# SCATTERING AND LOCALIZATION OF CLASSICAL WAVES IN RANDOM MEDIA

World Scientific Series on Directions in Condensed Matter Physics — Vol. 8

# SCATTERING AND LOCALIZATION OF CLASSICAL WAVES IN RANDOM MEDIA

Edited by

## Ping Sheng

**World Scientific**

*Singapore • New Jersey • London • Hong Kong*

*Published by*

World Scientific Publishing Co. Pte. Ltd.

5 Toh Tuck Link, Singapore 596224

*USA office:* 27 Warren Street, Suite 401-402, Hackensack, NJ 07601

*UK office:* 57 Shelton Street, Covent Garden, London WC2H 9HE

**Library of Congress Cataloging-in-Publication Data**
Scattering and localization of classical waves in random media / edited by Ping Sheng.
   (World Scientific series on directions in condensed matter physics: vol. 8)
  ISBN-13 978-9971-5-0539-4 -- ISBN-10 9971-5-0539-8
  ISBN-13 978-9971-5-0540-0 (pbk) -- ISBN-10 9971-5-0540-1 (pbk)
  1. Condensed matter. 2. Order-disorder models. 3. Waves.
  4. Scattering (Physics) 5. Multiple scattering (Physics)
  I. Sheng, Ping, 1946–  . II. Series: World Scientific series on directions in
  condensed matter physics; v. 8.
  QC173.4.C65S23    1990
  530.4'1--dc20
                                 90-30206
                                     CIP

**British Library Cataloguing-in-Publication Data**
A catalogue record for this book is available from the British Library.

The publisher would like to thank the authors concerned and the following publishers for their permission to reproduce the reprinted figures found in this volume: Academic Press Inc.; American Physical Society (*Phys. Rev. Lett.*).

To those who have not granted us permission prior to publication, we have taken the liberty to reproduce their figures without consent. We shall however acknowledge them in future editions of this work.

# PREFACE

The study of scattering and propagation of classical waves, i.e.
electromagnetic and elastic waves, is a topic familiar to a broad cross
section of scientists and engineers. This is so because waves are the
principal means of communication, detection, and measurement. As
a condensed-matter physicist, I am no exception to having a general
knowledge base about wave scattering. However, it was only during
the past decade, through the consideration of problems in oil exploration and subsurface imaging, that I learned about the fascinating
aspects of multiply-scattered wavefield. The timing of this learning experience was especially fortunate, since it coincided with breakthroughs
in the understanding of the localization phenomenon, a concept
originally proposed by P. W. Anderson in 1958 in the context of electron diffusion in random potentials. These insights, achieved through a
combination of phenomenological theories, experiments, developments
in mathematical techniques, and large-scale numerical simulations, were
soon realized to imply important predictions about the statistical
characteristics of multiply-scattered classical waves. It was during my
participation in the subsequent flurry of research activities that
I became keenly aware of the need for a central reference where these
exciting new developments can be quickly assimilated by a beginner as
well as by experts in related fields. This is the source of my primary
motivation that gives rise to the present volume.

The collection of papers contained in this volume is roughly
balanced between theory and experiment. They have the common

theme of being related, in one way or the other, to the characteristics of classical waves under the condition of strong multiple scattering. The phenomenon of localization, being the single most prominent theoretical prediction for the strong scattering regime, is examined from different perspectives by a number of authors. It should be noted that each article is self-contained and therefore can be read independently. In that sense the ordering of the chapters is immaterial. However, I have grouped papers on electromagnetic waves or general properties of scalar waves towards the front of the volume and those dealing with acoustic waves or specific topics towards the rear of the volume.

I wish to thank every contributing author for the effort in writing an up-to-date and pedagogical account of his/her topic. I also very much appreciate the interest and technical help of the World Scientific staff. In particular, the care and attention of Miss P. H. Tham is essential in catching many mistypes I would had overlooked. Finally I wish to thank my wife, Deborah, for the support and understanding of a husband who had on many nights stayed at home but away from the family.

Ping Sheng
*Annandale, NJ*
*1990*

# CONTENTS


Preface   v

The Localization of Waves in Disordered Media   1
    *Sajeev John*

Experiments on Weak Localization of Light and Their
Interpretation   97
    *M. P. van Albada, M. B. van der Mark, and*
    *A. Lagendijk*

Wave Diffusion and Localization in Random Composites   137
    *Zhao-Qing Zhang and Ping Sheng*

Novel Correlations and Fluctuations in Speckle Patterns   179
    *Shechao Feng*

Fluctuations, Correlation and Average Transport of
Electromagnetic Radiation in Random Media   207
    *Azriel Z. Genack*

Dynamical Correlations of Multiply Scattered Light   312
    *D. J. Pine, D. A. Weitz, G. Maret, P. E. Wolf,*
    *E. Herbolzheimer, and P. M. Chaikin*

Anderson Localization of the Classical Electromagnetic
Waves in a Disordered Dielectric Medium   373
    *Karamjeet Arya, Zhao-Bin Su, and Joseph L. Birman*

Optical Localization: Calculational Techniques and Results     404
     *E. N. Economou and C. M. Soukoulis*

Localization of Acoustic Waves     423
     *C. A. Condat and T. R. Kirkpatrick*

Localization of Surface Gravity Waves on a
Random Bottom     541
     *Max Belzons, Elisabeth Guazzelli, and*
     *Bernard Souillard*

Wave Localization and Multiple Scattering in
Randomly-Layered Media     563
     *Ping Sheng, Benjamin White, Zhao-Qing Zhang, and*
     *George Papanicolaou*

Subject Index     620

# THE LOCALIZATION OF WAVES
# IN DISORDERED MEDIA[*]


SAJEEV JOHN[†]

*Joseph Henry Laboratories of Physics*
*Jadwin Hall, Princeton University*
*Princeton, NJ 08544*



---

[*]This work was supported by the National Science Foundation under Grant DMR
85 18163.

[†]Present address: Dept. of Physics, University of Toronto, Toronto, Ontario,
Canada M5S1A7.

# Contents


1. Introduction     4

2. Photon Localization: The Physical Picture     9
   2.1. Independent scatterers     12
   2.2. Coherent scatterers     15

3. Diffusion and Coherent Backscattering     21
   3.1. Coherent backscattering     25

4. Field Theoretical Formulation of Diffusion:
Phonons in a Random Bond Model     30
   4.1. Gaussian path integrals and replicas     31
   4.2. The energy diffusion coefficient     33
   4.3. Field theory     36
   4.4. Discussion     42

5. Wave Propagation and Localization in a Long
Range Correlated Random Potential     43
   5.1. Introduction     43
   5.2. The model     44
   5.3. Field theory formulation     46
   5.4. Spontaneous symmetry breaking and the
       averaged one particle Green's function     49
   5.5. Goldstone modes and the average diffusivity     53
   5.6. Spectral content of normal modes     59
   5.7. The nonlinear $\sigma$-model     61
   5.8. Summary and discussion     65
   5.A. Appendix A: Asymptotic behavior of the
       mean free path and the conductance     65
   5.B. Appendix B: Spectrum of the nonlinear $\sigma$-model     68

6. Mobility Edges, Localization and Absorption     70
   6.1. Weakly dissipative media     74

7. Renormalization and the Nonlinear $\sigma$-Model     76
   7.1. Dynamical response of a non-interacting electron
       gas near the localization transition     76

7.2. Nonlinear $\sigma$-model with dynamical symmetry breaking    77

7.A. Appendix A: Constraint on the nonlinear $\sigma$-model    84

7.B. Appendix B: Momentum shell integration
of the nonlinear $\sigma$-model    84

8. Conclusions    90

References    93

## 1. INTRODUCTION

Wave propagation in ordered and weakly disordered media is a subject with a long and rich history. It is a phenomenon which is common to everyday experience and as such would at first glance appear as an unlikely source for fundamentally new and unexplored physics. It is precisely such an unexplored regime which I will discuss in this article. The existence of propagating electromagnetic and thermal modes plays an important role in nearly all condensed matter systems. It is precisely this ubiquitous aspect of classical wave propagation which can lead to far reaching consequences if for some reason such modes were either absent or failed to propagate. The localization of classical waves and in particular light poses entirely new experimental and theoretical challenges, some unencountered and some overlooked in its electronic counterpart. It touches the very heart of our understanding of transport of wave-like phenomenon in disordered media whether quantum mechanical or classical, while at the same time offering the possibility of the most direct experimental test of these ideas. It is my aim in this article to describe some of the basic motivations for pursuing this goal, and to review in simple physical terms the underlying concepts. In addition I will present a pedagogical introduction to the formal mathematical description of localization and discuss some of the outstanding experimental and theoretical challenges, which have emerged in this field.

The theory of localization dates back to the classic paper of Anderson.[1] For a recent review of the rather extensive literature on electronic localization, the reader may wish to consult the article of Lee and Ramakrishnan.[2] The study of classical localization focussed initially on elastic waves in disordered solids. A first principles theory of localization of phonons in a disordered elastic medium in $2 + \epsilon$ dimensions was introduced by John, Sompolinsky and Stephen[3] based on the field theoretical formulation of electron localization by Wegner.[4] This theory has been recovered and extended by diagrammatic perturbation methods.[5,6]

In the case of electronic localization, nature provides a variety of readily available materials. It is convenient to describe these materials by simple model Hamiltonians such as disordered tight binding models and continuum white noise models.[7,8] Such an approach to classical localization, can in some instances "discard the baby with the bath water" as will be discussed at length in Sec. 2 of this article and can lead to quite misleading conclusions about the observability of the effect. This is particularly evident in the case of electromagnetic wave localization. In this regime, the precursor effect to localization has been experimentally observed[9-13] and quantitative agreement between theory[14-19] and experiment has been established. The precursor effect, referred to as coherent backscattering is illustrated in Fig. 6 and will be discussed at length in Sec. 3. Extrapolation of this weak scattering behavior also leads to the highly plausible prediction that an actual photon mobility edge separating extended states from localized states may be observable[20-24] for sufficiently high dielectric contrast materials in an intermediate frequency window separating extended states at both higher and lower frequencies. The approach to this issue has been largely divided into two fundamentally different, although in some ways complementary, lines of reasoning. These are *microscopic* or single particle scattering resonance models[22-24] on the one hand, and on the other hand *macroscopic* or Bragg-like resonance mechanisms.[25] Before discussing the relative merits of these two approaches I will review some of the basic motivations for studying classical localization and in particular optical localization:

(i) It provides for the first time a complete test of the scaling theory of localization.[26] Optical propagation in lossless dielectric media provide the ideal realization of a single excitation in a static random medium at room temperature and is not hampered by the inevitable presence of electron-electron interactions and electron-phonon interactions which occur in the study of electron localization. High resolution optical techniques also provide for the first time a detailed study of the angular, spatial and temporal dependence of the specific intensity of light and not simply the average diffusion coefficient. This will provide a more complete understanding of localization as a critical phenomenon.

(ii) Photon localization will provide a more detailed understanding of the role of different types of disorder, spatial correlations and short range order in the determination of transport properties of waves. As will be discussed, the details of the structure factor and in particular, short range order play a vital role in determining the nature of transport and the very existence of localization. This is in contrast to the simple view put forward by Ioffe and Regel[27] which suggests that localization will occur whenever the elastic mean free path becomes comparable to the inverse wave vector of the wave. Optical studies in systems with well characterized forms of disorder and short or long range order will strengthen our understanding of the existence of pseudogaps in the density of electronic states in amorphous semiconductors and the resulting transport properties. High dielectric contrast colloidal suspensions may be useful in this regard. In systems in which there is an interplay between order and disorder, the Ioffe-Regel condition[27] for localization depends sensitively on the underlying structure factor, as I will describe, and must accordingly be reformulated.

(iii) Optical propagation in disordered layered dielectric media provides valuable laboratory models of geophysical wave propagation. Such materials may be fabricated by molecular beam epitaxy. These studies are valuable in seismology and oil exploration. In the strong scattering limit they pose new theoretical problems such as the incorporation of macroscopic anisotropies into the theory of localization.

(iv) Optical propagation in the diffusive scattering regime has already proved to be a valuable new spectroscopic tool for studying the hydrodynamics of complex fluids. This has been termed "Diffusing Wave Spectroscopy".[28-30] The time autocorrelation of light intensity in the laser speckle pattern is a measure of the velocity autocorrelation function of scattering particles. This technique has already been applied successfully to weak scattering systems such as polyballs in suspension. Static and dynamic structure of complex fluids especially in the vicinity of phase transformations may be studied even in the absence of wave localization. If higher dielectric contrast materials are prepared this could in turn yield new insights concerning the effect of static and dynamic structure on localization.

(v) Periodic dielectric superlattice structures possessing photonic band gaps or pseudogaps in the optical density of states have recently been suggested[25] as candidates for strong localization of light. This also has important implications for the interaction of light with atoms and molecules. For instance it leads to the inhibition of spontaneous emission which is of importance in the design of new and more efficient lasers.[31] It will likewise affect photochemical rates and radiative transfer rates of molecules and radicals in solution.

(vi) Photon localization and band gaps also provide an important new avenue for the study of nonlinear optical effects and optical bistability. As in the case of polarons in a disordered medium[32,33] there is a substantial synergetic interplay between localization and self focussing nonlinearities. This is of importance for future optically based technologies such as bistable switching devices, optical transistors[34] and neural optical computers.[35]

The choice of photons as opposed to other classical waves as the focus of this discussion is largely motivated by this potential for important applications as well as the obvious availability of high resolution measurement techniques in optics. However, it should be mentioned that many of the ideas to be discussed apply equally well to the other systems. The most notable of these in terms of experimental realization are third sound localization in $^4$He films on rough substrates,[36,37] phonon localization in disordered solids as evidenced by the low temperature thermal conductivity plateau[38-40] and also surface plasmon localization on rough metal surfaces.[41,42] Under the more general heading of non-electronic localization I also mention the intriguing experiments by Tawel and Canter[43] on positron annihilation in dense helium gases at low temperatures. Here an anomalous peak in the annihilation rate is observed when the positron mean free path becomes comparable to the position de Broglie wavelength suggesting the possibility of a localization related phenomenon.

I conclude this section with an overview of the remaining sections in this article. A relatively complete qualitative picture of classical wave localization may be obtained from Secs. 2, 3, 6 and 8. The remaining Secs. 4, 5 and 7 provide an introduction to more formal mathematical tools which express more precisely the basic physics of the more heuristic discussions. Sections 2 and 3 deal almost exclusively with electromagnetic waves. In this case, the vector nature of the propagating field plays a crucial role in the analysis of some of the predicted and observed experimental results. A field theoretical formalism for describing the phenomena of these chapters is in principle straightforward but unnecessarily cumbersome for pedagogical purposes. Consequently the next two sections which introduce mathematical formalism consider only the case of scalar classical waves. Phonons with random bond disorder are discussed in Sec. 4 and phonons with random mass disorder are described in Sec. 5. The entirely new feature of Sec. 5 is the introduction of essentially arbitrary spatial correlations in the disorder. This physics which emerges from the introduction of strong spatial correlations is essential for the experimental observability of photon localization in three dimensions. The formalism is presented in the context of phonons with monotonically decaying correlations in the disorder. The most important application of this formalism is scattering structures with a static structure factor exhibiting short range order. This work is still in progress. It nevertheless is an important advance in formal mathematical capability which supercedes the formal tools introduced for the treatment of electron localization. This section has been reprinted in its entirety from Phys. Rev. **B28** (1983) 6358.

Critical behavior is discussed in Sec. 6. Since the relevant renormalization group picture is essentailly the same as in the case of electronic localization, the discussion is relatively brief. The main emphasis here is on the new experimental consequences of established scaling ideas in the context of electromagnetic propagation through dielectric slabs and weakly absorbing or nonlinear media.

Finally I present in Sec. 7 a pedagogical introduction to the use of the nonlinear $\sigma$-model in obtaining a renormalization group picture for wave diffusion near a mobility edge. This approach was originated by Wegner in the context of electronic localization. This section is aimed at the reader who is a non-expert in field theory and for whom the original literature is either too scattered or too terse. The derivation of nonlinear $\sigma$-model is given in its original context namely that of a non-interacting electron gas in a static uncorrelated random potential.

## 2. PHOTON LOCALIZATION: THE PHYSICAL PICTURE

In this section I derive a simple physical picture of classical wave propagation in non-dissipative random media exhibiting three fundamentally different regimes. For wavelengths $\lambda$ long compared to the scale $a$ of the scattering structures, wave propagation is dominated by Rayleigh scattering. In three dimensions the elastic mean free path $l$ diverges with wavelength as $\lambda^4$. All states are extended in this weak scattering regime. On the other hand if the wavelength is made very small compared to the scattering structures (or more precisely, the correlation length of the disorder) wave propagation may be described by classical geometric ray optics. Since the mean free path can never become shorter than the correlation length $a$, it follows that the transport of wave energy is diffusive in nature on very long length scales but that all states are extended. Our primary focus will therefore be on the intermediate frequency window in which the wavelength is comparable to the correlation length $a$. For optical waves it is shown that sufficiently high dielectric contrast random composites with a sufficient degree of short range order can exhibit a photon localization transition.

By far the most familiar example of localization is that of an electron in a disordered solid. In order to illustrate the similarities and differences between photons and electrons, I will begin by constructing a formal analogy between them. An electron in the conduction band of a disordered solid may be described by a continuum effective mass Schrödinger equation

$$\left[ \frac{-\hbar^2}{2m^*} \nabla^2 + V(x) \right] \psi(x) = E\psi(x) \ . \tag{2.1}$$

Here the random potential $V(x)$ is chosen to have zero mean value and has a spatial autocorrelation function $\langle V(x)V(0)\rangle_{\text{ensemble}} = V_{\text{rms}}^2 e^{-x^2/a^2}$ where the length scale $a$ is of the order of the interatomic spacing in the solid. The resulting one-electron density of states in three dimensions is depicted in Fig. 1. For weak disorder there is a tail of strongly localized states, referred to as the Urbach edge,[44] for $E < 0$ which is separated by a mobility edge from positive energy extended states in the square root continuum. As the disorder parameter $V_{rms}$ is increased the mobility edge eventually[45] moves into the positive energy continuum. The detailed trajectory of the mobility edge as a function of disorder has been the subject of extensive numerical studies.[46] The noteworthy point, however, for present considerations is that only in the limit of very strong disorder do states with $E > 0$ exhibit localization.

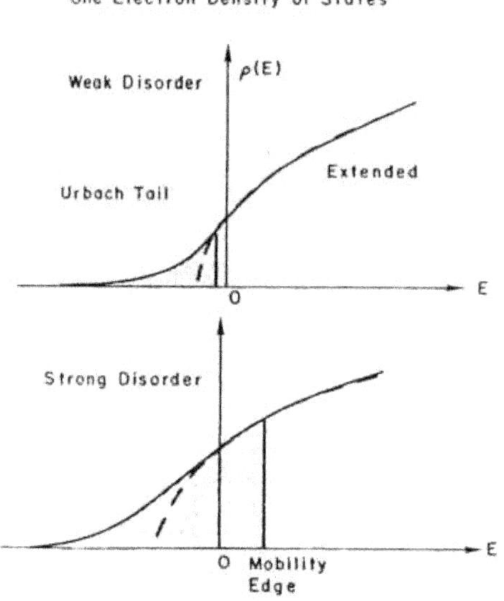


Fig. 1. (a) One electron density of states in a correlated Gaussian random potential. For weak disorder there is an Urbach band tail of localized states below the positive energy squareroot continuum of extended states which is shifted downward slightly by the disorder. (b) As the disorder is increased, the mobility edge eventually moves into the positive energy regime.

In the case of monochromatic electromagnetic waves of frequency $\omega$ propagating in a disordered nondissipative dielectric medium with dielectric constant

$$\epsilon(x) = \epsilon_0 + \epsilon_{fluct}(x) \,, \tag{2.2a}$$

the wave equation for the electric field vector **E** may be written in a form resembling the Schrödinger equation (2.1):

$$-\nabla^2 \mathbf{E} + \mathbf{\nabla}(\mathbf{\nabla} \cdot \mathbf{E}) - \frac{\omega^2}{c^2}\epsilon_{\text{fluct}}(x)\mathbf{E} = \epsilon_0 \frac{\omega^2}{c^2}\mathbf{E} \qquad (2.2b)$$

Here the randomly fluctuating part of the dielectric constant $\epsilon_{\text{fluct}}(x)$ plays the role of the random potential $V(x)$ with a mean value of zero and some appropriate spatial autocorrelation function. For nonmetallic scatterers with an everywhere real positive dielectric constant, it is evident that the energy eigenvalue $\epsilon_0\omega^2/c^2$ is always positive thereby precluding the possibility of band tail localization below the continuum. Herein lies the fundamental challenge of photon localization: in traditional electronic systems localization occurs at low energies in which the electron is trapped in potential wells or faces a number of large barriers. In the case of classical waves, however, as the energy eigenvalue is lowered by going to lower frequency $\omega$, the strength of the scattering potential $(\omega^2/c^2)\epsilon_{\text{fluct}}(x)$ also vanishes and the states are extended. In the opposite limit of very high frequencies, states are again extended irrespective of the strength of the disorder. Unlike the familiar picture of electronic localization, what we are really seeking in the case of light is an intermediate frequency window of localization within the positive energy continuum! The situation is even further complicated by the restriction on the dielectric constant to be everywhere real and positive. This corresponds to looking for localization at an energy which is higher than the highest potential barrier which the particle faces in the Schrödinger equation analogy.

In classical wave systems in which there is a natural upper frequency cutoff, localization occurs predominantly at the upper band edge. This is illustrated in Fig. 2 for the case of phonons in an isotopically or structurally disordered solid. Excitations at the low frequency $\omega \rightarrow 0$ end correspond to uniform translation of the entire solid and are extended irrespective of the strength of the randomness. The factor $\omega^2/c^2$ multiplying $\epsilon_{\text{fluct}}(x)$ in Eq. (2.2) is simply an expression of the occurrence of such a Glodstone mode in classical wave systems.

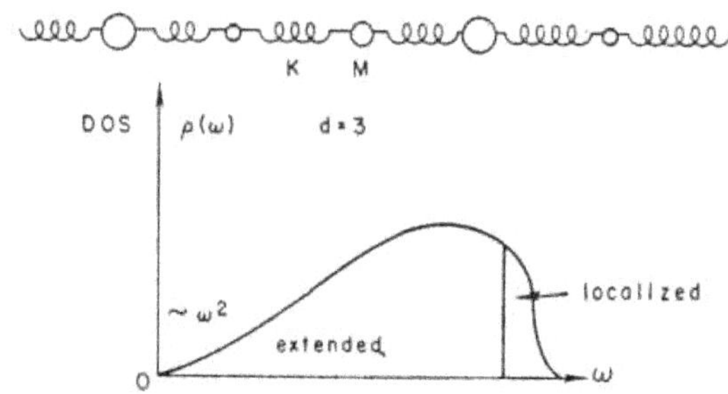


Fig. 2. Phonon localization for random masses or springs. Addition of defects leads to localization at the top of the phonon band. The $\omega = 0$ state corresponds to uniform translation of the entire medium and states in the vicinity are always extended for $d > 2$.

## 2.1. Independent Scatterers

The underlying physics of the high and low frequency limits in the case of photons which have no upper frequency cutoff can be made more precise by considering scattering from a single dielectric sphere. Consider a plane wave of wavelength $\lambda$ impinging on a small dielectric sphere of radius $a \ll \lambda$ of dielectric constant $\varepsilon_a$ embedded in a uniform background dielectric $\varepsilon_b$ in $d$-spatial dimensions. The scattered intensity $I_{\text{scatt}}$ at a distance $R$ from the sphere can be a function of only the incident intensity $I_0$, the dielectric constants $\varepsilon_a$ and $\varepsilon_b$ and the lengths $R$, $\lambda$ and $a$. In particular $I_{\text{scatt}}$ must be proportional to the square of the induced dipole moment of the sphere which scales as the square of its volume $\sim (a^d)^2$ and by conservation of energy must fall off as $1/R^{d-1}$ with distance from the scattering center:

$$I_{\text{scatt}} = f_1(\lambda, \varepsilon_a, \varepsilon_b) \frac{a^{2d}}{R^{d-1}} I_0 \tag{2.3}$$

Since ratio $I_{\text{scatt}}/I_0$ is dimensionless, it follows that $f_1(\lambda, \varepsilon_a, \varepsilon_b) = f_2(\varepsilon_a, \varepsilon_b)/\lambda^{d+1}$ where $f_2$ is another dimensionless function of the dielectric constants. The vanishing of the scattering cross section for long wavelengths as $\lambda^{-(d+1)}$ is the familiar result for why the sky is blue in three dimensions. This generalization of Rayleigh scattering to $d$-dimensions also reveals the origin of diverging localization lengths for one and two-dimensional classical waves and of course the existence of extended states in $d = 3$. For a dense random collection of scatterers this behavior remains evident in the elastic mean free path $l$ which is proportional to $\lambda^{d+1}$ for long wavelengths.[3] A consequence of the scaling theory of localization[26,3] is that in one and two dimensions, all states are localized but with diverging localization lengths $\xi_{loc}$ behaving as $\xi_{loc} \sim l$ in one dimension and $\xi_{loc} \sim l \exp\left(\frac{\omega}{c} l\right)$ in two dimensions. A detailed derivation of these results may be found in Ref. 3.

It is likewise instructive to consider the opposite limit in which the wavelength of light is small compared to the scale of the scattering structures. For scattering from a single sphere it is well known[47] that for $\lambda \ll a$, the cross section saturates at a value $2\pi a^2$. This is the result of geometric optics. For a dense random collection of scatterers it is useful to introduce the notion of a correlation length $a$. On scales shorter than $a$, the dielectric constant does not vary appreciably and wave propagation is well described in a WKB approximation. The essential point is that the elastic mean free path never becomes smaller than the correlation length. If one adopts the most naive version of the Ioffe Regel condition $2\pi l/\lambda \simeq 1$ for localization, with $\lambda$ being the vacuum wavelength or even an effective medium wavelength of light, it follows that extended states are expected at both high and low frequencies. However as depicted in Fig. 3, for strong scattering, there arises the distinct possibility of localization within a narrow frequency window when the quantity $\lambda/2\pi \simeq a$. It is this intermediate frequency regime which we wish to analyze in greater detail.

For the point like scatterers, a straightforward first principles calculation[48] in $d = 2+\epsilon$ dimensions yields the criterion $(l\omega/c)^{d-1} \simeq 1/\epsilon$ for the occurrence of a mobility edge. This result is asymptotically exact as $\epsilon \to 0$. Extrapolating to $\epsilon = 1$ yields in three dimensions what I will refer to as the free-photon Ioffe-Regel condition. This particular result is based on perturbation theory about free photon states which undergo multiple scattering from point-like objects and a disorder average is performed over all possible positions of the scatterers. In effect, statistical weight is evenly distributed over all possible configurations of the scatterers and the medium has an essentially flat structure factor on average.

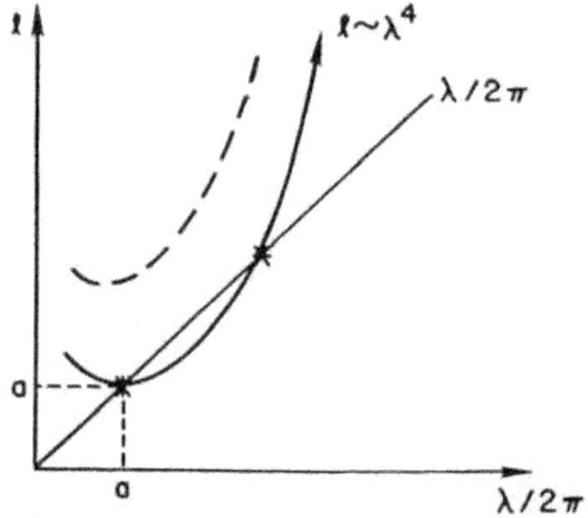


Fig. 3. Behavior of the elastic mean free path as a function of wavelength. In the long wavelength Rayleigh scattering limit $l\sim\lambda^4$. In the short wavelength limit $l\gtrsim a$, the correlation length. For a strongly disordered medium (solid curve) there may exist a range of wavelengths for which $2\pi l/\lambda\simeq 1$, exhibiting localization. This would not occur for dilute scatterers (dashed curve).

The first correction to this picture is to associate some nontrivial structure to the individual scatterers. The theory of Mie resonances[47] for scattering from dielectric spheres immediately tells us that this can have profound consequences on the elastic mean free path. For instance for spheres of radius $a$, of dielectric constant $\varepsilon_a$ embedded in a background dielectric $\varepsilon_b$, and for a ratio $\varepsilon_a/\varepsilon_b\simeq 4$, the Mie resonance which occurs at a frequency given by $\omega/c(2a)\simeq 1$, yields a scattering cross section $\sigma\simeq 6\pi a^2$. For a relatively dilute collection of spheres of number density $n$, the classical elastic mean free path becomes $l\sim\frac{1}{n\sigma}=\frac{2a}{9f}$. Here I have introduced the volume filling fraction $f$ of the spheres. Extrapolating this dilute scattering result to higher density, it is apparent that for a filling fraction $f\simeq 1/9$, the free-photon Ioffe Regel condition is satisfied on resonance. It is tempting to increase the density of scatterers so as to further decrease the mean free path. However, the fact that the cross section on resonance is 6 times the geometrical cross section indicates that a given sphere disturbs the wavefield over distances considerably larger than the actual sphere radius. The existence of the resonances requires that the "spheres of influence" of the scatterers do not overlap. Indeed for higher densities the reader may verify that the spheres become optically connected in this sense and that the mean free path in fact increases rather than decreases. From the single scattering or microscopic resonance point of view the free photon criterion for localization is a very delicate one to achieve. Similar considerations apply to other microscopic mechanisms such as resonance fluorescence in light scattering from atomic vapours or exciton fluorescence in solids.

## 2.2. Coherent Scatterers

The approach based on independent, uncorrelated scatterers however overlooks an important aspect of the problem which has far reaching consequences in the high density limit. In a sense it is the fundamental theorem of solid state physics: Certain geometrical arrangements of identical scatterers can give rise to large scale or macroscopic resonances. The most familiar example is the Bragg scattering of an electron in a perfectly periodic crystal. Such an effect is not given the required statistical weight by a disorder average which improperly averages over all positions of the scatterers.

Consider for instance a fluctuating dielectric constant $\epsilon(x) - \epsilon_0 \equiv \epsilon_{\text{fluct}}(x) = \epsilon_1(x) + V(x)$ where $\epsilon_1(x) = \epsilon_1 \Sigma_G U_G e^{iG \cdot x}$ is a perfectly periodic Bravais superlattice and $V(x)$ is a small perturbation arising from disorder. Here $\mathbf{G}$ runs over the appropriate reciprocal lattice and its value for the dominant Fourier component $U_\mathbf{G}$ is chosen so that the Bragg condition $\mathbf{k} \cdot \widehat{G} = (1/2)G$ may be satisfied for a photon of wavevector $\mathbf{k}$. Such a structure is attainable, albeit in a low dielectric contrast regime, with polyballs in suspension which exhibit *fcc* and *bcc* superlattice arrangements as well as a number of disordered phases. Setting $V(x) = 0$ for the time being, the effect of the periodic modulation or the photon spectrum may be estimated within a nearly-free-photon approximation. Unlike scalar electrons, there is a degeneracy between two possible photon helicity states, namely the right and left hand circularly polarized states, as well as the possibility of helicity flip scattering. For scattering by an angle $\theta$, the amplitude for helicity flip scattering is given by $(1 - \cos\theta)/2$. It follows that when the photon wavevector

16  *Sajeev John*

**k** lies along the Bragg plane defined by the reciprocal lattice vector **G**, the allowed photon frequencies are given by

$$\frac{\omega}{c} = \frac{k}{\sqrt{\epsilon_0 \pm \epsilon_1 U_G}} \tag{2.4a}$$

and

$$\frac{\omega}{c} = \frac{k}{\sqrt{(\epsilon_0 \pm \epsilon_1 U_G |1 - G^2/2k^2|)}} \,. \tag{2.4b}$$

Here $\epsilon_0$ is the average dielectric constant of the entire medium. The first pair of frequencies corresponds to the optical $s$-wave in which the polarization vector is perpendicular to the phase defined by the incident beam and its Bragg reflected partner. The second pair corresponds to the optical $p$-wave in which the polarization vector lies in the plane of scattering. The associated photon dispersion relations are depicted in Fig. 4.

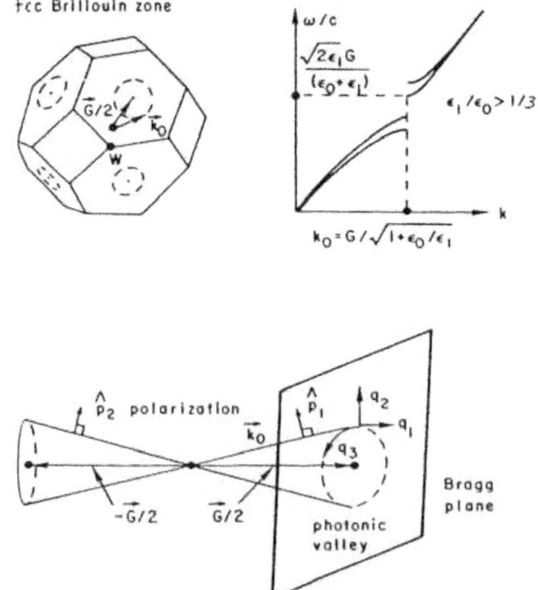


Fig. 4. Creation of a circular photonic valley near the band edge by the Bragg resonance of $p$-polarized light of wavevector $\mathbf{k}_0 + \mathbf{q}$ in a high dielectric contrast $\epsilon_1/\epsilon_0$ fcc superlattice. $q_1, q_2$ and $q_3$ are principal axes for photon dispersion.

The existence or near existence of a gap in the photon density of states is of paramount importance in determining transport properties and especially localization. Such a possibility was completely overlooked in the derivation of the free-photon Ioffe-Regel condition which assumed an essentially free-photon density of states. In the vicinity of a band edge the character of propagating states is modified. To a good approximation the electric field amplitude of the propagating wave is that of the free photon except modulated by an envelope Bloch wave whose wavelength $\lambda$ diverges as the mobility edge is approached. Under these circumstances the wavelength which must enter the localization criterion is that of the envelope. In the presence of even very weak disorder, the criterion $2\pi l/\lambda_{\mathrm{envelope}} \simeq 1$ is automatically satisfied as the photon frequency approaches the band edge frequency.

It is useful as a first step to estimate the dielectric contrast required to produce an optical band gap in three dimensions. For concreteness consider an $fcc$ superlattice. An estimate within the nearly-free-photon approximation follows from the condition that there exist frequencies $\omega$ which remain in the spectral gap defined by Eqs. (2.4a) and (2.4b) as the wavevector $\mathbf{k}$ is allowed to span the surface of the $fcc$ Brillouin zone. Setting $U_{\mathbf{G}} = 1$, it is apparent from these solutions that the lowest frequency state of the upper branch (-sign) occurs at the center of the Bragg plane $\mathbf{k} = \mathbf{G}/2$ provided $\epsilon_1/\epsilon_0 \leq 1/3$ whereas for higher dielectric contrast ($\epsilon_1/\epsilon_0 > 1/3$) the corresponding minimum occurs along a circle defined by the intersection of the Bragg plane with the sphere $k^2 = G^2/(1 + \epsilon_0/\epsilon_1)$. This is a new feature arising from the vector nature of light. Although the analysis based on Eqs. (2.4a) and (2.4b) ignores complications arising from the intersection of two or more Bragg planes, it leads to the estimate that for $\epsilon_1/\epsilon_0 \gtrsim .36$ a severe depression in the photon density of states (Fig. 5) and perhaps even a spectral gap occurs near the frequency $\omega/c = \sqrt{2\epsilon_1}\,G/(\epsilon_0 + \epsilon_1)$. The nature of light propagation just above this frequency is described by the associated photon dispersion relations. Defining $\mathbf{q}$ as a small deviation of the photon wavevector $\mathbf{k}$ from a point $\mathbf{k}_0$ on the Bragg plane in the circular photonic valley $k_0^2 = G^2/(1 + \epsilon_0/\epsilon_1)$, it follows that to quadratic order the lowest lying photon branch has frequency

$$\frac{\omega^2}{c^2} = E_c + A_1 q_1^2 + A_2 q_2^2 \qquad (\epsilon_1/\epsilon_0 > 1/3) \,. \qquad (2.5)$$

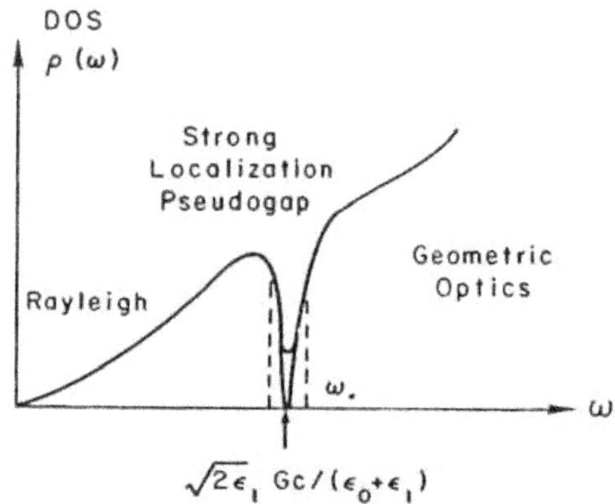


Fig. 5. Photon density of states in a disordered superlattice exhibiting low frequency Rayleigh scattering and high frequency geometric optics extended states separated by a pseudogap of strongly localized photons.

Here, the components of **q** along the principal axes perpendicular to the circular valley (Fig. 4) are labeled $q_1$ and $q_2$, whereas the third component $q_3$ tangent to the circle describes degenerate states and does not enter the dispersion relation (2.5) at quadratic order. The coefficients are given by $E_c = 2k_0^2/(\epsilon_0 + \epsilon_1)A_1 = (2/\epsilon_0)[(\epsilon_0^2 + \epsilon_1^2)/(\epsilon_0^2 + \epsilon_1^2)]$ and $A_2 = 2(3 - \epsilon_0/\epsilon_1)/(\epsilon_0 + \epsilon_1)$. The phase space available for photon propagation is accordingly restricted to a set of narrow symmetry related cones in k-space analogous to the pockets of electrons near a conduction band edge well known in semiconductor physics.

A detailed band structure calculation for a *fcc* crystal of dielectric spheres has been carried out by myself and R. Rangarajan using the Koringa-Kohn-Rostoker method. By evaluating the density of states of a classical scalar wave in a perfectly periodic system it is possible to ascertain the optimal volume filling fraction and minimum dielectric contrast required for the production of a total spectral gap. For the scalar case we find that for solid spheres in a low index (say unity) background the optimal volume filling fraction to be 10–15%. At 15%, the gap persists for refractive index ratios as low as 2.8. For higher contrasts a larger gap can be obtained but the optimal filling fraction decreases. Although the actual numerical limits will be slightly modified in the case of true (vector) electromagnetic waves, these numbers provide a very valuable guide for the fabrication of photonic materials capable of localizing light.

The perturbative introduction of randomness $V(x)$ leads to a mixing of all nearly degenerate photon branches. This invariably causes a smearing in wavevector space of any sharp Bragg resonances as well as a filling in of the pseudogap in the photon density of states (DOS). It is highly plausible nevertheless that photon near the band edge frequency retains certain general features of the dispersion relation (2.5) and that coherent backscattering of light occurs by disorder-induced scattering within and between such valleys in phase space. The dynamics of both inter and intravalley scattering are described approximately by an effective Schrödinger equation in wavevector space:

$$(A_1 q_1^2 + A_2 q_2^2)\Psi(\mathbf{k}_v, \mathbf{q}) + \frac{1}{\epsilon_0}\frac{\omega^2}{c^2} \sum_{\substack{\text{valleys} \\ v}} \oint_{\mathbf{k}_v'} \int d^2\mathbf{q}' \mathbf{V}(\mathbf{k}, \mathbf{k}')\Psi(\mathbf{k}_v', \mathbf{q})$$

$$= \left(\frac{\omega^2}{c^2} - E_c\right)\Psi(\mathbf{k}_v, \mathbf{q}) . \tag{2.6}$$

Here $\mathbf{k} = \mathbf{k}_v + \mathbf{q}$ (likewise for primed variables) and $\Psi(\mathbf{k}_v, \mathbf{q})$ in the q-Fourier transform of an envelope function $\Psi(\mathbf{k}_v, x)$ which multiplies the carrier wave electric field eigenvector

$$\mathbf{E}_0(\mathbf{k}_v, x) \equiv [\hat{p}_1 e^{i\mathbf{k}_v \cdot \mathbf{x}} + \hat{p}_2 e^{i(\mathbf{k}_v - \mathbf{G}) \cdot \mathbf{x}}]/\sqrt{2} ,$$

describing Bragg scattering at the bottom of a valley $v$. The polarization vectors $\hat{p}_1$ lie in the plane of Bragg scattering defined by $\mathbf{k}_v$ and $\mathbf{k}_v - \mathbf{G}$ and unit vectors normal to the surface of the wavevector cone depicted in Fig. 4. Integration over $\mathbf{k}_v'$ in (2.6) is along the intersection of this cone with the Bragg plane. $\tilde{V}$ is the matrix element of the random potential $V(x)$ between the actual electric fields (product of carrier wave and envelope) in the two states $\mathbf{k}$ and $\mathbf{k}'$.

The existence of the band gap guarantees the existence of strongly localized photonic band tail states at frequencies $\omega^2/c^2 \lesssim E_c$, analogous to the Urbach band tail in electronic systems. If the disorder is a weak perturbation on the underlying band structure, the position of the associated mobility edge is given by $ql \simeq 1$ where $q$ is defined in (2.6). The physical picture is entirely different from that described by the free-photon Ioffe-Regel condition in that it is now the wavelength of the envelope function $\Psi$ rather than that of the carrier wave which enters the localization criterion.

The occurrence of a photonic band gap guarantees the observability of localization. However, in real disordered systems it is likely that neither the band picture nor the single scattering approach provides a complete description. This is particularly evident in the fact that band gaps which may be produced for optical waves are at best very narrow and that in the presence of significant structural disorder what remains is merely a large depression or pseudogap in the DOS. In the limit of dilute uncorrelated scatterers, the microscopic resonance picture is entirely adequate whereas in the high density limit of highly correlated scatterers a macroscopic or band point of view is essential. Classical wave localization in three dimensions however occurs in an intermediate regime in which elements of both of these conceptually different approaches must enter but neither by itself is complete. The two approaches are in this sense complementary.

It is my primary hypothesis that photon localization is not simply the byproduct of a high degree of uncontrolled disorder but rather the result of a subtle interplay between order and disorder. This is in contradiction to effective medium approaches[22-24] which either "discard the baby with the bath water" or produce localization purely as an artifact of the approximation. The true criterion for localization in real systems depends strongly on the underlying static structure factor of the medium. What I have described so far are the two extreme limits of a structureless random medium for which the criterion $2\pi l/\lambda \simeq 1$ describes localization and that of a medium with nearly sharp Bragg peaks and a band gap for which $2\pi l/\lambda_{\text{envelope}} \simeq 1$ yields localization. It is highly plausible that a continuous crossover occurs between these two conditions as the structure factor of a suitably high dielectric contrast material is altered. The description of this crossover regime poses an important new theoretical problem unencountered in electronic systems, namely that of describing localization in a highly correlated scattering medium. A field theoretical formalism for describing wave propagation and localization in a medium with arbitrary spatial correlations has been developed by John and Stephen[49] and is presented in Sec. 5. It has been demonstrated that the lower critical dimension and critical exponents for the localization transition remain the same for long range correlated disorder. Further development of this formalism may lead to a complete description of classical wave localization in high dielectric contrast materials with short range order and pseudogaps in the density of states.

## 3. DIFFUSION AND COHERENT BACKSCATTERING

It is apparent from the discussion of the previous section that a precise mathematical description of classical wave localization requires important refinements of the tools used in electronic systems. I will give here only a brief outline of one formalism capable of recapturing the essential pieces of physics described above and postpone a detailed discussion of formal matters to Secs. 4, 5, and 7. The main point of this discussion is to give a more precise definition of the diffusion coefficient for wave energy which is valid even when the concept of a classical mean free path loses its utility. When the scattering is very weak it reduces to the familiar diffusion coefficient given by the product $(1/3)\,c_{\text{eff}}\,l$ of the effective medium speed $c_{\text{eff}}$ of wave propagation and the mean free path $l$. This new definition is essential because in the scaling theory of localization[26] the diffusion coefficient is no longer a local variable but is determined by macroscopic coherent wave interference throughout the entire disordered medium. In the vicinity of a mobility edge the diffusivity $D$ depends on the size of the entire sample. In addition I will present a physical picture of coherent backscattering of light in disordered media including two fundamental forms of broken symmetry, namely time-reversal noninvariance and parity nonconservation.

In order to determine whether the normal modes of a medium are extended or localized, it is useful to proceed by means of a simple thought experiment. Consider injecting a finite amount of energy into the disordered medium at some point in space over some very short interval of time. This might occur by means of a mechanical impulse given to an elastic medium or by a burst of electromagnetic radiation from a point source. We consider now the long time behavior of this energy. If the states are extended then random scattering leads to a diffusive spread of the locally conserved energy density $U(\mathbf{x}, t)$ but the energy eventually escapes to infinity. If on the other hand the modes are localized, the injected energy remains trapped in the vicinity of the initial impulse and energy diffusion coefficient in zero. More precisely, we may define the energy diffusion coefficient as

$$D \equiv \lim_{t \to \infty} \frac{1}{t} \frac{\int d^d x\, x^2 U(\mathbf{x}, t)}{\int d^d x\, U(\mathbf{x}, t)} \ . \tag{3.1}$$

The time evolution of the electric field $\mathbf{E}$ in response to the source $\mathbf{j}(x, t)$ is given by the equation of motion

$$\nabla^2 \mathbf{E}(x, t) - \boldsymbol{\nabla}(\boldsymbol{\nabla} \cdot \mathbf{E}) - \frac{\epsilon(x)}{c^2} \frac{\partial^2}{\partial t^2} \mathbf{E} = \mathbf{j}(x, t) \ . \tag{3.2}$$

The response to an impulse of the form $\mathbf{j}(\mathbf{x}, \mathbf{y}, t) = \hat{e}_0 \delta^d(\mathbf{x} - \mathbf{y}) \delta(t)$ is given by the retarded Green's function $G_{ij}^+(\mathbf{x}, \mathbf{y}, t)$ where $i$ and $j$ denote Cartesian components of the electric field. Since scattering causes a mixing of all polarization states it is sufficient to initially excite a single polarization state $\hat{e}_0$.

When dealing with a system consisting of a large number of scatterers, it is convenient to describe the response in an average sense. However, as in the case of electronic systems the averaged one-photon Green's function cannot distinguish between extended and localized states as illustrated by a simple one-dimensional example. In the absence of scattering the time Fourier transform of the Green's function takes the form $G^+(x) \sim \exp(i\sqrt{\epsilon_0 + \epsilon_{\text{fluct}}}\, \omega_+ |\mathbf{x}|)$ characteristic of wave propagation. Here, a small imaginary part has been added to the frequency $\omega_\pm \equiv \omega \pm i\eta$ to distinguish retarded and advanced propagators. Any statistical average over $\epsilon_{\text{fluct}}$ however leads to exponential decay of $G^+(x)$ with distance irrespective of the presence of scattering, arising purely from random phase cancellation between different members of the statistical ensemble. For instance if we take $\epsilon_{fluct}$ to be independent of $\mathbf{x}$ but distributed with probability weight $P(\epsilon_{fluct}) = (\gamma^2 + \epsilon_{fluct}^2)^{-1}$, it follows that $\langle G^+(x)\rangle_{ensemble} \sim \exp(i\sqrt{\epsilon_0 + i\gamma}\omega_+|x|)$ which decays with $x$ despite the fact that most members of the ensemble support unimpeded wave propagation. The occurrence of wave propagation in this model, however is manifest in the quantity $\langle|G^+(x)|^2\rangle_{ensemble} \sim \exp(-2\eta\sqrt{\epsilon}|\mathbf{x}|) \to 1$ in the limit as $\eta \to 0$.

The full form of the averaged two-photon Green's function is of considerable importance in optical localization. By imaging random media with monochromatic laser radiation the spatial, angular and frequency dependences of this correlation function as well as higher order correlation functions may be accurately measured.[51] It also determines the energy diffusion coefficient. Roughly speaking, the diffusivity is given by

$$D \simeq \lim_{\eta \to 0} \eta \frac{\int_{-\infty}^{\infty} d\omega \omega^2 \int d^d x\, x^2 \langle G(x,0;\omega_+)G(x,0;\omega_-)\rangle_{ens}}{\int_{-\infty}^{\infty} d\omega \omega^2 \int d^d x \langle G(x,0;\omega_+)G(x,0;\omega_-)\rangle_{ens}} . \qquad (3.3a)$$

In obtaining this result from (3.1) I have suppressed Cartesian indices on the Green's function and replaced the long time limit by a corresponding $\eta \to 0$ limit. Separated averages in the numerator and denominator are performed here since the denominator is simply the total energy injected into the system independent of the disorder. The frequency representation here suggests a spectral decomposition of the total diffusivity into its contributions $D(\omega)$ from individual normal modes of frequency $\omega$. That is to say

$$D \equiv \frac{\int_{-\infty}^{\infty} d\omega\, U(\omega) D(\omega)}{\int_{-\infty}^{\infty} d\omega\, U(\omega)} , \qquad (3.3b)$$

where $U(\omega)$ is the energy density injected into photons of frequency $\omega$ by the initial impulse and $D(\omega)$ is the diffusion coefficient for a photon of frequency $\omega$. It is the quantity $D(\omega)$ which we need to consider on longer and longer length scales in a renormalization group sense to determine whether states at the particular frequency $\omega$ are extended or localized.

This somewhat tortuous definition of $D(\omega)$ may seem surprising. In the weak scattering limit $(l \gg \lambda)$ it reduces to the well known classical result $D(\omega) \simeq 1/3\, c_{\text{eff}}\, l(\omega)$ where $c_{\text{eff}}$ is an effective medium speed of light and $l(\omega)$ is the frequency dependent elastic mean free path. The virtue of the definition (3.3b) lies in the fact that it is valid irrespective of the strength of the scattering and is unambiguous even near a mobility edge where the notion of a classical mean free path loses its utility. In this regime (3.3b) may be regarded as the defining equation of $l(\omega)$ via the familiar classical expression.

The existence of classical diffusion of waves in disordered media may be demonstrated by means of approximate evaluation of the ensemble averaged two photon Green's function in the long distance limit:

$$\int d^d x e^{i\mathbf{p}\cdot\mathbf{x}} \langle G^+(x,0;\omega_+)G^-(x,0;\omega_-)\rangle_{\text{ensemble}} \sim \frac{1}{\eta + D(\omega)p^2} \ . \quad (3.4)$$

In the case of point-like scatterers or a random medium with an essentially flat static structure factor, the derivation is identical to that in the electronic case. In terms of diagrammatic perturbation theory, this approximation corresponds to the summation of ladder diagrams. Alternatively this may be obtained from the replica functional integral representation[3,4,7,8] of the two-photon propagator within a saddle point approximation. In either case the coefficient of the $p^2$ term in the small wavevector expansion is identified with the frequency dependent diffusivity and exhibits the Rayleigh scattering singularity $D(\omega) \sim \omega^{-(d+1)}$ in the low frequency limit. The corresponding derivation for a medium with long range spatial correlations is more subtle but reveals the expected deviations from Rayleigh scattering when the correlation length of the disorder exceeds the wavelength. Further discussion of this formalism may be found elsewhere.[50]

The fact that energy spreads diffusively on sufficiently long length scales has important consequences for wave propagation in a random medium. As an illustration consider a classical random walker in $d$-spatial dimensions released at the origin at time $t = 0$. The probability that he will be found at a position $\mathbf{x}$ at a later time $t$ is given by

$$P(x,t) = (4\pi D t)^{-d/2} \exp(-x^2/(4\pi D t)) \ . \quad (3.5)$$

In particular, the probability of being found again at the origin is given by $P(0,t) \propto t^{-d/2}$. The integrated probability to return to the origin by some time $T$.

$$\int_{l/c}^{T} \frac{dt}{t^{d/2}}$$

is unbounded for $d \leq 2$ as $T$ increases. In the case of diffusing waves, rather than this classical random walker, it is the interference of these amplitudes returning to the origin which leads to localization. This argument also illustrates why the lower critical dimension for localization is in fact two. In dimensions less than or equal to two, wave interference always causes the classical diffusion coefficient $D(\omega)$ to be renormalized to zero on longer and longer length scales almost irrespective of the nature of the disorder.

## 3.1. Coherent Backscattering

In the case of optical waves propagating through a disordered dielectric medium, this interference effect has been vividly demonstrated in a beautiful series of recent experiments. This is the phenomenon of coherent backscattering. In these experiments,[9-13] incident laser light of frequency $\omega$ enters a disordered dielectric half space or slab and the angular dependence of the backscattered intensity is measured. For colloidal particles in suspension, the Brownian motion of the scatterers results in an adiabatic variation of the random potential. In any finite time interval Nature itself provides the desired ensemble averaging. In solid microstructures, essentially the same result may be obtained by rapidly spinning the sample during the experiment.[12] For circularly polarized incident light the intensity of the backscattering peak for the helicity preserving channel is a factor of two larger than the incoherent background intensity. Coherent backscattering into the reversed helicity channel however yields a considerably reduced backscattering intensity. The angular width of the peak in either case is roughly $\Delta\theta \sim \lambda/(2\pi l)$.

As shown in Fig. 6, one possible process is that in which incident light with wavevector $\mathbf{k}_i = \mathbf{k}_0$ is scattered at points $\mathbf{x}_1, \mathbf{x}_2, \ldots, \mathbf{x}_N$ into intermediate (virtual) states with wavevectors $\mathbf{k}_1, \mathbf{k}_2, \ldots, \mathbf{k}_{N-1}$ and finally into the state $\mathbf{k}_N = \mathbf{k}_f$ which is detected. For scalar waves undergoing an identical set of wavevector transfers, the scattering amplitudes at the points $\mathbf{x}_1, \ldots, \mathbf{x}_N$ are the same for the path $\gamma$ and the time reversed path $-\gamma$ (dashed line). The nature of interference between the two paths is determined entirely by their relative optical path lengths. The resulting relative phase is given by $\exp[i(\mathbf{k}_i + \mathbf{k}_f) \cdot (\mathbf{x}_N - \mathbf{x}_1)]$. In the exact backscattering direction $\mathbf{q} \equiv \mathbf{k}_i + \mathbf{k}_f = 0$ there is constructive interference and a consequent doubling of the intensity over the incoherent background. If the angle between $-\mathbf{k}_i$ and $\mathbf{k}_f$ is $\theta$, then the coherence condition for small $\theta$ becomes $\mathbf{q} \cdot (\mathbf{x}_N - \mathbf{x}_1) = 2\pi\theta|\mathbf{x}_n - \mathbf{x}_1|/\lambda < 1$. In the diffusion approximation $|\mathbf{x}_n - \mathbf{x}_1|^2 \simeq D(t_N - t_1) \simeq lL/3$, where $D$ is the photon diffusion coefficient and $L$ is the total length of path $\gamma$. Thus paths of length $L$ contribute to the coherent intensity for angles less than $\theta_m = \lambda/(2\pi\sqrt{lL/3})$.

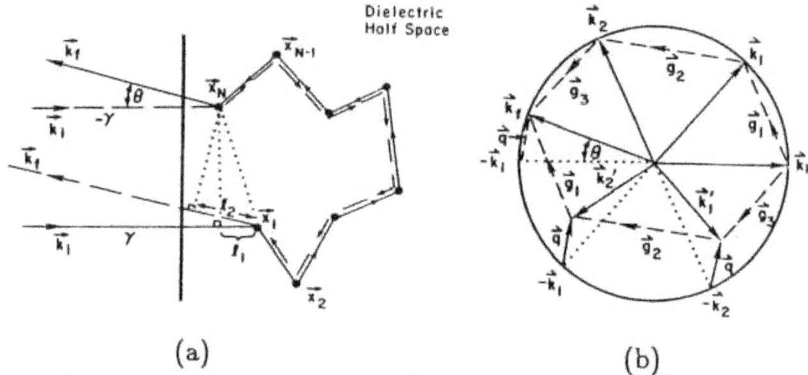


(a)           (b)

Fig. 6. (a) A typical scattering path $(\gamma)$ in real space and the time reverse $(-\gamma)$ with which it cohenrently interferes for small $\theta$. The phase difference between $\gamma$ and $-\gamma$ is simply proportional to the path length difference $l_2 - l_1$, where $l_1 = (\mathbf{x}_1 - \mathbf{x}_N) \cdot \mathbf{k}_i$ and $l_2 = (\mathbf{x}_N - \mathbf{x}_1) \cdot \mathbf{k}_f$. (b) A typical path drawn in momentum space. Equality of scattering amplitudes requires that the wave vector transfers $\mathbf{q}_j$ be the same but in the reversed order for the time-reversed path. Corresponding states differ in energy by an amount $E_{N-j} - E'_j \simeq c\hbar\mathbf{q} \cdot \hat{\mathbf{k}}_j$, which leads to a phase difference $(\Delta\phi)_{\mathrm{rms}} \simeq \sqrt{\frac{N}{3}}ql$.

To make contact with the phase space arguments of Sec. 2, it is useful to describe this process in wavevector space. This argument was originally given by Bergmann[51] in the context of electron localization.

Consider the path shown in Fig. 6b with wavevector transfers $\mathbf{g}_j = \mathbf{k}_j - \mathbf{k}_{j-1}$ $(j = 1, 2, \ldots, N)$. By time-reversing the sequence of transfers, it is apparent that the corresponding intermediate states $\mathbf{k}'_1, \mathbf{k}'_2, \ldots, \mathbf{k}'_{N-1}$ no longer lie on the energy shell for nonzero $\mathbf{q}$. This is allowed since these are virtual states with a lifetime $\tau = l/c$ and the energy shell is accordingly smeared by an amount $\hbar/\tau$. For small $\mathbf{q}$, the corresponding intermediate states differ in energy by an amount $E_{N-j} - E'_j \simeq c\hbar\mathbf{q} \cdot \widehat{k}_{N-j}$ $(j = 1, \ldots, N-1)$ where $\widehat{k}_{N-j}$ are unit vectors in the direction of propagation of the intermediate plane wave states. The resulting phase difference $\Delta\phi \simeq \widehat{k}_{N-j} \cdot \mathbf{q}l$. Since the direction of intermediate states is random, it follows that after an $N$ step random walk the accumulated root mean square phase difference $\Delta\phi_{\mathrm{rms}} \simeq \sqrt{N/3}(ql)$. Therefore only steps with $N \lesssim 3/(lq)^2$ contribute to the backscattering peak at the angle defined by $\mathbf{q}$. This accounts for the rapid decrease of coherent intensity for large $q$. A detailed derivation of the line shape for scalar waves has been given by Akkermans, Wolf and Maynard.[14]

The vector nature of electromagnetic waves leads to a nontrivial polarization content for backscattering. Stephen and Cwilich[15] have shown that for linearly polarized light incident on a scattering medium, the backscattering peak consists of a sharp narrow peak polarized parallel to the incident light as well as a broader depolarized peak. More recently MacKintosh and John[16] have shown that the fundamental symmetries relevant to photon localization are manifest in the helicity representation. In addition to time reversal invariance, there is parity of the right and left hand circular polarization states. These symmetries may be broken by the Faraday effect and by natural optical activity[52] respectively.

In the Faraday effect the application of a strong magnetic field $\mathbf{H}$ leads to a difference in the refractive indices for right $(R)$ and left $(L)$ hand circularly polarized photons depending on their direction of propagation $\widehat{k}$ given by $n_{R(L)} \simeq n_0 \pm \mathbf{g} \cdot \widehat{k}/(2n_0)$ where the gyration vector $\mathbf{g} = f\mathbf{H}$ and $f$ is the rotary power of the material. The resulting breakdown of time-reversal invariance, which is manifest here as breakdown of symmetry between light propagating in the $\widehat{k}$ direction and $-\widehat{k}$ direction, leads to a field dependent suppression of the coherent intensity in the helicity preserving channel but leaves the helicity flip channel essentially unaffected. Detailed lineshapes have been calculated and are shown in Fig. 7a. Depicted is the excess relative intensity with respect to the incoherent background. For instance an intensity level of 1.0 corresponds to a precise doubling of the light intensity in a particular direction relative to the diffuse background level which is labeled with excess intensity 0.0. The angle $\Theta$ (measured in radians) is the angle which the wavevector of the backscattered light makes with respect to the vector $-\mathbf{k}_i$, where $\mathbf{k}_i$ is the incident direction. Also in Fig. 7, $k_0 \equiv \omega/c = 2\pi/\lambda$ and thus the quantity $(gk_0)^{-1}$ defines a characteristic length over which Faraday rotation takes place. As the Faraday rotation per unit length is increased, the coherent intensity is suppressed more and more.

Natural optical activity occurs in the presence of certain helically shaped molecules and may be described phenomenologically as a difference in the refractive index for $R$ and $L$ states irrespective of the direction of propagation: $n_{R/L} \simeq n_0 \pm f/(2n_0)$. This maintains time reversal invariance but breaks parity. The result is a suppression of the backscattering peak in the helicity flip channel as shown in Fig. 7b whereas the helicity preserving channel is largely unaffected. The top curve in Fig. 7b depicts the backscattering peak in helicity flip channel in the absence of optical activity. As can be seen it is somewhat broader and much flatter than that of the helicity preserving channel. As in Fig. 7a, $k_0 \equiv \omega/c$ and $(fk_0)^{-1}$ represents a characteristic length over which rotation of the polarization state takes place. A detailed derivation of these lineshapes may be found elsewhere.[16] In all cases, the difference in lineshape of the two helicity channels arises from the noncommutativity of three dimensional rotations of the polarization state in a given scattering sequence and its time reverse.

In deriving the backscattering lineshapes we have assumed point like scatterers. As a result, no restriction is placed on direction of intermediate virtual states. However, in the vicinity of a mobility edge induced by a pseudogap in the photon DOS this assumption is drastically modified. Coherent backscattering is restricted to certain Bragg resonance channels and the wavevector space available for intermedi-

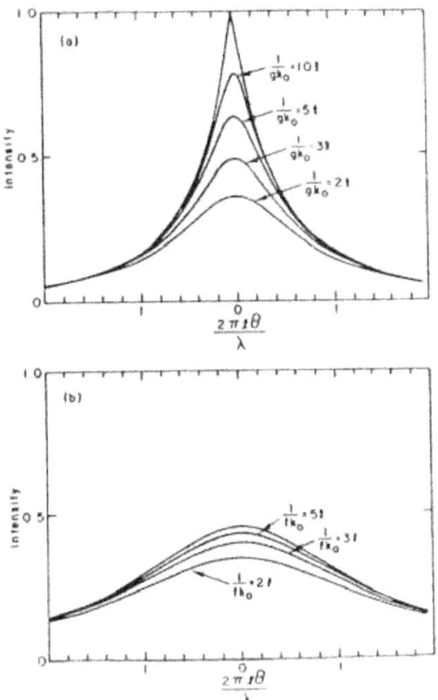

Fig. 7. (a) The effect of Faraday rotation on the peak line shape where $g = f\mathbf{B}$ is the gyration vector and $k_0 = \omega/c$. The curves shown are for incident and reflected light of the same circular polarization. The excess relative intensity with respect to the incoherent background is plotted as a function of the angle $\Theta$ (measured in radians) from the backscattering $-\mathbf{k}_i$ direction. The opposite helicity channel is unaffected by Faraday rotation. (b) The suppression of the intensity in the opposite helicity channel due to natural optical activity. The gyration vector $\mathbf{g} = f\hat{k}$ is proportional to the propagation direction. The helicity preserving channel is unaffected since time-reversal symmetry is retained. Both curves have been normalized by the isotropic background intensity in the helicity preserving channel.

ate states is greatly diminished. This may have observable effects on the backscattering lineshape and may be a direct probe of the structure factor of the disordered medium. The restriction of the phase space available for coherent backscattering provides a powerful mechanism for the occurrence of strong localization in three dimensions.[25] It is a much more likely source of localization than resonance modes of independent scatterers.[22-24]

## 4. FIELD THEORETICAL FORMULATION OF DIFFUSION: PHONONS IN A RANDOM BOND MODEL

In this section, I introduce the mathematical formalism upon which the heuristic discussion of the previous two sections is based. For the sake of simplicity, only the case of spatially uncorrelated disorder will be presented here. The effects of spatial correlations will be discussed formally in Sec. 5. There are a variety of models to which very similar mathematical approaches apply. Among these are noninteracting electrons in a static random potential, phonons in isotopically (random mass) disordered and structurally (random bonds) disordered solids, spin waves in a disordered magnet, classical scalar waves and physical electromagnetic waves in a random dielectric medium.

In the language of electronic localization which occurs in tight-binding models, random masses correspond to "diagonal disorder" whereas random bonds correspond to "off-diagonal disorder". In general, the discussion of "off-diagonal disorder" is mathematically more involved than that of "diagonal disorder" although the essential physics pertinent to localization is not qualitatively changed. In this section, I will discuss an example of the former, as illustrated by the random bond model for phonons in a structurally disordered solid. For simplicity attention will be restricted to scalar waves although the generalization to a vector displacement field is straightforward. In subsequent sections only "diagonal disorder" will be considered explicitly although the main results carry over to the random bond model as well. The main mathematical result of this section is to establish the precise relationship between wave diffusion and the two particle Green's function. The main physical result is that even for a dense random collection of scatterers, the bare frequency dependent diffusion coefficient $D(\omega)$ for phonons of frequency $\omega$ exhibits a Rayleigh scattering divergence in $d$-spatial dimensions of the form $\omega^{-(d+1)}$ as suggested by the single scattering result of Sec. 2. I begin this discussion with some necessary mathematical preliminaries. Readers who are familiar with the replica technique may wish to skip to the subsection entitled, "The Energy Diffusion Coefficient".

## 4.1. Gaussian Path Integrals and Replicas

The first mathematical preliminary, is the representation of the inverse of a matrix $A_{ij}$ by Gaussian integrals. Consider the "partition function"

$$Z_{\text{boson}} \equiv \int_{-\infty}^{\infty} \left( \prod_{m=1}^{N} dx_m \right) \exp\left( \frac{i}{2} x_j A_{jm} x_m \right) \tag{4.1}$$

for the real symmetric $N \times N$ matrix $A_{jm}$. Summation over repeated indices is implicit. Clearly $A$ may be diagonalized by an orthogonal transformation $O$:

$$O^T A O = D , \tag{4.2a}$$

where $D$ is the diagonal matrix whose entries $D_{ii} \equiv d_i \ i = 1, \ldots, N$ are the eigenvalues of $A$.

Since the Jacobian of the change of variables

$$y_i = O^T_{ij} x_j \tag{4.2b}$$

is unity, it follows that

$$Z_{\text{boson}} = \int_{-\infty}^{\infty} \left( \prod_{m=1}^{N} dy_m \right) \exp\left( \frac{i}{2} d_m y_m^2 \right) \tag{4.3a}$$

$$= \prod_{m=1}^{N} \left( \frac{-i d_m}{2\pi} \right)^{-1/2} = \det{}^{-1/2}\left( \frac{-iA}{2\pi} \right) . \tag{4.3b}$$

Similarly,

$$\langle x_j x_l \rangle \equiv \frac{1}{Z_{\text{boson}}} \int_{-\infty}^{\infty} \left( \prod_{m=1}^{N} dx_m \right) x_k x_l \exp\left( \frac{i}{2} x_j A_{jm} x_m \right) \tag{4.4a}$$

$$= A_{kl}^{-1} . \tag{4.4b}$$

For a more general Hermitian matrix $A_{jm}$ an analogous representation follows from the partition function

$$Z \equiv \int_{-\infty}^{\infty} \prod_{m=1}^{N} (dz_m \, dz_m^*) e^{-z_j^* A_{jm} z_m} , \tag{4.5}$$

where integration is over each complex $z_m$-plane. In this case there exists a unitary transformation $U$ on the $N$ complex variables $z_m$ for which $U^T A U$ is diagonal. It follows that

$$\langle z_k z_l^* \rangle \equiv \frac{1}{Z} \int \prod_{m=1}^{N} (dz_m \, dz_m^*) z_k z_l^* e^{-z_j^* A_{jm} z_m} , \qquad (4.6a)$$

$$= A_{kl}^{-1} . \qquad (4.6b)$$

The derivation of a renormalization group picture for the localization of waves is facilitated by the replica functional integral representation of the one and two particle Green's functions for the system. The Green's function for waves propagating in a disordered medium is the inverse of an operator containing a random element and the replica method provides an expression for this inverse which can be conveniently averaged.

The replica representation of $A_{kl}^{-1}$ follows by writting

$$A_{kl}^{-1} = \langle x_k x_l \rangle = \lim_{n \to 0} \langle x_k x_l \rangle Z_{\text{boson}} Z_{\text{boson}}^{n-1} \qquad (4.7a)$$

$$= \lim_{n \to 0} \int_{-\infty}^{\infty} \left[ \prod dx_m^\alpha \right] x_k^1 x_l^1 e^{(i/2) x_m^\alpha A_{mj} x_j^\alpha}, \quad \alpha = 1, \ldots, n . \qquad (4.7b)$$

Clearly, the form (4.7b) provides a convenient expression for cumulant averaging over the random matrix $A$. An alternative representation follows from the use of anticommuting Grassman integration variables $\{x_j, x_m\} = \{y_j, y_m\} = \{x_j, y_m\} = 0$. It follows from the rules of Grassman integration that

$$Z_{\text{fermion}} \equiv \int \prod_{i=1}^{N} (dx_i dy_i) e^{x_i A_{ij} y_j} \qquad (4.8a)$$

$$= \det{}^{1/2}(A) \qquad (4.8b)$$

leading to an expression analogous to (4.7) for $A_{kl}^{-1}$ in terms of $\langle x_k x_l \rangle_{\text{fermion}}$. This also suggests yet a third possible supersymmetric representation involving both classical and Grassman variables which avoids the use of replicas and the $n \to 0$ limit:

$$A_{kl}^{-1} = (2\pi i)^{-1/2} \langle x_k x_l \rangle Z_{\text{boson}} Z_{\text{fermion}} . \qquad (4.9)$$

Only the first of these possible representations, however, will be used in subsequent discussions. This leads to a field theory description of wave propagation in a disordered medium and a renormalization group analysis of the "phase transition" from extended to localized states. In analogy with the ferromagnetic phase transition in a Heisenberg spin system in $2 + \varepsilon$ dimension, localized states correspond to the high temperature paramagnetic phase and extended states to the low temperature phase. However, the nature of the spontaneously broken symmetry is entirely different from that of the simpler spin system.

## 4.2. The Energy Diffusion Coefficient

Consider an elastic medium in $d$-dimensions with a scalar displacement field $\phi$ of atoms from equilibrium described by the Lagrangian

$$L = \frac{1}{2} \int d^d x [\dot{\phi}^2(x,t) - V(x)|\nabla \phi(x,t)|^2] \; . \qquad (4.10)$$

Here, the atomic mass density has been taken to be unity and $V(x) = V_0 + V_1(x)$ is a randomly fluctuating nearest neighbor elastic coupling satisfying

$$\langle V_1(x) \rangle_{\text{ensemble}} = 0 \qquad (4.11a)$$

$$\langle V_1(x)V_1(y) \rangle_{\text{ensemble}} = \gamma^2 \delta^d(x - y) \; . \qquad (4.11b)$$

Elementary excitations in this disordered medium satisfy the associated Euler-Lagrange equation:

$$\ddot{\phi}(x,t) = \nabla \cdot (V(x)\nabla \phi) \; . \qquad (4.12)$$

This equation of motion leads to the local conservation law

$$\frac{\partial U(x,t)}{\partial t} + \nabla \cdot \mathbf{J}(x,t) = 0 \; , \qquad (4.13a)$$

where

$$U(x,t) = \frac{1}{2}\dot{\phi}^2 + \frac{1}{2}V(x)|\nabla \phi|^2 \qquad (4.13b)$$

is the phonon energy density and

$$\mathbf{J}(xt) = -V(x)\dot{\phi}\boldsymbol{\nabla}\phi \qquad (4.13c)$$

is the associated energy flux. The generalization to a vector displacement field $\boldsymbol{\phi}$ as discussed in the previous chapter is straightforward but does not alter the physics of localization.

The response of the system at point $x$ to an initial velocity impulse at $y$ is given by the retarded Green's function which satisfies the equation of motion (4.12) and the initial conditions

$$G(x, y, t) = 0 \quad t \leq 0 \qquad (4.14a)$$
$$\dot{G}(x, y, 0) = \delta^d(x - y) \ . \qquad (4.14b)$$

This leads to the equation

$$-\omega^2 G(x, y, \omega_+) = \nabla_x(V(x)\nabla_x G) + \delta^d(x - y) \ , \qquad (4.15a)$$
$$\omega_\pm \equiv \omega \pm i\eta \ . \qquad (4.15b)$$

The nature of normal modes is probed by giving the system at rest an initial kick and considering the long time behavior or the energy density (4.13b). In order to ensure that no net linear momentum is imparted to the system, the initial velocity distribution is chosen to be the divergence of a vector field:

$$\dot{\phi}(x, 0) = \boldsymbol{\nabla} \cdot \mathbf{u}(x) \ , \qquad (4.16)$$

where $\mathbf{u}(x) = \mathbf{u}(|x|)$ is highly peaked in the vicinity of the initial kick and zero elsewhere.

The long time behavior of any function $f(t)$ may be related to the value of its Laplace transform near the origin by the identity

$$\lim_{t \to 0} \frac{f(t)}{t^m} = \lim_{\eta \to 0}(2\eta)^{1+m} \int_0^\infty dt \, e^{-2\eta t} f(t) \quad m = 0, 1, 2 \dots \qquad (4.17)$$

choosing $f(t) = f_1 \cdot f_2$ where $f_1 \equiv e^{-\eta t} G(x, y, t)$ and $f_2 \equiv e^{-\eta t} G(x, y', t)$ and using Parseval's theorem which states that the $L^2$ Hilbert space scalar product is invariant under Fourier transformation:

$$\int_{-\infty}^{\infty} dt f_1^*(t) f_2(t) = \int_{-\infty}^{\infty} \frac{d\omega}{2\pi} \tilde{f}_1^*(\omega) \tilde{f}_2(\omega) \, ,$$

$$f_{1,2}(\omega) \equiv \int_{-\infty}^{\infty} dt \, e^{i\omega t} f_{1,2}(t) \tag{4.18}$$

it follows that

$$\lim_{t \to \infty} \frac{1}{t^m} V(x) |\nabla \phi|^2 = \lim_{\eta \to 0} (2\eta)^{1+m}$$

$$\times \int d^d y \, d^d y' \nabla \cdot u(y) \nabla \cdot u(y')$$

$$\times \int_{-\infty}^{\infty} \frac{d\omega}{2\pi} V(x) \nabla_x G(x, y, \omega_-) \nabla_x G(x, y', \omega_+) \, . \tag{4.19}$$

An analogous expression holds for the long time behavior of the kinetic energy density.

The averaged energy diffusion coefficient is related to the long time behavior of the energy density by

$$D \equiv \lim_{t \to 0} \frac{(1/t) \int d^d x \, x^2 \langle U(x, t) \rangle_{\text{ensemble}}}{\int d^d x \langle U(x, t) \rangle_{\text{ensemble}}} \tag{4.20}$$

Here, the ensemble average of the numerator and denominator may be performed separately since the denominator is simply the total energy injected into the medium by the initial kick and is independent of the particular realization of the disordered system. By the virial theorem, the kinetic and potential energy densities of this harmonic system contribute equally to the denominator of the above expression. Using Eq. (4.19) and the fact that $u(y)$ is highly peaked near $y = 0$, the potential energy contribution to the numerator may be approximated by

$$\frac{1}{2}(2\eta)^2 c^2 \int d^d x \, x^2 \int_{-\infty}^{\infty} \frac{d\omega}{2\pi} \left\langle V(x) \frac{\partial^2 G(x, 0, \omega_-)}{\partial x_i \partial y_1} \frac{\partial^2 G(x, 0, \omega_+)}{\partial x_i \partial y_1} \right\rangle_{\text{ensemble}} . \tag{4.21a}$$

Here, an integration to parts with respect to $y$ and $y'$ has been performed and the initial kick has been chosen such that

$$\int d^d y \, u_i(y) = c\delta_{il} \ . \tag{4.21b}$$

This differs from the corresponding contribution from the kinetic energy density by a term

$$(2\eta)^2 c^2 \int d^d x \int_{-\infty}^{\infty} \frac{d\omega}{2\pi} \left\langle V(x) \frac{\partial G(x,0,\omega_-)}{\partial y_1} x_i \frac{\partial^2 G(x,0,\omega_+)}{\partial x_i \partial y_1} \right\rangle \ . \tag{4.22}$$

However, if the spread of energy is in fact diffusive $(|x_i| \sim t^{1/2})$ this term should be negligible in the long time $(\eta \to 0)$ limit.

Since ensemble averaging restores isotropy to the system, (4.21a) may be symmetrized with respect to $y_i$, $i = 1, \ldots, d$ to give

$$D = \lim_{\eta \to 0} (2\eta) \frac{\int_{-\infty}^{\infty} d\omega \int d^d x \, x^2 \left\langle \frac{\partial^2 G(x,0,\omega_-)}{\partial x_i \partial y_j} \frac{\partial^2 G(x,0,\omega_+)}{\partial x_i \partial y_j} \right\rangle_{ensemble}}{\int_{-\infty}^{\infty} d\omega \int d^d x \left\langle \frac{\partial^2 G(x,0,\omega_-)}{\partial x_i \partial y_j} \frac{\partial^2 G(x,0,\omega_+)}{\partial x_i \partial y_j} \right\rangle_{ensemble}} \ . \tag{4.23}$$

Here, a further assumption of weak disorder has been made in the replacement of $V(x)$ by its average $V_0$.

### 4.3. Field Theory

The averaged Green's functions discussed in the previous section may be obtained as correlation functions in the $n_\pm \to 0$ limit from the replicated partition function

$$Z = \int D\phi \, e^{-(L_0^+ + L_0^- + L_{int})} \ , \tag{4.24a}$$

where

$$L_0^\pm = \frac{1}{2} \int d^d x \, \phi^\alpha(x)(\omega_\pm^2 + V_0 \nabla^2)\phi^\alpha(x); \quad \alpha = 1, \ldots, n_\pm \tag{4.24b}$$

and

$$L_{int} = -\frac{\gamma^2}{8} \int d^d x \, \partial_i \phi^\alpha \, \partial_i \phi^\alpha \, \partial_j \phi^\beta \, \partial_j \phi^\beta \quad \alpha, \beta = 1, \ldots, n_+ + n_- \ . \tag{4.24c}$$

Here, the abbreviated notation $\partial_i \equiv \frac{\partial}{\partial x_i}$ has been used and the contour of functional integration is from $-(e^{i\pi/4})_\infty$ to $(e^{-i\pi/4})_\infty$ for the $n_+$ retarded fields and from $-(e^{i\pi/4})_\infty$ to $(e^{i\pi/4})_\infty$ for the $n_-$ advanced fields.

The interaction term (4.24c) may be decoupled by the Hubbard-Stratonovitch transformation:

$$e^{-L_{int}} = \int DQ_{ij}^{\alpha\beta}$$
$$\times \exp\left\{ -\frac{1}{2} \int d^d x [Q_{ij}^{\alpha\beta} Q_{ij}^{\alpha\beta} + \gamma Q_{ij}^{\alpha\beta}(x)\partial_i \phi^\alpha(x)\partial_j \phi^\beta(x)] \right\} . \tag{4.25}$$

As may easily be verified by introducing a source term to (4.24a) the $\phi$ and $Q$ correlation functions are related by:

$$\langle Q_{ij}^{\alpha\beta}(x) \rangle = -\frac{\gamma}{2} \langle \partial_i \phi^\alpha(x)\partial_j \phi^\beta(x) \rangle \tag{4.25a}$$

and

$$\langle Q_{ij}^{+-}(x)Q_{ij}^{+-}(0) \rangle = \frac{\gamma^2}{4} \langle \partial_i \phi^+(x)\partial_j \phi^-(x)\partial_i \phi^+(0)\partial_j \phi^-(0) \rangle . \tag{4.25b}$$

The ensemble average defined by (4.8a) is given by

$$\langle G(x, y, \omega_+) \rangle_{ensemble} = -\langle \phi^+(x)\phi^+(y) \rangle . \tag{4.25c}$$

Integrating over the $\phi$-fields in (4.24a) with the substitution (4.25) yields a field theory in $Q$:

$$Z = \int DQ e^{-L[Q]} , \tag{4.26a}$$

where

$$L[Q] = \frac{1}{2} \, \text{tr} \, [\ln \, A(Q)] + \frac{1}{2} \int d^d x Q_{ij}^{\alpha\beta} Q_{ij}^{\alpha\beta} \tag{4.26b}$$

and

$$A(Q) = \begin{bmatrix} \omega_+^2 + V_0 \nabla^2 & 0 \\ 0 & \omega_-^2 + V_0 \nabla^0 \end{bmatrix} - \gamma \partial_i Q_{ij} \partial_j \ . \qquad (4.26c)$$

The Lagrangian $L[Q]$ possesses a saddle point $Q^0$ which is diagonal in both the replica indices $\alpha, \beta$ and Cartesian indices $i, j$ defined by

$$Q^0 = \delta_{ij} \left\{ \mathrm{Re}\ Q^+ \begin{pmatrix} I & 0 \\ 0 & I \end{pmatrix} + i\ \mathrm{Im}\ Q^+ \begin{pmatrix} I & 0 \\ 0 & -I \end{pmatrix} \right\} , \qquad (4.27a)$$

where

$$Q^+ = -\frac{\gamma}{2d} \int \frac{d^d k}{(2\pi)^d} k^2 G^+(k) \qquad (4.27b)$$

and

$$G^+(k) = \frac{1}{\omega_+^2 - k^2 (V_0 - \gamma Q^+)} \ . \qquad (4.27c)$$

The latter two equations define a coherent potential approximation (CPA) to the averaged one-phonon Green's function.

Expanding $L[Q]$ to quadratic order in the fluctuations $\widehat{Q} \equiv Q - Q^0$ transverse to the direction of broken symmetry defined by (4.27a) yields

$$L[\widehat{Q}] = \frac{1}{2} \int d^d x\, d^d y\ \widehat{Q}_{ij}^{+-}(x) C_{ijkl}(x - y) \widehat{Q}_{kl}^{+-}(y) + \ldots \qquad (4.28a)$$

where

$$C_{ijkl}(x) = \delta_{ik} \delta_{jl} \delta^d(x) - \frac{\gamma^2}{2} \frac{\partial^2 G^+(x, 0)}{\partial x_i \partial x_k} \frac{\partial^2 G^-(0, x)}{\partial x_j \partial x_l} \qquad (4.28b)$$

and $G^\pm$ are retarded and advanced one phonon Green's functions defined by the coherent potential approximation (4.27b) and (4.27c):

$$G^+(x, 0) = \int \frac{d^d k}{(2\pi)^d} e^{-ik \cdot x} G^+(k) \ . \qquad (4.28c)$$

The energy diffusion coefficient is then given by:

$$D = \lim_{\eta \to 0} (2\eta) \frac{\int_{-\infty}^{\infty} d\omega \int d^d x\, x^2 \langle \widehat{Q}_{ij}^{+-}(x) \widehat{Q}_{ij}^{+-}(0) \rangle}{\int_{-\infty}^{\infty} d\omega \int d^d x \langle \widehat{Q}_{ij}^{+-}(x) \widehat{Q}_{ij}^{+-}(0) \rangle} \qquad (4.29a)$$

$$= -\lim_{\eta \to 0} (2\eta) \frac{\int_{-\infty}^{\infty} d\omega \frac{\partial}{\partial p_i} \frac{\partial}{\partial p_i} \mathrm{Tr}\ C^{-1}(p)|_{p=0}}{\int_{-\infty}^{\infty} d\omega\ \mathrm{Tr}\ C^{-1}(0)} \ , \qquad (4.29b)$$

where

$$C_{ijkl}(p) = \int d^d x \, e^{ip \cdot x} C_{ijkl}(x) \qquad (4.29c)$$

and the Trace involves the contraction of the first two Cartesian indices with the last two in the form $ijij$. Clearly the diffusivity is determined by the low momentum expansion coefficients

$$C_{ijkl}(p) = C_0^{ijkl} + C_{2,m}^{ijkl} p_m^2 + \cdots , \qquad (4.30a)$$

where

$$C_0^{ijkl} = \delta_{ij}\delta_{jl} - \frac{\gamma^2}{2} \int d^d p \, p_i p_j p_k p_l |G^+(p)|^2 \qquad (4.30b)$$

and

$$C_{2,m}^{ijkl} = \gamma^2 \int d^d p \, p_i p_j p_k p_l \frac{\partial G^+}{\partial p_m} \frac{\partial G^-}{\partial p_m} . \qquad (4.30c)$$

Here, the summation over the repeated index $m$ in (4.30a) does not occur in (4.30c). By parity, both the coefficients $C_0$ and $C_{2,m}$ have a Cartesian index structure of the form const $\delta_{ik}\delta_{jl}$ + const$'$ $\delta_{ij}\delta_{kl}$ + const$''$ $\delta_{il}\delta_{jk}$ and therefore may be simultaneously diagonalized. Denoting the eigenvalues of $C_0$ and $C_{2,m}$ by $e_0^{ij}(\omega)$ and $e_m^{ij}(\omega)$ respectively it follows that the eigenvalue $e_0^{11}(\omega)$ corresponding to the eigenvector $\delta_{kl}$ is given by

$$e_0^{11}(\omega) = 1 - \frac{\gamma^2}{2d} \int d^d p (p^2)^2 |G^+(p)|^2 \qquad (4.31a)$$

$$= \eta \frac{2\pi\omega N(\omega^2)}{d(\text{Im } Q^+)^2} , \qquad (4.31b)$$

where

$$N(\omega^2) = -\frac{1}{\pi} \int \frac{d^d k}{(2\pi)^d} G^+(k) \qquad (4.31c)$$

is the phonon density of states. In obtaining the form (4.31b) use has been made of the CPA equation (4.27b). Since $e_0^{11}(\omega)$ vanishes as $\eta \to 0$, this demonstrates the existence of a Goldstone mode in the field theory and a nonzero diffusivity given by

$$D = \lim_{\eta \to 0} (2\eta) \frac{\int_{-\infty}^{\infty} d\omega \frac{C_2(\omega)}{C_0^2(\omega)}}{\int_{-\infty}^{\infty} d\omega \frac{1}{C_0(\omega)}} , \tag{4.32a}$$

where

$$C_0(\omega) = e_0^{11}(\omega) \tag{4.32b}$$

and

$$C_2(\omega) = \sum_{m=1}^{d} e_m^{11}(\omega) \tag{4.32c}$$

is the sum of the eigenvalues of $C_{2,m}^{ijkl}$ corresponding to eigenvector $\delta_{kl}$. Using the CPA equation (4.27b) this may also be evaluated:

$$C_2(\omega) = \gamma^2 \int d^d p \frac{\partial}{\partial p_l} (p_i p_k G^+(p)) \frac{\partial}{\partial p_l} (p_i p_k G^-(p)) \tag{4.33a}$$

$$= \gamma^2 \int d^d p \left\{ 2(d+1) p^2 |G^+(p)|^2 + (p^2)^2 |\nabla G^+(p)|^2 \right.$$

$$\left. + 2p^2 \left[ G^+(p)(\mathbf{p} \cdot \nabla) G^-(p) + G^-(p)(\mathbf{p} \cdot \nabla) G^+(p) \right] \right\} \tag{4.33b}$$

$$= \frac{1}{\mathrm{Im}\, Q^+} \left\{ 2(d+6)\pi N(\omega^2) - 4d\, \mathrm{Im} \left[ V_0 - \gamma Q^+ \frac{\partial}{\partial \omega^2} Q^+ \right] \right. \tag{4.33c}$$

$$\left. + \left[ \frac{|V_0 - \gamma Q^+|^2}{\gamma\, \mathrm{Im}\, Q^+} \frac{2\pi N(\omega^2)}{\mathrm{Im}\, Q^+} - 4d\, \mathrm{Re}\, \frac{\partial Q^+}{\partial \omega^2} \right] \right\} .$$

In the weak scattering limit, the CPA equation may be solved by making the approximation

$$\mathrm{Im}\, G^+(k) \simeq -\pi \delta(\omega^2 - V_0 k^2) . \tag{4.34}$$

It follows

$$N(\omega^2) \simeq \frac{1}{2} \frac{S_d}{(2\pi)^d} \frac{\omega^{d-2}}{V_0^{d/2}} \tag{4.35a}$$

and

$$\text{Im } Q^{+} \simeq \frac{\pi}{4d} \frac{S_d}{(2\pi)^d} \frac{\omega^d}{V_0^{d/2}} , \tag{4.35b}$$

where $S_d$ is the surface area of the unit $d$-dimensional sphere. Clearly in the weak scattering or low frequency limit, the dominant contribution to $C_2(\omega)$ in (4.33c) is given by

$$C_2(\omega) \simeq \frac{2\pi V_0^2 N(\omega)^2}{\gamma(\text{Im } Q^{+})^3} \sim \frac{1}{\gamma^4 \omega^{2d+2}} (\omega \to 0) . \tag{4.36a}$$

Using (4.31b) it also follows that

$$C_0(\omega) \sim \frac{\eta}{\gamma^2 \omega^{d+1}} (\omega \to 0) . \tag{4.36b}$$

Comparison of (4.32) with the spectral decomposition of the energy diffusion coefficient

$$D = \frac{\int_{-\infty}^{\infty} d\omega \, E(\omega) D(\omega)}{\int_{-\infty}^{\infty} d\omega \, E(\omega)} \tag{4.37}$$

suggests the identification of $C_0^{-1}(\omega)$ with the energy $E(\omega)$ injected into normal modes in the frequency range $\omega$ to $\omega + d\omega$ by the initial kick and $C_2(\omega)/C_0(\omega)$ with the spectral component $D(\omega)$ of the diffusion coefficient. In units in which the velocity of sound is unity, $D(\omega)$ is precisely the elastic mean free path of phonons of frequency $\omega$.

Clearly,

$$D(\omega) \sim \frac{1}{\gamma^2 \omega^{d+1}} \tag{4.38}$$

exhibits the same Rayleigh scattering behavior as in a random mass model of a disordered elastic medium.

The detailed derivation of the consequences of this behavior will be taken up in Secs. 6 and 7. For the moment I will state the main results. As in the case of photons, coherent backscattering gives rise to a renormalization of the bare diffusion coefficient $D(\omega)$. In both one and two dimensions $D(\omega)$ renormalizes to zero on long length scales so that all states are localized. It may be shown that for $d = 1$ the localization length, $\xi_{\mathrm{loc}}$ is simply a multiple of the elastic mean free path $l$. It follows from (3.15) that the effect of Rayleigh scattering is to cause $\xi_{\mathrm{loc}}$ to diverge as $\omega \to 0$.

$$\xi_{\mathrm{loc}} \sim l \sim \frac{1}{\omega^2} \quad (d = 1) \ . \tag{4.39a}$$

In two dimensions the corresponding divergence is exponentially fast:

$$\xi_{\mathrm{loc}} \sim l \exp\left(\frac{\omega}{c} l\right) \sim \frac{1}{\omega^3} e^{1/\omega^2} \ . \tag{4.39b}$$

Here $c$ is the speed of sound. In three dimensions, for localization to occur, the bare diffusion coefficient must fall below a certain critical value. Setting $D(\omega)$ equal to its critical value gives the Ioffe-Regel condition. Since $D(\omega)$ becomes arbitrarily large for small $\omega$, it follows that long wavelength phonons diffuse faster than the value required for localization. Low frequency states are extended in $d = 2 + \epsilon$ for arbitrary disorder. For $d = 3$, in a real solid phonon localization typically occurs only at very high frequencies near the Debye cutoff.

### 4.4. Discussion

The potential energy term in the Lagrangian (4.24) is similar in form to that of a Heisenberg ferromagnet with a weakly fluctuating exchange coupling $J(x)$ between nearest neighbor spins $\mathbf{S}(x)$:

$$H = \int d^d x J(x) |\nabla \mathbf{S}|^2, \quad \mathbf{S} \cdot \mathbf{S} = 1 \ . \tag{4.39c}$$

For example in $d = 3$, this leads to linearized equations of motion for small fluctuations $\mathbf{S} = (\psi_x, \psi_y, 0)$ from a direction $\hat{z}$ of spontaneous magnetization of the form:[53]

$$\frac{\partial \psi_x}{\partial t} = \mathrm{const}\ \boldsymbol{\nabla} \cdot \left(J(x)\boldsymbol{\nabla}\psi_y\right) \tag{4.40a}$$

$$\frac{\partial \psi_y}{\partial t} = -\,\mathrm{const}\ \boldsymbol{\nabla} \cdot \left(J(x)\boldsymbol{\nabla}\psi_x\right) \ . \tag{4.40b}$$

Spin waves of very long wavelength couple very weakly to disorder as in the case of phonons suggesting a qualitatively similar localization behavior.

The model presented in this section considers phonons in a harmonic approximation in the limit of weak disorder. Recent computer simulations by Nagel *et al.*[54] suggest that a qualitatively similar behavior may be found in the strong disorder limit of a glass. Grzonka and Moore[55] have also studied the nature of spin wave excitations in a planar spin glass by computer. The measured inverse participation ratios of these excitations likewise reveal that all states are localized in one and two dimensions with diverging localization lengths in the low frequency limit for the case of isotropic nearest spin interactions. In $d = 3$ there is a single mobility edge separating low frequency extended states from high frequency localized states. However, the addition of an anisotropic perturbation which breaks the global rotational symmetry of the system gives rise to localized states at both the high and low (non zero) frequency band edges.

## 5. WAVE PROPAGATION AND LOCALIZATION IN A LONG RANGE CORRELATED RANDOM POTENTIAL*

### 5.1. Introduction

The phenomenon of Anderson localization[1] is a fundamental property of waves in a disordered medium. The nature of this transition from extended to localized states has been studied extensively for elementary excitations in systems in which the disorder is spatially uncorrelated. In this section, we consider the transport properties of waves in a medium with long range correlated disorder. For exponentially decaying correlations, we examine the nature of normal modes of the disordered system as the range of correlations is made long compared to the wavelength of the excitation. The infinite-range case of power law decaying correlations is also discussed in detail. This latter case may be of importance for phonons in an elastic medium possessing vacancies, dislocations or other topological defects in the crystalline strucutre.[56] The long range strain field associated with a quenched distribution of such

---

*This section is reprinted from S. John and M. Stephen, *Phys. Rev.* **B 28** (1983) 6358.

defects gives rise to a scattering potential for phonons with power law decaying correlations.[57] Another example is that of electronic conduction in solids possessing a quenched distribution of defects with power law decaying impurity potentials. Also, in a polar semiconductor power law correlations may be realized by means of structural disorder. Here, the local charge imbalances associated with the deviation of atoms from their crystalline position produce long range correlated random electric fields familiar from the problem of the Urbach optical absorption edge.[58,59]

We show from first principles using the methods of field theory and renormalization that the localization transition for waves in a long range correlated random potential is in the same universality class as the spatially uncorrelated case. All states are localized in two dimensions and below. The crossover in $d \leq 2$ from localized to extended states as the range of correlations is made longer, and hence the potential smoother, is found to occur only in the limit of a completely flat potential. It is shown that in the presence of arbitrarily weak disorder, increasing the smoothness of the potential results simply in a corresponding increase of the localization lengths. Above two dimensions the mobility edge separating low frequency extended from high frequency localized states is shown to be characterized by critical exponents in accord with the theories of Wegner[60] and Abrahams *et al.*[26] of electronic conduction in a short range correlated random potential. As the range of correlations is made longer, this mobility edge moves to higher frequencies.

## 5.2. The Model

We consider the general problem of wave propagation in a medium with a correlated spatially varying index of refraction. For simplicity we consider a scalar wave equation in the continuum field theory limit. For concreteness we choose a model for elastic waves described by a displacement field $\phi$, in a medium consisting of atoms with spatially varying random masses $m(x)$ and constant "spring stiffness" $V_0$.

$$m(x)\frac{\partial^2 \phi}{\partial t^2} - V_0 \nabla^2 \phi = 0 \; . \tag{5.1}$$

The random masses here fluctuate about a mean value $m_0$

$$m(x) = m_0 + m'(x) \qquad (5.2a)$$

and are characterized by spatial correlations which we write as

$$\langle m'(x)m'(y)\rangle = B(x - y) \qquad (5.2b)$$
$$\langle m'(x)\rangle = 0 \ . \qquad (5.2c)$$

We assert that the critical properties of the present scalar model near the localization transition are identical to those of a general vector field propagating in an isotropic medium with inhomogeneous index of refraction.

As in the case of spatially uncorrelated disorder we probe the nature of normal modes of this system by introducing a spatially localized zero momentum phonon wavepacket and examining its subsequent time evolution as it scatters from random mass fluctuations in its environment. It is convenient to define an average diffusion coefficient in terms of the long time behavior of the vibrational energy $E(x,t)$:

$$D \equiv \lim_{t \to \infty} \frac{1}{t} \frac{\int d^d x\, x^2 \langle E(x,t)\rangle_{ensemble}}{\int d^d x \langle E(x,t)\rangle_{ensemble}} \ . \qquad (5.3a)$$

Here, the energy density takes the form

$$E(x,t) = \frac{1}{2} m(x)\dot{\phi}^2 + \frac{V_0}{2}|\nabla \phi|^2 \ . \qquad (5.3b)$$

The angular brackets denote an average over all possible realizations of the random mass field compatible with the conditions (5.2). By standard methods,[3] the diffusivity can be expressed in terms of a product of the retarded and advanced phonon Green's functions in the weak scattering limit

$$B(x - y)/m_0^2 \ll 1 \qquad (5.4a)$$

$$D = \lim_{\eta \to 0} \eta\, \frac{\int_{-\infty}^{\infty} d\omega\, \omega^2 \int d^d x\, x^2 \langle G(x,0;\omega_+)G(x,0;\omega_-)\rangle_{ensemble}}{\int_{-\infty}^{\infty} d\omega\, \omega^2 \int d^d x \langle G(x,0;\omega_+)G(x,0;\omega_-)\rangle_{ensemble}} \ . $$
$$(5.4b)$$

Here, $\omega_\pm \equiv \omega \pm i\eta$ denotes the addition of an infinitesimal imaginary part to the phonon frequency in the retarded and advanced Green's functions respectively. This latter form suggests a spectral decomposition of the diffusivity into contributions $D(\omega)$ from normal modes of a given frequency $\omega$.

$$D \equiv \frac{\int_{-\infty}^{\infty} d\omega \, D(\omega) E(\omega)}{\int_{-\infty}^{\infty} d\omega \, E(\omega)} \; . \tag{5.5}$$

Here, $E(\omega)$ is the spectral density of energy excited in the medium by the initially injected phonon wavepacket. The behavior of $D(\omega)$ on sufficiently long length scales, in the renormalization group sense, determines whether normal modes of a given frequency $\omega$ are extended or localized. If states of frequency $\omega$ are localized, then $D(\omega)$ will renormalize to zero whereas extended states of frequency $\omega$ gives rise to a finite value of the diffusivity at that frequency.

### 5.3. Field Theory Formulation

In order to calculate the diffusivity $D(\omega)$ we utilize the replica field representation[3,7,61] of the two particle phonon Green's function

$$G(x, y, \omega_+) G(x, y, \omega_-)$$
$$= \lim_{\substack{n_+ \to 0 \\ n_- \to 0}} \int D\phi \, \phi^{1+}(x) \phi^{1+}(y) \phi^{1-}(x) \phi^{1-}(y) e^{(L^+ + L^-)} \; , \tag{5.6a}$$

where

$$L^\pm = \frac{1}{2} \int d^d x \, \phi^\alpha (\omega_\pm^2 \, m(x) + V_0 \nabla^2) \phi^\alpha \quad \alpha = 1, \ldots, n_\pm \; . \tag{5.6b}$$

Here we have introduced two sets of replica indices denoted by superscripts $\pm$ for the retarded and advanced Green's functions respectively. Accordingly, the contours of integration are chosen for the $\pm$ replicas as shown in Fig. 8 so that the infinitesimal imaginary parts $\pm i\eta$ ensure convergence of the functional integrals. The symbol $D\phi$ denotes functional integration over all of the replica fields.

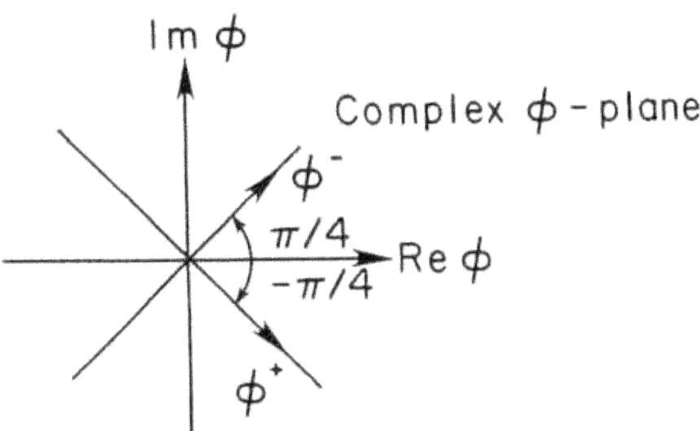



Fig. 8. Contour of Integration for the retarded (+) and advanced (-) replica fields.

The conditions (5.2) may be implemented by the probability distribution for the fluctuating part of the mass density

$$P\{m'(x)\} = \text{const } e^{-\frac{1}{2}\int d^d x d^d y \, m'(x) B_{0p}^{-1}(x-y) m'(y)} \ , \qquad (5.7\text{a})$$

where $B_{0p}^{-1}$ refers to an operator inverse of $B(x-y)$ with respect to its two coordinate space arguments:

$$\int d^d z B(x-z) B_{0p}^{-1}(z-y) = \delta^d(x-y) \ . \qquad (5.7\text{b})$$

Averaging over all possible realizations of the random mass field we obtain

$$\langle |G(x,y,\omega_+)|^2 \rangle_{\text{ensemble}} = \lim_{\substack{n_+ \to 0 \\ n_- \to 0}} \int D\phi \, \phi^{1+}(x)\phi^{1+}(y)$$
$$\times \, \phi^{1-}(x)\phi^{1-}(y) e^{-(L_0^+ + L_0^- + L_{\text{int}})} \ , \qquad (5.8\text{a})$$

where $L_0^{\pm}$ are obtained from $L^{\pm}$ by replacing $m(x)$ by its mean value $m_0$ and

$$L_{\text{int}} = \frac{-\omega^4}{8} \int d^d x d^d y \, \phi^\alpha(x)\phi^\alpha(x) B(x-y)\phi^\beta(y)\phi^\beta(y) \ . \qquad (5.8\text{b})$$

It is convenient to reduce this quartic replica field coupling, generated by averaging, through the introduction of a set of intermediate bilocal fields $Q^{\alpha\beta}(x,y)$, $\alpha,\beta = 1,\dots,n_+ + n_-$. The interacting part of the field theory (3.3b) can then be expressed as a functional integral over this set of fields:

$$e^{-L_{\mathrm{int}}} = \mathrm{const}\int DQ\, \exp\left\{-\frac{1}{2}\int d^dx\,d^dy[B^{-1}(x-y)Q^{\alpha\beta}(x,y)\right.$$
$$\left. \times Q^{\alpha\beta}(x,y) + \omega^2 Q^{\alpha\beta}(x,y)\phi^\alpha(x)\phi^\beta(y)]\right\}\ . \tag{5.9a}$$

Here, and in subsequent discussions $B^{-1}$ refers to the reciprocal of the correlation function

$$B^{-1}(x-y) \equiv \frac{1}{B(x-y)}\ . \tag{5.9b}$$

In terms of the intermediate fields, the generating functional for the averaged Green's functions may be expressed as a quadratic form on the $\phi$-fields:

$$Z(J) \equiv \lim_{\substack{n_+\to 0\\ n_-\to 0}}\int D\phi\, DQ$$
$$\times\, e^{-\frac{1}{2}(\phi,A(Q)\phi)-\frac{1}{2}\int d^dx\,d^dy[B^{-1}(x-y)Q^{\alpha\beta}Q^{\alpha\beta}+2J^{\alpha\beta}(x,y)Q^{\alpha\beta}(x,y)]}\ , \tag{5.10a}$$

where

$$\langle\phi,A(Q)\phi\rangle = \int d^dx\,d^dy\,\phi^\alpha(x)A^{\alpha\beta}(Q)\phi^\beta(y) \tag{5.10b}$$

and

$$A(Q) \equiv \begin{bmatrix} (m_0\omega_+^2 + V_0\nabla^2)\delta^d(x-y)\underline{I} & 0 \\ 0 & (m_0\omega_-^2 + V_0\nabla^2)\delta^d(x-y)I \end{bmatrix}$$
$$+\,\omega^2\begin{bmatrix} Q^{++}(x,y) & Q^{+-}(x,y) \\ Q^{-+}(x,y) & Q^{--}(x,y) \end{bmatrix}\ . \tag{5.10c}$$

Here $\underline{I}$ denotes the $n_\pm \times n_\pm$ identity matrix in replica space. Differentiation with respect to the source $J^{\alpha\beta}(x,y)$ before and after integration over the $\phi$-fields yields the following relations between the averaged Green's functions and the $Q$-correlation functions:

$$\langle Q^{\pm\pm}(x,y)\rangle = -\frac{\omega^2}{2}B(x-y)\langle G(x,y,\omega_\pm)\rangle_{\text{ensemble}}$$

$$(5.11a)$$

$$\langle Q^{+-}(x_1,y_1)Q^{+-}(x_2,y_2)\rangle = \frac{\omega^4}{4}B(x_1-y_1)B(x_2-y_2)$$
$$\times \langle G(x_1,x_2,\omega_+)G(y_1,y_2,\omega_-)\rangle_{\text{ensemble}} \cdot$$

$$(5.11b)$$

These are quantities relevant to the calculation of the average density of states and energy diffusivity respectively.

## 5.4. Spontaneous Symmetry Breaking and the Averaged One Particle Green's Function

Straightforward integration over the replica $\phi$-fields in (5.10) yields a field theory in the bilocal fields $Q$ of the form

$$Z(0) = \lim_{\substack{n_+ \to 0 \\ n_- \to 0}} \int DQ e^{-L[Q]}, \qquad (5.12)$$

where

$$L[Q] \equiv \frac{1}{2}\ln[\det A(Q)] + \frac{1}{2}\int d^d x d^d y\, Q^{\alpha\beta}(x,y)Q^{\alpha\beta}(x,y)B^{-1}(x-y).$$

$$(5.12b)$$

Evaluation of the expectation value $\langle Q\rangle$ with respect to the above action in a saddle point approximation yields the generalization to the case of spatially correlated disorder of the coherent potential approximation[62] (CPA) for the one particle Green's function. Expanding the Lagrangian $L[Q]$ about a point $Q_0$ and requiring that there be no linear terms in the small fluctuation $\widehat{Q} = Q - Q_0$ gives the condition that $Q_0$ be the required saddle point of the field theory. A straightforward calculation[63] yields

$$Q_0^{++}(xy) = \frac{-\omega^2}{2} B(x-y) G_0(x,y,\omega_+) \tag{5.13a}$$

$$Q_0^{--}(xy) = \frac{-\omega^2}{2} B(x-y) G_0(x,y,\omega_-) \tag{5.13b}$$

$$Q_0^{+-} = Q_0^{-+} = 0 , \tag{5.13c}$$

where

$$G_0(x,y,\omega_\pm) \equiv [(m_0 + Q_0^{\pm\pm})\omega_\pm^2 + V_0\nabla^2]_{xy}^{-1} . \tag{5.13d}$$

The condition (5.13c) corresponds to our choice of a replica diagonal
saddle point. Since averaging restores translational invariance, the CPA
Green's function $G_0$ depends only on the difference of its coordinate
space arguments. Defining the Fourier transform

$$Q^\pm(k) \equiv \int d^d(x-y) e^{ik(x-y)} Q_0^{\pm\pm}(x-y) \tag{5.14}$$

the condition (5.13d) on the one particle CPA Green's function may be
rewritten as

$$G_0^+(k) = \frac{1}{(m_0 + Q^+(k))\omega_+^2 - V_0 k^2} \tag{5.15a}$$

$$Q^+(k) = \frac{-\omega^2}{2} \int d^d q \, \tilde{B}(k-q) G_0^+(q) . \tag{5.15b}$$

In the last equation we have introduced the Fourier transform of the
correlation function:

$$\tilde{B}(k) \equiv \int d^d x \, e^{ik \cdot x} B(x) . \tag{5.15c}$$

As noted by Schäfer and Wegner,[61] the Lagrangian $L[Q]$ possesses
an internal global symmetry among the replicas. In the limit as $\eta \to 0$,
the symmetry between the retarded and advanced replica fields may be
expressed as

$$Q \to UQU^T . \tag{5.16a}$$

This leaves $L[Q]$ invariant for the set of "pseudo-orthogonal" matrices

$$U = \begin{pmatrix} \cosh\theta & i\sinh\theta \\ -i\sinh\theta & \cosh\theta \end{pmatrix} . \qquad (5.16b)$$

The saddle point solution (5.13) corresponds to a state of the field $Q$ in which this symmetry is spontaneously broken. This broken symmetry is manifest in the imaginary part of $Q_0$ for frequencies at which the density of states and hence $\mathrm{Im}\,Q^+(k)$ is non-zero

$$Q_0(k) = \mathrm{Re}\,Q^+(k)\begin{pmatrix} \underline{I} & 0 \\ 0 & \underline{I} \end{pmatrix} + i\,\mathrm{Im}\,Q^+(k)\begin{pmatrix} \underline{I} & 0 \\ 0 & -\underline{I} \end{pmatrix} . \qquad (5.17)$$

As usual,[3,7,61] by applying the symmetry operation $U$ to this solution we may generate a continuous manifold of distinct saddle points of the Langrangian $L[Q]$.

The CPA equations (5.15) must in general be solved self-consistently since the self energy (5.15b) of the Green's function involves the full propagator itself. We present approximate solutions to these equations for some particular choices of the correlation function $B(x)$ in the weak scattering limit (5.4b). Since the average mean free path for waves propagating in the disordered medium is related to the imaginary part of the self energy, we concentrate on the evaluation of this quantity:

$$\mathrm{Im}\,Q^+(k) = \frac{-\omega^2}{2}\int d^d q\,\tilde{B}(q-k)\,\mathrm{Im}\,G_0^+(q) . \qquad (5.18)$$

In the weak scattering limit, $\mathrm{Im}\,Q^+(k)$ is correspondingly small and so the imaginary part of the propagator (5.15a) takes the form of a delta function

$$\mathrm{Im}\,G_0^+(q) \rightarrow -\pi\delta(\omega^2 - q^2) , \quad (m_0 = V_0 = 1) . \qquad (5.19)$$

In this limit, the correlation function appearing in (5.18) may be evaluated for $|\mathbf{q}| = \omega$ and pulled out of the integral:

$$\mathrm{Im}\,Q^+(k) \simeq \frac{\omega^2}{2\pi m_0}\langle\tilde{B}(\mathbf{k} - \omega\hat{q})\rangle_{\hat{q}}\; N(\omega^2) . \qquad (5.20a)$$

Here we have introduced an angular average of the correlation function over the directions of the unit vector $\hat{q}$:

$$\langle \tilde{B}(\mathbf{k} - \omega\hat{q})\rangle_{\hat{q}} \equiv \frac{\int d\Omega\hat{q}\,\tilde{B}(\mathbf{k} - \omega\hat{q})}{\int d\Omega\hat{q}} \tag{5.20b}$$

and

$$N(\omega^2) \equiv \frac{-m_0}{\pi} \int d^d q \; \mathrm{Im}\; G_0^+(q) \tag{5.20c}$$

is the phonon density of states which has an asymptotic behavior $\sim \omega^{d-2}$ in the low frequency limit.[3]

We examine the dependence of the phonon phase coherence length

$$\Lambda(\omega) = \left[\frac{\omega^2}{V_0} \; \mathrm{Im}\; Q^+(\omega)\right]^{-\frac{1}{2}} \tag{5.21}$$

on the length scale of correlations in $d$-dimensions by means of the normalized, exponentially decaying correlation function

$$B(x) = \frac{\gamma^2}{a^d} e^{-x/a} \; . \tag{5.22}$$

Here $\gamma^2$ and $a$ determine the scattering strength and range of the correlations respectively. It is shown in Appendix A that in the limit of correlation length, $a$, very large compared to the phonon wavelength, the angular average in (5.20a) has the asymptotic behavior

$$\langle B\rangle \equiv \langle \tilde{B}(\mathbf{k} - \omega\hat{q})\rangle_{\hat{q}}\big|_{k=\omega} \sim \frac{\gamma^2}{(\omega a)^{d-1}}(\omega a \gg 1) \; . \tag{5.23a}$$

It follows that the phase coherence length diverges in the same limit as

$$\Lambda(\omega) \sim \left[\frac{m_0 V_0}{N(\omega^2)}\right]^{\frac{1}{2}} \frac{1}{\gamma\omega^2}(\omega a)^{(d-1)/2} \; . \tag{5.23b}$$

We consider also the behavior of the self energy (5.20a) in the limit of very long wavelength phonons. It is also shown in Appendix A that for any exponential correlation of the type of (5.22) or any power law decaying correlation of the form

$$B(x) = \frac{\gamma^2}{(x^2 + a^2)^m}, \quad (2m > d) \tag{5.24}$$

the low frequency $(\omega \to 0)$ asymptotic behavior of the self energy is the same as for uncorrelated disorder:

$$\text{Im } Q^+(\omega) \sim \omega^d, \quad (2m > d) . \tag{5.25a}$$

For longer range correlations $(0 < 2m \le d)$, however, this becomes

$$\text{Im } Q^+(\omega) \sim \begin{cases} -\omega^d \log \omega & (2m = d) \\ \omega^{2m} & (1 < 2m < d) \\ -\omega \log \omega & (2m = 1) \\ \omega^{m/(1-m)} & (m < \frac{1}{2}) \end{cases} . \tag{5.25b}$$

The stronger scattering in the low frequency limit apparent in this latter case is associated with the non-integrability of the correlation function (5.24) for $2m \le d$. Although the overall potential is smoother, the long range nature of the scattering centers gives rise to scattering which is stronger than the Rayleigh type (5.25a).

Finally in the limit of an ensemble of completely flat potentials $B(x) = \gamma^2$, the CPA equation (5.15) reduces to a simple quadratic equation with solution

$$G_0^\pm(k) = \frac{1}{\gamma^2 \omega^4} \left[ (m_0 \omega^2 - V_0 k^2) \mp \sqrt{(V_0 k^2 - m_0 \omega^2)^2 - 2\gamma^2 \omega^4} \right] . \tag{5.26}$$

We now proceed to examine the effect of these various spatial correlations on the average energy diffusivity.

## 5.5. Goldstone Modes and the Average Diffusivity

In a field theory with a spontaneously broken symmetry there is an associated Goldstone mode. In the present section we demonstrate the existence of a Goldstone mode associated with the symmetry (5.16) and discuss its relation to the average energy diffusivity.

We consider the expansion of the Lagrangian (5.12b) to second order in the fluctuations about the saddle point $Q_0$. Retaining only those fluctuations $\widehat{Q}^{+-}$ transverse to the direction of spontaneous symmetry breaking, which enter the expression (5.4b) for the diffusivity (see also (5.11b)) we may write

$$L[\widehat{Q}] = \frac{1}{2} \int d^d x_1 d^d x_2 d^d x_3 d^d x_4 \widehat{Q}^{+-}(x_1, x_2) C(x_1, x_2, x_3, x_4) \widehat{Q}^{-+}(x_3, x_4),$$

$$(5.27a)$$

where

$$C(x_1, x_2, x_3, x_4) \equiv B^{-1}(x_1 - x_2)\delta^d(x_1 - x_3)\delta^d(x_2 - x_4)$$
$$- \frac{\omega^4}{2} G_0^+(x_1 - x_3) G_0^-(x_2 - x_4) .$$

$$(5.27b)$$

By transforming to "center of mass" and relative coordinates

$$R_1 = \frac{x_1 + x_2}{2} \quad R_2 = \frac{x_3 + x_4}{2}$$
$$s_1 = x_1 - x_2 \quad s_2 = x_3 - x_4$$

$$(5.28)$$

and introducing the Fourier transform

$$C(K_1, K_2, k_1, k_2) \equiv \int d^d R_1 d^d R_2 d^d s_1 d^d s_2 e^{i(K_1 \cdot R_1 + K_2 \cdot R_2 + k_1 \cdot s_1 + k_2 \cdot s_2)}$$
$$\times C(R_1, R_2, s_1, s_2) \quad (5.29a)$$
$$\equiv C(K_1, k_1, k_2)\delta^d(K_1 - K_2) \quad (5.29b)$$

the Lagrangian for quadratic fluctuations may be expressed as

$$L[\widehat{Q}] = \frac{1}{2} \int d^d K d^d k_1 d^d k_2 \widehat{Q}^{+-}(K, k_1) C(K, k_1, k_2) \widehat{Q}^{-+}(K, k_2) ,$$

$$(5.30a)$$

where

$$C(K, k_1, k_2) \equiv \tilde{B}^{-1}(k_1 - k_2) - \frac{\omega^4}{2}\delta^d(k_1 - k_2) G_0^+\left(k_1 - \frac{K}{2}\right)$$
$$\times G_0^-\left(k_1 + \frac{K}{2}\right)$$

$$(5.30b)$$

and

$$\int d^d k_3 \, \tilde{B}(k_1 - k_3) \tilde{B}^{-1}(k_3 - k_2) = \delta^d(k_1 - k_2) \; . \qquad (5.30c)$$

The diagonal nature of this operator in $K$, expresses simply the translational symmetry of the theory resulting from averaging over the disorder.

A Goldstone mode in the field theory corresponds to a zero eigenvalue of the operator

$$C(0, k_1, k_2) = \delta^d(k_1 - k_2)[B^{-1}(i\nabla_{k_1}) - \frac{\omega^4}{2}|G_0^+(k_1)|^2] \; . \qquad (5.31a)$$

In order to obtain this differential form of the operator we have replaced the coordinate space argument of the reciprocal correlation function by $i\nabla_k$:

$$B^{-1}(i\nabla_k) \equiv \frac{1}{B(x \to i\nabla_k)} \; . \qquad (5.31b)$$

It is straightforward to verify, using the CPA equation (5.15) that there is in fact one zero eigenvalue in the limit as $\eta \to 0$ for which the associated eigenvector is Im $Q^+(k_2)$:

$$\int d^d k_2 \, C(0, k_1, k_2) \, \text{Im} \, Q^+(k_2) = -\eta m_0 \omega^3 |G_0^+(k_1)|^2 \to 0 \quad (\eta \to 0) \; . \qquad (5.32)$$

Since the diffusion coefficient is related to the operator $C$ by the expression

$$D = \lim_{\eta \to 0} \eta \frac{\int_{-\infty}^{\infty} \frac{d\omega}{\omega^2} \int d^d k_1 d^d k_2 \frac{d^2}{dK^2} C^{-1}(K, k_1, k_2)|_{K=0}}{\int_{-\infty}^{\infty} \frac{d\omega}{\omega^2} \int d^d k_1 d^d k_2 C^{-1}(0, k_1, k_2)} \qquad (5.33)$$

the existence of the Goldstone mode guarantees that the diffusivity is non-zero.

For the special case of a correlation function

$$B(x) = \frac{\gamma^2}{a^2 + x^2} \qquad (5.34a)$$

the operator $C \equiv C(0, k_1, k_2)$ takes the form of a Schrödinger operator

$$\gamma^2 C = -\nabla_k^2 + a^2 - \frac{\omega^4 \gamma^2}{2} |G_0^+(k)|^2 \; . \tag{5.34b}$$

We may use the relation

$$|G_0^+(k)|^2 = \frac{-\,\mathrm{Im}\; G_0^+(k)}{\omega^2\,\mathrm{Im}\; Q^+(k)} \tag{5.35}$$

to rewrite this operator in the weak scattering limit (5.19) as

$$\gamma^2 C = -\nabla_k^2 + V(k) \; , \tag{5.36a}$$

where

$$V(k) \equiv a^2 - g\delta(k - \omega), \quad (m_0 = V_0 = 1) \tag{5.36b}$$

and $g$ is a frequency dependent coupling constant. In $d = 3$, for instance, it is shown in Appendix B that

$$\frac{1}{g\omega} \equiv e^{-\omega a} \frac{\sinh \omega a}{\omega a} \; . \tag{5.36c}$$

This is a delta shell potential used sometimes in low energy nuclear scattering theory.[64] It is shown in Appendix 5.B, that there is precisely one $l = 0$ bound state for which the corresponding eigenvalue $\lambda_0$ of the operator $C$ is zero. This establishes the spherically symmetric Goldstone mode $\mathrm{Im}\; Q^+(k)$ as the ground state of this operator. Also for sufficiently high frequencies

$$g\omega > 2l + 1 \tag{5.37}$$

the delta shell potential admits bound states up to and including angular momentum $l$. In the geometrical optics limit $\omega a \gg 1$ the condition (5.37) is simply $\omega a > \frac{1}{2}(2l + 1)$. In this limit the disorder is extremely smooth on the scale of the phonon wavelength and the scattering is correspondingly weak. As a result the higher angular momentum eigenvalues $\lambda_l$ of the operator $C$ approach the Goldstone mode $\lambda_0 = 0$.

$$\lambda_l \sim \frac{2l(l+1)}{\gamma^2 \omega^2} \quad (\omega a \gg 1) \; . \tag{5.38}$$

This behavior is illustrative of the general case of correlations $B(x)$. For shorter range correlations of the form (5.22), the "kinetic energy" of the associated operator $C$, rather than taking the usual form $-\nabla_k^2$, will be exponentially large. By analogy with the spectrum of a quantum particle in a box, we expect the separation between bound states to be correspondingly large. In the opposite limit of longer range correlations than (5.34a), the "kinetic energy" of $C$ will be less dominant and so the separation between bound states smaller. Nevertheless in the presence of arbitrarily smooth disorder there is a gap between the Goldstone mode $\lambda_0$ and the first excited state $\lambda_1$. It is only in the limit of an ensemble of completely flat mass densities $B(x) = \gamma^2$, that (5.31) has no kinetic energy term:

$$\gamma^2 C(0, k_1, k_2) = \delta^d(k_1 - k_2) \left[ 1 - \frac{\gamma^2 \omega^4}{2} |G_0^+(k_1)|^2 \right] . \qquad (5.39\text{a})$$

It follows from the CPA equations (5.15) and (5.26) that for $(k^2 - \omega^2)^2 < 2\omega^4 \gamma^2$,

$$|G_0^+(k)|^2 = \frac{2}{\gamma^2 \omega^2} \qquad (5.39\text{b})$$

so that the continuous spectrum of (5.39a) includes an infinite set of Goldstone modes (see Fig. 9). An immediate consequence of this is that the energy diffusivity is infinite, as can be easily verified by direct evaluation of (5.33). Since in any particular realization of the mass density, there is no disorder, all phonon modes freely propagate.

For the case of decaying correlations such as (5.34a) we may replace $C^{-1}$ appearing in the diffusivity (5.33) by its projection onto the Goldstone subspace:

$$C^{-1}(K, k_1, k_2) \to \frac{1}{\lambda_0(K)} v_0^*(K, k_1) v_0(K, k_2) . \qquad (5.40)$$

Here, $v_0(K, k_2)$ is the eigenvector which as $K \to 0$ corresponds to the Goldstone mode Im $Q^+(k_2)$ and the associated eigenvalue may be expanded for small $K$ as

$$\lambda_0(K) = \lambda_0 + K^2 \lambda_0''(0) + \dots , \qquad (5.41\text{a})$$

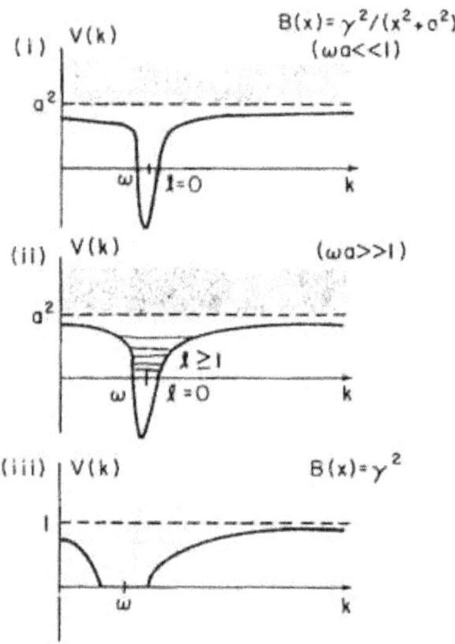



Fig. 9. Spectrum of the operator $C(0, k_1, k_2)$. For cases (i) and (ii) of correlations $B(x) = \gamma^2/(x^2 + a^2)$, the Goldstone mode $(l = 0)$ is separated from higher modes $(l \geq 1)$ by a gap. The continuous spectrum begins at $a^2$. In the long wavelength limit (i), there is only a single bound state, whereas in the geometrical optics limit (ii) the higher angular momentum bound states approach the Goldstone mode. For case (iii) $B(x) = \gamma^2$, the spectrum is that of a particle in a classical potential. There is no gap between the continuous spectrum and the infinite set of Goldstone modes in range $(k^2 - \omega^2)^2 < 2\omega^4 \gamma^2$.

where

$$\lambda_0''(K) \equiv \frac{1}{2} \frac{\partial^2}{\partial K^2} \lambda_0(K) . \tag{5.41b}$$

The diffusivity can then be expressed as

$$D = - \lim_{\eta \to 0} (2\eta) \frac{\int_{-\infty}^{\infty} d\omega \omega^2 N^2(\omega^2) \frac{\lambda_0''(0)}{\lambda_0^2}}{\int_{-\infty}^{\infty} d\omega \omega^2 N^2(\omega^2) \frac{1}{\lambda_0}} , \tag{5.42}$$

where $N(\omega^2)$ is the density of states (4.9c). Using (5.32) we may write

$$\lambda_0 = -\eta m_0 \omega^3 \int d^d k |G_0^+(k)|^2 \, \mathrm{Im} \, Q^+(k)$$

$$= \frac{-\eta}{\pi} m_0 \omega N(\omega^2) . \tag{5.43}$$

The coefficient $\lambda_0''(0)$ may be obtained by applying lowest order Rayleigh-Scrödinger perturbation theory to the operator $C(K, k_1, k_2)$, treating the terms arising from an expansion in small $K$ as a perturbation to the operator $C(0, k_1, k_2)$. This yields to leading order in the scattering strength $\gamma^2$ of the mass disorder:

$$\lambda_0''(0) = \frac{-\omega^4}{16d} \int d^d k (\mathrm{Im}\, Q^+(k))^2$$
$$\times \left[ \nabla_k^2 |G_0^+(k)|^2 - 4 \frac{\partial}{\partial k_i} G_0^+(k) \frac{\partial}{\partial k_i} G_0^-(k) \right] . \tag{5.44}$$

By substitution of (5.43) into (5.42) and comparison with (5.5) it is apparent that $\lambda_0''(0)$ is the conductance $\Sigma(\omega)$ associated with phonons of frequency $\omega$:

$$\Sigma(\omega) \equiv \rho(\omega) D(\omega) \tag{5.45a}$$
$$= \frac{4\pi}{m_0} \lambda_0''(0) . \tag{5.45b}$$

Here we have introduced the density of states normalized in the variable $\omega$:

$$\rho(\omega) \equiv 2\omega N(\omega^2) . \tag{5.46}$$

## 5.6. Spectral Content of Normal Modes

The eigenfunctions $v_l(0, k_1)$ of the operator $C(0, k_1, k_2)$ have a simple physical interpretation which we now proceed to describe. The Goldstone mode $v_0(0, k) = \mathrm{Im}\, Q^+(k)$ is associated with an exact "pseudo-orthogonal" symmetry (5.16) of the replica field Lagrangian. Physically, this symmetry expresses energy conservation for excitations in the disordered medium and leads to the existence of an isotropic diffusion mode in the two particle Green's function. As the disorder is made smoother and smoother a geometric optics limit is achieved in which there is also approximate momentum conservation over short length scales. The field theory realizes this approximate conservation law by means of an approximate translational symmetry in the interaction Lagrangian (5.8b). Namely, the replica fields $\phi^\alpha(x)$ and $\phi^\beta(y)$ can be independently translated on length scales over which $B(x - y)$ remains relatively flat. In the limit $B(x - y) = \gamma^2$ this symmetry becomes exact, leading to an infinite number of Goldstone modes (see Fig. 9 (iii)). The low lying eigenvalues $(l \geq 1)$ of the operator $C(0, k_1, k_2)$ for $\omega a \gg 1$ are an expression of this approximate symmetry and the propagation of plane wave-like excitations in the disordered medium over short distances.

If we introduce a transmitter of frequency $\omega$ into the disordered medium at a location $x = 0$, the response far away may be characterized by a coherence function[65]

$$\Gamma(x, x') \equiv \langle \phi(x, t)\phi^*(x', t) \rangle_{\text{ensemble}} \tag{5.47a}$$

$$= \langle G(x, 0, \omega_+)G(x', 0, \omega_-) \rangle_{\text{ensemble}} \ . \tag{5.47b}$$

This may be related to the present replica field theory by means of (5.11b)

$$\Gamma(R, s) = \frac{4}{\omega^4} B^{-1}(s)B^{-1}(0)\langle Q^{+-}(R, s)Q^{+-}(0, 0) \rangle \tag{5.48a}$$

where

$$R = \frac{x + x'}{2} \quad s = x - x' \ . \tag{5.48b}$$

Introducing the Fourier transform as in (5.29)

$$\Gamma(K, q) = \frac{\text{const}}{\omega^4} \int d^d k \, \tilde{B}^{-1}(q - k)\langle Q^{+-}(K, k)Q^{+-}(0, 0) \rangle \tag{5.49}$$

and expanding

$$Q^{+-}(K, k) = \sum_l Q_l(K)v_l(0, k) \tag{5.50}$$

in terms of the eigenfunctions of $C(0, k_1, k_2)$, the coherence for small $K$ (large $R$) may be expressed as:

$$\Gamma(K, q) = \frac{\text{const}}{\omega^4} \sum_l \frac{v_l \psi_l(q)}{\lambda_l + K^2 \lambda_l'' + \ldots} \ , \tag{5.51a}$$

where

$$v_l \equiv \int d^d k\, v_l(0,k) \qquad (5.51b)$$

and

$$\psi_l(q) \equiv \int d^d k\, \tilde{B}^{-1}(q-k) v_l(0,k) \ . \qquad (5.51c)$$

$\lambda_l''$ is the expansion coefficient of the $l^{\text{th}}$ eigenvalue $\lambda_l(K)$ of the operator $C(K,k_1,k_2)$.

At large distances $R$ from the transmitter, only the spherically symmetric ($l=0$) component has an appreciable amplitude, since $\lambda_0 = 0$. Higher angular momentum components are exponentially damped on a length scale determined by $(\lambda_l/\lambda_l'')^{-\frac{1}{2}}$. At high frequencies,

$$\lambda_l^{-\frac{1}{2}} \sim \frac{\gamma\omega}{\sqrt{2l(l+1)}} \qquad (\omega a \gg 1) \ . \qquad (5.52)$$

Clearly $\psi_l(q)$ represents the spectral content of the $l$th angular momentum component of the transmitted disturbance. For the isotropic diffusion mode ($l=0$),

$$\psi_0(q) = \int d^d k\, \tilde{B}^{-1}(k-q)\, \text{Im}\, Q^+(k) = \frac{-\omega^2}{2}\, \text{Im}\, G_0^+(q) \ , \qquad (5.53)$$

indicating that sufficiently far away from an arbitrary transmitter of frequency $\omega$, the set of wave-vectors excited in the medium is isotropic and highly peaked at $|\mathbf{q}| = \omega$ in the weak scattering limit.

## 5.7. The Nonlinear $\sigma$-Model

The nature of normal modes of frequency $\omega$ is determined by the manner in which the conductance $\Sigma(\omega)$ renormalizes as we integrate over short wavelength fluctuations of the field theory and rescale to longer lengths. For frequencies at which $\Sigma(\omega)$ renormalizes to zero we deduce that normal modes of frequency $\omega$ are localized. Extended states of frequency $\omega$ are characterized by corresponding $\Sigma(\omega)$ which renormalizes to a nonzero value in the long distance limit. Expanding the field $\widehat{Q}^{+-}$ in (5.30) in terms of the eigenfunctions $v_l(K,k_1)$ of

$C(K, k_1, k_2)$ (see (5.50)); the quadratic fluctuations about the saddle point may be written

$$L[Q] = \sum_l \int d^d K \, \lambda_l(K) Q_l(K) Q_l(-K) . \tag{5.54}$$

Since there is a gap between the Goldstone mode ($l = 0$) and other low lying modes of $C(0, k_1, k_2)$ for arbitrarily smooth disorder, the critical behavior of the field theory is determined by retaining the $l = 0$ term alone. The fluctuations about the saddle point manifold which we shall consider tending to restore the spontaneously broken symmetry (5.17) may be written as

$$Q(R, k) = i \, \text{Im} \, Q^+(k) U(R) \begin{pmatrix} I & 0 \\ 0 & -I \end{pmatrix} U^T(R) . \tag{5.55}$$

Comparison with the eigenfunction expansion (5.50) and using $v_0(0K) = \text{Im} \, Q^+(k)$ yields a nonlinear $\sigma$-model identical to that for uncorrelated disorder:[3,7,61]

$$L[Q] = \Sigma(\omega) \int d^d K \, K^2 Q_0^{\alpha\beta}(K) Q_0^{\alpha\beta}(-K) , \tag{5.56a}$$

where

$$Q_0(R) = iU(R) \begin{pmatrix} I & 0 \\ 0 & -I \end{pmatrix} U^T(R) . \tag{5.56b}$$

Here we have made use of the low momentum expansion (5.41a) of $\lambda_0(K)$ and the relation (5.45) between expansion coefficient $\lambda_0''(0)$ and the conductance $\Sigma(\omega)$. As in the uncorrelated case,[3] all normal modes are localized in two dimensions and below. In $d = 2 + \varepsilon$ there is a mobility edge $\omega_*$ separating low frequency extended states from high frequency localized states which is characterized by a localization length $\xi$ diverging as $|\omega - \omega_*|^{-1/\varepsilon}$. The divergence of the localization length as $\omega \to 0$ for $d \leq 2$ is governed by the corresponding asymptotic behavior of the bare conductance $\Sigma(\omega)$. This is evaluated in Appendix A using the solutions (5.25) of the CPA equation for correlations of the form (5.24).

$$\Sigma(\omega) \sim \begin{cases} \frac{1}{\omega^2} & (2m > d) \\[2mm] -\frac{1}{\omega^2 \log \omega} & (2m = d) \\[2mm] \omega^{d-2} - \frac{1}{\omega^{2m}} & (1 < 2m < d) \\[2mm] -\omega^{d-2} \frac{1}{\omega \log \omega} & (2m = 1) \\[2mm] \omega^{d-2} \frac{1}{\omega^{m/(1-m)}} & (0 < 2m < 1) \ . \end{cases} \qquad (5.57)$$

This leads to the following localization lengths in $d = 1$ and 2

$$d = 1 \quad \xi \sim \begin{cases} \frac{1}{\omega^2} & \left(\frac{1}{2} < m\right) \\[2mm] \frac{1}{\omega^2 \log \omega} & \left(m = \frac{1}{2}\right) \\[2mm] \frac{1}{\omega^{1/(1-m)}} & \left(0 < m < \frac{1}{2}\right) \end{cases} \qquad (5.58)$$

$$d = 2 \quad \xi \sim \begin{cases} e^{\frac{1}{\omega^2}} & (m > 1) \\[2mm] e^{-\frac{1}{\omega^2 \log \omega}} & (m = 1) \\[2mm] e^{\frac{1}{\omega^{2m}}} & \left(\frac{1}{2} < m < 1\right) \\[2mm] e^{-\frac{1}{\omega \log \omega}} & \left(m = \frac{1}{2}\right) \\[2mm] e^{-\frac{1}{\omega^{m/(1-m)}}} & \left(0 < m < \frac{1}{2}\right) \ . \end{cases} \qquad (5.59)$$

Similarly, the problem of electron localization in a long range correlated random potential is described within the same universality class. For the case of a Schrödinger equation

$$(-\nabla^2 + V(x))\psi(x) = E\psi(x) \qquad (5.60a)$$

with a random potential satisfying the conditions

$$\langle V(x)V(y)\rangle = B(x - y) \qquad (5.60b)$$

$$\langle V(x)\rangle = 0 \qquad (5.60c)$$

the quantity analogous to the energy diffusivity (5.3) is the diffusion coefficient for the locally conserved electron probability density:

$$D \equiv \lim_{t \to 0} \frac{\frac{1}{t} \int d^d x \, x^2 \langle |\psi(x, t)|^2 \rangle_{\text{ensemble}}}{\int d^d x \langle |\psi(x, t)|^2 \rangle_{\text{ensemble}}} \ . \qquad (5.61)$$

Here $\psi(x,t)$ is the electron wavefunction. The spectral decomposition of the diffusivity into contributions $D(E)$ from electron eigenstates of energy $E$ (compare with (5.5)) now takes the form

$$D \equiv \frac{\int_{-\infty}^{\infty} dE D(E)\rho(E)}{\int_{-\infty}^{\infty} dE \rho(E)} \, , \qquad (5.62)$$

where $\rho(E)$ is the electron density of states at energy $E$. The derivation of a nonlinear $\sigma$-model (5.56) follows in an analogous manner to that presented for the wave equation (5.1). The coupling constant $\Sigma(\omega)$ is now replaced by the d.c. electrical conductance

$$\Sigma(E) \equiv D(E)\rho(E) \, . \qquad (5.63)$$

Again, there is a gap between the Goldstone mode and higher angular momentum modes of the nonlinear $\sigma$-model, for arbitrarily long range correlations $B(x - y)$, leading to a lower critical dimension of two. All electron eigenstates are localized for $d \leq 2$. For $d = 2 + \varepsilon$, there are two mobility edges $\pm E_*$ characterized by localization lengths diverging as $\xi \sim |E \pm E_*|^{-1/\varepsilon}$. As in the case of uncorrelated disorder, the conductance vanishes linearly at the mobility edges:

$$\Sigma(E \simeq \pm E_*) \sim (E \pm E_*)^t \, , \quad t = 1 \, . \qquad (5.64a)$$

For the case of a system of non-interacting electrons in a long range correlated random potential filled to a fermi level $E_F$, the response to an external electric field of frequency $\nu$ is characterized by the a.c. conductance $\Sigma(E_F, \nu)$. The critical exponents are the same as those described in previous theories.[66,67,68]

$$\Sigma(E_F = E_*, \nu) \sim \nu^{(d-2)/d} (\nu \to 0) \, . \qquad (5.64b)$$

Also the static electrical polarizability diverges from the insulator side of the transition:

$$\alpha(E_F \simeq \pm E_*, \nu = 0) \sim |E_F \pm E_*|^{-2/\varepsilon} \qquad (5.64c)$$

## 5.8. Summary and Discussion

We have shown that the Anderson localization transition for waves in a disordered medium with spatial correlations as long as $B(x) = \gamma^2/(x^2 + a^2)^m$ $(m > 0)$ can be described as a phase transition in an appropriate nonlinear $\sigma$-model in $2+\varepsilon$ dimensions. Due to the existence of a gap between the Goldstone mode of the field theory and higher modes, this nonlinear $\sigma$-model is in the same universality class as that of uncorrelated disorder. Since the Goldstone mode and higher modes can be interpreted as the bound states of a quantum particle in a suitable potential well with a generalized "kinetic energy" we argue that for arbitrarily smooth disorder, the low lying spectrum of the nonlinear $\sigma$-model is discrete and hence leads to the same critical behavior as discussed in the theories of Wegner[60] and Abrahams *et al.*[26] The effect of long range correlations in the geometrical optics limit $\omega a \gg 1$ is to produce a corresponding increase in the mean free path or for $d \leq 2$ an increase in the localization lengths.

This is in contrast to the critical behavior of a Heisenberg ferromagnet with power law decaying exchange interactions. For example, in the case of spins with interactions falling off with spatial separation as $R^{-(d+\sigma)}$ and $(0 < \sigma < 2)$, the associated nonlinear $\sigma$-model has a weak singularity of the form $K^\sigma$ leading to the existence of a phase transition between the paramagnetic and ferromagnetic phases for all dimensions $d > \sigma$, (Ref. 69). In the present case the eigenvalue $\lambda_0(K)$ of the operator $C(K, k_1, k_2)$ is analytic near $K = 0$ leading to the usual $K^2$ term in the nonlinear $\sigma$-model (7.3). The only way that a lowering of the lower critical dimension could arise in the present model is by the existence of a continuous spectrum of $C(0, k_1, k_2)$ touching the Goldstone mode $\lambda_0 = 0$. This however, occurs only in the limit $B(x) = \gamma^2$ of perfect order within each realization of the mass field $m(x)$.

## 5.A. Appendix A: Asymptotic Behavior of the Mean Free Path and the Conductance

The behavior of the mean free path, the energy diffusivity and hence the localizations lengths in the asymptotic regimes $\omega \to 0$ and $\omega a \gg 1$ are governed by the solutions of CPA equation (5.15) in these same limits. In the weak scattering limit these follow from the evaluation of the angular average $\langle B \rangle \equiv \langle \tilde{B}(\mathbf{k} - \omega \hat{q}) \rangle_{\hat{q}}|_{k=\omega}$. We consider first the case of exponentially decaying correlations (4.11). In $d = 1$, this average is trivial and yields

$$\operatorname{Im} Q^+(k)|_{k=\omega} = \frac{\gamma^2 \omega}{2}\left[1 + \frac{1}{1 + (2\omega a)^2}\right] \to \gamma^2 \omega \quad \omega a \gg 1 \ . \quad (5.65)$$

For $d = 2$ we may write

$$\langle B \rangle = \frac{1}{2}\int_0^\infty dx\, x J_0^2(\omega x) B(x) \quad (5.66\text{a})$$

$$= \frac{\gamma^2}{2}\frac{\partial}{\partial a}\int_0^\infty dx e^{-x/a} J_0^2(\omega x) \ , \quad (5.66\text{b})$$

where $J_0$ is a Bessel function. The integral here may be expressed as a complete elliptic integral $K(s)$ of the first kind[70]

$$\langle B \rangle = \frac{\gamma^2}{\pi}\frac{\partial}{\partial s}(sK(s))\frac{\partial s}{\partial y} \ , \quad (5.66\text{c})$$

where

$$s \equiv y/\sqrt{1 + y^2} \ , \quad y \equiv 2\omega a \ . \quad (5.66\text{d})$$

In the limit of the correlation length much longer than the phonon wavelength $\omega a \gg 1$, we may use the asymptotic form of the elliptic inetgral[71]

$$K(s) \sim \frac{1}{2}\ln\frac{16}{1 - s} \ , \quad s \to 1 \quad (5.66\text{e})$$

leading to the required asymptotic behavior

$$\langle B \rangle \sim \frac{\gamma^2}{2\pi(\omega a)} \quad (\omega a \gg 1) \ . \quad (5.66\text{f})$$

Finally for $d = 3$, the angular average may be expressed in terms of the spherical Bessel functions $j_0$:

$$\langle B \rangle = \frac{\gamma^2}{2\pi^2 a^3}\int_0^\infty dx\, x^2 j_0^2(\omega x) e^{-x/a} \quad (5.67\text{a})$$

$$= \frac{\gamma^2}{4\pi^2}\frac{1}{1 + (a\omega)^2} \ . \quad (5.67\text{b})$$

We consider also the low frequency ($\omega \to 0$) limit. From (5.20a).

$$\text{Im } Q^+(\omega) \sim \omega^d \langle B \rangle \ . \tag{5.68}$$

For any integrable $\int d^d x B(x) < \infty$ correlation function $\langle B \rangle$ approaches a finite constant as $\omega \to 0$. For power law correlations of the form (5.24) in the non-integrable range $1 < 2m \leq d$,

$$\langle B \rangle \sim \begin{cases} -\log \omega a & 2m = d \\ \\ \omega^{2m-d} & 1 < 2m < d \ . \end{cases} \tag{5.69}$$

If $2m \leq 1$, the de function approximation (5.19) to the CPA Green's function is no longer self consistent. Long range power laws of this form result in a singularity in the angular average of the Fourier transformed correlation function

$$\langle \tilde{B}(\mathbf{k} - \mathbf{q}) \rangle_{\hat{q}} \sim \begin{cases} -\log |k - q| & (2m = 1) \\ \\ |k - q|^{2m-1} & (2m < 1) \ . \end{cases} \tag{5.70}$$

This leads to a divergent self energy if the approximation (5.19) is used. More generally, we may write

$$\text{Im } Q^+(k) \sim \omega^4 \int_0^\infty \frac{dq\, q^{d-1}}{(kq)^{(d-1)/2} |k-q|^{1-2m}}$$
$$\times \frac{\text{Im } Q^+(q)}{(\omega^2 - q^2)^2 + \omega^4 (\text{Im } Q^+(q))^2} \ . \tag{5.71}$$

It follows that

$$[\text{Im } Q^+(\omega)]^{2-2m} \sim \begin{cases} \omega^{2m} & (2m < 1) \\ \\ -\omega \log \omega & (2m = 1) \end{cases} \tag{5.72}$$

yielding the required result for the low frequency asymptotic behavior of the self energy (5.25b) for $2m \leq 1$.

The low frequency behavior of the conductance (5.44) is governed by the term

$$\Sigma(\omega) \sim \frac{\omega^4}{4} \int d^d k (\mathrm{Im}\, Q^+(k))^2 \frac{\partial}{\partial k_i} G_0^+(k) \frac{\partial}{\partial k_i} G_0^-(k) \,. \qquad (5.73\mathrm{a})$$

Using

$$\frac{\partial}{\partial k_i} G_0^+(k) = -(G_0^+(k))^2 \left( -2k_i + \omega^2 \frac{\partial}{\partial k_i} Q^+(k) \right) \qquad (5.73\mathrm{b})$$

the most singular contribution to the condustance as $\omega \to 0$ may be written (see also Ref. 3, Appendix B)

$$\Sigma(\omega) \sim \int d^d k \, k^2 (\mathrm{Im}\, G_0^+(k))^2 \qquad (5.73\mathrm{c})$$

$$\sim \frac{\omega^{d-2}}{\mathrm{Im}\, Q^+(\omega)} \,. \qquad (5.73\mathrm{d})$$

Using the asymptotic behavior of the self energy $\mathrm{Im}\, Q^+(\omega)$ from (5.68), (5.69) and (5.72) we obtain the required result (5.57).

## 5.B.  Appendix B: Spectrum of the Nonlinear $\sigma$-Model

We consider the spectrum of the Schrödinger equation

$$[-\nabla_k^2 + V(k)]v_l = \gamma^2 \lambda_l v_l \qquad (5.74\mathrm{a})$$

for the delta shell potential

$$V(k) \equiv a^2 - \frac{\gamma^2 \delta(k^2 - \omega^2)}{N(\omega^2)\langle B \rangle} \,. \qquad (5.74\mathrm{b})$$

For $d = 3$,

$$N(\omega^2)\langle B \rangle = \frac{\gamma^2 \omega}{\pi} \int_0^\infty dx \frac{x^2 j_0^2(\omega x)}{x^2 + a^2} \qquad (5.75\mathrm{a})$$

$$= \frac{\gamma^2}{2} e^{-\omega a} \frac{\sinh \omega a}{\omega a} \,. \qquad (5.75\mathrm{b})$$

This establishes the result (5.36c). The potential may be rewritten in the form

$$V(k) = a^2 - g\delta(k - \omega) , \qquad (5.76a)$$

where

$$\frac{1}{-g\omega} = \omega a j_0(i\omega a) h_0(i\omega a) . \qquad (5.76b)$$

Here, the spherical Bessel and Hankel functions have a pure imaginary argument. It is possible to show[64] that the condition for a bound state of angular momentum $l$ is given by

$$-\frac{1}{g\omega} = \beta j_l(i\beta) h_l(i\beta) , \qquad (5.77a)$$

where $\beta$ is related to the eigenvalue $\gamma^2 \lambda_l$ of the operator (5.74a) by

$$\gamma^2 \lambda_l = a^2 - \beta^2/\omega^2 . \qquad (5.77b)$$

Clearly there is precisely one solution for $l = 0 (\beta = \omega a)$ for which the corresponding eigenvalue $\lambda_0 = 0$. For sufficiently large values of $\omega a$ the equation (5.77a) admits higher angular momentum bound states. In the geometric optics limit $\omega a \gg 1$, these excited states ($l \geq 1$) approach the ground state $l = 0$. Using the asymptotic (large $\beta$) forms of the Bessel function,[71] Eq. (5.77a) becomes

$$-\frac{1}{2\omega a} \sim -\frac{1}{2\beta}\left[1 - \frac{l(l+1)}{\beta^2}\right] . \qquad (5.78a)$$

It follows that

$$\beta \sim \omega a - \frac{l(l+1)}{\omega a} . \qquad (5.78b)$$

Substitution into (5.77b) yields the asymptotic behavior of the $l^{\text{th}}$ bound state

$$\lambda_l \sim \frac{2l(l+1)}{\gamma^2 \omega^2} \quad (\omega a \gg 1) . \qquad (5.78c)$$

## 6. MOBILITY EDGES, LOCALIZATION
## AND ABSORPTION

The incorporation of wave interference and, in particular, cohenrent backscattering into the theory of diffusion of wave energy leads to a simple renormalization group picture of transport (see Fig. 10). The careful definition of the diffusion coefficeint $D(\omega)$, for waves of frequency $\omega$ developed in Sec. 3 is of particular importance here. In a situation where wave interference plays an important role in determining transport, the spread of wave energy is not diffusive at all in the sense that a photon performs a classical random walk. This presents a very complicated situation. Fortunately, there is a way of applying the concept of classical diffusion here provided we make one major concession in our classical way of thinking: the diffusion coefficient is no longer a local quantity determined by a classical mean free path and a speed of propagation but depends on the macroscopic coherence properties of the entire illuminated sample. In a random medium it is reasonable to expect that scatterers which are very far apart do not on, *average*, cause large interference corrections to the classical diffusion picture. (The word average here is very important. Changes in distant scatterers *can* give rise to significant fluctuations about the average.) It follows that there exists a coherence length $\xi_{coh} \gtrsim l$ which represents a scale on which we must very carefully incorporate interference effects in order to determine the effective diffusion coefficient at any point within the coherence volume. To put it in other words, the possible amplitudes for a photon to diffuse from point $A$ to point $B$ within a coherence volume $\xi_{coh}^d$ interfere significantly with each other. Depending on the distance between the point $A$ at which the photon is injected into the medium and the point $B$ at which it is detected, the effective diffusion coefficient of the photn is strongly renormalized by the wave interference. Another example is that of a finite size sample of linear size $L$. By changing the scale of the sample, the number of diffusion paths which can interfere changes giving rise to an effective diffusion coefficient $D(L)$ at any point within the sample which depends on the macroscopic scale $L$ of the sample. (Here the dependence of $D$ on $\omega$ is suppressed for convenience).

There are many formal mathematical ways of describing this physical picture.[3,4,6−8,26] In the nonlinear $\sigma$-model approach it is expressed in terms of a dimensionless scale dependent resistance variable $g(L)$ which is closely related to the diffusion coefficient $D(L)$. Here again $L$ denotes the linear scale of the sample and on the scale of the elastic mean free path, the resistance $g(l)$ is equal to the bare resistance given by $[\rho(\omega)D(\omega)l^{d-2}]^{-1}$ where $\rho(\omega)$ is the photon DOS. For point-like scatterers, this reduces to $g(l) \sim (\frac{\omega}{c}l)^{1-d}$. This is just the classical diffusion result which incorporates no wave interference corrections. In the vicinity of a mobility edge, there arises a new coherence length $\xi_{\mathrm{coh}} \geq l$ such that on length scales $L$ in the range $l < L < \xi_{\mathrm{coh}}$ the spread of energy is subdiffusive in nature as a result of coherent backscattering which gives a significant wave interference correction to classical diffusion. In this range, the spread of wave energy may be interpreted in terms of a scale dependent diffusion coefficient which behaves roughly as $D(L) \simeq \frac{cl}{3}\left(\frac{l}{L}\right)$. On length scales long compared to $\xi_{\mathrm{coh}}$, the photon resumes its diffusive motion except with a lower or renormalized value $\frac{cl}{3}\left(\frac{l}{\xi_{\mathrm{coh}}}\right)$ of the diffusion coefficient.

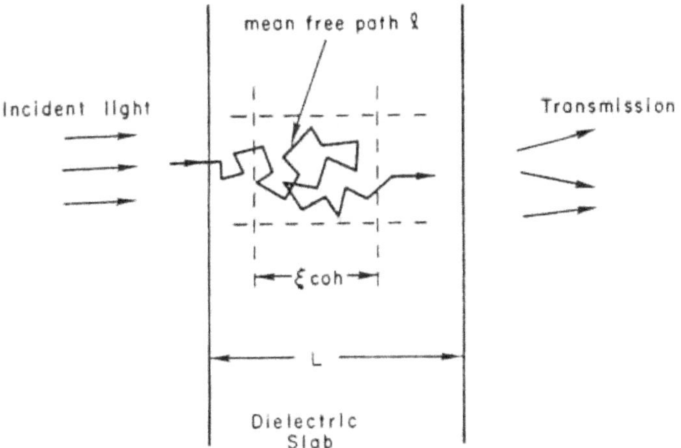


Fig. 10. Physical picture of optical transport at incipient localization. On scales short compared to the coherence length $\xi_{\mathrm{coh}}$, the spread of electromagnetic energy is subdiffusive in nature due to coherent backscattering. On scales longer than $\xi_{\mathrm{coh}}$, the photon resumes the diffusive behavior except with a renormalized diffusion coefficient $D \simeq \frac{cl}{3}(\frac{l}{\xi_{\mathrm{coh}}})$.

Strictly speaking this physical picture is an extrapolation of a renormalization group analysis in $d = 2 + \epsilon$ dimensions which becomes asymptotically exact as $d \to 2$ and for $g \ll 1$ or $l \gg \lambda$. As in the scaling theory of electron localization it is the best available first principles result for three dimensions. A self consistent theory of localization which interpolates between the weak scatterers and strong localization regimes has been derived by Vollhardt and Wölfle.[74] This may be particularly useful for quantitative comparison between theory and experiment. Optical systems provide a unique opportunity to experimentally test nearly every aspect of this interpolation and extrapolation. The result in $d = 2 + \epsilon$ dimensions is identical to that for electrons and may be summarized by a simple differential equation for the resistance first obtained by Abrahams *et al.*[26]

$$\frac{dg(L)}{d\ln L} = (2 - d)g + g^2 + \dots . \tag{6.1}$$

In three dimensions the solution of this equation yields $\xi_{\mathrm{coh}} \sim |\omega - \omega_*|^{-1}$ where $\omega_*$ is the mobility edge frequency and[21,25]

$$D(L) \simeq \frac{cl}{3}\left( \frac{l}{\xi_{\mathrm{coh}}} + \frac{l}{L} \right) . \tag{6.2}$$

The relevance of this result to an optical transmission experiment in the absence of dissipation or absorption is depicted in Fig. 11. Consider first the case in which the coherence length is short compared to the slab thickness. The time required for an incident photon to traverse the thickness $L$ is given by $\tau(L) = L^2/D(L)$. For $l \lesssim \xi_{\mathrm{coh}} \ll L$, the average displacement $R$ of the photon as a function of time is that of classical diffusion $R \sim t^{1/2}$. In the case of incipient localization $l \ll L \ll \xi_{\mathrm{coh}}$, the diffusion coefficient has the value $D(L) = \frac{cl}{3}(\frac{l}{L})^{d-2}$. The transit time from one face of the slab to the other now scales as $\tau(L) \sim L^d$. In other words the mobility edge regime is characterized by a critical slowing down of the photon which now traverses a distance $R \sim t^{1/d}$ rather than that of a classical random walker. The critical slowing down of light is also likely to be apparent in long time tails associated with picosecond pulse propagation of essentially monochromatic light through the slab.[72,73] For classical diffusion, the transmitted intensity at times $t \gg \tau(L)$ the average time of flight decays with time as $\exp(-\pi^2 Dt/L^2)$. Since the classical diffusion coefficient enters directly into this expression, it follows that in the case of incipient localization this expression must be generalized by introducing either a scale or time dependence to the parameter $D$.

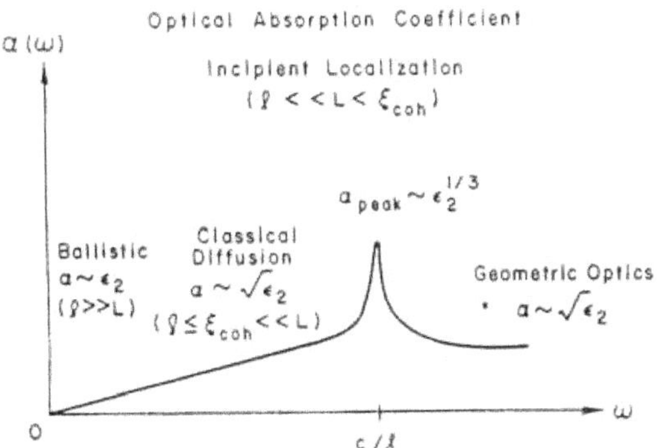


Fig. 11. Scaling of the absorption coefficient $\alpha$ with the imaginary part $\epsilon_2$ of the dielectric constant. As $\omega \to 0$, propagation is ballistic $l \gg L$ the sample size and $\alpha \sim \epsilon_2$. In diffusive scattering regime $l \ll L$ $\alpha \sim \sqrt{\epsilon_2}$. At incipient localization $l \ll L < \xi_{coh}$, $\alpha \sim \epsilon_2^{1/3}$. For very high frequencies the geometric optics regime is reached and $\alpha \sim \sqrt{\epsilon_2}$ again. All results are for weakly dissipative ($l \ll l_{inel}$) systems.

Recently Anderson[21] has suggested that anomalies associated with incipient localization may appear in the total transmitted intensity through a disordered dielectric slab illuminated by a steady state monochromatic plane wave source. For the case of classical diffusion the transmission coefficient $T$ defined as the ratio of the total transmitted intensity to the total incident intensity is given by the relation $T = l/L$ where $l$ is the classical elastic mean free path. This may also be written as $T = \frac{3D}{cL}$. Anderson's conjecture is that more generally one may replace the classical diffusion coefficient in this expression by the scale dependent function (6.2). This then leads to a transmission coefficient $T \sim l^2/(\xi_{coh} L)$ for $l \lesssim \xi_{coh} \ll L$ but to a new scale dependence $T \sim l^2/L^2$ in the incipient localization regime $l \ll L \ll \xi_{coh}$. This conjecture is appealing because of its simplicity but is by no means *a priori* unique. Further theoretical work is essential here. A complete

                    *Sajeev John*

first principles theory of the wavevector dependence of the two photon
Green's function in the critical regime is required to fully resolve this
issue.

## 6.1. Weakly Dissipative Media

In the case of electrons there is a conservation law which prevents
their total number from changing. Photons on the other hand can be
absorbed. The discussion of classical localization is therefore not complete without the analysis of wave propagation in a weakly dissipative
disordered medium. By weak dissipation I mean that the inelastic mean
free path or typical distance between absorption events is large compared to $l$ but nevertheless smaller than the sample size $L$. This may
be introduced by means of a small imaginary part $\epsilon_2$ to the dielectric
constant $\epsilon(x) = \epsilon_0 + \epsilon_{\text{fluct}}(x) + i\epsilon_2$. Within a classical diffusion approximation the ensemble averaged two-photon propagator takes the
form[48]

$$\int d^d x e^{i\mathbf{p}\cdot\mathbf{x}} \langle |G(x,0,\omega_+)|^2 \rangle_{\text{ensemble}} \sim \frac{1}{\epsilon_2 + D(\omega)p^2 + \ldots} . \tag{6.3}$$

The optical absorption coefficient $\alpha$ is defined as the decay constant
for the intensity from a source of intensity $I_0$: $I = I_0 e^{-\alpha x}$. It follows
from (6.3) that $\alpha \sim \sqrt{\epsilon_2/D}$. The effects of coherent wave interference may be incorporated into this picture by means of a straightforward renormalization group analysis. From (6.2) it follows that for an
infinite medium ($L = \infty$) the diffusion coefficient vanishes as the coherence length diverges. If $l_{\text{inel}} > \xi_{\text{coh}}$, it follows that the absorption
coefficient increases in the same manner that the diffusion coefficient
$D(\omega) \sim \omega - \omega_*$ decreases as the mobility edge frequency $\omega_*$ is ap-
proached from the extended state side: $\alpha(\omega) \sim \sqrt{\epsilon_2/|\omega - \omega_*|}$. On the
other hand, if the coherence length exceeds the inelastic length, then
$l_{\text{inel}}$ acts as a long distance cutoff for coherent wave interference. It
follows that even at the mobility edge frequency $\omega_*$ there is a residual
diffusivity given by $D(\omega^*) \simeq \frac{cl}{3}\left(\frac{l}{l_{\text{inel}}}\right)^{d-2}$. The inelastic mean free path
which enters here must be determined self consistently from the actual
diffusion coefficient: $l_{\text{inel}} = \sqrt{D\tau_{\text{inel}}}$. Here $\tau_{\text{inel}}$ is the inelastic mean

free time and is related to the imaginary part of the dielectric constant as $\tau_{\text{inel}} \sim 1/(\epsilon_2 \omega)$. Solving the two equations relating the residual diffusivity to $\epsilon_2$ yields $D(\omega^*) \sim \epsilon_2^{(d-2)/d}$.

This result for the residual diffusivity is analogous to the critical behavior of the a.c. conductivity of non-interacting electron gas as the Fermi level passes through the electronic mobility edge.[66] Substituting the value of the residual diffusivity into the expression for $\alpha$ reveals that the absorption coefficient as a function of frequency exhibits a peak at the mobility edge $\omega^*$ which scales as $\alpha(\omega^*) \sim \sqrt{\epsilon_2/D(\omega^*)} \sim \epsilon_2^{1/d}$. The physical origin of the critical exponent $1/d$ is the critical slowing down of the photon as it approaches localization. This is precisely the exponent which describes time of flight $(R \sim t^{1/d})$ at incipient localization. The behavior of the absorption coefficient in various scattering regimes is depicted schematically in Fig. 11.

The anomalies in absorption associated with localization are a general indication of enhanced coupling of the electromagnetic field to other degrees of freedom. It is almost certain to enhance nonlinear effects and especially those of a self focussing nature. A strong localization pseudogap in the photon DOS is likely to be accompanied by various forms of optical bistability as the light intensity is increased. Such effects may have useful technological applications. If for instance, a change from localized to extended state behavior at a particular laser frequency can be produced in a suitable medium by small variations in the incident light intensity, this may be useful as a switching device. If the light intensity can be varied by using a second beam which simultaneously propagates through the medium, the large differential gain which is likely to occur near a mobility edge may provide a useful transistor effect.

# 7. RENORMALIZATION AND THE NONLINEAR $\sigma$-MODEL[*]

## 7.1. Dynamical Response of a Non-interacting Electron Gas Near the Localization Transition

The linear response of a non-interacting electron gas in a disordered solid is determined by the relative position of the fermi level $E_F$ with respect to the mobility edge $E_*$ separating localized from extended one-electron states. If $E_F$ lies in the region of extended states the solid exhibits metallic behavior and has a nonzero d.c. electrical conductivity. If $E_F$ lies in the region of localized states the d.c. conductivity is zero and the solid has a finite electrical polarizability characteristic of an insulator. It has been shown by Wegner[60] in $d = 2 + \varepsilon$ dimensions and independently by Abrahams *et al.*[26] in $d = 3$ that the d.c. conductivity vanishes as

$$\sigma_{d\cdot c} \sim \left(E_F - E_*\right)^{(d-2)\nu}, \quad \nu = 1/\varepsilon, \tag{7.1}$$

when the Fermi level approaches the mobility edge from the metallic side.

In this section, I derive from first principles the critical behavior of the a.c. conductivity at the localization transition

$$\sigma_{a\cdot c}(\omega, E_F = E_*) \sim \omega^{(d-2)/d}, \quad (\omega \to 0) \tag{7.2}$$

and the divergence of the static electrical polarizability as $E_F$ approaches the mobility edge from the insulating side:

$$\alpha(\omega = 0) \sim \frac{1}{|E_F - E_*|^{2/\epsilon}} \tag{7.3}$$

using the $\eta \to 0$ replica field representation of the electron-hole propagator and momentum shell renormalization scheme. These results have also been obtained by Shapiro and Abrahams[66] using a physical argument.

---

[*]Reprinted from Sajeev John Ph.D. thesis, Harvard University, Chap. 7, (1984).

## 7.2. Nonlinear $\sigma$-model with Dynamical Symmetry Breaking

Consider a gas of non-interacting electrons at zero temperature in a random potential $V(x)$ described by the Hamiltonian

$$H = \int d^d x\, \psi^+(x)[-\nabla^2 + V(x)]\psi(x), \quad \frac{h^2}{2m} = 1 , \qquad (7.4)$$

where $\psi(x)$ is the second quantized electron annihilation field operator. The dynamical linear response of this system to a perturbing field of frequency $\omega$ is given by

$$\chi(x, x'; \omega) \equiv i \int_0^\infty dt e^{i\omega_+ t} \langle 0|[\rho(r,t), \rho(r',0)]|0\rangle \qquad (7.5)$$

$$\omega_\pm = \omega \pm i\eta ,$$

where $\rho \equiv \psi^+ \psi$ is the electron density operator and $|0\rangle$ denotes the ground state of the electron gas. The ensemble averaged response (averaged over all possible realizations of $V(x)$) for sufficiently low frequencies $\omega$ may be expressed in terms of the advanced and retarded one-electron Green's function as follows (see for instance Vollhardt and Wolfle[74])

$$\chi(q, \omega) \equiv \int d^d(x - x') e^{iq\cdot(x-x')} \langle \chi(x, x'; \omega)\rangle_{\text{ensemble}} \qquad (7.6a)$$

$$\simeq \omega\phi(q,\omega) + N(E_F) , \qquad (7.6b)$$

where $N(E_F)$ is the electronic density of states at the Fermi level,

$$\phi(q,\omega) = -\frac{1}{2\pi i} \int d^d x e^{iq\cdot x} \left\langle G^R\left(x, 0; E_F + \frac{\omega}{2}\right) \right.$$
$$\left. \times\, G^A\left(x, 0; E_F - \frac{\omega}{2}\right) \right\rangle_{\text{ensemble}} \qquad (7.6c)$$

and in terms of the one electron eigenfunctions $\phi_n(x)$ and energy levels $E_n$,

$$G^{R,A}(x, x'; E) = -\sum_n \frac{\phi_n(x)\phi_n^*(x')}{E - E_n \pm i\delta} . \qquad (7.6d)$$

The hydrodynamic response function (7.6a) is related to the wave-vector and frequency dependent diffusion coefficient $D(q,\omega)$ for density fluctuations in the electron gas (see for instance Forster[75]) by

$$\chi(q,\omega) = \frac{iq^2 D(q,\omega)}{\omega + iq^2 D(q,\omega)} N(E_F) \,. \tag{7.7}$$

From (6b) it follows that

$$\phi(q,\omega) = \frac{N(E_F)}{\omega + iq^2 D(q,\omega)} \,. \tag{7.8}$$

For the case of a random potential $V(x)$ satisfying

$$\langle V(x) \rangle_{\mathrm{ensemble}} = 0 \tag{7.9a}$$

and

$$\langle V(x)V(y) \rangle_{\mathrm{ensemble}} = \gamma^2 \delta^d(x-y) \tag{7.9b}$$

the averaged electron-hole propagator appearing in (7.6c) has the replicated functional integral representation

$$\left\langle G^R\left(x,0;E_F + \frac{\omega}{2}\right) G^A\left(x,0;E_F - \frac{\omega}{2}\right) \right\rangle$$
$$= \lim_{\eta_\pm \to 0} \int D\phi \, \phi_+^1(x)\phi_+^1(0)\phi_-^1(x)\phi_-^1(0)e^L \,, \tag{7.10a}$$

where

$$L = L_0^+ + L_0^- + L_{\mathrm{int}} + i\frac{\omega}{4}\int d^d x \, \phi^\alpha(x)\phi^\alpha(x) \quad \alpha = 1,\dots,n_+ + n_- \tag{7.10b}$$
$$L_0^\pm = \pm\frac{i}{2}\int d^d x \, \phi^\alpha(x)(-\nabla^2 - E_F \pm i\eta)\phi^\alpha(x) \tag{7.10c}$$

and

$$L_{\mathrm{int}} = -\frac{\gamma^2}{8}\int d^d x (\phi_+^\alpha \phi_+^\alpha - \phi_-^\alpha + \phi_-^\alpha)^2 \quad \alpha = 1,\dots,n_\pm \,. \tag{7.10d}$$

Here the contour of functional integration is along the real $\phi$-axis. The structure of the field theory is similar[48] to that for waves propagating in a disordered medium with a small imaginary part to the dielectric constant. Accordingly, the two particle Green's function (7.10a) may be re-expressed in terms of a quadropole field $Q_{(x)}^{\alpha\beta}$:

$$\left\langle G^R\left(x,0;E_F+\frac{\omega}{2}\right)G^A\left(x,0;E_F-\frac{\omega}{2}\right)\right\rangle$$
$$=\frac{4}{\gamma^2}\lim_{n_\pm\to 0}\int DQ\, Q_{+-}^{11}(x)Q_{+-}^{11}(0)e^{L[Q]}\ , \tag{7.11a}$$

where

$$L[Q]=L_0[Q]-\frac{\omega}{2\gamma}\int d^d x(Q_{++}^{\alpha\alpha}-Q_{--}^{\alpha\alpha}) \tag{7.11b}$$

and

$$L_0[Q]=-\frac{1}{2}\operatorname{tr}\ln\left\{\begin{bmatrix}-\nabla^2-E_F-i\eta & 0 \\ 0 & -\nabla^2-E_F+i\eta\end{bmatrix}+\gamma\underline{Q}\right\}$$
$$-\frac{1}{2}\int d^d x\, Q^{\alpha\beta}Q^{\alpha\beta}\ . \tag{7.11c}$$

Here, I have made use of the identity:

$$e^{L_{\mathrm{int}}}=\int DQ e^{-\frac{1}{2}\int d^d x\{\operatorname{tr} Q^2+\gamma(\phi,\underline{C}Q\underline{C}\phi)\}}\ , \tag{7.12a}$$

where $(\ ,\ )$ denotes the $n_++n_-$ dimensional scalar product in replica space and

$$\underline{C}\equiv\begin{pmatrix}e^{i\pi/4}\underline{I}_{n_+} & 0 \\ 0 & e^{-i\pi/4}\underline{I}_{n_-}\ .\end{pmatrix} \tag{7.12b}$$

$\underline{I}_{n_\pm}$ denote $n_\pm$-dimensional identity matrices. In the absence of a perturbing external field of frequency $\omega$, it is evident from (7.10) that the Lagrangain $L$ is invariant under transformation

$$\phi\to U\phi\ , \tag{7.13}$$

where $U$ is any member of the group $\mathbf{O}(n_+, n_-)$ and $\phi$ denotes an $n_+ + n_-$ dimensional column vector.

The domain of integration for the field $Q$ in (7.11a) is dictated by the requirements that the integral (7.12a) converges for any real $\phi$ and that upon substitution of (7.12a) for (7.10d) the functional integral over $\phi$ in (7.10a) converges uniformly in $Q$. The latter criterion would not be satisfied for instance by choosing the domain of $Q$ integration to be the group of real symmetric matrices. The correct domain may be obtained by noting that a symmetry transformation $\phi \to U\phi$ in (7.12a) is equivalent to a transformation $Q \to \tilde{U}^T Q \tilde{U}$ where $\tilde{U} = \underline{C} U \underline{C}^*$. Accordingly, the domain of $Q$-integration is the group of matrices which may be expressable in the form[61]

$$Q(x) = \tilde{U}^T(x) D_0(x) \tilde{U}(x) , \qquad (7.14)$$

where $D_0$ is an arbitrary real diagonal $n_+ + n_-$ dimensional matrix and $\tilde{U}$ is an arbitrary pseudo-orthogonal matrix as defined above ($\tilde{U}^T \tilde{U} = I$). The identity (7.12a) may be verified by deformation of the hypercontour (7.14) onto the group of real symmetric matrices and integrating over $Q$.

The nonlinear $\sigma$ model arises from the deformation of the hypercontour (7.14) to pass through the saddle point $Q^0$ of $L_0[Q]$ defined by

$$Q^0 = \mathrm{Re}\ Q^+ \begin{pmatrix} I_{n_+} & 0 \\ 0 & I_{n_-} \end{pmatrix} + i\ \mathrm{Im}\ Q^+ \begin{pmatrix} I_{n_+} & 0 \\ 0 & -I_{n_-} \end{pmatrix} \qquad (7.15a)$$

where $Q^+$ is the averaged one-electron Green's function in a coherent potential approximation:

$$Q^+ = -\frac{\gamma}{2} \int \frac{d^d k}{(2\pi)^d} G^+(k) , \qquad (7.15b)$$

where

$$G^+(k) = \frac{1}{k^2 - E_f - i\eta + Q^+} . \qquad (7.15c)$$

Expanding $L_0[Q]$ to quadratic order in the fluctuation $\widehat{Q} \equiv Q - Q^0$ leads to the nonlinear $\sigma$-model

$$L_{\text{eff}}[Q] = -\frac{1}{2} \int \frac{d^d k}{(2\pi)^d} C(k) Q^{\alpha\beta}(k) Q^{\alpha\beta}(-k) - \frac{\omega}{2\gamma} \int d^d x (Q^{\alpha\alpha}_{++} - Q^{\alpha\alpha}_{--}) ,$$
$$(7.16\text{a})$$

where

$$C(k) \equiv 1 - \frac{\gamma^2}{2} \int d^d q\, G^+(k+q/2) G^-(k-q/2) = \frac{2}{\pi\gamma^2} k^2 \frac{D(k,0)}{N(E_F)}$$
$$(7.16\text{b})$$

with the constraint on the field $Q$ defined by (14), that $D_0(x) = Q^0$. Since $\operatorname{Im} Q^+ = -\frac{\pi}{2}\gamma N(E_F) < 0$ the distortion of the $D_0$ contour to pass through the saddle point $Q^0$ avoids the singularities of the logarithm in (11c) as required. The nonlinear $\sigma$-model makes the further approximation of replacing the contour of integration for each of the elements of $D_0$ by the single point $Q^0$. Fluctuations in the eigenvalues of $Q$ about $Q^0$ correspond to massive modes of the field theory and are therefore irrelevant in the renormalization group sense. The effective Lagrangian (7.16a) retains only the leading quadratic term in massless modes of the field theory. Higher order terms may be obtained by direct substitution of (7.14) into (7.11c) and then cumulant averaging terms involving the massless fluctuations $\tilde{U}(x)$ with respect to the massive modes by integrating over the massive modes $D(x)$ (see for instance Pruisken[76]).

Rescaling the field $Q^{\alpha\beta}(x)$ by a factor $\operatorname{Im} Q^+$ and coordinate and momentum variables by the elastic electron mean free path $l$ as discussed previously[48] leads to a nonlinear $\sigma$-model in terms of the new dimensionless variables $Q, x$ and $k$:

$$L_{\text{eff}}[Q] = -\frac{1}{2g} \int \frac{d^d k}{(2\pi)^d} k^2 Q^{\alpha\beta}(k) Q^{\alpha\beta}(-k) + h \int d^d x (Q^{\alpha\alpha}_{++} - Q^{\alpha\alpha}_{--})$$
$$|k| < 1 \qquad\qquad (7.17\text{a})$$

with the constraint

$$\underline{Q}(x) = i\tilde{U}^T(x) \begin{pmatrix} \underline{I}_{n_+} & 0 \\ 0 & -\underline{I}_{n_-} \end{pmatrix} \underline{\tilde{U}}(x) , \qquad (7.17\text{b})$$

where the dimensionless coupling constant is

$$\frac{1}{g} \equiv \frac{\pi}{2} N(E_F) D(E_F) l^{d-2} \, , \, D(E_F) \equiv D(k=0, \omega=0) \qquad (7.17\text{c})$$

and the dimensionless symmetry breaking field is

$$h \equiv \frac{\pi}{4} \omega N(E_F) l^d \, . \qquad (7.17\text{d})$$

It is shown in Appendix A that the constraint (7.17b) is equivalent to writing

$$\underline{Q}(x) = \begin{pmatrix} i\sqrt{1 + \underline{\Pi}\,\underline{\Pi}^T} & \underline{\Pi} \\ \underline{\Pi}^T & -i\sqrt{1 + \underline{\Pi}^T\underline{\Pi}} \end{pmatrix} \, , \qquad (7.18)$$

where $\underline{\Pi}(x)$ is a general real matrix. Substituting this into (7.17a) gives

$$L_{\text{eff}}[\Pi] = -\frac{1}{g} \int d^d x \left[ \operatorname{tr} \partial_\mu \Pi \partial_\mu \Pi^T \right.$$

$$\left. - \frac{1}{2} \operatorname{tr} \partial_\mu \Pi \Pi^T \Pi \partial_\mu \Pi^T - \frac{1}{2} \operatorname{tr} \partial_\mu \Pi^T \Pi \partial_\mu \Pi^T \Pi + \dots \right]$$

$$+ ih \int d^d x \left[ \operatorname{tr} \Pi \Pi^T - \frac{1}{4} \operatorname{tr} \Pi^T \Pi \Pi^T \Pi + \dots \right] \, . \qquad (7.19)$$

As shown in Appendix B, a straightforward momentum shell integration and rescaling leads to the following recursion relations at the one loop level in the $n \to 0$ limit:

$$\frac{dg}{d\ln L} = -\varepsilon g + \frac{1}{8\pi} \frac{g^2}{1 - igh} \qquad (7.20\text{a})$$

$$\frac{dh}{d\ln L} = dh \, . \qquad (7.20\text{b})$$

Linearization about the Anderson-Wegner fixed point $(g^*, h^*) = (8n'\varepsilon, 0)$ yields

$$\frac{d\Delta g}{d\ln L} = \varepsilon \Delta g, \quad \Delta g \equiv g - g^* \sim (E_F - E_*) . \tag{7.20c}$$

The quantities of physical interest, the dynamical conductivity $\sigma(\omega)$ and the electrical polarizability $\alpha(\omega)$ are given by

$$\sigma(\omega) = -i\omega\alpha(\omega) = e^2 \lim_{q \to 0} \left(\frac{-i\omega}{q^2}\right) \chi(q, \omega) . \tag{7.21}$$

In the nonlinear $\sigma$-model the electron-hole propagator determining $\chi(q, \omega)$ is a function of only the two scaling variables $g$ and $h$. For fixed values of the coupling constants, if one regards the physical system on a length scale larger than the original by a factor $b$, it follows from (7.21) and (7.7) on dimensional grounds and $\sigma(\omega)$ is augmented by a factor $b^{d-2}$. However, in the vicinity of the mobility edge the linearized recursion relations (7.20b) and (7.20c) suggest that this is equivalent to the conductivity on the original length scale except for the new values of the coupling constants $b^\varepsilon \Delta g$ and $b^d h$. That is to say,

$$\sigma(\Delta g, h) = b^{2-d} \sigma(b^\varepsilon \Delta g, b^d h) \tag{7.22a}$$

For the d.c. conductivity $(h = 0)$, setting $b = (\Delta g)^{-1/\varepsilon}$ in this homogeneity relation yields the result of Wegner, Eq. (7.1). At the mobility edge $(\Delta g = 0)$ the vanishing of the a.c. conductivity with frequency follows by setting $b = h^{-1/d}$:

$$\sigma(\omega)|_{E_F = E_*} \equiv \sigma(0, h) = h^{(d-2)/d}\sigma(0, 1) \sim \omega^{(d-2)/d} . \tag{7.22b}$$

Using (7.21) and (7.17d), the electrical polarizability $\alpha(\omega)$ may also be expressed in terms of the scaling variables:

$$\alpha(\Delta g, h) = i\frac{\pi}{4} N(E_F) l^d \frac{\sigma(\Delta g, h)}{h} . \tag{7.23a}$$

Using (7.22a) yields the homogeneity relation

$$\alpha(\Delta g, h) = b^2 \alpha(b^\varepsilon \Delta g, b^d h) . \tag{7.23b}$$

In the static limit $(h = 0)$ and setting $b = (\Delta g)^{-1/\epsilon}$ it follows that zero frequency polarizability diverges as the mobility edge is approached from the insulator side:

$$\alpha(\omega = 0) = \alpha(\Delta g, 0) = (\Delta g)^{-2/\epsilon}\alpha(1,0) \sim \frac{1}{|E_F - E_*|^{2/\epsilon}} \, . \qquad (7.23c)$$

## 7.A. Appendix A: Constraint on the Nonlinear $\sigma$-Model

*Theorem:*

The set of matrices $S_1$ such that $Q = i\tilde{U}^T \begin{pmatrix} I & 0 \\ 0 & -I \end{pmatrix} \tilde{U}$, $\tilde{U} = CUC^*$ and $U\epsilon O(n,n)$ is identical to the set of matrices $S_2$ which are expressible as

$$Q = \begin{pmatrix} i\sqrt{1 + \Pi\Pi^T} & \Pi \\ \Pi^T & -i\sqrt{1 + \Pi^T\Pi} \end{pmatrix}$$

for some real matrix $\Pi$.

*Proof:*

(i) Suppose $Q\epsilon S_1$. Then $Q = -C^*\chi C^*$ where $\chi \equiv U^T U$. Since $\chi = \chi^T$ it may be written as $\chi = \begin{pmatrix} A & \Pi \\ \Pi^T & B \end{pmatrix}$ with $A = A^T$ and $B = B^T$. Since $\chi\epsilon O(n,n)$, it follows that $A^2 - \Pi\Pi^T = B^2 - \Pi^T\Pi = I$ and $A\Pi = \Pi B$. Therefore $Q\epsilon S_2$.

(ii) Suppose $Q\epsilon S_2$. Let $B = \frac{\Pi}{\sqrt{2}}(1 + \sqrt{1 + \Pi^T\Pi})^{1/2}$. Therefore $\Pi = 2B\sqrt{1 + B^T B}$. It follows that $Q = CU^TUC$ where $U = \begin{pmatrix} \sqrt{1 + BB^T} & B \\ B^T & \sqrt{1 + B^T B} \end{pmatrix} \epsilon \mathbf{O}(n,n)$. Inserting the identity $iC^* \begin{pmatrix} I & 0 \\ 0 & -I \end{pmatrix} C^* = I$ between $U^T$ and $U$, it follows that $Q\epsilon S_1$.

## 7.B. Appendix B: Momentum Shell Integration of the Nonlinear $\sigma$-Model

The diagrams associated with each of the integrals appearing in Eq. (7.10) are given in Fig. 12.

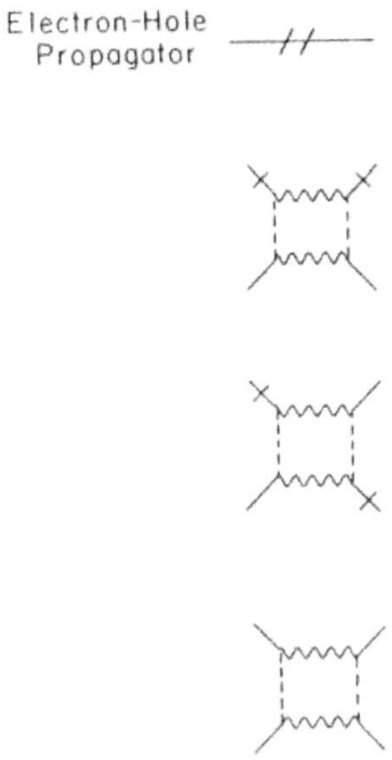

Electron-Hole
Propagator

$$(7.24)$$

$$(7.25)$$

$$(7.26)$$

$$(7.27)$$

Fig. 12. Quartic Terms in Expansion of the Nonlinear $\sigma$-Model. Solid lines correspond to electron-hole propagators in a ladder (diffusion) approximation. Slashes on solid lines represents spatial derivatives (a factor of $\mathbf{q}$). Dashed and wavy lines correspond to electron propagators and hole propagators respectively.

(i) Propagator:

$$-\frac{1}{g}\int d^d x\left(\operatorname{tr}\partial_\mu\Pi\partial_\mu\Pi^T - ih\,\operatorname{tr}\Pi\Pi^T\right) = -\frac{1}{g}\int_q\left(q^2 - ihg\right)\operatorname{tr}\Pi_q\Pi_{-q}^T \,,$$

$$(7.24a)$$

where

$$\int_q \equiv \int_{|q|<1}\frac{d^d q}{(2\pi)^d} \tag{7.24b}$$

and

$$\Pi_q^{\alpha\beta} \equiv \int d^d x\, e^{iq\cdot x}\Pi^{\alpha\beta}(x)\quad \alpha,\beta = 1,\dots,n_\pm \tag{7.24c}$$

 *Sajeev John*

(ii)

$$\frac{1}{2g} \int d^d x \, \mathrm{tr} \, \partial_\mu \Pi \Pi^T \Pi \partial_\mu \Pi^T$$

$$= -\frac{1}{2g} \int_{q_1 q_2 q_3} (\mathbf{q}_1 \cdot \mathbf{q}_2) \Pi_{q_1}^{a\alpha} \Pi_{q_2}^{a\beta} \Pi_{q_3}^{b\beta} \Pi_{-q_1-q_2-q_3}^{b\alpha} \tag{7.25}$$

(iii)

$$\frac{1}{2g} \int d^d x \, \mathrm{tr} \, \partial_\mu \Pi^T \Pi \partial_\mu \Pi^T \Pi$$

$$= -\frac{1}{2g} \int_{q_1 q_2 q_3} (\mathbf{q}_1 \cdot \mathbf{q}_3) \Pi_{q_1}^{a\alpha} \Pi_{q_2}^{a\beta} \Pi_{q_3}^{b\beta} \Pi_{-q_1-q_2-q_3}^{b\alpha} \tag{7.26}$$

(iv)

$$-\frac{ih}{4} \int d^d x \, \mathrm{tr} \, \Pi^T \Pi \Pi^T \Pi$$

$$= -\frac{ih}{4} \int_{q_1 q_2 q_3} \Pi_{q_1}^{a\alpha} \Pi_{q_2}^{a\beta} \Pi_{q_3}^{b\beta} \Pi_{-q_1-q_2-q_3}^{b\alpha} \, . \tag{7.27}$$

Integrals contributing to the one loop renormalization of $g$ and $h$ are depicted in Fig. 13. For the dimensionless conductance $g^{-1}$ these are:

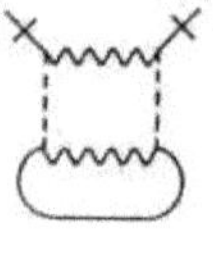

$$\tag{7.28}$$

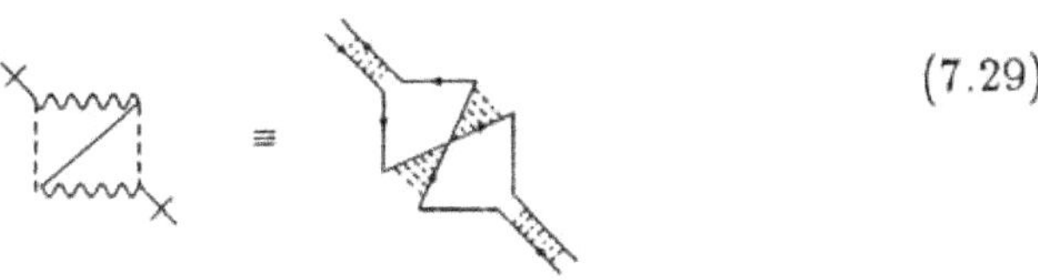

$$(7.29)$$

$$(7.30)$$

Fig. 13. Graphs contributing to one-loop renormalization of the conductance $g^{-1}$ and the symmetry breaking field $h$. Graph (7.29) contributes to a renormalization of $g$ and is equivalent as indicated to a maximally crossed impurity scattering diagram of Langer and Neal. Graph (7.28) and the first and third graphs of (7.30) vanish in the $n \to 0$ limit. The second the fourth garphs of (7.30) precisely cancel so that at the one-loop level there is no renormalization of $h$.

(i)

$$-\frac{1}{2g} \int\limits_{\substack{q_1 \cdot q_2 < b^{-1} \\ 1 > q_3, |q_1 + q_2 + q_3| > b^{-1}}} (\mathbf{q}_1 \cdot \mathbf{q}_2) \Pi_{q_1}^{\alpha\alpha} \Pi_{q_2}^{\alpha\beta} n \delta_{\beta\alpha} \frac{g}{2} (2\pi)^d \frac{\delta^d(\mathbf{q}_1 + \mathbf{q}_2)}{q_3^2 - ihg}$$

$$= \frac{n}{4} b^{-(d+2)} \varsigma^2 \int_q q^2 \Pi_q^{\alpha\alpha} \Pi_{-q}^{\alpha\alpha} f_0(b, h) \,, \tag{7.28a}$$

where $\varsigma$ is a "spin" rescaling factor and

$$f_m(b, h) \equiv \int_{1 > |q| > b^{-1}} \frac{q^{2m}}{q^2 - ihg} \,. \tag{7.28b}$$

Here $n = n_+ = n_-$ and similarly

(ii)

$$\frac{1}{4} b^{-(d+2)} \varsigma^2 \int_q q^2 \Pi_q^{a\alpha} \Pi_{-q}^{a\alpha} f_0(b, h) \ . \tag{7.29}$$

The sum of the graphs contributing to the renormalization of $h$ gives:

$$b^{-d} \varsigma^2 \frac{n+1}{4} \int_q \Pi_q^{a\alpha} \Pi_{-q}^{a\alpha} f_2(b, h)$$

$$- \frac{ih}{4} (4n + 2) \frac{g}{2} \int_q \Pi_q^{a\alpha} \Pi_{-q}^{a\alpha} f_0(b, h) \ . \tag{7.30a}$$

The part of the above sum which does not vanish in the limit at $h \to 0$:

$$b^{-d} \varsigma^2 \frac{(n+1)}{4} \int_q \Pi_q^{a\alpha} \Pi_{-q}^{a\alpha} \int_{b^{-1} < |q| < 1} 1 \tag{7.30b}$$

cannot contribute to the renormalized field $h'$ since the renormalization procedure must preserve the original symmetry of $L_0[Q]$ under pseudo-orthogonal transformations. The term (7.30b) is in fact cancelled by the Jacobian associated with the transformation (7.18) to the new variable of integration $\Pi$.[77] For example in the case $n_+ = n_- = 1$, this constraint may be implemented by the delta functions $\delta(Q_{+-} - Q_{-+})\delta(Q_{++} + Q_{--})\delta(Q_{++}^2 + \Pi^2 + 1)$. The Jacobian of this transformation is

$$\prod_x \left\{ \frac{1}{2} (1 + \Pi^2(x))^{-1/2} \right\} = \text{const } e^{-1/2 \int d^d x \ln(1 + \Pi^2)} \ , \tag{7.31a}$$

where

$$\rho \equiv \int_{|q| < 1} 1 \tag{7.31b}$$

is the number of degrees of freedom per unit volume. After length and spin rescaling, the leading term in the expansion of the logarithm in (7.31a) may be written

$$-b^{-d} \frac{\varsigma^2}{2} \rho \int_q \Pi_q \Pi_{-q} \ . \tag{7.31c}$$

Clearly the shell part $(b^{-1} < |q| < 1)$ of the integral (7.31b) precisely cancels $(n = 1)$ the term (7.30b).

Retaining only those terms that vanish in the limit $h \to 0$ in (7.30) yields for the renormalized field $h'$:

$$h' = b^{-d}\varsigma^2 h \left[1 - \frac{ng}{4} f_0(b,h)\right] . \tag{7.32a}$$

Combining (7.28a) and (7.29) and inverting yields the renormalized resistance $g'$:

$$g' = b^{d+2}\varsigma^{-2} \left[1 + \frac{(n+1)}{4} g f_0(b,h)\right] . \tag{7.32b}$$

But since $h$ couples linearly to $(Q^{++}_{q=0} - Q^{--}_{q=0})$ in the Lagrangian (17a) it follows that

$$h' = \varsigma h . \tag{7.33}$$

(All components of the matrix $Q$ have the same spin rescaling factor in order to preserve the symmetry of $L_0[Q]$ under the renormalization group transformation). Solving for $\varsigma$ from (7.31a) and (7.32) gives

$$\varsigma^{-1} = b^{-d} \left(1 - \frac{ng}{4} f_0(b,h)\right) . \tag{7.34}$$

Substituting into (7.31a) and (7.31b) yields

$$g' = b^{-\epsilon} g \left[1 - \frac{(n-1)}{4} g f_0(b,h)\right] \tag{7.35a}$$

$$h' = b^d h \left[1 + \frac{n}{4} g f_0(b,h)\right] . \tag{7.35b}$$

Letting $b = L/l$ and differentiating the above with respect to $\ln(L/l)$ leads to the following differential recursion relations:

$$\frac{dg}{d\ln(L/l)} = -\varepsilon g - \frac{(n-1)}{4} \frac{S_d}{(2\Pi)^d} \frac{hg}{1 - igh} \tag{7.36a}$$

$$\frac{dh}{d\ln(L/l)} = dh + \frac{n}{4} \frac{S_d}{(2\Pi)^d} \frac{hg}{1 - igh} . \tag{7.36b}$$

Here $S_d$ is the surface area of the unit sphere in $d$-dimensions and use has been made of the fact that

$$\frac{\delta f_0}{\delta b}\bigg|_{b=1} = \frac{S_d}{(2\pi)^d}\frac{1}{1 - igh} .$$

Taking the $n \to 0$ limit and evaluating $S_d$ in two dimensions $(S_2 = 2\pi)$ yields the recursion relations (7.20a) and (7.20b).

The momentum shell renormalization scheme used here is equivalent to the procedure of Langer and Neal[78] and Abrahams *et al.*[79] for the case of the random impurity problem. The graph (7.29) which contributes in the $n \to 0$ limit to the renormalized resistance corresponds precisely to the maximally crossed impurity scattering diagram arising in the Edwards[80] model as shown in Fig. 13.

## 8. CONCLUSIONS

The outlook for the observation of photon localization is promising in view of recent development in the manufacture and use of lossless fiber optic quality materials. In light of recent theoretical developments, the present challenge is an experimental one, namely that of preparing materials which satisfy the predicted conditions for localization. There are three basic conditions that must be satisfied for experimental observation to be feasible. There are many materials in which one or the other of these conditions is met. The fact that photon localization has not been previously observed is due to the fact that naturally occuring materials have not simultaneously satisfied all three criteria. The first of these is high dielectric contrast between scatterers and background. A dielectric ratio of about 2.5 would be quite adequate in this regard. $TiO_2$ is one such material which is currently being tried.[72] What is needed is a relatively monodisperse collection of $TiO_2$ spheres of diameter comparable to the inverse wavevector $k_0^{-1}$ of light which could be suspended in a low dielectric constant background such as water. In general, semiconductors such as Si and Ge exhibit high dielectric constant at wavelengths just below their fundamental (Urbach) absorption edge. For instance pellets of Si might provide suitable scatterers in the infrared spectrum. Highly reflective metal balls of diameter $k_0^{-1}$ are also a possibility provided that the absorption can be decreased sufficiently. The absorption criterion is the second fundamental requirement. To be of much value, the inelastic mean free path should be of the order of 100 times the classical elastic mean free path. A detailed analysis of the possible use of metal particles has recently been given by Arya, Su and Birman.[41]

The third and perhaps least recognized criterion in choosing a material for the study of photon localization is the ability to control macroscopic structure and thereby diminish the phase space available for optical propagation. One system which is ideal from the structural point of view are polyballs[81] (charged polystyrene spheres in solution). These form *bcc* and *fcc* colloidal crystals as well as glassy phases. The spheres currently in use have a refractive index of 1.5. With a background refractive index of 1.3 (water in this case) the dielectric contrast is sufficient to be of value from the point of view of classical localization. However, there appears no fundamental reason why higher dielectric contrast colloidal crystals cannot be made. It is tempting to simply dry out these polyball systems since the dielectric contrast between air and polystyrene is close to 2.3. Although the dried out phase continues to exhibit crystalline order, the spheres now touch, diminishing the relevant Fourier amplitude for Bargg scattering. Coating the spheres with a low index material might help in this regard. Another advantage of polyball systems is that disorder can be introduced in a highly controlled manner. For example shaking a polyball suspension with unltrasound of varying intensity introduces structural disorder which on the time scale of optical propagation is essentially static. Also applying shear[82] to such a system causes the colloidal crystal to melt. The realization of glassy phases in high dielectric contrast materials can allow detailed studies of the influence of the structure factor on localization.

I have stressed the importance of structure a great deal in this article. Some numerical support for this point of view is available from molecular dynamics studies of electronic localization in liquid and amorphous Si. A recent study by Car and Parinello[83] on the influence of the structure factor on electronic transport is particularly suggestive. These authors numerically evaluated the one-electron density of states

(DOS) for liquid Si. It showed no particular structure at the electronic fermi level $E_f$ and all states were extended at this energy despite the disordered liquid like arrangement of atoms. Upon quenching the system into an amorphous Si phase the DOS rapidly developed a large gap-like structure at $E_f$ and the states became localized. This illustrates very vividly the high degree of sensitivity of transport properties to the underlying structure factor of the material. It is very likely that many systems currently under investigation[54,55] as candidates for photon localization are near the borderline and with better control over macroscopic structure can very easily exhibit the effects described in this article.

Another possibility which is beginning to receive attention is the use of microwaves. Here, very high dielectric constant spheres are available on the centimeter wavelength scale. The construction of large superlattices here is cumbersome but potentially very fruitful in terms of providing insight for optical systems. Recently Yablonovitch[84] has reported the existence of a photonic band gap in a sample consisting of an $8 \times 8 \times 8$ *fcc* array of spherical cavities drilled in a background dielectric of refractive index close to 3.5.

On the theoretical side, the most significant challenge is the accurate description of strongly correlated scatterers. Unlike the situation in electron systems where this could be simplified by considering tight binding model Hamiltonians with narrow bands, and large gaps in the one-electron density of states, optical systems in which photon localization is possible have very broad bands and narrow gaps. Treating a dense collection of strong scatterers of this type by means of a coherent potential approximation or any other effective medium theory which only accounts accurately for single scattering is quite misleading from the point of view of classical localization.

Since the free-photon Ioffe-Regel condition is barely attainable in the case of even an almost structureless random medium, nearly any incorporation of short range order is likely to be a significant step toward observing optical localization. The existence of pseudogaps in the electronic density of states in amorphous semiconductors provides ample support for this hypothesis. Achieving the photonic analog of amorphous semiconductors is, in my judgement, the most promising direction in systematically demonstrating the strong localization of light. The benefits of such an achievement are two folds. Not only will it deepen our understanding of localization and other transport phenomena in disordered systems, but the materials produced in the process constitute an entirely new class of photonic materials which may prove to be of significant technological value.

## REFERENCES


1. P. W. Anderson, *Phys. Rev.* **109** (1958) 1492.
2. P. A. Lee and T. V. Ramakrishnan, *Rev. Mod. Phys.* **57** (1985) 287.
3. S. John, H. Sompolinsky and M. J. Stephen, *Phys. Rev.* **B27** (1983) 5592.
4. See, for instance, F. Wegner "Anderson Localization", eds. Y. Nagaoka and H. Fukuyama (Springer, New York, 1982) and L. Schafer and F. Wegner, *Z. Phys.* **B38** (1980) 113.
5. T. R. Kirkpatrick, *Phys. Rev.* **B31** (1985) 5746.
6. E. Akkermans and R. Maynard, *Phys. Rev.* **B32** (1985) 7850.
7. A. J. McKane and M. Stone, *Ann. Phys., (N.Y.)* **131** (1981) 36.
8. S. Hikami, *Phys. Rev.* **B24** (1981) 2671.
9. Y. Kuga and A. Ishimaru, *J. Opt. Soc. Am.* **A1** (1984) 831.
10. M. van Albada and A Lagendijk, *Phys. Rev. Lett.* **55** (1985) 2692.
11. P. F. Wolf and G. Maret, *Phys. Rev. Lett.* **55** (1985) 2696.
12. S. Etemad, R. Thompson and M. J. Andrejco, *Phys. Rev. Lett.* **57** (1986) 575.
13. M. Kaveh, M. Rosenbluh, I. Edrei and I. Freund, *Phys. Rev. Lett.* **57** (1986) 2049.
14. E. Akkermans, P. E. Wolf and R. Maynard, *Phys. Rev. Lett.* **56** (1986) 1471.
15. M. J. Stephen and G. Cwilich, *Phys. Rev.* **B39** (1986) 7564.
16. F. C. MacKintosh and S. John, *Phys. Rev.* **B37** (1988) 1884.
17. M. Rosenbluh, I. Edrei, M. Kaveh and I. Freund, *Phys. Rev.* **A** (1987).
18. M. B. van der Mark, M. P. van Albada and Ad Lagendijk, preprint.
19. E. Akkermans, P. E. Wolf, R. Maynard and G. Maret, *J. Physique* **49** (1988).
20. S. John, *Phys. Rev. Lett.* **53** (1984) 2169.
21. P. W. Anderson, *Phil. Mag.* **B52** (1985) 505.
22. K. Arya, Z. B. Su and J. L. Birman, *Phys. Rev. Lett.* **57** (1986) 2725.
23. P. Sheng and Z. Zhang, *Phys. Rev. Lett.* **57** (1986) 1879.

24. C. A. Condat and T. R. Kirkpatrick, *Phys. Rev. Lett.* **58** (1987) 226.

25. S. John, *Phys. Rev. Lett.* **58** (1987) 2486.

26. E. Abrahams, P. W. Anderson, D. C. Licciardello and T. V. Ramakrishnan, *Phys. Rev. Lett.* **42** (1979) 673.

27. A. F. Ioffe and A. R. Regel, *Prog. Semicond.* **4** (1960) 237.

28. G. Maret and P. E. Wolf, *Z. Phys.* **B65** (1987) 409.

29. I. Freund, M. Kaveh and M. Rosenbluh, *Phys. Rev. Lett.* **60** (1988) 1130; D. J. Pine, D. A. Weitz, P. M. Chaikin and E. Herbolzheimer, *Phys. Rev. Lett.* **60** (1988) 1134.

30. J. M. Drake and A. Z. Genack, unpublished.

31. E. Yablonovitch, *Phys. Rev. Lett.* **58** (1987) 2059.

32. M. H. Cohen, E. N. Economou and C. M. Soukoulis, *Phys. Rev. Lett.* **51** (1983) 1202.

33. S. John, *Phys. Rev.* **B35** (1987) 9291; C. H. Grein and S. John, *Phys. Rev.* **B36** (1987) 7457.

34. H. Gibbs, "Optical Bistability: Controlling Light with Light", (Academic Press, New York, 1985).

35. Y. S. Abu-Mostafa and D. Psaltis, "Optical Neural Computers", (Scientific American, 1987).

36. S. M. Cohen and J. Machta, *Phys. Rev. Lett.* **54** (1985) 2242.

37. S. M. Cohen, M. Machta, T. R. Kirkpatrick and C. A. Condat, *Phys. Rev. Lett.* **58** (1987) 785.

38. J. Jackie, *Solid State Comm.* **39** (1981) 1261.

39. J. E. Graebner, B. Golding and L. C. Allen, *Phys. Rev.* **B34** (1986) 5696; J. E. Graebner and B. Golding, **B34** (1986) 5788.

40. H. Ikari, *Phys. Lett.* **A123** (1987) 179.

41. K. Arya, Z. B. Su and J. L. Birman, *Phys. Rev. Lett.* **54** (1985) 1559.

42. A. R. McGurn, A. A. Maradudin and V. Celli, *Phys. Rev.* **B31** (1985) 4866.

43. R. Tawel and K. F. Canter, *Phys. Rev. Lett.* **56** (1986) 2322.

44. S. John, C. Soukoulis, M. H. Cohen and E. N. Economou, *Phys. Rev. Lett.* **57** (1986) 1777.

45. For very weak disorder, the trajectory of the mobility edge is dominated by the shift of the squareroot continuum which moves to lower energy with increasing disorder. Only for stronger disorder does the mobility edge begin to move to positive energies.

46. E. N. Economou, C. M. Soukoulis and A. D. Zdetsis, *Phys. Rev.* **B30** (1984) 1686.

47. See for instance M. Kerker, "The Scattering of Light and Other Electromagnetic Radiation", (Academic Press, New York, 1969) or J. D. Jackson, "Classical Electrodynamics", (John Wiley and Sons, New York,

1975) pp. 449–452.

48. S. John, *Phys. Rev.* **B31** (1985) 304.

49. S. John and M. J. Stephen, *Phys. Rev.* **B28** (1983) 6358.

50. S. John, Ph.D. thesis, (Harvard University, 1984).

51. G. Bergmann, *Phys. Rep.* **107** (1984) 1.

52. A. Yariv and P. Yeh, "Optical Waves in Crystals", (John Wiley and Sons, New York, 1984).

53. See for instance: R. P. Feynman, "Statistical Mechanics", (Benjamin Cummings, 1972).

54. S. R. Nagle, A. Rahman and G. S. Grest, *Phys. Rev. Lett.* **47** (1981) 1665.

55. R. B. Grzonka and M. A. Moore, *J. Phys.* **C16** (1983) 1109.

56. L. D. Landau and E. M. Lifshitz, "Theory of Elasticity", (Pergamon, New York, 1970).

57. P. Carruthers, *Rev. Mod. Phys.* **33** (1) (1961) 92.

58. H. Sumi, *J. Phys. Soc. Japan* **32** (3) (1972) 616.

59. J. D. Dow and D. Redfield, *Phys. Rev.* **B5** (2) (1972) 594.

60. F. Wegner, "Anderson Localization", eds. Y. Nagaoka and H. Fukuyama (Springer-Verlag, 1982).

61. L. Scafer and F. Wegner, *Z. Phys.* **B38** (1980) 113.

62. R. J. Elliott, J. A. Krumhansl and P. L. Leath, *Rev. Mod. Phys.* **46** (3) (1974) 465.

63. This is done in complete analogy with the case of uncorrelated disorder discussed in Ref. 3.

64. K. Gottfried, "Quantum Mechanics", (W. A. Benjamin, 1966).

65. V. I. Tatarski, "Wave Propagation in a Turbulent Medium", (McGraw-Hill, 1961).

66. B. Shapiro and E. Abrahams, *Phys. Rev.* **B24** (1981) 4889.

67. R. Oppermann and F. Wegner, *Z. Phys.* **B34** (1979) 327.

68. S. Hikami, *Phys. Rev.* **B24** (5) (1981) 2671.

69. E. Brézin and J. Zinn-Justin, *Phys. Rev.* **B14** (7) (1976) 3110.

70. I. S. Gradshteyn and I. M. Ryzhik, "Table of Integrals, Series, and Products", (Academic Press, 1980).

71. "Handbook of Mathematical Functions", eds. M. Abramowitz and I. A. Stegun (National Bureau of Standards, 1964).

72. A. Z. Genack, *Phys. Rev. Lett.* **58** (1987) 2043.

73. G. Watson, P. Fleury and S. McCall, *Phys. Rev. Lett.* **58** (1987) 945.

74. D. Vollhardt and P. Wölfle, *Phys. Rev. Lett.* **48** (1982) 699; *Phys. Rev.* **B22** (1980) 4666.

75. D. Forster, "Hydrodynamic Fluctuations, Broken Symmetry and Correlation Functions", (W. A. Benjamin, 1975).

76. A. M. M. Pruisken, *Nucl. Phys.* B (to be published).

77. D. R. Nelson and R. A. Pelcovits, *Phys. Rev.* **B16** (1977) 2191.

78. J. S. Langer and T. Neal, *Phys. Rev. Lett.* **16** (1966) 984.

79. E. Abrahams and T. V. Ramakrishnan, *J. Non Crystalline Solids* **35** (1980) 15.

80. S. F. Edwards, *Phil. Mag.* **3** (1958) 1020.

81. P. M. Chaikin *et al.* "Physics of Complex and supermolecular Fluids", eds. S. A. Safran and N. A. Clark (John Wiley and Sons, New York, 1987).

82. B. J. Ackerson and N. A. Clark, *Phys. Rev.* **A30** (1984) 906.

83. R. Car and M. Parinello, *Phys. Rev. Lett.* **60** (1988) 204.

84. E. Yablonovitch, *Bull. Am. Phys. Soc., March meeting* (1988).

# EXPERIMENTS ON WEAK LOCALIZATION OF LIGHT AND THEIR INTERPRETATION


M. P. VAN ALBADA,* M. B. VAN DER MARK* and
A. LAGENDIJK*†

*Natuurkundig Laboratorium der Universiteit
van Amsterdam, Valckenierstraat 65,
1018 XE Amsterdam
The Netherlands

†FOM-Instituut voor Atoom- en Molecuulfysica
Kruislaan 407, 1098 SJ Amsterdam
The Netherlands



In this paper we discuss light scattering experiments on random media consisting of dielectric particles ($TiO_2$, polystyrene) in a medium (air, water) having another refractive index. A rigorous scalar wave theory is treated as a model to represent the enhanced backscattering (weak localization) of light for finite slabs, and a diffusion approximation to this theory is discussed. Experiments and theories are compared, and effects which are the consequence of the vector character of light are explained. We comment in some detail upon the backscattering-enhancement factor. On the basis of experimental results we speculate about the possibility to observe strong localization of light.

  *M. P. van Albada, M. B. van der Mark & A. Lagendijk*

## Contents


1. Introduction   99
    1.1. Interference and weak localization   100
    1.2. Enhanced backscattering   102
    1.3. Coherent backscattering   102
    1.4. Shape and width of the cones of coherent backscattering   103

2. Experimental Techniques   104

3. Some Results   106

4. Interpretation of the Vector Character Effects   107

5. The Parallel Cone, Scalar Theory, and Its Verification   111
    5.1. Rigorous scalar theory for point-scatterers   112
    5.2. Scalar diffusion approximation   115
    5.3. Time-resolved coherent backscattering   118
    5.4. Comparison with experiment   120

6. On the Factor of 2   124

7. On the Road to Strong Localization   129
    7.1. Dependent scattering and correlated scatterers   129
    7.2. Very strongly-scattering media and localization   131

Acknowledgement   134

References   134

# 1. INTRODUCTION

Localization of light is a fascinating new field in which modern developments of condensed matter physics are applied to the well-established field of linear optics. Localization as introduced by Anderson[1] refers to a dramatic change in the propagation of an electron when it is subjected to a spatially random potential.[2]

Several approaches can be used to describe the phenomenon of localization in a disordered medium. One method is to look at the transport properties of the electron. In this picture localization is concerned with the vanishing of the diffusion coefficient. This view of localization was given a thorough foundation by Götze,[3] and later workers,[4] who used transport equations of the mode-coupling type. At localization the interference of scattered waves cannot be neglected any more, but indeed its influence becomes essential ("there is life after a mean free path").

The dimension of the disordered medium is a crucial parameter. In 1-D and 2-D any degree of disorder will lead to a finite localization length while in 3-D a certain critical degree of disorder is needed before localization will set in. In the latter case the heuristic Ioffe-Regel criterion

$$\lambda_{\mathrm{MF}} \leq \lambda/(2\pi) , \tag{1}$$

(in which $\lambda_{\mathrm{MF}}$ is the mean free path of the electron, and $\lambda$ is its wavelength) applies, as may be seen as follows: If this condition is met the wave character of the wave (function) is completely lost and localization is established.

Although the majority of theories of localization has been developed for the Schrödinger wave equation, it is quite clear that the concept is much broader. In principle for almost any wave equation localized solutions can be obtained when solved for a random medium. This becomes particularly clear when one realizes that the Ioffe-Regel criterion can be applied to all wave phenomena. Localization of electromagnetic waves is of particular interest because one deals with localization of *vector* waves, described by a well-established and well-studied set of equations basically different from the Schrödinger wave equation.

## 1.1. Interference and Weak Localization

If the mean free path is not short enough to make the diffusion constant vanish as a result of Anderson localization, one may still observe interference effects that affect this constant. Let us look at interference effects in multiple scattering: A well known phenomenon in the scattering of coherent light by a rigid random medium is "laser speckle", which results from interference between scattered waves that traveled through the sample along different light paths (by "light path" we mean a sequence of scattering events $s_1, s_2, \ldots, s_f$). If the scatterers are allowed to move over distances of the order of the wavelength of the light or more at the time scale of the measurement, the spatial distribution of the scattered intensity will be the average of a rapidly changing speckle pattern, and therefore essentially flat. Liquid random samples in which the scatterers are subject to Brownian motion, or rigid random samples that are spun, yield scattered-intensity patterns that almost look as they would if interference between light paths did not exist. One type of interference however survives: In the direction of pure backscattering, the waves that travel along the same light path in opposite directions (time-reversed paths) will always have the same phase and interfere constructively (actually this holds only for the backscattered light component polarized parallel to the incident beam, but for the moment we neglect polarization effects). Moving away from this direction, a difference in phase $\Delta\phi$ develops, which increases with the angle $\theta$ between the incoming and outgoing wave vectors and further depends on the relative positions of the first and last scatterer $s_1$ and $s_f$, according to

$$\Delta\phi = \frac{r}{\lambda}\{\sin(\alpha) - \sin(\alpha - \theta)\} = 2\frac{r}{\lambda}\cos(\alpha - \theta/2)\sin(\theta/2) , \qquad (2)$$

where $r$ is the distance between $s_1$ and $s_f$ (c.f. Fig. 1). With increasing $\theta$ the interference for an individual light path will oscillate between constructive and destructive. At $\theta = 0$ all time-reversed pairs interfere constructively, at small $\theta$ only light paths with large values of $r$ will interfere destructively, and at large $\theta$ constructive and destructive interference will average out (c.f. Fig. 2). Thus, within a certain solid angle

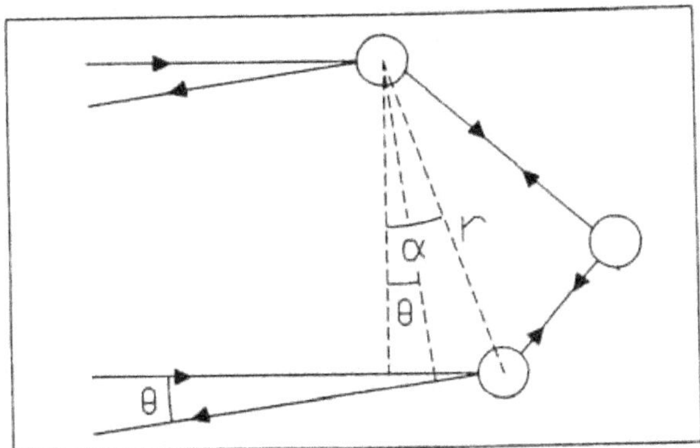

Fig. 1. Time-reversed pair of light paths. In case of exact backscattering the paths are equally long. In all other directions there will be a phase difference between the time reversed waves.

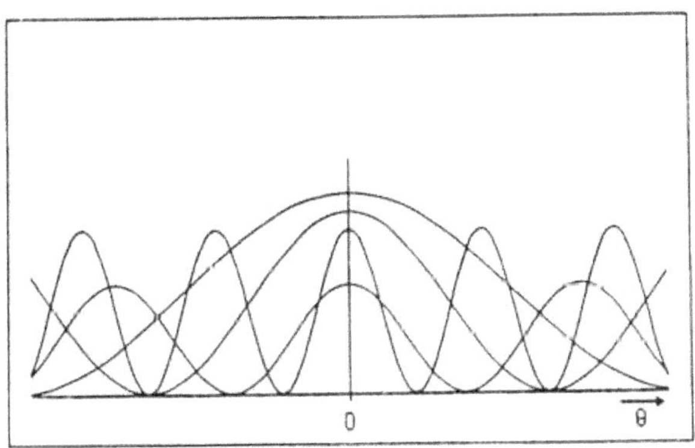

Fig. 2. A number of two-point-source interference patterns belonging to different time-reversed pairs of waves.

around the direction of pure backscattering there will be an enhanced intensity (c.f. Fig. 3) or, interference increases the returning probability and hence reduces the diffusion coefficient. This phenomenon is known as weak localization.

Since the displacement $r$ scales with the mean free path, the width of the cone of enhanced backscattering will increase and the diffusion constant will decrease with increasing randomness of the medium. The width of the cone is expected to be inversely proportional to the mean free path.

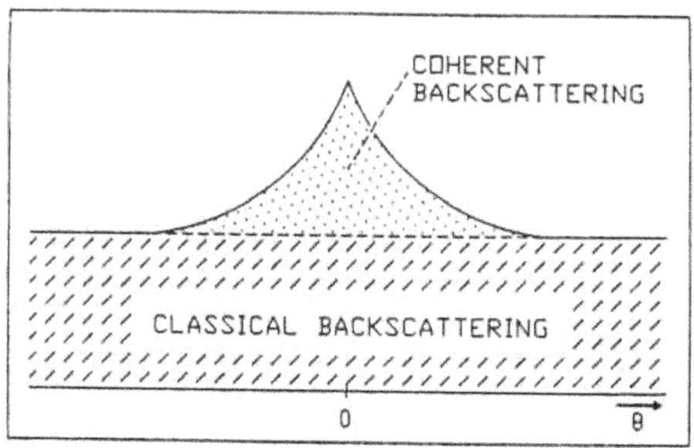


Fig. 3. Sum of the two-point-source interference patterns of all contributing time-reversed pairs of light paths.

## 1.2. Enhanced Backscattering

From the former paragraphs one might conclude that in order to detect weak localization, one should merely look for enhanced backscattering. Enhanced backscattering has been known for long. An early reference to it was made by the 16th century italian artist Benvenuto Cellini[5] who interpreted a halo that he observed just around the shadow of his own head among the shadows of a group of people as a sign of genius. There is also more recent experimental work,[6] and a number of possible explanations not involving interference have been very thoroughly discussed.[7] Enchanced backscattering may result for instance from shadow working (only in the direction of pure backscattering one will not see scatterers that are located in the shadow of other scatterers) lens action, (dew drops may focus sunlight onto a plant leaf and refract much of the scattered light into the backward direction), or retroreflection (e.g. by cubic crystals). We may conclude that once enhanced backscattering is detected, one still needs to prove that it results from interference before it may be attributed to weak localization. The inverse proportionality between the mean free path and the width of the cone of enhanced backscattering seems to be a good criterion.

## 1.3. Coherent Backscattering

By coherent backscattering we will mean enhanced backscattering as a result of interference.

Enhanced backscattering from a random sample (consisting of polystyrene micro spheres in water)[8] was for the first time interpreted in terms of interference between time-reversed paths by Tsang and Ishimaru.[9] At first, this work was not noticed by researchers in the field of localization, probably because the authors did not make the link with localization theory. The first to make this link were the present authors[10] and Wolf and Maret.[11] Both these groups reported the observation of cones of enhanced backscattering of light from polystyrene sphere-water samples, found qualitatively the expected relationship between width of the cone and mean free path, and in addition observed a marked polarization effect: the enhanced intensity in the backscattered light is much more pronounced in the component with polarization parallel to that of the incident beam than in the perpendicularly polarized component.

Enhanced backscattering from rigid random samples was later reported by Etemad c.s.,[12] who obtained the cone by adding many speckle patterns that were recorded for slightly different orientations of the sample with respect to the incident beam, and by Kaveh c.s.[13] who obtained ensemble averaging by spinning the sample. In these experiments the influence of the mean free path was not investigated, but the different enhancement factors in the parallel and perpendicular components were confirmed in Kaveh's work.

## 1.4. Shape and Width of the Cones of Coherent Backscattering

As the cones are made up of the interference patterns of individual light paths, their shapes may be expected to reflect the characteristics of these light paths. For instance: The longer the paths, the larger will be the average value of $|r|$, so that long light paths may be expected to yield a narrow contribution to the cone. The relative contributions of lower- and higher-order scattering will depend on the scattering matrix of the particles used, so the shape of the cone may be expected to depend on e.g. particle size.

Detailed studies have been made. Coherent backscattering from polystyrene-sphere suspensions in water has been studied as a function of concentration, particle size, sample-thickness, polarization of the backscattered light with respect to incident polarization, and of the spatial orientation of the scan.[14] Suspensions of $TiO_2$ particles in 2-methylpentane-2,4-diol have been studied as a function of concentration and sample-thickness.[15] The cutoff of long light paths has also been studied by addition of absorbing dye to the samples.[16] The width of the contribution to the cone as a function of the length of the light paths has been studied in a time-resolved experiment on femtosecond scale.[17]

## 2. EXPERIMENTAL TECHNIQUES

The present authors recorded cones of coherent backscattering using a set up of the type drawn schematically in Fig. 4. A linearly polarized laser beam was expanded and then reflected from a beam splitter onto the sample cell. The intensity backscattered through the beamsplitter was recorded as a function of the scattering angle, using a pinhole-detector assembly mounted on a stepper-motor driven translation stage and positioned with the pinhole in the focal plane of the lens $L$. (The transmission characteristics of the beam splitter are polarization- and angle-dependent, and recorded intensity profiles need to be corrected for these effects). The cell was tilted off-axis so as to keep its window reflections well away from the detector. In scans of high-viscosity samples the cell was spun around its axis to average out the speckle that results from interference between different light paths. Using a difference technique, the width and the enhancement factor of the contribution to the backscatter cone were studied as a function of the depth in the sample: subtracting the backscattering patterns of slabs of thickness $d_1$ and $d_2$ ($d_2 > d_1$), the scattering contributions coming from the front layer of the sample ($z < d_1$) cancel out, and what remains is the contribution of light that has "seen" the deeper part ($d_1 < z < d_2$) of the slab (c.f. Fig. 5). In this way the single-scattering contribution (which has no interference term) can be eliminated. It is also possible to exclude the contribution of the very long lightpaths that give rise to the narrow top of the cone by choosing $d_2$ relatively small.

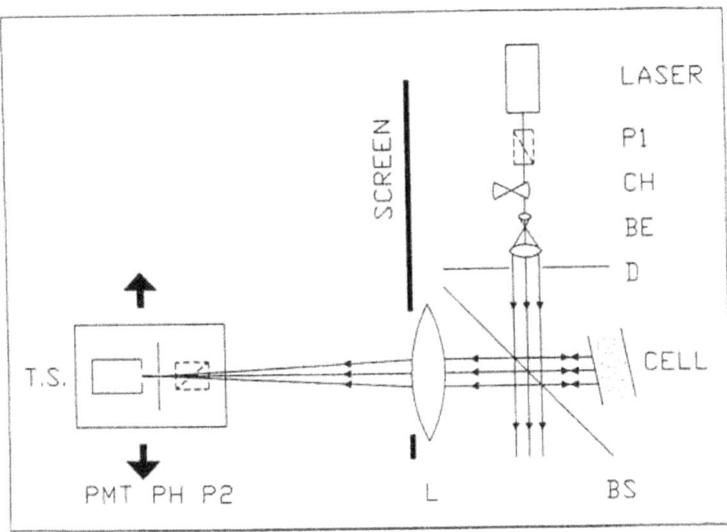


Fig. 4. Experimental setup for the observation of backscattering enhancement. P1, P2, polarizers; CH, chopper; BE, beam expander; BS, beamsplitter; L, lens; PH, pinhole; TS, translation stage.

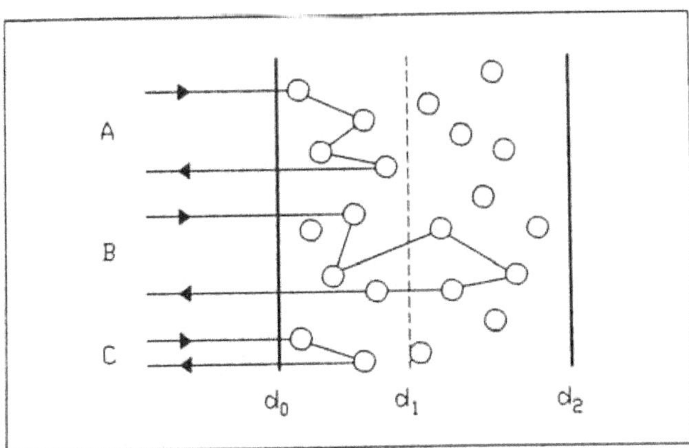


Fig. 5. The intensity pattern of the "full slab" $(d_0 - d_2)$ contains contributions of A, B, and C; that of the "front slab" $(d_0 - d_1)$ contains contributions of A and C. Subtraction of these patterns yields the pattern of the "deep slab" $(d_1 - d_2)$ containing only the "long path" contribution of B.

Etemad c.s.[16] studied the effect of the cutting off of long light paths both by limiting the sample-thickness and by addition of absorbing dye to the samples. In this work, a $\lambda/4$ plate was used in front of the sample cell, which leads to elimination of a.o. the single scattering contribution to the backscattered intensity.

In a study of solid bariumsulphate-air samples, Kaveh c.s.[13] eliminated the complication of polarization- and angle-dependent transmission coefficients of the beam splitter by using a mirror instead. In this way, the very top of the cone could not be observed, but the "blind angle" was kept small by locating both the mirror and the detector at a large distance from the sample.

Vreeker *et al.*[17] studied the evolution in time of the cone of coherent backscattering, that results from very short (100 fs) laser-pulses using a colliding-pulse mode-locked laser as a source and a light-gating detection technique based upon second-harmonic generation.

## 3. SOME RESULTS

In Fig. 6 (taken from Ref. 15) the measured width $W$ of the cone in the parallel light component is plotted as a function of $\lambda_{tr}$, the transport mean free path. The transport mean free path is related to the scattering mean free path $\lambda_{sc}$ by $\lambda_{tr} = \lambda_{sc}/(1 - \overline{\cos\theta})$, with $\overline{\cos\theta}$ the average angle of scattering of light by a single particle. Clearly, for isotropic scatterers $\lambda_{tr} = \lambda_{sc}$. We use $\lambda_{tr}$ instead of $\lambda_{sc}$ in order to map a theory for isotropic scatterers (see Sec. 5) on results that were obtained for non-isotropic (Mie) scatterers. This is only justified if a certain number of scattering events is involved, for instance, single scattering is only poorly described. The expected inverse proportionality of $W$ as a function of $\lambda_{tr}$ is seen to hold over the accessible range of nearly three decades. The enhancement factors $I_{top}/I_{background}$ were found to depend slightly on the type of scatterers and ranged from 1.5–1.8 for the parallel component (not corrected for single scattering), and from 1.05–1.3 for the perpendicular component. For very small (Rayleigh-like) particles, the parallel cone was found[14] to be spatially anisotropic. Figure 7 shows intensity profiles as recorded using a 0.215 $\mu$m polystyrene-sphere suspension in water as sample. The "cone" of a "full" slab is actually elliptically shaped, its width being abnormally large in the direction of incident polarization, whereas in the perpendicular direction it is consistent with the data obtained for suspensions of somewhat larger (Mie-type) particles, which give isotropic cones (c.f. Fig. 6).

Figure 8 shows intensity profiles belonging to a "deep" slab, that were obtained by subtracting the intensity profiles recorded for a (thinner) "front" slab of the same sample from the ones of the "full" slab. The two peaks in Fig. 8 are equally wide, showing that the "difference cone" of the deep slab is spatially isotropic. We therefore conclude that the spatial anisotropy of the full cone results from lower-order multiple scattering processes, taking place in the front slab. The curves in Fig. 9 were found for deep slabs, scanned parallel to the direction of incident polarization, the top curve gives the intensity profile in the parallel light component, and the bottom one gives that of the perpendicular light component. It is seen that there is no enhancement in the perpendicular component. The (low) enhancement factors observed in the perpendicular component for full slabs are therefore also a result of lower-order multiple scattering contributions.

## 4. INTERPRETATION OF THE VECTOR CHARACTER EFFECTS

The occurrence of backscattering enhancement in the perpendicularly polarized component, and of spatial anisotropy for Rayleigh-like scatterers may be understood in the following way:

Let light with polarization vector $\mathbf{P}_i$ be incident on a random medium in the $z$ direction, and let $s_1, s_2, \ldots s_{n-1}, s_n$ be a "light path" formed by $n$ scattering centers, along which a wave travels. A scattering event will normally change the polarization vector of the travelling wave. We observe the wave that is scattered from $s_n$ using a detector that is located in the $-z$ direction and denote the polarization vector of this wave by $\mathbf{P}_f^+$. Since $\mathbf{P}_i$ and $\mathbf{P}_f^+$ are both parallel to the $z = 0$ plane, they are related by $\mathbf{P}_f^+ = M^+ \mathbf{P}_i$, $M^+$ being a $(2 \times 2)$ matrix. For the wave that travels along the same path but in the opposite direction (the time-reversed path) it holds $\mathbf{P}_f^- = M^- \mathbf{P}_i$, and $M^+$ and $M^-$ are

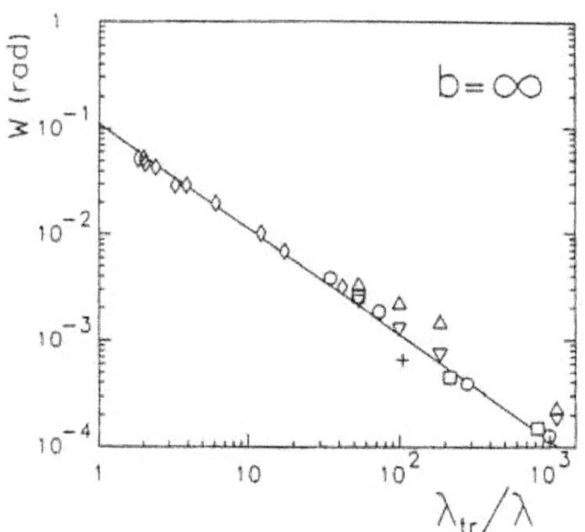


Fig. 6. Full width $W$ of the cone of enhanced backscattering vs $\lambda_{tr}/\lambda$ for very thick slabs. $\diamond$ = 0.22 $\mu$m TiO$_2$ in 2-methylpentane-2,4-diol; $\bigcirc$ = 1.091 $\mu$m polystyrene speres in water; $\square$ = 0.482 $\mu$m polystyrene spheres in water; $\triangle$ = 0.214 $\mu$m polystyrene spheres in water (// scan); $\triangledown$ = 0.214 $\mu$m polystyrene spheres in water ($\perp$ scan); + 2.02 $\mu$m polyvinyltoluene spheres in water. line: $W = \frac{(0.7)}{(2\pi)} \, (\lambda/\lambda_{tr})$, the extrapolated result from the exact theory for $b = \infty$ (see Sec. 5).

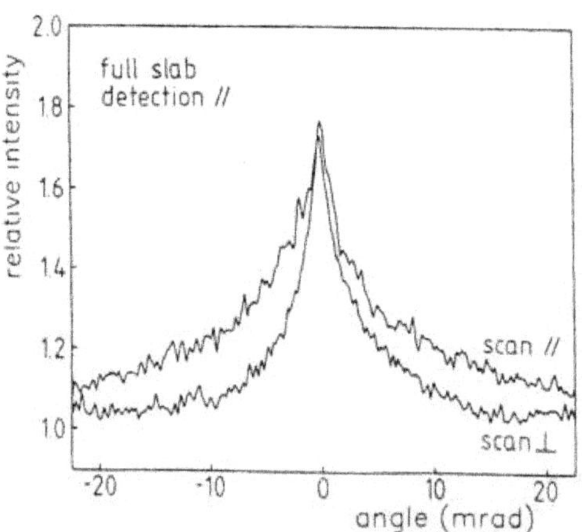


Fig. 7. Backscattered intensity from a 1500 $\mu$m slab of 9.6 volume % of 0.215 $\mu$m polystyrene particles in water as a function of angle. Upper curve: spatial scan // polarization vector; lower curve: spatial scan $\perp$ polarization vector. The two curves differ approximately by a factor of two in width. ($\lambda_{tr}$ = 26 $\mu$m).

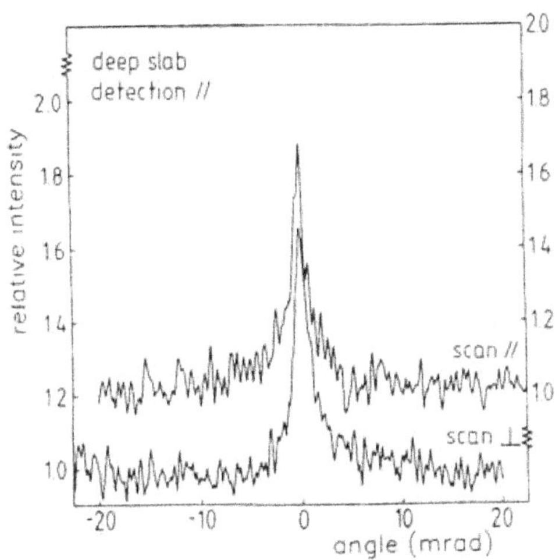


Fig. 8. Difference backscatter cones from the slab between 60 $\mu$m and 1500 $\mu$m of the suspension of Fig. 7. The mutually perpendicular scans now give the same width.

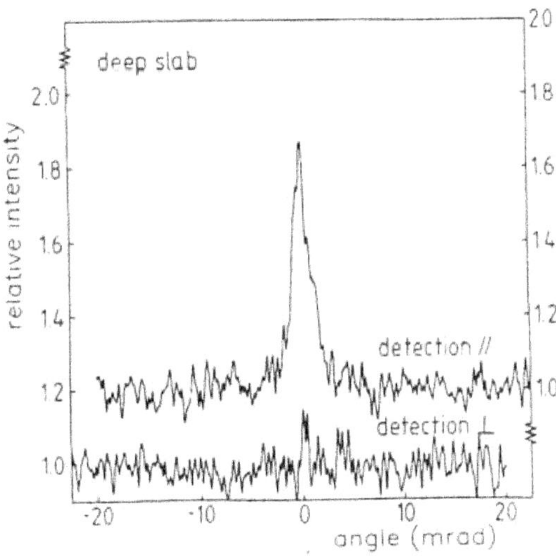


Fig. 9. Parallel- and perpendicularly-polarized difference critical backscattering from the slab between 70 $\mu$m and 1500 $\mu$m of 9.6 volume % of 1.091 $\mu$m polystyrene in water. Upper curve: detected light polarized parallel to incident light; lower curve: detected light polarized perpendicular to incident light.

related by $M^- = (M^+)^T$. For directions slightly different from $-z$, the relationship $M^- = (M^+)^T$ may be thought to remain valid, because over a small angle the amplitudes of the waves scattered from $s_1$ and $s_n$ will change very little. The phases of a time-reversed pair of waves are correlated and the waves will interfere: For the polarized component the interference will be constructive in the $-z$ direction, and in directions different from $-z$ the phase shift between the waves traveling in opposite directions along an individual path will depend on the angle $\theta$ between $-z$ and the direction of observation and on the relative coordinates of $s_1$ and $s_n$: if $s_1$ is situated in the origin and $x_n$ and $y_n$ are the $x$ and $y$ coordinates of $s_n$, then the phase shifts that result from a displacement of the detector over an angle $\theta$ in the $x$-$y$ plane ($// \mathbf{P}_i$) and in the $y$-$z$ plane ($\perp \mathbf{P}_i$) will be $\simeq x_n \sin\theta/\lambda$ and $\simeq y_n \sin\theta/\lambda$ respectively. If $\theta$ is sufficiently large, the phase shift for individual light paths will be random and the detected polarized and depolarized intensities will be $2\langle (M_{11})^2 \rangle$ and $\langle (M_{12})^2 \rangle + \langle (M_{21})^2 \rangle = 2\langle (M_{12})^2 \rangle$ respectively. (The triangular brackets represent summing over all light paths). The intensities observed in the $-z$ direction will be $\langle (2M_{11})^2 \rangle$ and $\langle (M_{12} + M_{21})^2 \rangle$. For the enhancement factors in the polarized and depolarized components $E_{//}$ and $E_\perp$ it will hold

$$E_{//} = \frac{\langle (2M_{11})^2 \rangle}{2\langle (M_{11})^2 \rangle} = 2; \qquad E_\perp = 1 + \frac{\langle M_{12} M_{21} \rangle}{\langle (M_{12})^2 \rangle} . \tag{3}$$

It is easily seen that for 2nd order backscattering $M_{12} = M_{21}$, and this already suffices to explain that $E_\perp > 1$. Likewise, the spatial anisotropy of the parallel polarized "cone" may be understood. If in 2nd order backscattering the angle between the incident polarization vector ($\hat{\mathbf{x}}$) and the normal to the scattering plane is $\varphi$, the backscattered polarized amplitude will have maxima for $\varphi = 0$, $\pi$, and minima for $\varphi = \pi/2$, $3\pi/2$, so that for the transport of the parallel polarized component it will hold that the average light path has $y_n > x_n$. Consequently a displacement of the detector over an angle $\theta$ in the $y$-$z$ plane will affect the average phase shift to a larger extent than would an equal displacement in the $x$-$z$ plane or, the 2nd order contribution to the parallel polarized "cone" is broader in the $x$- than in the $y$-direction. A computer simulation of weak localization in Rayleigh scattering[18] fully confirms this interpretation: Simulated cones of coherent backscattering showing both a perpendicular enhancement factor slightly larger than 1 and spatial anisotropy in the parallel cone, are presented in Fig. 10.

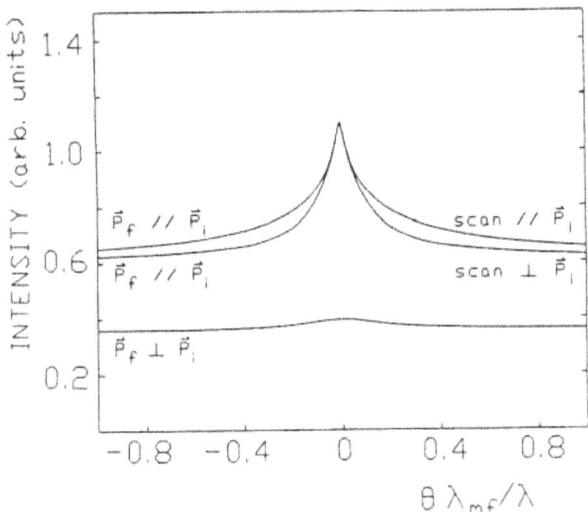


Fig. 10. Anisotropy in the enhanced backscattering of light from a random medium containing Rayleigh scatterers: total cones as found by summing the orders 1-1000 (corresponding to 94% of the total backscattering). The curves are normalized with respect to 100% of the total (// + ⊥) background intensity. Upper curve: scan //$P_i$,$P_f$//$P_i$ middle curve: scan ⊥$P_i$,$P_f$//$P_i$; lower curve: $P_f$⊥$P_i$ (curves for scans //$P_i$ and ⊥$P_i$ coincide).

## 5. THE PARALLEL CONE, SCALAR THEORY, AND ITS VERIFICATION

In the case of *scalar* waves, the amplitudes of a time-reversed pair of waves are equal. For *vector* waves, the same holds (in the direction of pure backscattering) for the amplitude component parallel to the polarization of the incident wave (and generally not for the perpendicular component). One might therefore try to apply scalar theory to coherent backscattering in the parallel light component. Clearly, with scalar theory one will not be able to describe the spatial anisotropy of the parallel cone, but this effect was found to be important only for very small scatterers (near the Rayleigh limit). In this chapter both a rigorous theory for isotropic point scatterers and a scalar diffusion approximation will be described and verified against experimental results.

## 5.1. Rigorous Scalar Theory for Point-Scatterers

Systematic perturbation theory in terms of diagrammatic expansions is a fruitful way of describing wave propagation in random media. A very useful review has been published by Frisch.[19]

The system we will study is a collection of identical random point scatterers in a slab. The direction perpendicular to the slab is the $z$ direction and the slab is confined within $[0, d]$, so $d$ is the thickness of the slab. The quantity $a$ is the albedo, the ratio of the scattering- and total cross-section of a scatterer.

After solving the summation of the ladder and most crossed diagrams[9,15,19,20] using the assumption of isotropic point scatterers, we find an integral equation, the solution of which can be used to obtain the incoherent and time-reversed interference contribution to the intensity. We find

$$F(\mathbf{r}_1, \mathbf{r}_2) = a^2 \kappa^2 A(\mathbf{r}_1 - \mathbf{r}_2) + \frac{a}{4\pi\kappa} \int_0^d A(\mathbf{r}_1 - \mathbf{r}')F(\mathbf{r}',\mathbf{r}_2)dr' , \quad (4\mathrm{a})$$

in which no diffusion approximation is made, and where $\kappa \equiv \lambda_{\mathrm{MF}}^{-1}$ is the extinction rate or turbidity, with $\lambda_{\mathrm{MF}}$ the elastic mean free path and $F(\mathbf{r}_1, \mathbf{r}_2)$ is the propagator for the intensity, which describes the energy density at $\mathbf{r}_1$ due to a point source at $\mathbf{r}_2$, and with

$$A(\mathbf{r}_1 - \mathbf{r}_2) = \frac{\exp(-\kappa|\mathbf{r}_1 - \mathbf{r}_2|)}{|\mathbf{r}_1 - \mathbf{r}_2|^2} , \quad (4\mathrm{b})$$

the averaged amplitude Green's function squared.

Using the translational invariance of the slab in the $(x, y)$ plane, Eq. (4a) can be Fourier transformed in two dimensions

$$\Gamma(r_1, r_2; \alpha) \equiv \frac{a}{4\pi\kappa^2} \int F(\mathbf{r}_\perp, z_1, z_2)\exp(-i\mathbf{r}_\perp \cdot \mathbf{k}_\perp)dr_\perp , \quad (5)$$

where $\mathbf{r}_\perp = (x_1 - x_2, y_1 - y_2)$, the dimensionless optical depth $\tau \equiv \kappa z$, and in which

$$\alpha \equiv \frac{k_\perp}{\kappa} = \frac{k_0}{\kappa} \left\{ (\sin\theta_i \cos\varphi_i + \sin\theta_s \cos\varphi_s)^2 \right.$$

$$\left. + (\sin\theta_i \sin\varphi_i + \sin\theta_s \sin\varphi_s)^2 \right\}^{\frac{1}{2}} , \tag{6}$$

where $k_0 = (2\pi)/\lambda$ is the magnitude of the wave vector associated with the empty (that is without scatterers) medium, and the polar angles refer to the incoming $\widehat{K}_i$ and to the scattered direction $\widehat{K}_s$. As a consequence of this we have, with $b = \kappa d$ the optical thickness,

$$\Gamma(\tau_1, \tau_2; \alpha) = \frac{a^2}{2} W\left(|\tau_1 - \tau_2|; \alpha\right)$$

$$+ \frac{a}{2} \int_0^b W\left(|\tau_1 - \tau'|; \alpha\right) \Gamma(\tau', \tau_2; \alpha) d\tau' . \tag{7}$$

which has the form of the Milne equation,[21] and with

$$W(\tau, \alpha) \equiv \int_1^\infty \frac{\exp\left[-\tau(t^2 + \alpha^2)^{\frac{1}{2}}\right]}{(t^2 + \alpha^2)^{\frac{1}{2}}} dt . \tag{8}$$

Now, after calculating $\Gamma(\tau_1, \tau_2; \alpha)$, we can find the scattering intensity $I(\mathbf{r})$ with the help of the so-called bistatic scattering coefficient,[9,22] which can be partitioned into the coefficients for single scattering, $\gamma_s$, incoherent- (ladder terms) $\gamma_l$, and coherent-intensities $\gamma_c$ (cyclical terms), and is given by

$$\gamma(\mu_s, \mu_i) \equiv \frac{4\pi r^2}{A\mu_i} I(\mathbf{r}) = \gamma_s(\mu_s, \mu_i) + \gamma_l(\mu_s, \mu_i) + \gamma_c(\mu_s, \mu_i) , \tag{9a}$$

where $A$ is the area of the target, $\mu_{i,s} \equiv \cos\theta_{i,s}$, and

$$\gamma_s(\mu_s,\mu_i) = \frac{a\mu_s}{\mu_i + \mu_s}\left[1 - \exp\left\{-b\left(\frac{1}{\mu_i} + \frac{1}{\mu_s}\right)\right\}\right], \tag{9b}$$

$$\gamma_l(\mu_s,\mu_i) = \frac{1}{\mu_i}\int_0^b\int_0^b \Gamma(\tau_1,\tau_2;\alpha=0)\exp\left\{-\left(\frac{\tau_1}{\mu_s} + \frac{\tau_2}{\mu_i}\right)\right\}d\tau_1\,d\tau_2, \tag{9c}$$

$$\gamma_c(\mu_s,\mu_i) = \frac{1}{\mu_i}\int_0^b\int_0^b \Gamma(\tau_1,\tau_2;\alpha)\cos\left\{\frac{k_0}{\kappa}(\mu_i - \mu_s)(\tau_1 - \tau_2)\right\}$$

$$\times \exp\left\{-\frac{1}{2}\left(\frac{1}{\mu_s} + \frac{1}{\mu_i}\right)(\tau_1 + \tau_2)\right\}d\tau_1\,d\tau_2. \tag{9d}$$

To allow for easy formal manipulations it is very useful to introduce the following matrix multiplication. For matrices $F$ and $G$ it is defined as

$$FG = \int_0^b F(\tau)G(\tau)d\tau. \tag{10}$$

Notice that if the matrices depend on two $\tau$ coordinates the multiplication only involves the first coordinate. With

$$M(\tau_1,\tau_2;\alpha) \equiv \frac{1}{2}W\left(|\tau_1 - \tau_2|;\alpha\right). \tag{11}$$

Equation (7) can be written in matrix form

$$\Gamma = S + aM\Gamma, \tag{12}$$

where the initial source $S$ represents the lowest order of scattering to be accounted for, for instance single- or double-scattering. The formal solution of Eq. (12) is

$$\Gamma = (1 - aM)^{-1}S, \tag{13}$$

so the solution of the total scattering problem can be obtained by inverting the matrix $(1 - aM)$. It may also be developed into a power series, in which case the solution takes the form

$$\Gamma = S + aMS + a^2M^2S + a^3M^3S + \ldots, \tag{14}$$

where each term is found by multiplying the preceding one with the matrix $aM$. The physical significance of the power series is that each term corresponds to a separate order of scattering. So if we let $\Gamma_n$ represent the contribution for the $n$th order of scattering, we can use the iterative formula

$$\Gamma_{n+1} = aM\Gamma_n , \qquad (15)$$

to calculate the successive orders of scattering, starting with the two particle scattering term

$$S = a^2 M = \frac{a^2}{2} W(|\tau_1 - \tau_2|; \alpha) \equiv \Gamma_2 . \qquad (16)$$

Notice that this term is unbound for $\tau_1 \rightarrow \tau_2$.

The convergence of our iterative procedure (15) to the total solution $\Gamma$ is, except for numerical problems that might occur due to build up of inaccuracy for very high orders of scattering, determined by the albedo $a$ and the optical thickness of the slab $b$. For a semi-infinite slab, the iteration converges as $a^n n^{-3/2}$. In practice this means that it is possible to calculate the whole solution for $b = \infty$ and $a < 0.9$ within 1% accuracy in less than 20 iterations. For the albedo $a = 1$ only slabs with $b < 2$ can be treated with about the same accuracy.

Of course we are also interested in thick slabs ($b > 2$) with the albedo close or equal to 1. The solution to this problem is obtained by solving Eq. (12) by diagonalization. With the diagonalization procedure we can handle slabs up to 32 optical depths with albedo equal to 1, which already gives 95% of the backscattering intensity of the semi-infinite slab. So the only problem left, are very thick slabs ($32 < b < \infty$) in case $a = 1$. However, thick slabs and dominance of high-order scattering processes are precisely the conditions for which a diffusion approximation should hold. Therefore we are also interested in a diffusion theory.

## 5.2. Scalar Diffusion Approximation

In order to obtain a diffusion theory, consider a scattering medium of infinite size and take advantage of the translational invariance in all directions, then include boundary conditions to handle the finite slab problem.

From the literature[22-26] we know that the boundary conditions for a random walk can be given in the form of trapping planes in $-\tau_0$ and $b + \tau_0$ respectively. It appears that $\tau_0 \simeq 0.7104$ for $a = 1$. The solution of Eq. (4a) for a finite slab in the diffusion approximation is

$$\Gamma(\tau_1, \tau_2; \alpha) = \sum_{n=-\infty}^{\infty} \{\Gamma_0(\tau_1 - \tau_2 - 2nB; \alpha)$$

$$- \Gamma_0(\tau_1 + \tau_2 + 2nB + 2\tau_0; \alpha)\}$$

$$= \frac{3a^2}{2c\sinh(cB)}[\cosh\{c(B - |\tau_1 - \tau_2|)\}$$

$$- \cosh\{c(b - [\tau_1 + \tau_2])\}], \tag{17}$$

with $\Gamma_0(\tau; \alpha)$ the infinite-space Green's function and

$$c \equiv [3(1 - a) + \alpha^2]^{\frac{1}{2}}, \tag{18a}$$

$$B \equiv b + 2\tau_0 . \tag{18b}$$

The summation in this equation results from the multiple reflection in both mirror planes. For a semi-infinite slab, only the term for $n = 0$ remains.

This result can be used to calculate both the incoherent and the coherent (interference) contribution to the scattering in the diffusion approximation. We define

$$u \equiv \frac{k_0}{\kappa}(\mu_i - \mu_s), \tag{19a}$$

$$v \equiv \frac{1}{2}\left(\frac{1}{\mu_i} + \frac{1}{\mu_s}\right). \tag{19b}$$

After tedious algebra we find for the cyclical part

$$\gamma_c(\mu_s,\mu_i) = \frac{3a^2 \exp(-vb)}{2\mu_i cv \sinh(cB)}$$

$$\times \left\{ \left[ \frac{1}{(v-c)^2 + u^2}\{v - (v-c)\cosh(2c\tau_0)\} \right. \right.$$

$$+ \left. \frac{1}{(v+c)^2 + u^2}\{v - (v+c)\cosh(2c\tau_0)\} \right] \cosh(ub)$$

$$+ \left[ \frac{u}{(v+c)^2 + u^2} - \frac{u}{(v-c)^2 + u^2} \right] \sinh(2c\tau_0)\sin(ub)$$

$$+ \frac{1}{(v-c)^2 + u^2}[(v-c)\cosh\{(v-c)b - 2c\tau_0\}$$

$$- (v)\cosh\{(v-c)b\}]$$

$$+ \frac{1}{(v+c)^2 + u^2}[(v+c)\cosh\{(v+c)b + 2c\tau_0\}$$

$$\left. - (v)\cosh\{(v+c)b\}] \right\} . \tag{20}$$

Note that for pure backscattering, $\mu_i = \mu_s$, $\gamma_l = \gamma_c$ in all cases.

For $b \to \infty$ expression (68b) reduces to

$$\gamma_{c,\infty} = \frac{3a^2[c + v\{1 - \exp(-2c\tau_0)\}]}{2\mu_i cv[(v+c)^2 + u^2]} , \tag{21}$$

which for an albedo of 1 agrees with the result of Ref. 28.

In Fig. 11 we compare the lineshapes of the exact theory and the diffusion approximation for two different optical thicknesses, with the albedo $a = 1$. We see, that although their shapes are comparable, the diffusion approximation shows a small underestimation for all angles with respect to the exact theory. It appears, as we will see, that this is caused by a small underestimation of the lowest orders of scattering. This also explains why the backscattering for the thin slab ($b = 1$) (for which the lower-order scattering contributions are dominant) suffers more from this underestimation than that for the thick slab ($b = 32$).

In order to decompose the results of the diffusion theory in a summation of terms in which each term describes the next order of

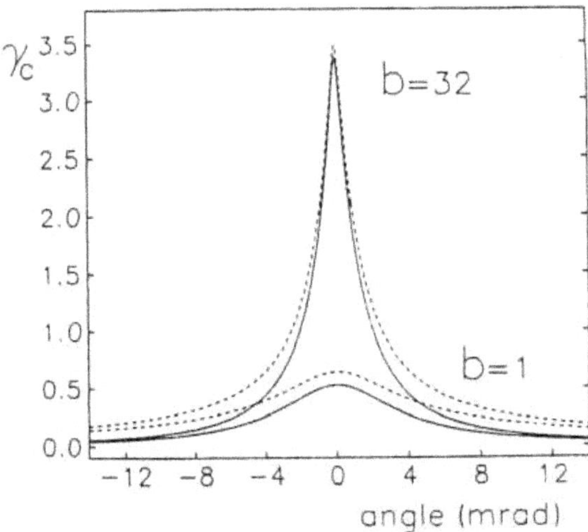

Fig. 11. Comparison of the absolute lineshapes, $\gamma_c(\mu_i,\mu_s)$, as found from the exact theory (dashed lines) and the diffusion approximation (solid lines) for slab-thickness $b = 32$ and $b = 1$, $\lambda_{MF}/\lambda = 53.5$, and albedo $a = 1$.

scattering, we have to express the bistatic coefficient in a Taylor series of the albedo. Expansion of (21) in the albedo gives the separate contribution $\gamma_n$ for each order of scattering for the case $b = \infty$. The results of this expansion for $\mu_i = \mu_s = 1$ are shown in Fig. 12 and Table 1. It is seen that only for the very lowest orders of scattering there is some discrepancy between the values for $\gamma_{c,\infty}$ as calculated from the exact theory and the diffusion approximation. This was to be expected since a wave is only propagating diffusively after it has scattered several times. From Fig. 12 it is also seen that the contribution of each order of scattering is proportional to $n^{-3/2}$ for higher order scattering in case $b = \infty$.

## 5.3. Time-Resolved Coherent Backscattering

Akkermans *et al.*[27] developed an approximate diffusion theory in which the shape of the backscattering cone is found by integration over all contributing path lengths, neglecting the depth of the first and last scatterer. This theory can be used to find the time dependence of the

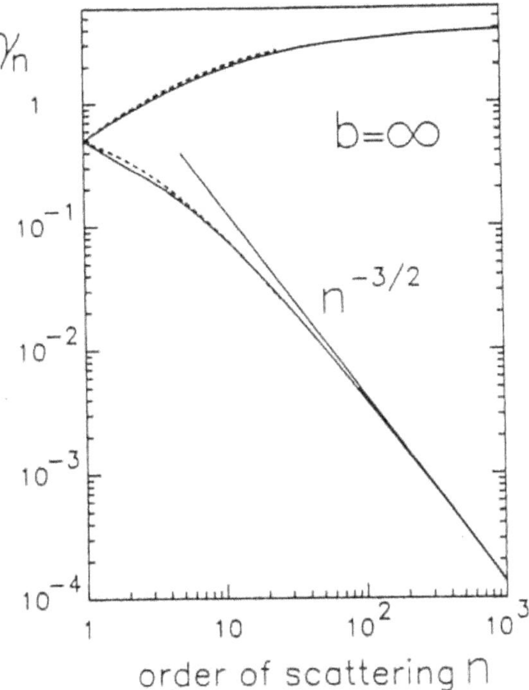


Fig. 12. The $n$th order scattering contribution to the total amount of scattering $\gamma$ in case $b = \infty$, $a = 1$ and $\mu_i = \mu_s = 1$ for both the exact theory (dotted lines) and the diffusion approximation (solid lines). The convergence of the high-order scattering to the $n^{3/2}$ asymptote is clear. It is also seen, that the lower-order scattering is underestimated by the diffusion approach. The cumulative curve $\sum_n \gamma_n$ is also given. See also Table 1.

bistatic coefficient $\gamma(\theta, t)$:

$$\gamma(\theta, t) \propto t^{-3/2}[1 + \exp\{-4\pi^2 ct\lambda_{\mathrm{MF}}\theta^2/(3\lambda^2)\}] \,, \qquad (22)$$

where $c$ is the speed of light. The equation describes a Gaussian shaped backscattering cone with a width of the order $\lambda/(\lambda_{\mathrm{MF}}ct)^{\frac{1}{2}}$. In principle the equation is only valid for times $t$ long enough for the scattering to have reached the diffusive limit, which is only after a certain number of scattering events. As will be seen however, this condition was apparently met within the error of our experiment, since our data are well

Table 1. The $n$th order scattering contribution to the total amount of scattering $\gamma$ in case $b = \infty$, $a = 1$ and $\mu_i = \mu_o = 1$ for both the exact theory and the diffusion approximation. It is seen, that the lower-order scattering is underestimated by the diffusion approach. The cumulative curve $\sum_n \gamma_n$ is also given. See also Fig. 12.

| $n$ | $\gamma_{n,\text{diff}}$ | $\gamma_{n,\text{exa}}$ | $\sum_n \gamma_{n,\text{diff}}$ | $\sum_n \gamma_{n,\text{exa}}$ |
|---|---|---|---|---|
| 1 | 0.5000 | 0.5000 | 0.5000 | 0.5000 |
| 2 | 0.3071 | 0.3465 | 0.8071 | 0.8465 |
| 3 | 0.2356 | 0.2611 | 1.0426 | 1.1076 |
| 4 | 0.1903 | 0.2066 | 1.2330 | 1.3142 |
| 5 | 0.1582 | 0.1692 | 1.3912 | 1.4834 |
| 6 | 0.1343 | 0.1421 | 1.5254 | 1.6254 |
| 7 | 0.1159 | 0.1214 | 1.6413 | 1.7469 |
| 8 | 0.1013 | 0.1055 | 1.7426 | 1.8524 |
| 9 | 0.0896 | 0.0928 | 1.8323 | 1.9452 |
| 10 | 0.0801 | 0.0825 | 1.9124 | 2.0277 |
| $\infty$ | 0.0000 | 0.0000 | 4.1313 | 4.2277 |

described by the theory. The observed intensity at a particular angle $\theta$ still has to be convolved with the pulse shape. We then get

$$\tilde{\gamma}(\theta, t) = \iint \gamma(\theta, r - t) I_p(t) I_p(t' - \tau) dt dt' , \qquad (23)$$

in which the first integral gives the backscattered signal as a convolution of the impulse response function of the scattered $\gamma(\theta, t)$, and the incident pulse shape $I_p(t)$. The second integral gives the measured signal as a convolution of the backscattered signal with a light-gating pulse, which is used in case of ultra fast detection.

## 5.4. Comparison with Experiment

We will now test the theories against experimentally obtained cones of parallel polarization. We first note that the predicted enhancement factor for the multiple scattering contribution to the backscattered light is 2 in all cases. As the single scattering contribution to the incoherent background cannot be very much larger than $\simeq 10\%$ for thick slabs, one would expect enhancement factors of about 1.9. Experimental enhancement factors may be significantly lower. We will return to this point later.

Even if the predicted relative intensities of background and cone do not fully agree with the experimentally determined values, scalar theory may still correctly predict the *width* and the *shape* of the cone, provided that the latter is indeed mainly due to most-crossed-diagram type of interference. Our rigorous isotropic theory predicts the width of the cone to relate with the transport mean free path $\lambda_{tr}$ according to

$$W \simeq \frac{0.7}{2\pi}(\lambda/\lambda_{tr}) \ . \tag{24}$$

This relationship is in excellent agreement with the experimental data (c.f. Fig. 6).

In Fig. 13, the backscattering pattern as recorded using a thick slab ($b \simeq 240$) of 1.05-vol % and diameter $\phi \simeq 0.22\mu$m rutile particles in 2-methylpentane-2,4-diol as a sample, is plotted together with the intensity profile as calculated for this sample from diffusion theory, in which for $\lambda_{MF}$ the value of $\lambda_{tr}$ that was obtained from transmission experiments[15] was used. The theoretical curve was fitted to the experimental data in the following way: the calculated intensity profile was convolved with the instrumental resolution of .5 mrad. (The advantage of convolving the calculated curve instead of deconvolving the experimental one is that no assumptions regarding the shape of the latter are needed). The tops of the resulting theoretical and experimental curves were then superimposed and the vertical scale of the calculated curve was adapted so as to make the outermost parts of its wings coincide with those of the experimental cone. The shape and width of the calculated cone is found to fit perfectly to the experimental data. At the same time we see that the observed intensity profile has an offset with respect to the calculated one that amounts to *ca.* 30% of the total background. In Fig. 13, we denoted the bistatic scattering coefficient of all angle-independent (ai) terms by $\gamma_{ai}$. The latter is too big to be explained in terms of just single scattering by an ideal sample. The large experimental value of $\gamma_{ai}$ may be (at least partly) explained by multiple scattering contributions without interference terms, or with angle-independent interference terms (c.f. Sec. 6). More trivial explanations should also be considered: Suppose that the $TiO_2$-pigment tends to adhere to the cell-window. The concentration of scatterers in the very front slab of a real sample will then be higher than everywhere else, and this would result in a relative increase of the single-scattering contribution. However, "deep difference slabs" (see below) should then exhibit enhancement factors near 2, and this we do not find experimentally.

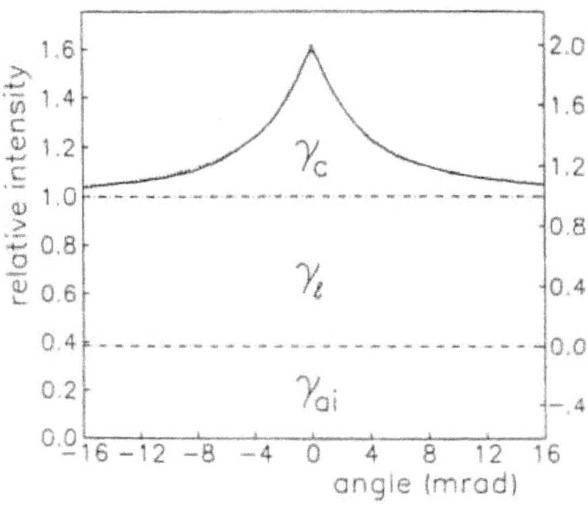


Fig. 13. Enhanced backscattering of the parallel light component from a 2300 $\mu$m slab of a 1.05-vol % sample of $\phi \simeq$ 220 nm particles of $TiO_2$ in 2-methylpentane-2,4-diol at $\lambda_{vac} = 514.5$ nm.
Solid line: experimental curve, dotted line: curve, calculated from diffusion theory, using $\lambda_{tr} = 9.7$ $\mu$m.
The left-hand vertical axis corresponds to the experimental curve, while the right-hand axis corresponds to the theoretical decomposition in contributions from the ladderterms, $\lambda_l$, the interference terms, $\gamma_c$, and the angle-independent terms, $\gamma_{ai}$.

As explained in Sec. 2, we used a difference technique (see Ref. 14) to probe the width and the enhancement factor of the contribution to the backscatter cone as a function of the depth in the sample.

Figure 14a shows difference cones that were obtained with this technique, using the sample that gave the full cone of Fig. 13. In order to present several curves in the same figure without loosing clarity, a

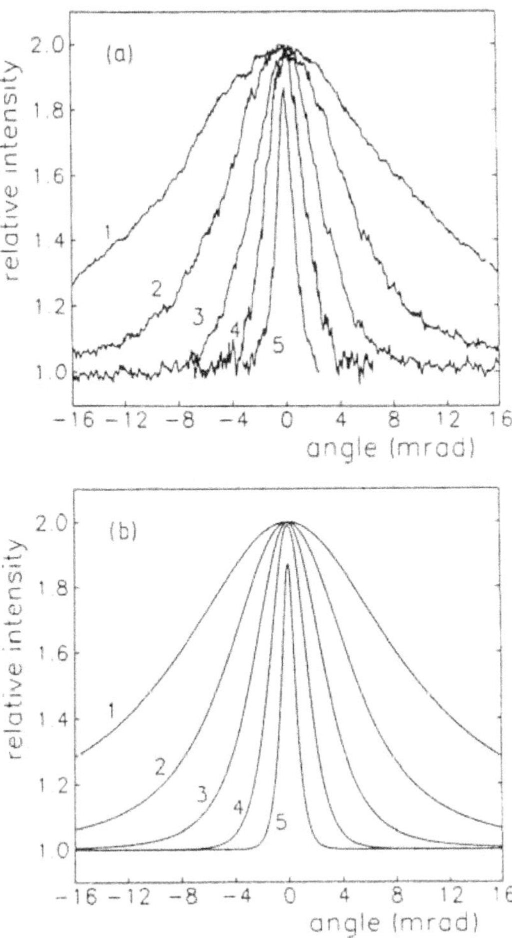

Fig. 14(a). Normalized (see text) difference cones recorded for the parallel light component using the sample of Fig. 13. Difference slabs: curve 1: 0–13 $\mu$m; curve 2: 13–25 $\mu$m; curve 3: 25–50 $\mu$m; curve 4: 50–100 $\mu$m; curve 5: 100–1000 $\mu$m.
(b) Normalized backscattering curves, calculated from diffusion theory and convolved with the resolving power of .5 mrad for the difference slabs of (a), using $\lambda_{tr}$ - 9.7 $\mu$m.

normalization procedure was performed that is more conveniently outlined if we first discuss the corresponding calculated cones. The latter are shown in Fig. 14b and were obtained in the following way: the intensity profiles were calculated from diffusion theory using experimentally obtained $\lambda_{tr}$ values, and convolved with the instrumental resolution. The resulting (convolved) curves were normalized, each one with respect to its own background level. We return now to the experimental curves: the vertical scales of the cones in Fig. 14a were adapted so as to match their heights (intensities) to those of their calculated counterparts. The experimental and calculated widths and relative intensities and the experimental enhancement factors for series of difference cones that were recorded using samples with different volume fractions of $TiO_2$, are listed in Table 2. (The values listed for the 1.05-vol % sample correspond to the cones of Figs. 14.) It is concluded that scalar theory correctly describes the width and shape of these parallel polarized cones.

From Figs. 14 and Table 2 it is further seen that the deeper a difference slab is situated in the sample, the narrower its contribution to the cone. (This was to be expected, because light paths visiting deep slabs are necessarily long.)

The precise relationship between path length and width of the contribution to the cone was studied in a time resolved experiment. Figure 15 shows cones recorded at 30 fsec and 330 fsec after the CPM-laser pulse respectively, using a 6.8 vol% $TiO_2$/2-methylpentane-2,4-diol sample ($\lambda_{tr} \simeq 2\mu m$). The smooth lines are theoretical curves, calculated from Eq. (23). The agreement is excellent. Figure 16 shows the time decay of pulses scattered from the $TiO_2$ sample, as measured at fixed angles with respect to the direction of pure backscattering. It is seen that for all delay times the scattered intensity is largest in the direction of exact backscattering. At small angles with the backscattering direction, the phase-difference between time-reversed paths increases with $t^{\frac{1}{2}}$, and a coherent contribution to the scattered intensity will only occur during a short time following the laser flash.

## 6. ON THE FACTOR OF 2

Basically the theory of weak localization is well understood. The shape and width of the backscattered intensity profile in a variety of experimental situations are reasonably well described. A disturbing fea-

Table 2. Experimental values for width, relative intensity and enhancement factor, and values calculated for width and relative intensities from diffusion theory for difference slabs as a function of $\lambda_{tr}$ and optical depth. Relative intensities were normalized with respect to the total intensities of the difference slabs given for each sample.

| Concentration | | Optical | $\Delta(\tau_d)$ (mrad) | | Relative intensity | | Enhancement factor (expt) |
|---|---|---|---|---|---|---|---|
| (vol %) | $\lambda_{tr}$ ($\mu$m) | depth $\tau_d$ | expt. | calc. | expt. | calc. | |
| 0.36 | 28.2 | 0.35- 0.9 | 7.8 | 7.5 | 0.09 | 0.13 | 2.0 |
| | | 0.9 - 1.6 | 4.6 | 5.0 | 0.26 | 0.19 | 1.83 |
| | | 1.6 - 4.6 | 2.4 | 2.6 | 0.45 | 0.46 | 1.69 |
| | | 4.6 - 8.9 | 1.1 | 1.2 | 0.21 | 0.22 | 1.60 |
| 1.05 | 9.68 | 0. - 1.3 | 21. | 20. | 0.24 | 0.22 | 1.38 |
| | | 1.3- 2.6 | 10.3 | 10.4 | 0.21 | 0.21 | 1.87 |
| | | 2.6- 5.2 | 5.4 | 5.9 | 0.21 | 0.23 | 1.71 |
| | | 5.2- 10.3 | 3.1 | 3.1 | 0.17 | 0.17 | 1.69 |
| | | 10.3-103. | 1.2 | 1.1 | 0.16 | 0.17 | 1.65 |
| 2.50 | 4.06 | 2.0- 6.1 | 17. | 15. | 0.54 | 0.57 | 1.71 |
| | | 6.1- 9.8 | 8.0 | 7.0 | 0.17 | 0.17 | 1.63 |
| | | 9.8- 49.1 | 3.9 | 3.0 | 0.28 | 0.27 | 1.61 |
| 9.81 | 1.04 | 5.8- 12.5 | 29. | 25. | 0.53 | 0.56 | 1.67 |
| | | 12.5- 24. | 15. | 13. | 0.21 | 0.27 | 1.54 |
| | | 24. -48. | 7.9 | 6.8 | 0.26 | 0.16 | 1.65 |

ture however is the enhancement factor, defined as the ratio of the total backscatter intensity and the incoherent background intensity. An accurate experimental determination of the factor is difficult for technical reasons. Some experimentalists maintain that they found evidence for an enhancement factor differing from two, whereas others claim that an exact factor of two is being measured. So far, all theories add up to a description of backscattering in terms of exclusively self-avoiding light paths (ladder diagrams) and interference exclusively between time-reversed pairs of such paths (most-crossed diagrams). This description

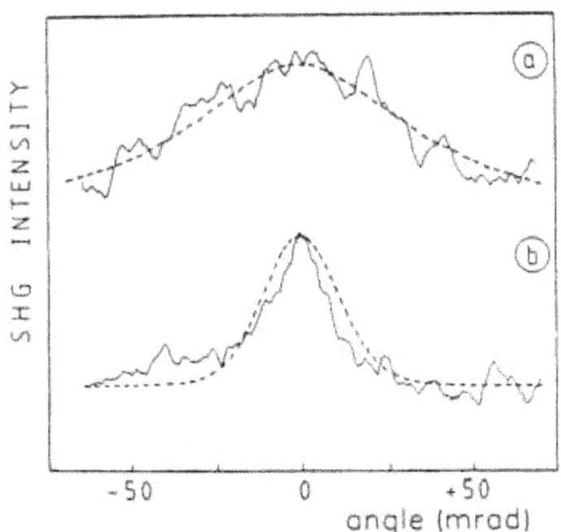

Fig. 15. Enhanced backscattering cones measured for a 6.8 vol% suspension of $TiO_2$ particles in 2-methylpentane-2,4-diol $(\lambda_{MF}=2\mu m)$ at different delay times after the laser pulse: (a) 30 fs; (b) 330 fs. Dashed curves represent results of calculations based on Eq. (23).

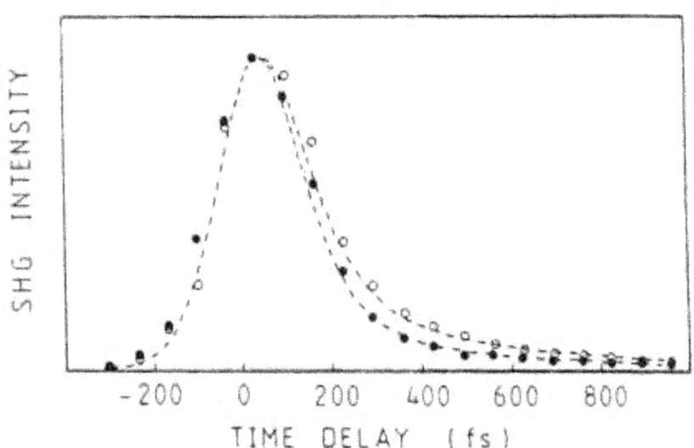

Fig. 16. Backscattered light pulses from the sample of Fig. 15 measured for $\theta = 0$ mrad (open circles) and $\theta = 27$ mrad (full circles). Dashed curves represent results of calculations based on Eq. (23).

leads to a doubling of the intensity of all multiple-scattering contributions in the 180° direction both for scalar waves and for the parallel polarized component in the case of vector waves, since each multiple-scattering ladder diagram has a most-crossed counterpart. The first-order ladder diagram however, is identical to its most-crossed counterpart (time-reversal symmetry) and does not yield an interference term. The single scattering contribution therefore is not doubled in the 180° direction. In the case of isotropic scattering of scalar waves by point-scatterers, the single scattering contribution is about 12% of the incoherent background intensity.

Recently it was stated that the phenomenon of weak localization for scalar waves is principally different from that of electromagnetic waves in the sense that electromagnetic theory in contrast to scalar theory would allow for an enhancement factor differing from two.[28] This statement is based upon a misinterpretation of theoretical results of Stephen and Cwillich,[24] who calculated the enhancement factors in the backscattering of scalar- and vector-waves, neglecting single scattering in the scalar case but not in the vector case.

If in addition to the self-avoiding light paths also non-self-avoiding paths are considered, classes of diagrams may be identified that will further decrease the predicted enhancement factor.[29] One class corresponds to "loop" type of light paths and is pictured in Fig. 17. The corresponding interference terms are isotropic (angle-independent) and therefore apparently contribute to the incoherent background. Another class corresponds to time-reversal-invariant multiple-scattering processes. Figure 18 shows a two-particle third-order scattering event that is identical to its time-reversed version. In the same figure the square of this diagram, relevant for the calculation of the intensity Green's function, is displayed. Restricting ourselves to two-particle scattering it is possible to sum all time-reversal-invariant diagrams rigorously as they are part of the general intensity diagram of Fig. 19. These diagrams can be described as comprising all intensity contributions brought about by two-particle scattering events in all orders starting and ending on the same scattering center. A calculation shows that they renormalize the first-order scattering efficiency, and their contribution may be up to a few percent of the total first-order scattering, depending on the density of scatterers. In addition one can easily indicate time-reversal invariant more-particle scattering events.

So far, neither the multiple-scattering time-reversal invariant diagrams nor isotropic interference contributions have been included in the theoretical work. We showed that in addition to the well-known

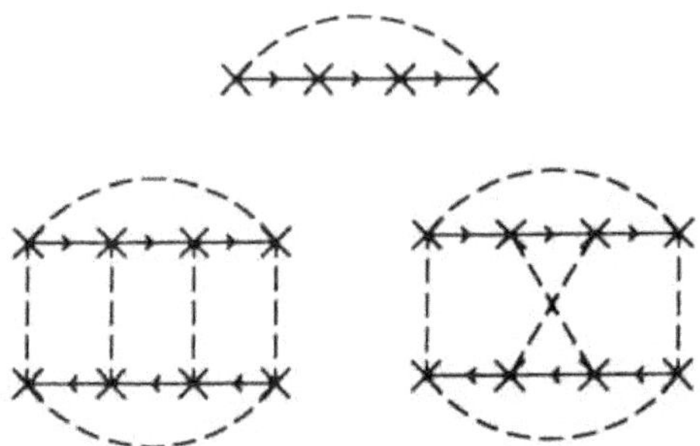

Fig. 17. A time-reversal-variant three-particle amplitude diagram (upper),
the incoherent contribution of this diagram to the intensity (left), and the intensity contribution of the interference of this diagram with its time-reversed
version (right). All loop-type diagrams, both the incoherent and the time-
reversed interference versions, will cause an *angle-independent* backscatter intensity. (Diagrammatic rules: (crosses denote *t*-matrix of scatterer, solid lines
represent full amplitude Green's function, and dashed lines connect identical
particles.)

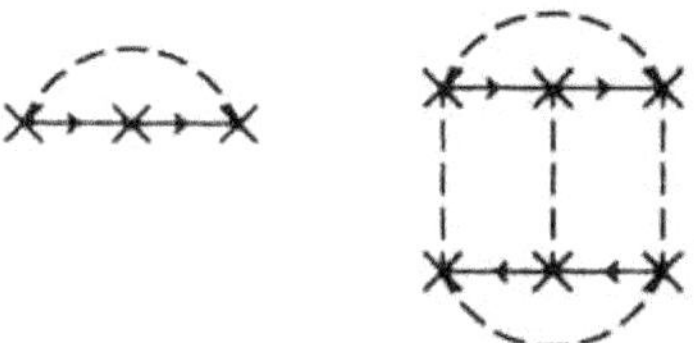

Fig. 18. A time-reversal-invariant third order scattering diagram both for
the amplitude (left) and for the intensity (right).

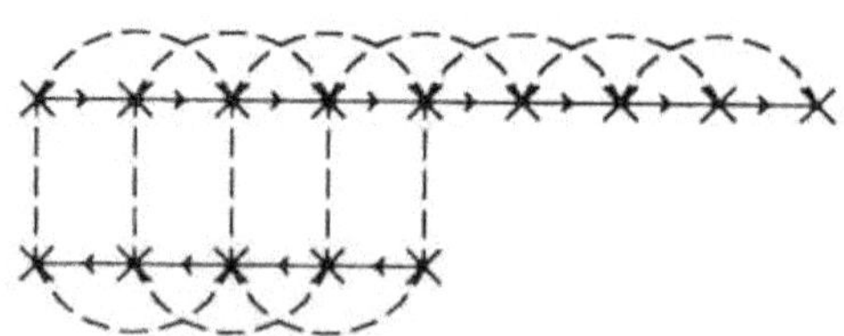

Fig. 19. A general two-particle multiple-scattering intensity diagram starting
and ending in both channels on the same particle.

incoherent background (ladder diagrams) and interference contribution between time-reversed events (cyclical diagrams) other contributions to the backscattering from a random medium exist. The experimental observation of these events would make this area of research unique in the field of physics of scattering as to the experimental possibility of discrimination between various types of multiple-scattering events. We conclude that the question of the enhancement factor deserves much more study both experimentally and theoretically.

## 7. ON THE ROAD TO STRONG LOCALIZATION

### 7.1. Dependent Scattering and Correlated Scatterers

Many theories of elastic multiple scattering in random media are based on two assumptions that become highly questionable at high densities of scatterers. These assumptions are (i) that particle-particle correlations are absent, and (ii) that scattering paths which go through the same scatterer more than once may be neglected: Independent-Scattering Approximation (IPA).

Inclusion of particle-particle correlations is desirable at high densities, but interaction between scatterers complicates matters enormously. Even with the simplest possible forms such as hard-sphere interaction, the static behavior of a collection of scatterers, expressed in terms of density-density correlation functions, becomes an insolvable many-body system. Approximative methods exist to obtain reliable values for density-density correlation functions. The problem with multiple scattering is that one needs higher-order density correlation functions, that are difficult to obtain. Usually one relies on decoupling procedures to express high-order density correlation functions in terms of products of lower-order correlation functions. Anyhow, the introduction of even approximate particle-particle correlations makes the multiple-scattering problem much more difficult. Now without knowing the precise magnitude of the corrections induced by particle-particle correlations, it is expected that in general the trend will be to reduce the scattering efficiency or, starting from a weakly-scattering medium, increasing the density of scatterers will result in a reduction of the mean free path until the effect of particle-particle correlations balances that of the increasing density and saturation occurs.

For very high densities, mean-field approximate theories exist to calculate amplitude Green's functions. These methods are known under names as average-$t$-matrix approximation (ATA), and coherent-potential approximation (CPA).[30] They find useful application in the theory of electronic structure of disordered composites, for instance: In an $AB$ disordered crystalline system, both limits (100% $A$ and 100% $B$) should be described correctly and the mean-field theories are tailored to do so. The theories also include some dependent-scattering corrections. The mean-field theories are usually optimized to calculate the real part of the self energy of the amplitude Green's function, which is expressed as an effective dielectric constant. We are more interested in the mean free path, which is associated with the imaginary part of the self energy of the amplitude Green's function.

The independent-scattering approximation (IPA) is expected to break down when the scatterers get very close. For instance for two particles the IPA implies that there are contributions to the $t$-matrix of the type $t_1, t_2, t_1 t_2$, and $t_2 t_1$, excluding all higher-order multiple scattering between the two particles. However, when the particles are close compared to the wavelength, multiple scattering between the two particles may become very important. One sees that in this situation the particle correlations may also become important. To get more insight in the effect of dependent scattering without the simultaneous influence of particle correlations Reiter and Lagendijk studied *point* scatterers in an isotropic scalar theory.[31] This theory yields the total $t$-matrix as the inversion of a matrix with a dimension equal to the number of particles. Their $t$-matrix is not averaged yet over the realizations of the sample and is therefore a configuration-dependent $t$-matrix. The inversion allows for an exact numerical diagonalization for a finite but large (in the order of 1000) number of particles. This numerical result can then be compared with the approximate theories to check their quality. In addition the theory of Reiter and Lagendijk can be used to average over the particle realizations and obtain dependent-scattering corrections to the amplitude Green's function.

Before discussing the implications of the effects of particle correlations and dependent scattering on localization we want to point out that in principle localization is connected with unaveraged amplitude Green's functions or configurationally averaged intensity Green's functions. In the preceding paragraph we discussed mainly the results for the averaged amplitude Green's function. In principle, no information on localization is contained in these results, as is for instance illustrated very well by the Lloyds model.[32] Of course, the results for the intensity Green's function depend on those for the amplitude Green's function. In practice it is very likely that if corrections of the type just discussed make the scattering (described by the scattering mean free path $\lambda_{sc}$) less efficient, the scattering in the intensity Green's function, as described by the transport mean free path $\lambda_{tr}$, will also be less efficient. In Fig. 20 the results of a renormalized theory including all types of two-particle dependent scattering on the mean free path and the effective wave vector are displayed as a function of the disorder parameter $\alpha$, defined as

$$\alpha \equiv 4\pi n_0 \mathrm{Im}\,(f)k_0^{-2} \, , \tag{25}$$

in which $n_0 = N/V$ is the density of scatterers, $f$ is the scattering amplitude of the scatterers, and $k_0$ is the bare wave vector of the medium.

We see that indeed the dependent-scattering corrections make the scattering mean free path longer. We expect that this will also make the transport mean free path longer and as such make the possible observation of localization of waves more difficult. In the next section we will discuss the experimental observation of scattering and transport mean free paths for very strongly scattering media in order to find out whether an indication of particle-particle correlations or dependent-scattering effects can be found.


## 7.2. Very Strongly-Scattering Media and Localization

Now that the occurrence of weak localization has been proved, the question as to whether it will be possible to observe strong localization of light should be considered. The crucial parameter is the transport mean free path $\lambda_{tr}$, which should be as short as possible.

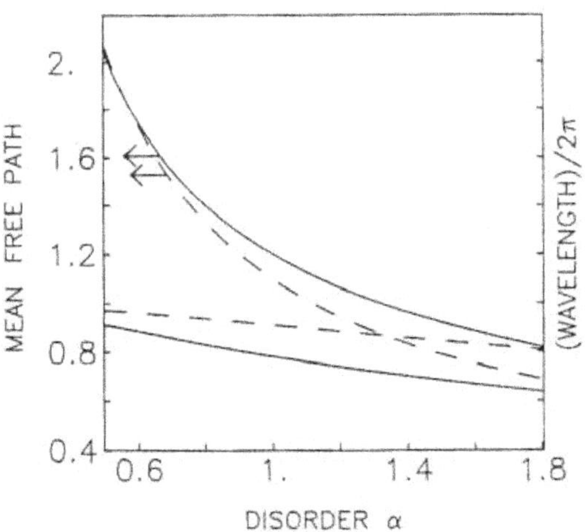



Fig. 20. Mean free path and wavelength as a function of the disorder parameter $\alpha$ for resonating particles. Dashed lines refer to the independent scattering result and solid lines are the result of a self-consistent theory which incorporates dependent-scattering corrections.

It seems that a suspension of $TiO_2$ in air or in some low-refractive-index substance is the system in which the shortest $\lambda_{tr}$-values for visible light may be realized in practice. Once $\lambda_{tr} < \lambda_{cr}$ (where $\lambda_{cr}$ is some critical mean free path) the system will be localized. The Ioffe-Regel criterion gives $\lambda_{cr}$ as $\lambda/(2\pi)$, and indeed the existence of delocalized states at $\lambda_{tr} < \lambda/(2\pi)$ seems to be very unlikely. On the other hand it cannot be excluded beforehand that the localization transition might occur earlier. To detect strong localization one should study light that traveled through the medium over at least the localization length $\lambda_{loc}$, and near the localization transition the latter may be larger than $\lambda_{tr}$ by orders of magnitude. Since in backscattering lower-order scattering contributions are dominant, transmission experiments seem to be indicated for the detection of strong localization. Whereas in the regime of classical diffusion the transmission decays with increasing length $L$ of the sample according to $L^{-1}$, in the localized regime this decay is expected to be faster: possibly starting as $L^{-2}$ for $L < \lambda_{loc}$ and exponential for $L > \lambda_{loc}$. Absorption by the sample is a complicating factor, as it also leads to a faster than $L^{-1}$ decay.

Watson c.s.[33] performed time-resolved experiments on the diffusion of light through scattering media. For polystyrene-sphere suspensions in water with $2\pi\lambda_{\mathrm{MF}}/\lambda \simeq 360$, they found no signs of departure from classical diffusive behavior. In $TiO_2$-water and $TiO_2$-air samples, they could not determine $\lambda_{\mathrm{MF}}$, because of absorption.

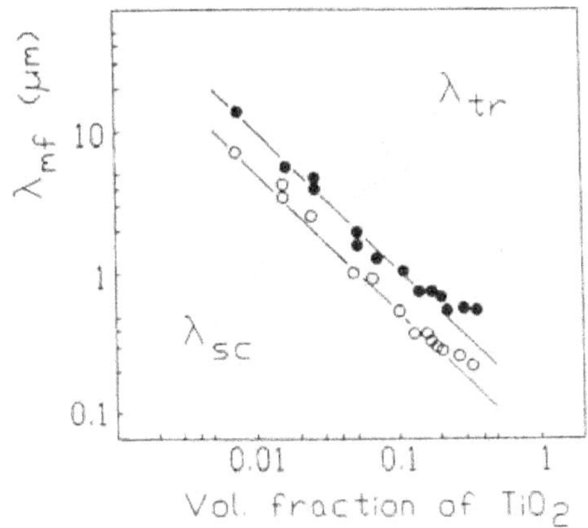


Fig. 21. Scattering mean free path (lower plot) and transport mean free path (upper plot) as a function of the volume fraction $\phi$ of 220 nm $TiO_2$ particles.

The present authors determined $\lambda_{\mathrm{sc}}$ and $\lambda_{\mathrm{tr}}$ for $TiO_2$ suspensions in 2-methylpentane-2,4-diol. Results are shown in Fig. 21. It is seen that, upon increasing the volume fraction of $TiO_2$ above $\simeq 15\%$, the transport mean free path seems to saturate at a value that is slightly larger than the wavelength of the light in the medium. (The width of the cone of enhanced backscattering equally saturates.) It is difficult to tell whether particle-particle correlations or dependent scattering is the main cause of this effect, but since the average interparticle separation at this volume fraction is roughly equal to the wavelength in the medium, dependent scattering may be expected to at least play a part. Thus, in the samples with strongest randomness, $\lambda_{\mathrm{tr}}$ is still by a factor of 8 higher than the Ioffe-Regel value for $\lambda_{\mathrm{cr}}$. A systematic study of the diffusion coefficient as a function of the sample-thickness might reveal how near these samples come to the localization transition.

## ACKNOWLEDGEMENT

This work is part of the research program of the "Stichting voor Fundamenteel Onderzoek der Materie" (FOM), which is financially supported by the "Nederlandse Organisatie voor Wetenschappelijk Onderzoek" (NWO).

## REFERENCES

1. P. W. Anderson, "Absence of diffusion in certain random lattices", *Phys. Rev.* **109** (1958) 1492.
2. For a recent collection of papers on Anderson localization, see: *Anderson Localization*, ed. Y. Nagaoka and H. Fukuyama (Springer, Berlin, 1982).
3. W. Götze, "A theory for the conductivity of a fermion gas moving in a strong three-dimensional random potential", *J. Phys.* **C12** (1979) 1279; W. Götze, "The mobility of a quantum particle in a three-dimensional random potential", *Phil. Mag.* **B43** (1981) 219.
4. D. Vollhardt and P. Wölfle, "Diagrammatic, self-consistent treatment of the Anderson localization problem in $d \leq 2$ dimensions", *Phys. Rev.* **B22** (1980) 4666.
5. B. Cellini, *Autobiography*, through M. Minnaert, *Natuurkunde van het vrije veld* (Thieme, Zutphen) Vol. I, p. 267.
6. W. W. Montgomery and R. H. Kohl, "Opposition effect experimentation", *Opt. Lett.* **5** (1980) 546.
7. T. S. Trowbridge, "Retroreflection from rough surfaces", *J. Opt. Soc. Am.* **68** (1978) 1225.
8. Y. Kuga and A. Ishimaru, "Retroreflection from a dense distribution of spherical particles", *J. Opt. Soc. Am.* **A1** (1984) 831.
9. L. Tsang and A. Ishimaru, "Backscattering enhancement of random discrete scatterers", *J. Opt. Soc. Am.* **A1** (1984) 836; "Theory of backscattering enhancement of random discrete isotropic scatterers based on the summation of all ladder and cyclical terms", *ibid.* **2** (1985) 1331; "Radiative wave and cyclical transfer equations for dense nontenuous media", *ibid.* **2** (1985) 2187.
10. M. P. van Albada and A. Lagendijk, "Observation of weak localization of light in a random medium", *Phys. Rev. Lett.* **55** (1985) 2692.
11. P. E. Wolf and G. Maret, "Weak localization and coherent backscattering of photons in disordered media", *Phys. Rev. Lett.* **55** (1985) 2696.

12. S. Etemad, R. Thompson and M. J. Andrejco, "Weak localization of photons: universal fluctuations and ensemble averaging", *Phys. Rev. Lett.* **57** (1986) 575.

13. M. Kaveh, M. Rosenbluh, I. Edrei and I. Freund, "Weak localization and light scattering from disordered solids", *Phys. Rev. Lett.* **57** (1986) 2049.

14. M. P. van Albada, M. B. van der Mark and A. Lagendijk, "Observation of weak localization of light in a finite slab; anisotropy effects and light-path classification", *Phys. Rev. Lett.* **58** (1987) 361.

15. M. B. van der Mark, M. P. van Albada and A. Lagendijk, "Light scattering in strongly scattering media: multiple scattering and weak localization", *Phys. Rev.* **B37** (1988) 3575.

16. S. Etemad, R. Thompson, M. J. Andrejco, Sajeev John and F. C. MacKintosh, "Weak localization of photons: termination of coherent random walks by absorbtion and confined geometry", *Phys. Rev. Lett.* **59** (1987) 1420.

17. R. Vreeker, M. P. van Albada, R. Sprik and A. Lagendijk, "Femtosecond time-resolved measurements of weak localization of light", *Phys. Lett.* **A132** (1988) 51.

18. M. P. van Albada and A. Lagendijk, "Vector character of light in weak localization: spatial anisotropy in coherent backscattering from a random medium", *Phys. Rev.* **B36** (1987) 2353 (rapid communication).

19. U. Frisch, *Wave propagation in random media* in "Probabilistic methods in applied mathematics", ed. A. T. Barucha-Reid, (Academic Press, New York, 1968) Vol. I, p. 76.

20. G. Bergmann, "Weak localization in thin films; a time of flight experiment with conduction electrons", *Phys. Rep.* **107** (1984) 1.

21. K. E. Atkinson, *A survey of numerical methods for the solution of Fredholm integral equations of the second kind* (Society for Industrial and Applied Mathematics, Philadelphia, 1976).

22. A. Ishimaru, *Wave propagation and scattering in random media*, (Academic, New York, 1978) Vols. I and II.

23. H. C. van de Hulst, *Multiple light scattering*, (Academic, New York, 1980) Vols. I and II.

24. M. J. Stephen and G. Cwilich, "Rayleigh scattering and weak localization: effects of polarization", *Phys. Rev.* **B34** (1986) 7564.

25. A. A. Golubentsev, "Suppression of interference effects in multiple scattering of light", *Sov. Phys. JETP* **59** (1984) 26 [*Zh. Eksp. Teor. Fiz.* **86** (1984) 47].

26. M. N. Barber and B. W. Ninham, *Random and Restricted Walks* (Gordon and Breach, New York, 1970).

27. E. Akkermans, P. E. Wolf and R. Maynard, "Coherent backscattering of light by disordered media: analysis of the peak line shape", *Phys. Rev. Lett.* **56** (1986) 1471.

28. M. Rosenbluh, I. Edrei, M. Kaveh and I. Freund, "Precision determination of the line shape for coherently backscattered light from disordered solids: comparison of vector and scalar theories", *Phys. Rev.* **A35** (1987) 4458.

29. A. Lagendijk, "Interference in multiple scattering of scalar and vector waves", unpublished.

30. V. A. Davies and L. Schwartz, "Electromagnetic propagation in close-packed disordered suspensions", *Phys. Rev.* **B31** (1985) 5155.

31. G. F. Reiter and A. Lagendijk, "Dependent-scattering effects in the determination of the criterion for the localization of waves", to be published.

32. P. Lloyd, "Exactly solvable model of electronic states in a three-dimensional disordered Hamiltonian: non-existence of localized states", *J. Phys.* **C2** (1969) 1717.

33. G. H. Watson, Jr., P. A. Fleury and S. L. McCall, "Search for photon localization in the time domain", *Phys. Rev. Lett.* **58** (1987) 945.

# WAVE DIFFUSION AND LOCALIZATION IN RANDOM COMPOSITES


ZHAO-QING ZHANG* and PING SHENG
*Exxon Research and Engineering Company*
*Route 22 East, Annandale, NJ 08801*



By starting from the effective medium description for a two-component random composite, we delineate the wave transport behavior in both the propagating regime, valid for scale smaller or comparable to the mean free path, and the diffusion/localization regime, valid on a scale large compared to the mean free path. The accompanying mathematical framework is built up step-by-step in conjunction with the development of the physical concepts. It is shown that the wave diffusion is distinguished from the ordinary diffusion by a phase memory effect in the backscattering direction, which leads to a renormalization of the diffusion constant in a global sense. The resulting scaling behavior of the diffusion constant near a mobility edge is described in some detail by using the analogy with electronic systems. Calculation of the localization phase diagram as a function of wave frequency, relative volume fraction of the two components, and the impedance ratio between the



*Present address:* Institute of Physics, Academia Sinica, Beijing, People's Republic of China.

components indicates that the occurrence of localization requires a minimum impedance ratio whose value is on the order of 2.5. Furthermore, localization is shown to be closely tied to resonant scattering, and there is a multitude of mobility edges and localization regions associated with resonance harmonics. Pulse propagation behavior, as well as problems for further study, are also discussed.

# Contents


1. Introduction     140

2. Model and Its Effective-Medium Characteristics     141
    2.1. Model description     141
    2.2. Wave scattering and the T operator     142
    2.3. Effective medium description     145
    2.4. Interpolation scheme for $c_0$     147
    2.5. Beyond the effective medium description     147

3. Wave Diffusion and Localization     149
    3.1. Motivation     149
    3.2. The vertex function     150
    3.3. Derivation of the wave diffusion behavior     152
    3.4. Wave localization due to coherent backscattering     156
    3.5. Scaling behavior near the mobility edge     163
    3.6. Localization phase diagrams     168

4. Pulse Propagation Behavior     173

5. Concluding Remarks     176

References     177

## 1. INTRODUCTION

As a phenomenon generic to waves in random media, localization has many general features common to both quantum particles, e.g. electrons,[1,2] and classical waves,[3-11] i.e. electromagnetic and elastic waves. For example, it is now known that all waves localize in two or one dimensional systems with an arbitrary amount of randomness, and that in three dimensions a wave localizes only in certain energy regimes that are separated from the delocalized regimes by so-called mobility edges. However, in spite of these general similarities one would expect that in terms of specific behaviors, such as the frequency dependence of the localization length and the localization phase diagram for waves in three dimensional systems, significant differences should exist due to the following factors. First, the dispersion relation for a quantum particle differs from that of the classical wave. A direct result is that whereas the scattering strength of a quantum particle (by an obstacle) generally increases as the energy decreases, the opposite is the case for the classical waves[10] as shown by the frequency dependence of the Rayleigh scattering, $\omega^{d+1}$, where $\omega$ denotes the frequency and $d$ the spatial dimension. Second, if we look at how the randomness affects the wave vector $k$ of a wave, it is seen that for a quantum particle $k \propto (E-V)^{\frac{1}{2}}$, where $E$ is the total energy of the particle and $V$ the random potential, whereas for the classical waves $k \propto n\omega/c$, where $n$ is the (spatially) random index of refraction and $c$ the wave speed. That is, the randomness enters the problem additively in the case of quantum particles but multiplicatively in the case of classical waves. Third, in the case of electron localization one usually deals with discrete models where the interatomic distance provides the natural scale for discretization. For classical waves, however, one can encounter either discrete systems, such as phonons in a random lattice, or continuous random systems (or randomnesses on a scale much larger than atomic distances). The difference between the discrete and the continuous systems can mean that whereas localization may always occur in discrete models for some energy regimes (by nucleating localized states near the band edge as a discrete lattice is randomized, for example), in continuous systems localization may be absent for all energies of the wave unless some other condition is satisfied.[6]

It is the purpose of this paper to delineate the localization behavior for classical waves in a disordered, two-component composite. To simplify our task, we will consider only scalar waves. That means effects arising specifically from the vector nature of elastic or electromagnetic waves will not be considered. Nevertheless, based on scalar wave's success in modeling diverse array of classical-wave phenomena, we expect the picture presented here to be at least qualitatively correct for the vector waves as well. In what ensues, Sec. 2 presents the model together with a discussion of its effective-medium wave characteristics. This is followed by the description of wave diffusion and localization in Sec. 3 together with the calculation of localization phase diagrams and discussion of their implications. Consideration of pulse propagation is given in Sec. 4, and Sec. 5 states some unresolved issues.

## 2. MODEL AND ITS EFFECTIVE-MEDIUM CHARACTERISTICS

### 2.1. Model Description

Let us consider a composite[6] consisting of two components with indices of refraction $n_1$ and $n_2 (> n_1)$. That is, component 1 will be the "fast" component and component 2 will be the "slow" component. We define the ratio $m = n_2/n_1$ $(> 1)$ to be the contrast of the composite medium. We will limit our consideration to cases where $n_1$ and $n_2$ are real, i.e. the wave is locally propagating in nature. This will help us in identifying the localized states since any imaginary part to the index of refraction (which can arise from either dissipation or metallic character of the material) can induce exponential decay in the wave amplitude and thereby mask the localization behavior. Volume fraction of component 1 is denoted by $1 - p$ and that of component 2 by $p$. At $p \to 0$ or $p \to 1$, the composite microstructure is assumed to consist of spheres of the minority component, with diameter $d$, randomly dispersed in the matrix of the majority component. In the intermediate range we assume the microstructure to be that resulting from a symmetric randomization of the two components.

## 2.2.  Wave Scattering and the T Operator

The scalar-wave amplitude $u$ satisfies the wave equation

$$[c^{-2}(\mathbf{r})\partial_t^2 - \nabla^2]u = 0\,, \tag{1}$$

where the wave speed $c$ takes the value $c_1$ in component 1 and the value $c_2$ ($< c_1$) in component 2. For time-harmonic wave with frequency $\omega$, Eq. (1) can be re-written as

$$[\nabla^2 + k_0^2 - \Sigma(\mathbf{r})]u = 0\,, \tag{2a}$$
$$\Sigma(\mathbf{r}) = k_0^2 - c^2(\mathbf{r})\omega^2\,, \tag{2b}$$

where $k_0 = \omega/c_0$, and $c_0$ is an "average" speed of the medium to be defined later. If we treat $\Sigma(\mathbf{r})$ as a perturbation to the operator $L_0 = [\nabla^2 + k_0^2]$ and let $G_0$ be the Green's function for $L_0$, then we can rewrite Eq. (2a) as

$$L_0 u = \Sigma u\,, \tag{3a}$$

and the solution $u$ may be written formally as

$$u = u_0 + G_0 \Sigma u\,, \tag{3b}$$

where $u_0$ denotes the solution to the homogeneous equation $Lu_0 = 0$. Equation (3b) may be iterated to obtain

$$u = u_0 + G_0 \Sigma u_0 + G_0 \Sigma G_0 \Sigma u_0 + \ldots$$
$$= u_0 + G_0 T u_0\,, \tag{4}$$

where the $T$ operator[12] is the sume of the infinite series $\Sigma + \Sigma G_0 \Sigma + \ldots$. Equation (4) expresses the exact result $u$ in terms of the solution $u_0 = \exp(i\,\mathbf{k}_0 \cdot \mathbf{r})$. The meaning of the $T$ matrix is that it embodies all the multiple scattering the wave experiences in the random medium. By writing out Eq. (4) explicitly, we get

$$u^+(\mathbf{r}) = \frac{\exp(i\mathbf{k}_0 \cdot \mathbf{r})}{(2\pi)^{3/2}}$$
$$+ \iint d\mathbf{r}_1 d\mathbf{r}_2 G_0^+(\mathbf{r},\mathbf{r}_1) T^+(\mathbf{r}_1,\mathbf{r}_2) \frac{\exp(i\,\mathbf{k}_0 \cdot \mathbf{r}_2)}{(2\pi)^{3/2}}\,, \tag{5a}$$

where the Green's function $G_0^+$ is easily verified to be

$$G_0^+ (\mathbf{r}, \mathbf{r}_1) = -\frac{1}{4\pi|\mathbf{r} - \mathbf{r}_1|} \exp(ik_0|\mathbf{r} - \mathbf{r}_1|) \ . \tag{5b}$$

The $+$ sign on $u, G_0$, and $T$ denotes the fact that in Eq. (5b) we have chosen the outgoing spherical wave instead of the incoming one. By noting that as $|\mathbf{r} - \mathbf{r}_1| \to \infty$, $k_0|\mathbf{r} - \mathbf{r}_1| \simeq k_0 r - k_0 \cos\theta r_1$ and $|\mathbf{r} - \mathbf{r}_1| \simeq r^{-1}$, with $\theta$ denoting the angle between $\mathbf{r}$ and $\mathbf{r}_1$, we get

$$\begin{aligned}
u^+ (\mathbf{r}) = {} & \frac{\exp(i\,\mathbf{k}_0 \cdot \mathbf{r})}{(2\pi)^{3/2}} - \frac{\exp(ik_0 r)}{4\pi r} \iint d\mathbf{r}_1 d\mathbf{r}_2 \\
& \times \exp(-i\,\mathbf{k}_f \cdot \mathbf{r}_1)\langle\mathbf{r}_1|T^+|\mathbf{r}_2\rangle \frac{\exp(i\,\mathbf{k}_0 \cdot \mathbf{r}_2)}{(2\pi)^{3/2}}
\end{aligned} \tag{6a}$$

Here $\mathbf{k}_f$ has magnitude $k_0$ and is along the direction of $\mathbf{r}$. By substituting $\langle\mathbf{r}|\mathbf{k}\rangle = (2\pi)^{-3/2} \exp(i\,\mathbf{k} \cdot \mathbf{r})$ into Eq. (6a), $u^+ (\mathbf{r})$ may be expressed succinctly as

$$u^+ (\mathbf{r}) = \frac{\exp(i\,\mathbf{k}_0 \cdot \mathbf{r})}{(2\pi)^{3/2}} - \frac{\pi}{(2\pi)^{1/2}} \frac{\exp(ik_0 r)}{r} \langle\mathbf{k}_f|T^+|\mathbf{k}_0\rangle \ . \tag{6b}$$

In the case of a single scatterer, we can compare Eq. (6b) to the familiar classical expression

$$u^+ (\mathbf{r}) = \frac{\exp(i\,\mathbf{k}_0 \cdot \mathbf{r})}{(2\pi)^{3/2}} + f_{k_0} (\theta) \frac{\exp(ik_0 r)}{(2\pi)^{3/2}\, r} \tag{6c}$$

and obtain

$$-2\pi^2 \langle\mathbf{k}_f|t^+|\mathbf{k}_0\rangle = f_{k_0} (\theta) \ , \tag{7}$$

where $f_{k_0} (\theta)$ is the scattering amplitude, and $t^+$ denotes the scattering matrix for a single scatterer.

In order to calculate the scattering amplitude, we start by considering the single scatterer problem as shown in Fig. 1. Inside the scatter $i$, we take the solution to the wave equation with speed $c_i$, and outside the scatterer we take the wave equation solution with the speed

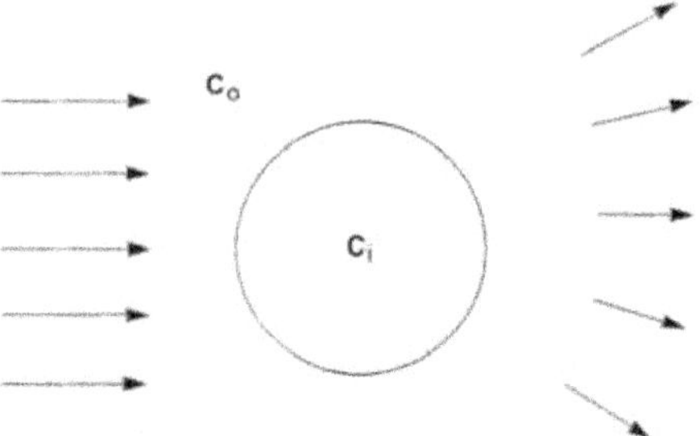


Fig. 1. The basis of the effective medium description for a two-component composite is the choice of a $c_0$ for the random medium in which the forward scattering amplitudes (the quantity $\Sigma$) from individual grains of component 1 $(c_1)$ and component 2 $(c_2)$ averages to zero.

$c_0$. The scattering amplitude $f_{k_0}^{(i)}(\theta)$ can be obtained by matching the wave-equation solution and its derivative inside the sphere,

$$u_{\text{in}}^{(i)} = (2\pi)^{-3/2} \sum_{l=0}^{\infty} A_l^{(i)} j_l(k_i r) P_l(\cos\theta), \quad r \le (d/2), \qquad (8\text{a})$$

to that outside the sphere,

$$
\begin{aligned}
u_{\text{out}}^{(i)} &= (2\pi)^{-3/2} \bigg[ \exp(i\,\mathbf{k}_0 \cdot \mathbf{r}) \\
&\quad + \sum_{l=0}^{\infty} D_l^{(i)} h_l^{(1)}(k_0 r) P_l(\cos\theta) i^l (2l+1) \bigg] \\
&= (2\pi)^{-3/2} \sum_{l=0}^{\infty} i^l (2l+1) P_l(\cos\theta) \\
&\quad \times [j_l(k_0 r) + D_l^{(i)} h_l^{(1)}(k_0 r)], \quad r > (d/2) .
\end{aligned}
\qquad (8\text{b})
$$

Here $j_l$ and $h_l^{(1)}$ are the spherical Bessel and Hankel functions of the first kind, respectively, $P$ denotes the Legendre polynomial, and $k_i = \omega/c_i$. The solution of the boundary value problem yields the expansion coefficients $A_l^{(i)}$ and $D_l^{(i)}$, and by noting that $h_l^{(1)}(k_0 r) \simeq \exp(ik_0 r)/r$ as $r \to \infty$, we get

$$f_{k_0}^{(i)}(\theta) = -\frac{i}{k_0} \sum_{l=0}^{\infty} D_l^{(i)} P_l(\cos\theta)(2l+1) . \qquad (9)$$

## 2.3. Effective Medium Description

In the above, the treatment of the medium outside the scatterer $i$ as homogeneous with speed $c_0$ is obviously an approximation since the medium is inhomogeneous in reality. However, at low frequencies, i.e. wavelength $\gg$ size of the inhomogeneities, this approximation is an excellent one since the wave not only scatters weakly but also cannot resolve the small inhomogeneities. As a result, the medium appears effectively homogeneous to the probing wave. The task is then to determine the value $c_0$ for the effective medium. A mathematical way to look at this problem is by examining the form of the Green's function $G$ for the full operator $L = L_0 - \Sigma$. That is,

$$G = L^{-1} = G_0 + G_0 \Sigma G_0 + G_0 \Sigma G_0 \Sigma G_0 + \cdots$$
$$= G_0 + G_0 T G_0 \, , \tag{10a}$$

or alternatively,

$$G = G_0 + G_0 \Sigma G \, . \tag{10b}$$

Equation (10b) is generally called the Dyson equation. From Eq. (10b) one can formally write

$$G = \frac{1}{G_0^{-1} - \Sigma} \, , \tag{11}$$

and by comparing Eq. (10a) with Eq. (10b) we get

$$\Sigma G = T G_0 \, , \tag{12a}$$

or

$$\Sigma = T(1 + T G_0)^{-1} \, . \tag{12b}$$

Looking at Eq. (11), we see that in order for the effective medium approximation to be valid, the condition $\Sigma \simeq 0$ must be satisfied through the choice of $c_0$. To be more precise, we note that if the medium is *macroscopically* homogeneous, we can always write, in wavevector representation

$$\langle G \rangle = G_e = \frac{\delta(\mathbf{p} - \mathbf{p}')}{k_0^2 - p^2 - \Sigma(\mathbf{p})} \, , \tag{13}$$

where $\langle\ \rangle$ denotes configurational averaging. From Eq. (12b),

$$\delta(\mathbf{p} - \mathbf{p}')\Sigma(\mathbf{p}) \cong \langle\langle\mathbf{p}|T^+|\mathbf{p}'\rangle\rangle$$

$$\cong n_1 \int d\,\mathbf{r}_i \exp[i\,\mathbf{r}_i \cdot (\mathbf{p} - \mathbf{p}')]\langle\mathbf{p}|t_1^+|\mathbf{p}'\rangle$$

$$+ n_2 \int d\,\mathbf{r}_j \exp[i\,\mathbf{r}_j \cdot (\mathbf{p} - \mathbf{p}')]\langle\mathbf{p}|t_2^+|\mathbf{p}'\rangle$$

$$= (2\pi)^3 \delta(\mathbf{p} - \mathbf{p}')[n_1\langle\mathbf{p}|t_1^+|\mathbf{p}\rangle + n_2\langle\mathbf{p}|t_2^+|\mathbf{p}\rangle]\,,$$

$$(14\mathrm{a})$$

where we have expressed $\langle\langle\mathbf{p}|T^+|\mathbf{p}'\rangle\rangle$ as the configurational average of $\langle\mathbf{p}|t^+|\mathbf{p}'\rangle$, $n_1$ and $n_2$ are the number densities of the two types of scatterers, $r_i$ and $r_j$ denote their positions, and $\mathbf{p}',\mathbf{p}$ are understood to have magnitude $k_0$. By using relation (7), we have

$$\Sigma(\mathbf{p}) = -4\pi[n_1 f_{k_0}^{(1)}(0) + n_2 f_{k_0}^{(2)}(0)]$$

$$= -24d^{-3}[(1 - p)f_{k_0}^{(1)}(0) + pf_{k_0}^{(2)}(0)]\,. \qquad (14\mathrm{b})$$

The condition $\Sigma(\mathbf{p}) = 0$ therefore reduces to

$$\sum_{l=0}^{\infty}(2l + 1)[(1 - p)D_l^{(1)} + pD_l^{(2)}] = 0 \qquad (15)$$

by using Eq. (9) and noting that $P_l(1) = 1$. Since $D_l^{(i)}$ is an implicit function of $c_0$ through the solution of the scattering boundary-value problem, Eq. (15) represents the effective-medium equation for the determination of $c_0$. At low frequencies, only the term $D_0^{(i)}$ is significant. That means only the static problem has to be sloved. By going through a little bit of algebra in matching boundary conditions, it can be shown that

$$c_e^{-2} = c_0^{-2}(\omega = 0) = (1 - p)c_1^{-2} + p\,c_2^{-2}\,. \qquad (16)$$

This equation is in fact well-known for acoustic wave in a composite medium (such as fluid with solid particle suspensions). If we recall that $c^{-2} = \rho/K$, where $\rho$ is the density and $K$ the bulk modulus, then for composites with $\rho_i = \rho_0$ Eq. (16) reduces to $K_0^{-1} = (1-p)K_1^{-1} + pK_2^{-1}$, which is the known effective-medium condition for bulk modulus in terms of component moduli.

## 2.4. Interpolation Scheme for $c_0$

At low frequencies, therefore, all that multiple-scattering does is to renormalize the medium speed, and this is true for all values of $p$. As frequency increases, however, Eq. (15) will not always have a well-defined and unique solution for all values of $p$. In those instances one notes that it is still possible to get a renormalized $\bar{c}_1$ or $\bar{c}_2$ for $p \simeq 0$ or $p \simeq 1$, respectively, by perturbation calculations. That is,

$$\bar{c}_1^{-1} = \mathrm{Re}\,[c_1^{-2} - (\Sigma/\omega^2)]^{1/2} \,, \tag{17a}$$

where

$$\Sigma = -24p\,d^{-3} f_{k_1}^{(2)}(0) \,, \tag{17b}$$

with $f_{k_1}^{(2)}(0)$ denoting the forward-scattering amplitude calculated from the scattering of a plane wave in medium 1 by a sphere of component 2. By exchanging the roles of two components, one can similarly obtain $\bar{c}_2$. $\bar{c}_1$ and $\bar{c}_2$ approaches values of $c_1$ and $c_2$ as frequency increases, and $\bar{c}_1 \simeq c_e \simeq \bar{c}_2$ at low frequencies. That means if we use the interpolation formula

$$c_0^{-1} = (1 - p)\bar{c}_1^{-1} + p\,\bar{c}_2^{-1} \,, \tag{18}$$

then $c_0$ would approach the correct low and high-frequency limits (at high frequencies the geometric-optics value of $c_0^{-1} = (1 - p)c_1^{-1} + pc_2^{-1}$) as well as the $p \simeq 0$ and $p \simeq 1$ limits. In Fig. 2 we plot the values of $c_0$, $\bar{c}_1$ and $\bar{c}_2$ as a function of reduced frequency for $n_1 = 1, n_2 = 2.11$, and $p = 0.4$. It is seen that $c_0$ tends to the correct $c_e$ value at low frequencies but displays oscillation at higher frequencies due to the existence of resonant scatterings (the positions of minima in $c_0$ correspond roughly to resonant scattering peaks).

## 2.5. Beyond the Effective Medium Description

While the value of $c_0$ determined from Eq. (18) obviously accounts for much of the multiple scattering effects responsible for the variation

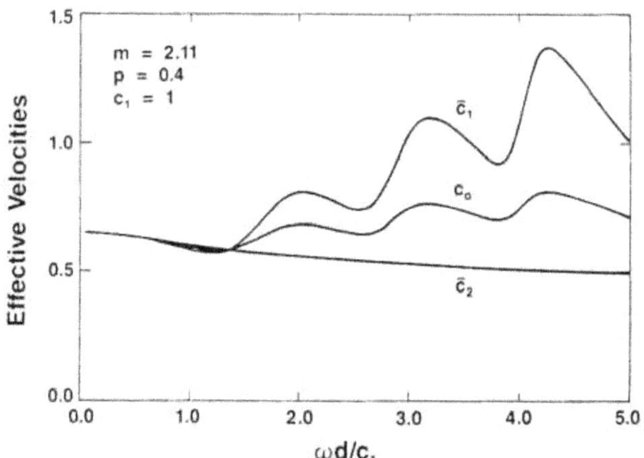



Fig. 2. A plot of the effective medium speed $c_0$ and the $\bar{c}_1, \bar{c}_2$ calculated from the perturbation theory as a function of frequency. The undulations are caused by scattering resonances.

of the medium characteristics in different frequency and compositional regimes, yet except for $\omega \to 0$, by no means can the choice of $c_0$ account for all the multiple scatterings. This is obvious since at frequencies where the wavelength becomes comparable to the inhomogeneities both intuition and experimental evidence would tell us that the wave is broken up and scattered in all directions in these situations. Mathematically, this is manifested by the non-existence of unique solutions to the effective medium equations, i.e. it becomes impossible to make $\Sigma = 0$ in a self-consistent manner.

The Green's function one has to use is therefore $G_e(\mathbf{p}) = (k_0^2 p^2 - \Sigma(\mathbf{p}))^{-1} \delta(\mathbf{p} - \mathbf{p}')$. However, even $G_e$ still differs from the exact $G$, and one write

$$G = G_e + G_e T' G_e , \qquad (19)$$

with the condition $\langle T' \rangle = 0$ so that $\langle G \rangle = G_e$. Here the meaning of Eq. (19) and $\langle T' \rangle = 0$ is that while there are scatterings caused by local deviations, yet overall there is no additional scattering (on the average) than those expressed by $\Sigma$. In other words, we can express $T'$ heuristically as $T - \langle T \rangle$, with $\langle T \rangle \simeq \Sigma$.

Since $\Sigma$ is generally complex, its imaginary part, related to the total scattering cross-section by the optical theorem, naturally introduces a spatially decaying character to the Green's function $G_e$ (in the r representation). Does that imply localization? The answer is no because the decay pertains only to an averaged quantity. By analogy, in Brownian motion the average displacement vector of a particle, $\langle \mathbf{r} \rangle$, is always zero and yet it does not imply that the particle is localized at the origin. In the present case, the decay of $G_e$ means the loss of the wave character, or phase coherence, for the transport of the excitation beyond the decay length. One therefore needs an alternative description for the transport behavior on a scale larger than the decay length. That is the topic of the next section.

## 3. WAVE DIFFUSION AND LOCALIZATION

### 3.1. Motivation

The study of wave transport in random media is divided into two regimes related to different spatial scales. On a scale smaller than the mean-free path of scattering, $l$, the behavior is wave-like. However, at scales much larger than $l$ the numerous scatterings make the energy transfer effectively diffusion-like. This picture is exactly analogous to that for the transport of a Brownian particle where the motion of a particle is ballistic between the scatterings but becomes diffusive when averaged over many collisions. The analogy also carries over to the mathematical analysis of the problem. In the study of random walks it is well known that the large-scale transport behavior is obtained not from the mean of its displacement vector, $\langle \mathbf{r} \rangle$, bur rather by the analysis of its variance, $\langle r^2 \rangle$. Similarly, for waves the study of their transport on a scale $\gg l$ requires an understanding of $\langle GG \rangle$, i.e. the correlation between two Green's functions. Since each Green's function can have its own frequency and wavevectors, we will let $\omega_\pm = \omega \pm (\Delta\omega/2)$ be the two frequencies and let $\mathbf{p}_\pm = \mathbf{p} \pm (\Delta \mathbf{k}/2)$, $\mathbf{p}'_\pm = \mathbf{p}' \pm (\Delta \mathbf{k}/2)$ be the wavevectors. For reasons that will be clear in Sec. 4, the precise form of $\langle GG \rangle$ we want is

$$C_{\mathbf{p}\mathbf{p}'}(\Delta \mathbf{k}, \Delta\omega | \omega) = \frac{(2\pi)^3}{V} \langle\langle \mathbf{p}_+ | G^+(\omega_+) | \mathbf{p}'_+ \rangle \langle \mathbf{p}'_- | G^-(\omega_-) | \mathbf{p}_- \rangle\rangle , \quad (20)$$

where $V$ is the spatial volume, and $G^\pm$ indicates that the frequency argument has a small but non-vanishing imaginary part, $\pm i\eta$, distinguishing between whether the Green's function is valid in time $t > 0(+i\eta)$ or $t < 0(-i\eta)$, with the source excitation at $t = 0$. Since $\Delta\omega$ represents the difference between the two frequencies, its conjugate time variable has the meaning of the mean travel time for the two waves as can be seen by writing the time variation of Eq. (20) as $\exp[i(\omega + \Delta\omega/2)t_1]\exp[-i(\omega - \Delta\omega/2)t_2] = \exp[i\omega(t_1 - t_2)]\exp[i\Delta\omega(t_1 + t_2)/2]$. Similarly, $\Delta k$ can be regarded as the conjugate variable to the travel distance. To obtain the long-time, large-distance behavior of wave transport therefore means analyzing $C_{\mathbf{pp'}}(\Delta\mathbf{k}, \Delta\omega|\omega)$ in the limit of $\Delta\omega \to 0$ and $\Delta k \to 0$.

### 3.2. The Vertex Function

Since the exact Green's function $G$ is not known, we would like to express $\langle GG \rangle$ in terms of $G_e$. Recalling that $G = G_e + G_e T' G_e$ with $T' \cong T - \langle T \rangle$, we get

$$\langle GG \rangle = G_e G_e + G_e \langle T' G_e G_e T' \rangle G_e \ , \tag{21a}$$

where the fact $\langle T' \rangle = 0$ has been used. Alternatively, one can rewrite Eq. (21a) in the form of

$$\langle GG \rangle = G_e G_e + G_e G_e \Gamma G_e G_e \ , \tag{21b}$$

where the object $\Gamma \simeq \langle T'T' \rangle$ is usually called the vertex function.[4,12] By writing $\Gamma = U + U G_e G_e \Gamma$ in terms of a new quantity $U$, one can further express $\langle GG \rangle$ as

$$\langle GG \rangle = G_e G_e + G_e G_e U(1 + G_e G_e \Gamma) G_e G_e \ ,$$
$$= G_e G_e + G_e G_e U \langle GG \rangle \ . \tag{21c}$$

Here $U$ is called the irreducible vertex function and plays the same role as the self-energy $\Sigma$ for the single-particle Green's functions. We note in passing that $G_e G_e$ always denotes $G_e^+(\mathbf{p}_+, \omega_+) G_e^-(\mathbf{p}_-, \omega_-)$. Equation (21c) is generally denoted as the Bethe-Salpeter equation. To get an understanding of the object $\Gamma$ (and $U$), we note that heuristically it is essentially the variance of $T$, i.e. $\Gamma \simeq \langle T'T' \rangle = \langle TT \rangle - \langle T \rangle^2$. If one writes $TT$ as $\langle \mathbf{p}_+|T^+(\omega_+)|\mathbf{p}'_+ \rangle\langle \mathbf{p}'_-|T^-(\omega_-)|\mathbf{p}_- \rangle$, it is clear that the scatterings represented by the two $T$'s can occur either independently or coincidentally. In the former, the spatial positions of the scatterers for $T^-$ bears no relation to those for $T^+$, and we may thus volume average the two $T$'s independently and obtain the square of the mean, exactly cancelling the $\langle T \rangle^2$ term. (We have ignored the differences between $T^+$ and $T^-$, but more careful treatment would yield the same result.) That leaves for $\Gamma$ the term in which the waves scatter in the same medium, i.e., the spatial positions of the scatterers for the two $T$'s are the same. Volume averaging yields

$$
\begin{aligned}
V\Gamma = n_1 \int d^3 r_i \, \exp[i\,\mathbf{r}_i \cdot (\mathbf{p}_+ - \mathbf{p}'_+ + \mathbf{p}'_- - \mathbf{p}_-)] \\
\times \langle \mathbf{p}_+ | t_1^+(\omega_+) | \mathbf{p}'_+ \rangle \langle \mathbf{p}'_- | t_1^-(\omega_-) | \mathbf{p}_- \rangle \\
+ n_2 \int d^3 r_j \, \exp[i\,\mathbf{r}_j \cdot (\mathbf{p}_+ - \mathbf{p}'_+ + \mathbf{p}'_- - \mathbf{p}_-)] \\
\times \langle \mathbf{p}_+ | t_2^+(\omega_+) | \mathbf{p}'_+ \rangle \langle \mathbf{p}'_- | t_2^-(\omega_-) | \mathbf{p}_- \rangle
\end{aligned}
$$

$$(22a)$$

or

$$
\begin{aligned}
\Gamma_{\mathbf{p}\mathbf{p}'}(\Delta\mathbf{k}, \Delta\omega | \omega) \simeq (2\pi)^3 [ & n_1 \langle \mathbf{p}_+ | t_1^+(\omega_+) | \mathbf{p}'_+ \rangle \langle \mathbf{p}'_- | t_1^-(\omega_-) | \mathbf{p}_- \rangle \\
& + n_2 \langle \mathbf{p}_+ | t_2^+(\omega_+) | \mathbf{p}'_+ \rangle \langle \mathbf{p}'_- | t_2^-(\omega_-) | \mathbf{p}_- \rangle ] \, .
\end{aligned}
$$

$$(22b)$$

One point to be noted about $\Gamma$ is that the four momenta involved always sum to zero as required by the condition of macroscopic spatial homogeneity. In the limit of $\Delta\omega, \Delta k \rightarrow 0$, we have

$$
\Gamma_B \simeq 48\pi^2 d^{-3} \sigma(\theta) \, ,
\tag{22c}
$$

with

$$
\sigma(\theta) = \frac{1}{4\pi 4}[(1-p)|f_{k_0}^{(1)}(\theta)|^2 + p|f_{k_0}^{(2)}(\theta)|^2] \, ,
\tag{22d}
$$

where the subscript $B$ denotes the present expression as the Boltzmann approximation for reasons that will become clear later. Since $U \simeq \Gamma$

in the lowest order, $U_B = \Gamma_B$. Now in terms of the irreducible vertex function $U$, $\Gamma$ may be expressed as an infinite series:

$$\Gamma = U + UG_eG_e\Gamma = U + UG_eG_eU + UG_eG_eUG_eG_eU + \dots \; . \quad (23)$$

If we let $U = U_B$ and represent it by a diagram as shown in Fig. 3a, then the series given by Eq. (23) may be represented as shown in Fig. 3b by so-called ladder diagrams. $U$ is noted to be the basic building block of the ladder diagrams, hence the description "irreducible".

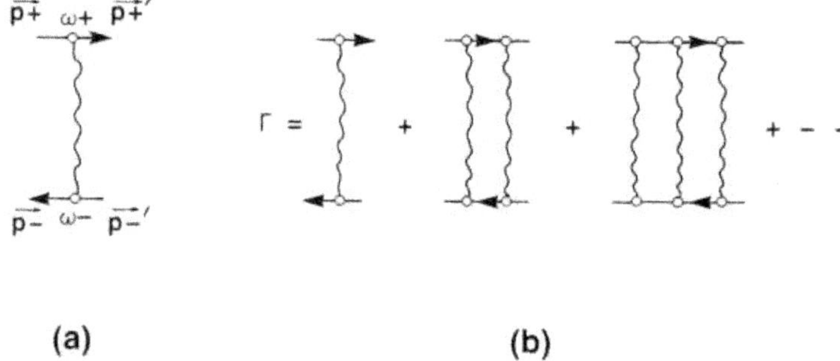


Fig. 3. (a) Diagram for the lowest order contribution to the irreducible vertex function $U$. The line on top indicates the momenta $\mathbf{p}_+$ and $\mathbf{p}'_+$ whereas the line at bottom indicates the momenta $\mathbf{p}_-$ and $\mathbf{p}'_-$. The arrows indicate the order of appearance, from left to right, of the momenta in the matrix elements, and the circle denotes the $t$ operator. The wiggly line connecting the two circles indicates that the scatterings represented by the two $t$'s are coincident, i.e. from the same scatterer. (b) So-called "ladder" diagrams for the vertex function $\Gamma$.

### 3.3. Derivation of the Wave Diffusion Behavior

To proceed further, we return to Eq. (21c):

$$C_{\mathbf{p}\mathbf{p}'}(\Delta\mathbf{k}, \Delta\omega|\omega)$$

$$= G_eG_e[\delta(\mathbf{p} - \mathbf{p}') + \int d\mathbf{p}_1 U_{\mathbf{p}\mathbf{p}_1}(\Delta\mathbf{k}, \Delta\omega|\omega)C_{\mathbf{p}_1\mathbf{p}'}(\Delta\mathbf{k}, \Delta\omega|\omega)] \; .$$

By integrating both sides with respect to $\mathbf{p}'$, denoting

$$C_{\mathbf{p}}(\Delta\mathbf{k}, \Delta\omega|\omega) = \int d\mathbf{p}' C_{\mathbf{p}\mathbf{p}'}(\Delta\mathbf{k}, \Delta\omega|\omega) \; ,$$

and using the identity $AB = (A - B)/(B^{-1} - A^{-1})$ for $G_e(\mathbf{p}_+,\omega_+)$ $\times G_e(\mathbf{p}_-,\omega_-)$, we get

$$\left[ -\frac{2\omega\Delta\omega}{c_0^2} + 2\Delta\mathbf{k}\cdot\mathbf{p} + \overset{+}{\underset{p+}{\Sigma}}(\omega_+) - \overset{-}{\underset{p-}{\Sigma}}(\omega_-) \right] C_{\mathbf{p}}(\Delta\mathbf{k},\Delta\omega|\omega)$$

$$= \Delta G_e(\mathbf{p}_+,\mathbf{p}_-,\omega_+,\omega_-)$$

$$\times \left[ 1 + \int d\mathbf{p}_1 U_{\mathbf{p}\mathbf{p}_1}(\Delta\mathbf{k},\Delta\omega|\omega)C_{\mathbf{p}_1}(\Delta\mathbf{k},\Delta\omega|\omega) \right] , \qquad (24)$$

where $\Delta G_e(\mathbf{p},\omega) = G_e(\mathbf{p}_+,\omega_+) - G_e(\mathbf{p}_-,\omega_-)$. From Eq. (24) we wish to derive two equations for the two averaged quantities:

$$S(\Delta k,\Delta\omega|\omega) = \int \frac{d\mathbf{p}}{(2\pi)^3} C_{\mathbf{p}}(\Delta\mathbf{k},\Delta\omega|\omega) \qquad (25a)$$

and

$$J(\Delta k,\Delta\omega|\omega) = \int \frac{d\mathbf{p}}{(2\pi)^3}(\Delta\mathbf{k}\cdot\mathbf{p})C_{\mathbf{p}}(\Delta\mathbf{k},\Delta\omega|\omega) . \qquad (25b)$$

Here $S$ is the correlation function $C_{\mathbf{p}\mathbf{p}'}$ averaged over all incident and scattering directions, and $J$ is the averaged "current density". To get the first equation relating these two quantities, we just integrate both sides of Eq. (24) with respect to $\mathbf{p}$ and use the so-called Ward identity

$$\int d\mathbf{p}\Delta G_e(\mathbf{p},\omega)U_{\mathbf{p}\mathbf{p}'}(\Delta\mathbf{k},\Delta\omega|\omega) = \overset{+}{\underset{p'_+}{\Sigma}}(\omega_+) - \overset{-}{\underset{p'_-}{\Sigma}}(\omega_-) \qquad (26)$$

to cancel terms. By further noting that in the limit of $\Delta k, \Delta\omega \to 0$, $\Delta G_e \simeq 2i\pi\delta(p^2 - k_0^2)$, we get from Eq. (24)

$$-\frac{2\omega\Delta\omega}{c_0^2}S + 2J = -\frac{i}{2\pi}k_0 . \qquad (27a)$$

While the derivation of the Ward identity can be found in standard textbooks, for the case of $\Delta k, \Delta\omega \to 0$, $U \sim U_B$, and $\Delta G$ approximated by a delta function, it is easy to see that Eq. (26) is exactly equivalent to the optical theorem relating the total scattering cross section to the imaginary part of the forward scattering amplitude.

For the second equation, we multiply both sides of Eq. (24) by $(\Delta\mathbf{k}\cdot\mathbf{p})$ and integrate with respect to $\mathbf{p}$:

$$-\frac{2\Delta\omega\Delta\omega}{c_0^2}J + \frac{2k_0^2(\Delta k)^2}{3}S - 2i\gamma(\omega)J$$
$$= \iint \frac{d\mathbf{p}\,d\mathbf{p}'}{(2\pi)^3}\Delta G(\mathbf{p},\omega)(\Delta\mathbf{k}\cdot\mathbf{p})U_{\mathbf{p}\mathbf{p}'}(\Delta\mathbf{k},\Delta\omega|\omega)C_{\mathbf{p}'}(\Delta\mathbf{k},\Delta\omega|\omega)\ . \tag{27b}$$

Here $\gamma(\omega)$ denotes the imaginary part of the self energy $\Sigma$, and the term $2k_0^2(\Delta k)^2 S/3$ is obtained by approximating $p^2$ by $k_0^2$ in the integral since $C_{\mathbf{p}}(\Delta\mathbf{k},\Delta\omega|\omega)$ is a function that is sharply peaked around $p = k_0$ (due to the existence of poles in the Green's functions). In the limit of $\Delta k \to 0$, $C_{\mathbf{p}}$ may be expanded to first order of $\Delta k$ as

$$C_{\mathbf{p}}(\Delta\mathbf{k},\Delta\omega|\omega) \simeq \Delta G_e(\mathbf{p},\omega)[\phi_{\mathbf{p}}^{(0)} + (\Delta\mathbf{k}\cdot\mathbf{p})\phi_{\mathbf{p}}^{(1)}]\ . \tag{28}$$

By again invoking the delta function approximation for $\Delta G_e$, Eq. (28) may be integrated with respect to $\mathbf{p}$ to get

$$\phi_{\mathbf{p}}^{(0)} = 2\pi i k_0^{-1} S\ . \tag{29a}$$

Multiplying both sides of Eq. (28) by $(\Delta\mathbf{k}\cdot\mathbf{p})$ and integrating with respect to $\mathbf{p}$ yields

$$\phi_{\mathbf{p}}^{(1)} = 2\pi i k_0^{-1}\frac{3}{(\Delta k)^2 p^2}J\ . \tag{29b}$$

Substituting Eqs. (28) and (29) back into Eq. (27b) finally gives (with the $S$ term integrating to zero)

$$\left[-\frac{2\omega\Delta\omega}{c_0^2} - 2i\gamma(\omega)\right]J + \frac{2k_0^2(\Delta k)^2}{3}S = AJ\ , \tag{30a}$$

where

$$A = \frac{6\pi i k_0^{-1}}{(\Delta k)^2}\iint \frac{d\mathbf{p}\,d\mathbf{p}'}{(2\pi)^3}\Delta G_e(\mathbf{p},\omega)$$
$$\times \Delta G_e(\mathbf{p}',\omega)(\Delta\mathbf{k}\cdot\mathbf{p})(\Delta\mathbf{k}\cdot\mathbf{p}')(p')^{-2}U_{\mathbf{p}\mathbf{p}'}(\Delta k,\Delta\omega|\omega)\ . \tag{30b}$$

By neglecting the $2\omega\Delta\omega/c_0^2$ term in Eq. (30a) since it is small compared to $\gamma(\omega)$ and $A$ in the limit of $\Delta\omega \to 0$, one can combine Eq. (30a) with Eq. (27a) to get

$$S(\Delta k, \Delta\omega|\omega) = \frac{c_0}{4\pi} \frac{1}{-(i\Delta\omega) + D(\omega,\Delta\omega)(\Delta k)^2} , \qquad (31a)$$

with

$$D(\omega,\Delta\omega) = \frac{2}{3}\omega\frac{1}{2\gamma - iA} . \qquad (31b)$$

Since we will see that the quantity $S(\Delta k, \Delta\omega|\omega)$ can be directly related to the time variation of an injected pulse, the form of the denominator in Eq. (31a) clearly indicates a diffusive behavior with diffusion constant $D$, i.e. $(\Delta k)^{-2} \simeq D(\Delta\omega)^{-1}$. We have thus shown that for waves in a random medium the long-time, large-distance behavior of the transport is diffusive in nature. The physical information of this transport regime is contained in the expression for $D$. To evaluate the diffusion constant, we recall that $\gamma$, the imaginary part of the self energy (related to the forward-scattering amplitude by Eq. (14b)), may be expressed in terms of the scattering cross section through the optical theorem:

$$2\gamma = 6d^{-3}(2\pi)^4 k_0 \int_{-1}^{1} d(\cos\theta)\sigma(\theta) , \qquad (32a)$$

where $\sigma(\theta)$ is given by Eq. (22d). On the other hand, by using $U_B$ for $U_{\mathbf{p}\mathbf{p}'}$ in Eq. (30b), we get

$$iA^{(B)} = 6d^{-3}(2\pi)^4 k_0 \int_{-1}^{1} d(\cos\theta)\sigma(\theta)\cos\theta , \qquad (32b)$$

where we have used the relation $\mathbf{p}\cdot\mathbf{p}' = pp'\cos\theta$, $(\Delta\mathbf{k}\cdot\mathbf{p})(\Delta\mathbf{k}\cdot\mathbf{p}') = pp'(\Delta k)^2\cos\theta/3$ after carrying out one of the $\mathbf{p}$ integrations, and the delta function approximation for $\Delta G$. Combining Eqs. (32a) and (32b) yields

$$D^{(B)} = \frac{c_0 d^3}{144\pi^4}\left[\int_{-1}^{1} d(\cos\theta)\sigma(\theta)(1-\cos\theta)\right]^{-1} . \qquad (33)$$

One recognizes from the integral of $\sigma(\theta)(1-\cos\theta)$ that Eq. (33) is precisely the Boltzmann equation relating the scattering cross section to the diffusion constant. This justifies our denoting the earlier approximations as the Boltzmann approximation. Classically, since the diffusion coefficient is directly related to the mean free path $l^*$ by $D^{(B)} = c_0 l^*/3$, Eq. (33) thus also offers a way to calculate the mean free path $l^*$ for wave diffusion. It should be noted that there is an alternative definition of the mean free path as $l = (n\sigma)^{-1}$, where $n$ is the density of scatterers and $\sigma$ the total scattering cross section. The two values of the mean free path are usually different, with $l^* > l$. This is due to the fact that the $(1 - \cos\theta)$ factor in Eq. (33) emphasizes the importance of non-forward scattering contribution to $l^*$ whereas for $l$ all scattering directions are treated equal. Therefore $l^*$ is a more accurate value for the distance over which the wave propagation direction, or phase, is randomized.

We have carried out evaluations of the wave diffusion coefficient for light scattering in a random medium consisting of polystyrene spheres. For wavelength in air $= 5896$Å, $d = 0.5$–$0.8$ $\mu$m, $m = 1.59$, and $p$ varying from 0.4 to 0.6, we get $D^{(B)} = (1.1$–$1.5) \times 10^6$ cm$^2$/sec. This is an excellent agreement with the measured value[14] of $D \simeq 1 \times 10^6$ cm$^2$/sec.

## 3.4. Wave Localization Due to Coherent Backscattering

While the derivation of the diffusive behavior confirms our intuitive expectation, it also raises some interesting questions in regard to localization of the wave. This is because any cursory inspection of Eq. (33) for $D^{(B)}$ will tell us that $D^{(B)}$ is always positive, i.e. there cannot be localization unless something else happens. In the past decade it is indeed discovered that wave diffusion differs from ordinary diffusion in one important aspect. The caveat is that while phase information is generally lost after many collisions, yet in the direction of $180°$ backscattering there can still be phase coherence regardless of the multiple scattering. The physics of this mechanism is easily demonstrated in Fig. 4, where any scattering path, as long as its final direction is exactly opposite to the incident direction, will have a time-reversed path possessing exactly the same phase delay as the original one. That means there can be constructive interference in the backscattering direction, resulting in a higher probability for the particle to be backscattered and thereby effectively reducing the diffusion coefficient. This mechanism, generally called the weak-localization phenomenon,[15-18] has been experimentally demonstrated. The question for us now is how to incorporate this backscattering phase-memory effect in the mathematical formalism. For this purpose let us return to the general expression for the diffusion constant, Eq. (31b), with $A$ given by Eq. (30b). It is clear that in our calculation of $D^{(B)}$ the use of $U_B$ for $U_{\mathbf{p}\mathbf{p}'}$ must be an over-simplification. To include more terms in the irreducible vertex function $U$, Vollhardt and Wölfle[13] have identified the maximally-crossed diagrams shown in Fig. 5a as the dominant higher-order contribution to $U$ that also correspond physically to the coherent backscattering effect. If we denote by $U^{(M)}$ the sum of all the maximally-crossed diagrams, then $A = A^{(B)} + A^{(M)}$, and from Eq. (31b)

$$D(\omega, \Delta\omega) = \frac{D^{(B)}}{1 - i\left[3A^{(M)}/(2\omega)\right] \cdot D^{(B)}} , \qquad (34a)$$

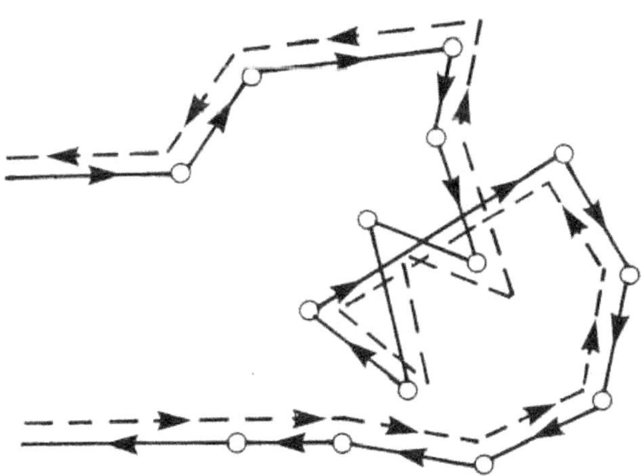

Fig. 4. Diagram illustrating the phase coherence effect in the backscattering direction. For any scattering path as indicated by the solid line, one can always find a time-reversed path as indicated by the dotted line which has exactly the same phase delay. As a result, they interfere constructively in the backscattering direction.

or

$$\frac{1}{D(\omega, \Delta\omega)} = \frac{1}{D^{(B)}} - i\frac{3}{2\omega}A^{(M)} \ ,$$

$$= \frac{1}{D^{(B)}} + \frac{9\,c_0^3}{8\pi^2\omega^4(\Delta k)^2} \iint d\mathbf{p}\,d\mathbf{p}'\Delta G_e(\mathbf{p},\omega)$$

$$\times \Delta G_e(\mathbf{p}',\omega)(\Delta\mathbf{k}\cdot\mathbf{p})(\Delta\mathbf{k}\cdot\mathbf{p}')U^{(M)}_{\mathbf{p}\mathbf{p}'} \ . \tag{34b}$$

In Eq. (34b) we have taken the factor $(p')^2$ outside the integral and expressed it as $(\omega/c_0)^2$ due to the $\Delta G_e(\mathbf{p}',\omega)$ delta function inside the integral. To evaluate $U^{(M)}$ and therefore $A^{(M)}$, it turns out that the maximally-crossed diagrams can be transformed into a ladder-diagram form by first turning the bottom line $180°$ as shown in Fig. 5b and then time-reversing the momenta as shown in Fig. 5c. The final step is to require the exit momentum $\mathbf{p}_2$ to be the opposite of the incident momentum, $\mathbf{p}_1$. That is, $\mathbf{p}_1 + \mathbf{p}_2 = \Delta\mathbf{k} \simeq 0$. In that case we may let $\mathbf{p} = -\mathbf{p}_2 + (\Delta\mathbf{k}/2)$ and $\mathbf{p}' = -\mathbf{p}_1 + (\Delta\mathbf{k}/2)$ where $\mathbf{p} = -\mathbf{p}'$ as shown in Fig. 5d. The last two steps are noted to be exactly those required for the appearance of weak localization described earlier, with the number of intervening $\mathbf{p}$'s as indicative of how many scatterings a particular diagram includes. The ladder diagrams are now exactly in the form as shown in Fig. 3 and therefore can be summed as in the Boltzmann approximation case. We recognize that the first term is $U_B G_e G_e U_B$, and the entire series may be written as

$$U^{(M)} = U_B G_e G_e U_B + U_B G_e G_e U_B G_e G_e U_B + \dots$$

$$= U_B [G_e G_e + G_e G_e U G_e G_e + \dots]U_B$$

$$= U_B \langle GG \rangle_B U_B \ .$$

By denoting $\langle GG \rangle_B$ as $C^{(B)}$, we finally get

$$U^{(M)} = U_B C^{(B)} U_B \tag{35}$$

with the superscript $(B)$ and subscript $B$ used to denote the quantity as calculated in the Boltzmann approximation. Writing out Eq. (35) in

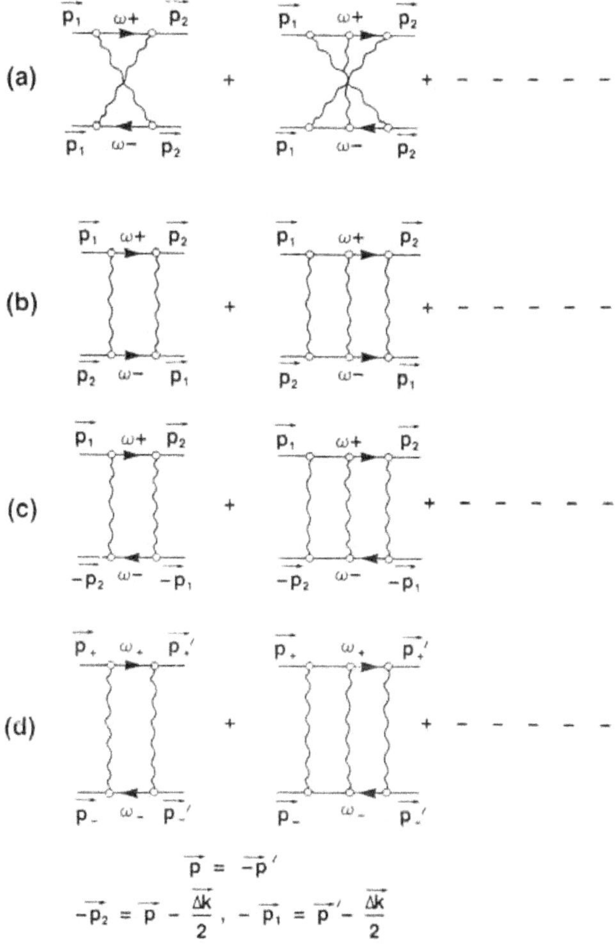



$$\vec{p} = -\vec{p}\,'$$
$$-\vec{p}_2 = \vec{p} - \frac{\vec{\Delta k}}{2}, \quad -\vec{p}_1 = \vec{p}\,' - \frac{\vec{\Delta k}}{2}$$

Fig. 5. Schematic illustration for the steps that convert the maximally-crossed diagrams into ladder diagrams. (a) The maximally-crossed diagrams. Each diagram represents a higher-order contribution to the irreducible vertex function $U$. (b) By turning the bottom line 180°, the diagram in (a) is made to look like a ladder. (c) By changing the order of appearance of the **p** states as well as reversing their momenta, the bottom line is now time-reversed. (d) The **p**'s are re-labeled with the provision that the incident and exit momenta are roughly opposite of each other, i.e. scattering is in the backward direction. In the above, steps (c) and (d) are noted to originate from the requirements of coherent backscattering.

full gives

$$U_{\mathbf{p}_1\mathbf{p}_2}^{(M)} = n^2(2\pi)^6 \iint d\mathbf{p}_3 d\mathbf{p}_4 \sigma(\mathbf{p}_1,\mathbf{p}_3) C_{\mathbf{p}_3\mathbf{p}_4}^{(B)}(\Delta\mathbf{k},\Delta\omega|\omega)\sigma(\mathbf{p}_4,\mathbf{p}_2),$$

$$\tag{36}$$

where we have used Eq. (22c) for $U_B$, with $n = 6\pi d^{-3}$ denoting the total density of scatterers. Since in the integral of Eq. (36) frequency of the wave is confined narrowly in the range of $\omega \pm (\Delta\omega/2)$, the magnitudes of $\mathbf{p}_3$ and $\mathbf{p}_4$ are held to a roughly constant value (mathematically this is achieved by the existence of poles in the Green's functions that determines the dispersion relation for the wave). As a result, integrations with respect to $\mathbf{p}_4$ and $\mathbf{p}_3$ may be regarded as essentially angular integrals. If we make the decoupling approximation and take $\sigma(\mathbf{p}_1,\mathbf{p}_3)$, $\sigma(\mathbf{p}_4,\mathbf{p}_2)$ outside the integral by giving them their average value, i.e.

$$\sigma(\mathbf{p}_1,\mathbf{p}_3),\sigma(\mathbf{p}_4,\mathbf{p}_2) \to (4\pi)^{-1}\int d\Omega\sigma(\mathbf{p}_1,\mathbf{p}_3)\,,$$

then Eq. (36) becomes

$$U_{\mathbf{p}_1\mathbf{p}_2}^{(M)} = \frac{2}{\pi}\gamma^2(\omega)k_0^{-2} \iint \frac{d\mathbf{p}_4 d\mathbf{p}_3}{(2\pi)^3} C_{\mathbf{p}_3\mathbf{p}_4}^{(B)}(\Delta\mathbf{k},\Delta\omega|\omega)\,, \tag{37a}$$

where we have applied the optical theorem in relating $\gamma$ to the integral of $\sigma(\theta)$. Recognizing that the integral now represents exactly $S(\Delta k,\Delta\omega|\omega)$, we use Eq. (31a) with $D = D^{(B)}$ to get

$$U_{\mathbf{p}_1\mathbf{p}_2}^{(M)} = \frac{\gamma^2(\omega)}{2\pi^2}c_0 k_0^{-2} \frac{1}{-i(\Delta\omega) + D^{(B)}(\Delta k)^2}\,, \tag{37b}$$

where $\Delta k = |\mathbf{p}_1 + \mathbf{p}_2|$ as specified earlier. We have thus expressed the contribution of the maximally-crossed diagrams in terms of the Boltzmann diffusion constant. Substitution of Eq. (37b) into Eq. (34b) yields

$$\frac{1}{D(\omega, \Delta\omega)} = \frac{1}{D^{(B)}} + \frac{9}{8\pi^2} \frac{c_0^3}{\omega^4 (\Delta k)^2} \iint d\mathbf{p}_1 d\mathbf{p}_2$$

$$\times \Delta G(\mathbf{p}_1, \omega) \Delta G(\mathbf{p}_2, \omega)(\Delta\mathbf{k} \cdot \mathbf{p}_1)(\Delta\mathbf{k} \cdot \mathbf{p}_2) U^{(M)}_{\mathbf{p}_1 \mathbf{p}_2}$$

$$= \frac{1}{D^{(B)}} - \frac{9\gamma^2(\omega) c_0^6}{16\pi^4 \omega^6 (\Delta k)^2} \iint d\mathbf{p} \, d(\Delta\mathbf{k})$$

$$\times \Delta G(\mathbf{p}, \omega) \Delta G(-\mathbf{p} + \Delta\mathbf{k}, \omega) \frac{(\Delta\mathbf{k} \cdot \mathbf{p})^2}{-i(\Delta\omega) + D^{(B)}(\Delta k)^2} ,$$

$$(38a)$$

where we have dropped a term $(\Delta\mathbf{k} \cdot \mathbf{p})(\Delta k)^2$ which is third order in $\Delta k$. By further neglecting $\Delta\mathbf{k}$ in $\Delta G(-\mathbf{p} + \Delta\mathbf{k}, \omega)$, we can break the $\mathbf{p}$ integration into two parts: the angular part over $(\Delta\mathbf{k} \cdot \mathbf{p})^2$ yields $4\pi(\Delta k)^2 p^2/3$, then by writing $\Delta G(\mathbf{p}, \omega)\Delta G(-\mathbf{p}, \omega) = [\Delta G(\mathbf{p}, \omega)]^2$ to the same level of approximation as that yielding the delta function for $\Delta G$:

$$(\Delta G)^2 = \left\{ \frac{-2i\gamma}{[k_0^2 - p^2]^2 + \gamma^2} \right\}^2 , \qquad (38b)$$

the $dp$ integration over $p^4(\Delta G)^2$ immediately gives[4]

$$\frac{1}{D(\omega, \Delta\omega)}$$

$$= \frac{1}{D^{(B)}} + \frac{3}{4} \frac{\gamma(\omega)}{\pi^2} \left(\frac{c_0}{\omega}\right)^3 \int d(\Delta\mathbf{k}) \frac{1}{-i(\Delta\omega) + D^{(B)}(\omega)(\Delta k)^2} .$$

$$(38c)$$

Here the $\Delta\mathbf{k}$ integration has to be limited to a sphere with radius $\Delta k \leq K/l^*(\omega) = q_c$ because the diffusive behavior is only valid on a spatial scale larger than $l^*(\omega)$. This leaves $K$ as a free parameter in the calculation. The effect of $K$ on localization will be investigated when we calculate the phase diagrams, but we note here that the maximum value of $K$ should be on the order of $2\pi$.

In Eq. (38c) we have used $D^{(B)}$ in the integral. Since now there is a more accurate value of $D$ as given by (38c), we can use that value in

162                           *Zhao-Qing Zhang & Ping Sheng*

the integral and thereby make Eq. (38c) a self-consistent expression for $D(\omega, \Delta\omega)$. By multiplying both sides of Eq. (38c) by $D(\omega, \Delta\omega)D^{(B)}$, we get

$$D(\omega, \Delta\omega) = D^{(B)}\left[ 1 - \frac{3\gamma(\omega)}{4\pi^2}\left(\frac{c_0}{\omega}\right)^3 \int\limits_{\Delta k < q_c} d(\Delta k) \right.$$

$$\left. \times \frac{1}{(\Delta k)^2 - iD^{-1}(\omega, \Delta\omega)(\Delta\omega)} \right]. \tag{39}$$

To examine whether a wave is localized or not, we are interested in the limit of $\Delta\omega \to 0$, i.e. infinite travel time behavior of $D$. In that case as long as $D \neq 0$ we have

$$D(\omega) = D^{(B)}(\omega)\left[ 1 - \frac{3\gamma(\omega)}{4\pi^2}\left(\frac{c_0}{\omega}\right)^3 \int_{\Delta k < q_c} d(\Delta k)\frac{1}{(\Delta k)^2} \right]$$

$$= D^{(B)}(\omega)\left[ 1 - \frac{3\gamma(\omega)}{\pi}\left(\frac{c_0}{\omega}\right)^3 \frac{K}{l^*(\omega)} \right]. \tag{40}$$

It is seen that the overall effect of coherent backscattering is to renormalize the Boltzmann diffusion constant by a factor that is always less than one as we have intuitively expected. Localization occurs when that factor is zero, i.e.

$$1 = \frac{3\gamma(\omega)}{\pi}\left(\frac{c_0}{\omega}\right)^3 \frac{K}{l^*(\omega)} \tag{41}$$

is the condition for calculating localization phase diagrams. The frequency $\omega^*$ at which Eq. (41) is satisfied is denoted as the mobility edge.

To actually use Eq. (40) with a given input frequency $\omega$ and volume fraction $p$, one first has to evaluate $c_0$ using the procedure that led to Eq. (18). The imaginary part of the forward scattering amplitude $\gamma(\omega)$ can then be evaluated by solving the boundary-value problem to get the scattering amplitudes $f_{k_0}^{(1)}(\theta)$ and $f_{k_0}^{(2)}(\theta)$ and let $\gamma(\omega) = \text{Im}\left[(1 - p)f_{k_0}^{(1)}(0) + pf_{k_0}^{(2)}(0)\right]$. From $f_{k_0}^{(1)}(\theta)$ and $f_{k_0}^{(2)}(\theta)$ one also obtains $\sigma(\theta)$ (Eq. (22d)) and $D^{(B)}$ (Eq. (33)) as well as the mean free path $l^* = 3D^{(B)}(\omega)/c_0$.

## 3.5. Scaling Behavior Near the Mobility Edge

While Eq. (41) gives the condition for wave localization, yet it does not give a physical picture for how it occurs. Obviously, the diffusion constant cannot just vanish everywhere since the wave still has to propagate on the scale of the mean free path. To get a better idea about the behavior near the mobility edge, we appeal to the scaling theory of localization[19] which has been formulated in the language of electronic systems. Although on the microscopic scale the electrons behave ballistically between collisions and therefore differ from waves, yet on the macroscopic scale where diffusion behavior dominates we expect the approach to localization to possess universal characteristics. Therefore below we will use the language of electronic systems and then make the connection to wave behavior via the proportionality between the conductivity $\sigma$ and the diffusion constant $D$.

It is a fact of common experience that the conductance $G$ of a conducting material, measured in units of $(\text{ohm})^{-1}$, is dependent on the sample size. So when one increases the size of a cubic sample, the conductance increases proportionally with its linear dimension $L$ since $G$ is proprotional to sample's cross-sectional area and inversely proportional to its length. A second fact, pointed out by Mott,[20] is that in order for a material to be metallic its conductivity, $\sigma = G/L$, has to be larger than a certain minimum value $\sigma_{\min}$ that is expressible in terms of a natural conductance unit, $e^2/\hbar = (4108 \text{ ohm})^{-1}$, divided by an atomic length $a$. Based on these two facts, the scaling theory makes the bold assumption that (1) the fundamental quantity in the conduction problem is the conductance, since it can be expressed in terms of the fundamental unit $e^2/\hbar$, and (2) the dimensionless conductance $g = G/(e^2/\hbar)$ varies with the linear dimension $L$ of a cubic sample in a general manner that is describable by a scaling relation:

$$g(bL) = f_b[g(L)] , \qquad (42)$$

where $b$ the dilatation factor and $f$ is some unknown scaling function. In words, what the relation Eq. (42) describes is that the conductance $g$ at a certain size $bL$ is completely determined by the conductance at another sample size $L$ and the ratio $b$ between these two linear dimensions. While not obvious at first sight, Eq. (42) directly implies the following differential relation. For $b \simeq 1$,

$$g(bL) \cong g(L) + (b-1)Lg'(L) \; ,$$

or

$$\frac{dg}{dL} = \frac{g(bL) - g(L)}{(b-1)L} = \frac{f_b[g(L)] - f_1[g(L)]}{(b-1)L} \; . \tag{43a}$$

Multiplying both sides by $L/g$ gives

$$\frac{d\ln g}{d\ln L} = \frac{(df/db)}{g} = \beta(g) \; . \tag{43b}$$

That is, the change of $g$ at a certain length scale is entirely determined by its value at that scale. From our discussion before we know that $g \propto L$ as $g \to \infty$ for conductors. That means $\beta(g) = 1$ for $g \to \infty$. On the other hand, for insulators $g \propto \exp[-L/\varsigma]$ due to the exponential decay of wave functions inside an insulator. That means $\beta(g) = \ln g$. For $g$ small, $\beta(g)$ is negative. Therefore, a plot of $\beta(g)$ vs. $\ln g$ would look roughly like that shown in Fig. 6, with $\beta(g)$ crossing zero at some value $g = g_c$. If now we couple the behavior of $\beta(g)$ with Eq. (43b), it is immediately obvious that for $\beta(g) > 0$, then the change in $\ln g$ is positive, resulting in an even faster increase for $\ln g$ since $\beta(g)$ is a monotonic function. However, if $\beta(g) < 0$, then the change in $\ln g$ is negative, resulting in an even faster decrease for $\ln g$. That is, one can attach arrows to the $\beta(g)$ vs. $\ln g$ curve as in Fig. 6 that clearly indicate the value $g_c$ as the dividing value between insulating and metallic behaviors, i.e. $g_c$ is a "minimum metallic conductance". One may be uncomfortable with the use of conductance as an indicator for physical characteristics since it is not an intensive quantity. For example, one may ask whether the sample size of a conductor can be shrunk to such a degree as to make its $g$ less than $g_c$ and thereby insulating. However, from the empirical evidences compiled by Mott in support of the existence of $\sigma_{\min}$ the sample size required in that case would be so small as to be less than atomic dimensions. Therefore, in practical terms this is unlikely. Furthermore, the scaling theory also tells us that it would be impossible to crossover from the conducting regime to the insulating regime (or vice versa) by just changing the sample size. This is due to the anomalous effects near the mobility edge which we will address presently.

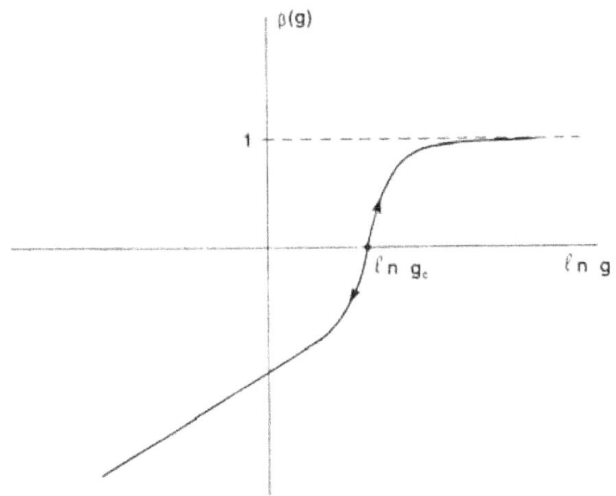


Fig. 6. The scaling function $\beta(g)$ plotted as a function of ln $g$. The value $g_c$ denotes the critical conductance at the mobility edge that separates the conducting regime $(\beta > 0)$ from the insulating regime $(\beta < 0)$.

For $g$ slightly larger than $g_c$, $\beta(g) \simeq 0$. Given that $\beta(g) = 1$ means $g \propto L$ as we have described earlier, $\beta(g) < 1$ implies that $g$ *does not scale linearly with $L$* as in the normal situation. It follows that if one decreases the conductance of a finite conducting sample by decreasing its size, then at some point before reaching $g_c$ the conductance will stay nearly constant as the sample is shrinked down further. Therefore, it would be impossible to reach $g_c$, or values below that, through sample size variation. A more important point to note, however, is that as a consequence of conductance's scale invariance near $g_c$, the usual intensive quantities, such as the conductivity and the diffusion constant, would become scale dependent since they are defined as proportional to $g/L$. To describe this effect more precisely, let us approximate $\beta(g)$ as $s(g - g_c)/g_c \simeq s \ln(g/g_c)$ for $g$ close to $g_c$. Then Eq. (43b) can be integrated to give

$$\frac{\ln(g_0/g_c)}{\ln(g/g_c)} = \left[\frac{L_0}{L}\right]^s , \tag{44a}$$

where we denote by $L_0, g_0$ the initial values of these quantities on the scale of the mean free path and $L, g$ their final values. We will specify $g$ to be the value at which $\beta(g) \simeq s \ln(g/g_c) \simeq 1$ so that the conductance scales as $g \propto L$ from that point on. The associated length scale $L = \varsigma$ at which this happens is then given by

$$\frac{L_0}{\varsigma} = [\ln(g_0/g_c)]^{1/s} \simeq \left(\frac{g_0 - g_c}{g_c}\right)^{1/s} , \tag{44b}$$

where we have used the fact that $s \sim 1$ and $\beta(g) \simeq 1 = \ln(g/g_c)$. Equation (44b) tells us that close to the mobility edge, the wave-field properties are homogeneous only on a scale larger than a length scale $\varsigma$. That is, only if one measures on a scale $> \varsigma$ would the diffusion constant and conductivity be intensive (i.e. scale independent) in nature. Moreover, the scale $\varsigma$, which we denote as the correlation length, is seen to diverge at the mobility edge according to Eq. (44b). For conductivity,

$$\sigma \propto g/\varsigma = (g/L_0)(L_0/\varsigma) \simeq (g_0/L_0)(L_0/\varsigma) , \tag{44c}$$

where we have noted that $g \simeq g_0$ due to conductance's relative size independence near $g_c$. Since $D \propto \sigma$, we finally get

$$D = D^{(B)}\left(\frac{l^*}{\varsigma}\right) . \tag{45}$$

Here $L_0$ is set equal to the mean free path $l^*$ and $D_0$ is identified as the Boltzmann diffusion constant. By combining with our discussion in the previous section, the effect of coherent backscattering is seen to create a new length scale $\varsigma$ near the mobility edge beyond which the diffusion constant has the value given by Eq. (45). Comparison of Eq. (45) with Eq. (40) yields

$$\frac{l^*}{\varsigma} = 1 - \frac{3\gamma(\omega)}{\pi}\left(\frac{c_0}{\omega}\right)^3 \frac{K}{l^*(\omega)} . \tag{46}$$

What happens for sample size $L < \varsigma$? Since $D^{(B)}l^* \propto g(l^*) \simeq G(L) \propto DL$ for $L < \varsigma$, we have

$$D(L < \varsigma) = D^{(B)}\left(\frac{l^*}{L}\right) . \tag{47}$$

That is, $\varsigma$ is simply replaced by $L$ for $l^* \leq L \leq \varsigma$. It follows that overall, the diffusion constant $D$ has the qualitative scale dependence shown in Fig. 7. Therefore, the physical picture of how the localization occurs is that of a global transformation where the correlation length $\varsigma$ diverges, thereby making the diffusion constant vanish on a macroscopic scale and yet still remains finite on any finite scale. Based on this picture, one way to detect the localization transition is to observe the so-called "anomalous diffusion" phenomenon[3] which can be explained as follows. For regular diffusion, we have

$$L^2 = Dt , \tag{48a}$$

where $t$ is the diffusion time. Near the mobility edge, when $\varsigma > L, D$ acquires a scale dependence given by Eq. (47). That is,

$$L^3 = D^{(B)}l^*t . \tag{48b}$$

The cubic $L$ variation as a function of $t$ is called anomalous diffusion and should be a sign for the onset of localization. According to our calculation based on Eq. (46), however, the frequency or composition range in which this occurs is rather narrow, making its observation probably difficult.

So far, all our discussion has focused on the delocalized side of the transition. On the localized side, we have to return to Eq. (39) and observe that since now $D = 0$, we can no longer neglect the $D^{-1}(\omega, \Delta\omega)(\Delta\omega)$ term in the integrand even when $\Delta\omega \to 0$. Instead, it is observed that $D^{-1} \cdot (\Delta\omega)$ has the units of $[\text{length}]^{-2}$. If we denote that length as $\varsigma_l = \kappa^{-1}$, then $S \sim (\kappa^2 + (\Delta k)^2)^{-1}$ implies an exponential-decay envelope for the Fourier transform of $S$. Since this is precisely the manifestation of a localized wave, we may identify $\varsigma_l$ as

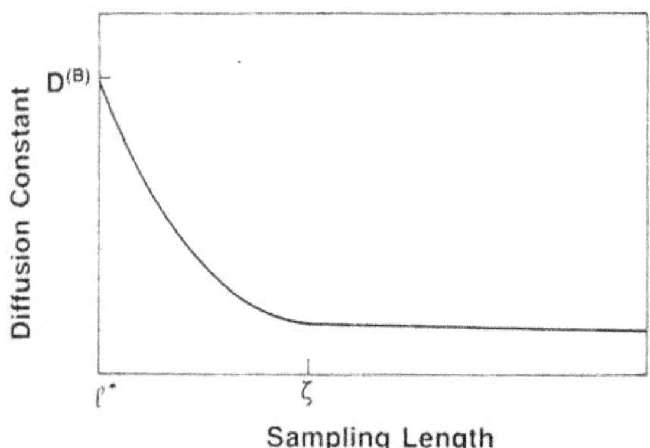


Fig. 7. Schematic illustration of diffusion constant's scale dependence near the mobility edge. The correlation length $\varsigma$ marks the scale above which the diffusion constant reverts to its normal scale-independent behavior.

the localization length. With $D = 0$ identically in the localized region, the equation for the determination of $\varsigma_l$ is

$$1 = \frac{3\gamma(\omega)}{\pi}\left(\frac{c_0}{\omega}\right)^3 \int_0^{q_c} \frac{(\Delta k)^2}{(\Delta k)^2 + \varsigma_l^{-2}}\, d(\Delta k)\, , \qquad (49)$$

The localization length may be regarded as the counterpart to the correlation length on the delocalized side of the transition. Similar to $\varsigma$, $\varsigma_l$ diverges as one approaches the mobility edge (from the localized size).

## 3.6. Localization Phase Diagrams

The frequency and composition requirement for the occurrence of wave localization in a two-component composite is a question of paramount importance to the experimental observation of the localization or the diffusion phenomenon. We have carried out numerical calculations based on Eq. (41) to delineate the phase diagrams[6] as a function of frequency, relative volume fraction, and the impedance ratio between the two components. Since there is a free parameter $K$ in the definition of $q_c = K/l^*$, we will repeat the calculation for two values of $K = 2\pi$ and 1. In Fig. 8 is shown the case for $K = 2\pi$. What is plotted is the mobility-edge contours as a function of $p$ and the dimensionless frequency $\omega d/c_1$. Each contour, which represents a single value of the impedance ratio $m$, separates the localized region inside the contour from the delocalized region outside. As seen in the figure, for impedance ratio up to $m = 3$ there are three large regions of localization centered roughly at $\omega d/c_1 \simeq 1.7, 2.9$, and $3.8$. As a function of increasing $m$, localization appears in the order marked by Roman numerals. The first appearance, in the form of almost a dot, occurs at $m = 2.107$. That means $m \simeq 2.11$ is the *minimum* impedance ratio required for wave localization.

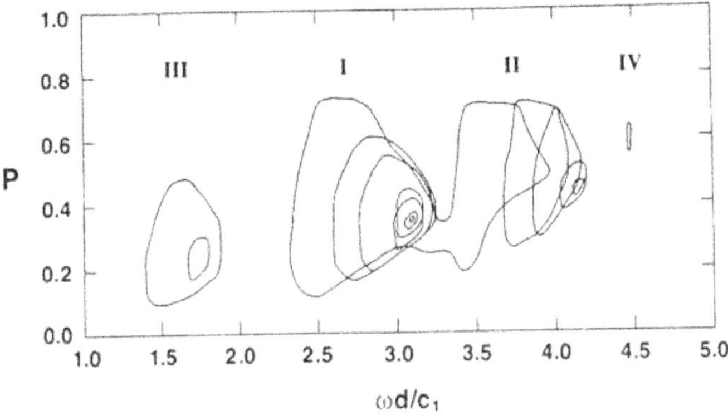


Fig. 8. Mobility-edge contours projected on the composition-frequency plane. The order of appearance of the three regions as $m$ is increased is noted by I, II, III, and IV. In each region the smaller contours always appear before the larger ones. In I the first six contours correspond to $m = 2.107, 2.11, 2.12, 2.14, 2.22$, and $2.3$. The seventh one, $m = 2.5$, extends over to region II. In II the first four contours correspond to $m = 2.12, 2.14, 2.22$, and $2.3$. In III the two contours correspond to $m = 2.3$ and $2.5$. In IV the narrow strip of localized region correspond to $m = 2.5$. The value of $K$ used in the calculation is $2\pi$.

Examination of the scattering cross sections in the three regions of localization shows that they are related to the $i = 1, 2, 3$ resonances of the component-2 spheres (slow medium), where $i$ denotes the angular degeneracy of the resonant mode. The first occurrence of localization, centered around $\omega d/c_1 = 3$, is associated with the $i = 2$ resonance. Why $i = 2$ appears first, but not $i = 1$, has to do with the fact that there is an optimal frequency window for the occurrence of localization.

If the frequency is too low, the scattering would be weak and $\gamma(\omega)$ small, and therefore the renormalization of the diffusion constant is minimal. At high frequencies, on the other hand, the decrease of wavelength inevitably limits the growth of the coherence length $\varsigma$ and thereby also inhibits localization. This leaves the intermediate frequency range, which has a wavelength in the fast medium (component 1) 1.5–3 times that of $d$, the most susceptible to localization. Localization first occurs when the enhancement of the scattering cross section by a particular resonance is just enough to bring $D$ to zero. As the impedance ratio $m$ is increased, the resonant scattering effect increases, and this broadens the localization region.

As a function of relative volume fraction, since both $p \simeq 0$ and $p \simeq 1$ regions have relatively homogeneous microstructures and therefore weak scattering effects, localization occurs usually in the intermediate range of $p \simeq 0.3$–0.6. The phase diagram is not symmetric relative to $p = 0.5$, however, because we found that the wave scattering is much stronger for a scatterer of the slower component embedded in a fast medium than the reverse configuration. This is perhaps due to the focusing and total reflection effects of the former configuration tends to trap the wave energy inside the scatterer and thereby producing resonant scattering effect; whereas in the latter the scatterer would act more as a diverging lens for the wave.

This simple asymmetry has some interesting experimental implications. For light, the slow medium is usually a solid and the fast medium is either the vacuum or a fluid. That means the optimal geometry for light localization would be hard to achieve since the fast medium cannot act as a rigid matrix to hold the scatterers. Exactly the reverse is the case for acoustic waves. In that case the solid medium is usually the fast medium and the voids or fluids the slower ones. Producing a strongly-scattering medium is therefore fairly natural for the acoustic wave.

Figure 9 shows the localization phase diagram for the case of $K = 1$. While the contours look very different, yet qualitative features remain the same. Localization now occurs first near $\omega d/c_1 = 2.6$, with a minimum impedance ratio of $m = 2.5$. As $m$ increases, there is a tendency for localization region to shift down in frequency. Due to the fact that the contours now spread all over the place, we have to label each one individually.

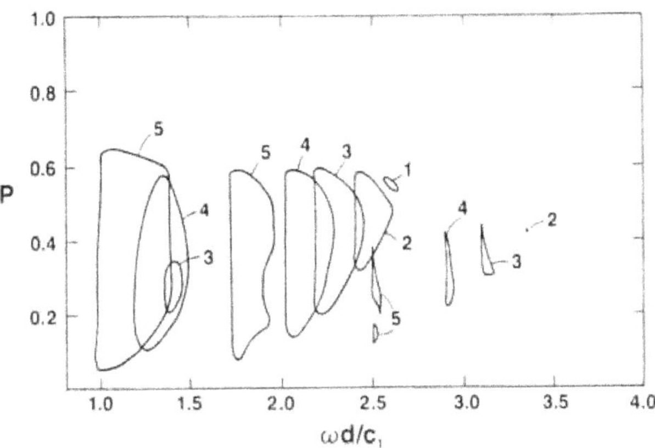

Fig. 9. Same as Fig. 8 but calculated with $K = 1$. The numbered label for each contour has the following correspondence: 1 means $m = 2.5$, 2 means $m = 2.6$, 3 means $m = 2.8$, 4 means $m = 3.0$, and 5 means $m = 3.5$.

An important conclusion of our numerical calculation is therefore the existence of a minimum impedance ratio, in the range of $m \simeq 2.5$ (we regard $K = 1$ as a more reasonable value), required for the onset of classical wave localization.[6] Moreover, resonance scattering plays an important role in enhancing coherent backscattering effect and in inducing localization. These features should be contrasted with those of electron localization, e.g. electrons in a random lattice always have a mobility edge at some energy near the band edge regardless of the amplitude of potential fluctuations. Since our calculation is by no means rigorous, one may wonder about the accuracy of our predictions. While numerical simulation in the present model would be difficult due to the computational magnitude of such an effort, it should be noted that the accuracy of the general approach has indeed been verified in the electronic tight-binding model where a comparison between theory and numerical simulation yields excellent agreement.[21] Therefore, it is expected that the present predictions should generally be on the right track.

From Fig. 8 or Fig. 9 one may fix either $p$ or $\omega$ and calculate the correlation and localization length at a horizontal or vertical slice of the phase diagram for one value of $m$. In Fig. 10 we show the case of $K = 2\pi$, $m = 2.5$, and $p = 0.55$. Three points should be noted. First, at low frequencies we have $\varsigma \simeq l^*$, so the increase in $\varsigma$ reflects the increase of the mean free path $l^*$. Also, the undulation seen in the low-frequency region is the manifestation of a scatterer resonance. Second, the divergences of $\varsigma$ and $\varsigma_l$ at the mobility edges are verified to have the form $(\omega - \omega^*)^{-1}$. Third, due to the uncertainty in the value of $K$, the values of $\varsigma$ and $\varsigma_l$ should only be viewed as an order-of-magnitude estimate. At some frequencies $\varsigma_l$ is seen to be less than $d$. This is in fact the result of using too large a value for $K$. A smaller $K$ would shift the mobility edges and increase the minimum values of $\varsigma_l$. If we require $\varsigma_l$ to be greater than $d$, then the maximum value of $K$ is limited to be less than $\pi/2$. However, the qualitative behavior would still remain the same.

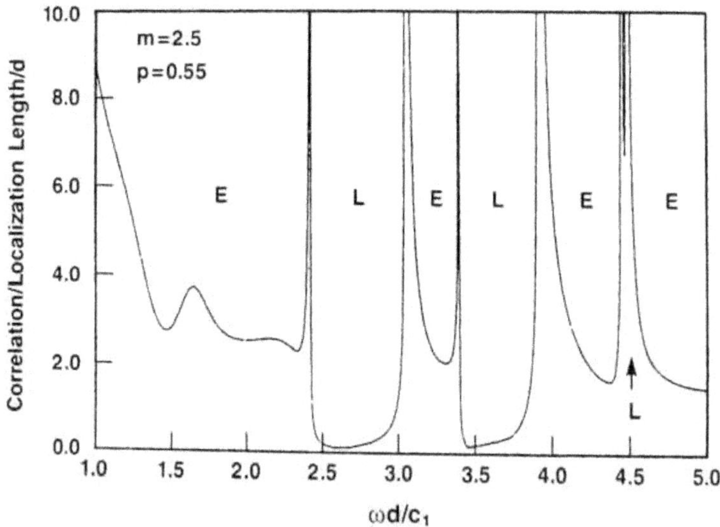


Fig. 10. Correlation length and localization length plotted as a function of the dimensionless frequency variable $\omega d/c_1$. $E$ denotes extended region. The curves in extended regions represent the correlation length $\varsigma$. $L$ denotes the localized region. The curves in the localized regions represent the localization length $\varsigma_l$. Values of $m$ and $p$ are given in the figure.

The divergence behavior of $\varsigma, \varsigma_l \sim (\omega - \omega^*)^{-1}$ is modified when the fixed $p$ (or $\omega$) slice is tangential to the contour so that one never crosses the mobility edge but only touches it tangentially at a point. In that case we have $\varsigma, \varsigma_l \sim (\omega - \omega^*)^{-2}$. The same thing happens at the isolated critical point when the contour shrinks down to a point.

## 4. PULSE PROPAGATION BEHAVIOR

Whereas in previous sections we looked at localization from the point of view of a single-frequency wave, in this section we wish to consider the behavior of pulse propagation. That is, with the wave Eq. (1), we specify the initial conditions to be

$$G^+\left(\mathbf{r}, t \le 0 | \mathbf{r}'\right) = 0 , \tag{50a}$$

$$\partial_t G^+\left(\mathbf{r}, t = 0 | \mathbf{r}'\right) = -c_0^2 \delta(\mathbf{r} - \mathbf{r}') . \tag{50b}$$

The time and spatial evolution of the injected energy pulse is then given by

$$P(\mathbf{r}, t | \mathbf{r}') = \left[\partial_t G^+\left(\mathbf{r}, t | \mathbf{r}'\right)\right]^2 . \tag{51}$$

What we want to show now is that $P(\mathbf{r}, t | \mathbf{r}')$ can be related to $S$ in Eq. (31a) through the following set of equations.

$$P(\mathbf{r}, t | \mathbf{r}') = \frac{1}{2\pi} \int d(\Delta\omega) \exp(-i\Delta\omega t) \bar{P}(\mathbf{r}, \Delta\omega | \mathbf{r}') , \tag{52a}$$

$$\langle \bar{P}(\mathbf{r}, \Delta\omega | \mathbf{r}') \rangle = \int_{-\infty}^{\infty} \frac{d\omega}{2\pi} \left(\omega + \frac{\Delta\omega}{2\pi}\right) \left(\omega - \frac{\Delta\omega}{2}\right) R(\mathbf{r}, \omega, \Delta\omega | \mathbf{r}') , \tag{52b}$$

$$
\begin{aligned}
R(\mathbf{r}, & \omega, \Delta\omega | \mathbf{r}') \\
&= \frac{1}{(2\pi)^3} \int d(\Delta\mathbf{k}) \exp[i(\Delta\mathbf{k}) \cdot (\mathbf{r} - \mathbf{r}')] S(\Delta k, \Delta\omega | \omega) \\
&= \frac{1}{(2\pi)^3} \int_{\Delta k \le q_c} d(\Delta\mathbf{k}) \frac{(c_0/4\pi) \exp[i(\Delta\mathbf{k}) \cdot (\mathbf{r} - \mathbf{r}')]}{-i(\Delta\omega) + D(\omega, \Delta\omega)(\Delta k)^2} .
\end{aligned}
\tag{52c}
$$

Due to the fact that the diffusion form for $S$ is valid only for $\Delta k < q_c$ and $\Delta\omega$ small, the pulse propagation behavior calculated from Eq. (52)

should be regarded as descriptive of the actual behavior only in the limit of long propagation time $t$ and large travel distance $|\mathbf{r} - \mathbf{r}'|$.

To prove the relations given by Eq. (52), let us first define a function $f(\mathbf{r}, t|\mathbf{r}')$ as

$$G^+(\mathbf{r}, t|\mathbf{r}') = \theta(t) f(\mathbf{r}, t|\mathbf{r}') . \tag{53a}$$

It follows that

$$G^-(\mathbf{r}, t|\mathbf{r}') = -\theta(-t) f(\mathbf{r}, t|\mathbf{r}') , \tag{53b}$$

and we can assume $f(\mathbf{r}, -t|\mathbf{r}') = -f(\mathbf{r}, t|\mathbf{r})$. The Fourier transform of $f$ is denoted as $\bar{f}(\mathbf{r}, \omega|\mathbf{r}')$:

$$\bar{f}(\mathbf{r}, \omega|\mathbf{r}') = \int_{-\infty}^{\infty} f(\mathbf{r}, t|\mathbf{r}') \exp(i\omega t) dt . \tag{54}$$

Since $f$ is real, $\bar{f}$ must be imaginary and $\bar{f}(\mathbf{r}, \omega|\mathbf{r}') = -\bar{f}(\mathbf{r}, -\omega|\mathbf{r}')$. Then $G^+(\mathbf{r}, \Delta\omega|\mathbf{r}')$ may be written as

$$
\begin{aligned}
G^+(\mathbf{r}, \Delta\omega|\mathbf{r}') &= \int_{-\infty}^{\infty} G^+(\mathbf{r}, t|\mathbf{r}') \exp(i\Delta\omega t) dt \\
&= \int_{0}^{\infty} f(\mathbf{r}, t|\mathbf{r}') \exp(i\Delta\omega t) dt \\
&= \int_{0}^{\infty} dt \exp(i\Delta\omega t) \int_{-\infty}^{\infty} \frac{d\omega}{2\pi} \bar{f}(\mathbf{r}, \omega|\mathbf{r}') \exp(-i\omega t) \\
&= \frac{i}{2\pi} \int_{-\infty}^{\infty} \frac{\bar{f}(\mathbf{r}, \omega|\mathbf{r}')}{\Delta\omega - \omega + i\eta} d\omega , 
\end{aligned} \tag{55a}
$$

or

$$G^+(\mathbf{r}, \Delta\omega|\mathbf{r}') = \frac{i}{2\pi} \int_{-\infty}^{\infty} d\omega \frac{\omega \bar{f}(\mathbf{r}, \omega|\mathbf{r}')}{(\Delta\omega + i\eta)^2 - \omega^2} . \tag{55b}$$

For $G^-(\mathbf{r}, \Delta\omega|\mathbf{r}')$ we can get a similar expression except now $\Delta\omega + i\eta$ is replaced by $\Delta\omega - i\eta$. Now

$$
\begin{aligned}
\bar{P}(\mathbf{r}, \Delta\omega|\mathbf{r}') &= \int_{0}^{\infty} [\partial_t f(\mathbf{r}, t|\mathbf{r}')]^2 \exp(i\Delta\omega t) dt \\
&= \frac{-i}{(2\pi)^2} \iint_{-\infty}^{\infty} d\omega_1 d\omega_2 \frac{\omega_1 \omega_2 \bar{f}(\mathbf{r}, \omega_1|\mathbf{r}') \bar{f}(\mathbf{r}, \omega_2|\mathbf{r}')}{\Delta\omega - \omega_1 - \omega_2 + i\eta} \\
&= \frac{i}{2(2\pi)^2} \iint_{-\infty}^{\infty} d\omega_1 d\omega_2 \, \omega_1 \omega_2 \bar{f}(\mathbf{r}, \omega_1)|\mathbf{r}') \bar{f}(\mathbf{r}', \omega_2|\mathbf{r}) .
\end{aligned}
$$

$$\left[ \frac{\omega_1 - \Delta\omega}{(\omega_1 - \Delta\omega - i\eta)^2 - \omega_2^2} - \frac{\omega_1 + \Delta\omega}{(\omega_1 + \Delta\omega + i\eta)^2 - \omega_2^2} \right] , \qquad (56a)$$

where in the last step we have used the fact that $\bar{f}(\mathbf{r}, \omega | \mathbf{r}') = -\bar{f}(\mathbf{r}, -\omega | \mathbf{r}')$ and $f(\mathbf{r}, \omega | \mathbf{r}') = f(\mathbf{r}', \omega | \mathbf{r})$. At this point it is noted that the quantity inside the square bracket on last line of Eq. (56a) is equal to

$$\frac{i}{\pi} \int_{-\infty}^{\infty} d\omega \frac{(\omega + (\Delta\omega/2))(\omega - (\Delta\omega/2))}{\left[(\omega + (\Delta\omega/2) + i\eta)^2 - \omega_1^2\right]\left[(\omega - (\Delta\omega/2) - i\eta)^2 - \omega_2^2\right]} ,$$

so that

$$\begin{aligned}
\bar{P}(\mathbf{r}, \Delta\omega | \mathbf{r}') = -\frac{1}{(2\pi)^3} &\int_{-\infty}^{\infty} d\omega \left(\omega + \frac{\Delta\omega}{2}\right)\left(\omega - \frac{\Delta\omega}{2}\right) \\
&\times \int_{-\infty}^{\infty} d\omega_1 \frac{\omega_1 \bar{f}(\mathbf{r}, \omega_1 | \mathbf{r}')}{\left(\omega + (\Delta\omega/2) + i\eta\right)^2 - \omega_1^2} \\
&\times \int_{-\infty}^{\infty} d\omega_2 \frac{\omega_2 \bar{f}(\mathbf{r}', \omega_2 | \mathbf{r})}{\left(\omega - (\Delta\omega/2) - i\eta\right)^2 - \omega_2^2} \\
= &\int_{-\infty}^{\infty} \frac{d\omega}{2\pi} \left(\omega + \frac{\Delta\omega}{2}\right)\left(\omega - \frac{\Delta\omega}{2}\right) \\
&\times G^+\left(\mathbf{r}, \omega + \frac{\Delta\omega}{2} \Big| \mathbf{r}'\right) G^-\left(\mathbf{r}', \omega - \frac{\Delta\omega}{2} \Big| \mathbf{r}\right) \qquad (57)
\end{aligned}$$

by using Eq. (55b). Comparison with Eq. (52b) shows that $\langle G^+\left(r, \omega + \frac{\Delta\omega}{2} | \mathbf{r}'\right) G^-\left(\mathbf{r}', \omega - \frac{\Delta\omega}{2} | \mathbf{r}\right)\rangle$ is denoted by $R(\mathbf{r}, \omega, \Delta\omega | \mathbf{r}')$ in that equation. The condition of macroscopic homogeneity implies $R$ has translational invariance, i.e. the spatial dependence of $R$ is a function of $(\mathbf{r} - \mathbf{r}')$ only so that the Fourier transform of $R$ (denoted by $\bar{R}$) is given by

$$\begin{aligned}
\bar{R}(\Delta k, &\Delta\omega | \omega) \\
&= \frac{1}{V} \iint d\mathbf{r} \, d\mathbf{r}' \exp[-i(\Delta \mathbf{k}) \cdot (\mathbf{r} - \mathbf{r}')] \langle\langle \mathbf{r} | G^+ | \mathbf{r}'\rangle \langle \mathbf{r}' | G^- | \mathbf{r}\rangle\rangle .
\end{aligned}$$

$$(58a)$$

Below we show that $\bar{R}(\Delta k, \Delta\omega | \omega)$ is exactly $S(\Delta k, \Delta\omega | \omega)$ as given by Eq. (52c). Now $\langle \mathbf{r} | G^+ | \mathbf{r}'\rangle \langle \mathbf{r}' | G^- | \mathbf{r}\rangle$ may be transformed into the $p$

representation as $\iiiint d\mathbf{p}_1\, d\mathbf{p}_2\, d\mathbf{p}_3\, d\mathbf{p}_4\ \langle\mathbf{r}|\mathbf{p}_1\rangle\ \langle\mathbf{p}_1|G^+|\mathbf{p}_2\rangle\ \langle\mathbf{p}_2|\mathbf{r}'\rangle$ $\times\langle\mathbf{r}'|\mathbf{p}_3\rangle\ \langle\mathbf{p}_3|G^-|\mathbf{p}_4\rangle\langle\mathbf{p}_4|\mathbf{r}\rangle$. By writing $\langle\mathbf{p}|\mathbf{r}\rangle = \exp(-i\mathbf{p}\cdot\mathbf{r})/(2\pi)^{3/2}$, substituting the transformed expression back into Eq. (58a), and integrating with respect to $\mathbf{r},\mathbf{r}'$, as well as two of the four $\mathbf{p}$'s, we get

$$
\begin{aligned}
\bar{R}(\Delta k, \Delta\omega|\omega) &= \frac{1}{V}\iint d\mathbf{p}\, d\mathbf{p}'\langle\langle\mathbf{p}_+|G^+(\omega_+)|\mathbf{p}'_+\rangle\langle\mathbf{p}_-|G^-(\omega_-)|\mathbf{p}_-\rangle\rangle \\
&= \iint \frac{d\mathbf{p}\, d\mathbf{p}'}{(2\pi)^3} C_{\mathbf{p}\mathbf{p}'}(\Delta k, \Delta\omega|\omega) \\
&= S(\Delta k, \Delta\omega|\omega),
\end{aligned}
\tag{58b}
$$

where $\mathbf{p}_\pm = \mathbf{p} \pm (\Delta\mathbf{k}/2)$ and $\omega_\pm = \omega \pm (\Delta\omega/2)$ as defined in Sec. 3. This proves the relations as given by Eq. (52).

It is expected that in the time domain, an observer at a fixed distance from the source point should observe a long time tail which is proportional to $t^{-3/2}$, characteristic of diffusion. Close to the mobility edge or inside the localized regime, however, the decay rate of the tail should become slower. This is indeed observed in our rough evaluations using Eq. (52).

## 5. CONCLUDING REMARKS

In this article we have delineated in some detail the mathematical framework of wave diffusion and localization together with its physical underpinnings. While much about the localization problem have been understood in the past decade, yet much more remain to be studied. In particular, three issues concerning the present theory need further clarification:

1) The transition between the propagating regime and the diffusion regime. We have described the wave transport behavior on both the small scale (less than the mean free path) and the large scale. How is the transition between these two regimes achieved? To understand that one needs a better solution of the Bethe-Salpeter equation for all values of $\Delta\omega$ and $\Delta k$.

2) Multiple scattering effects in the propagating regime. So far all the scattering effects as expressed by the self energy $\Sigma$ have been calculated on the basis of the single particle scattering in an effective

medium. While the choice of $c_0$ inherently contains some multiple-scattering effects, yet this may not go far enough in accounting for those specific scatterings arising from strong short-range correlations between the scatterers. Better calculations could improve the predictive accuracy of the theory.

3) Determination of the free parameter $K$. This may be achieved by either comparing the prediction of the theory with experiments or simulation results, or by extending the evaluation of the two-particle Green's function to the transition regime.

It is our hope that besides complementing other articles in this volume, the present work can stimulate further research in the clarification of the above issues.

# REFERENCES


1. Many articles and references on electron localization can be found in *Percolation, Localization, and Superconductivity*; eds. A. M. Goldman and S. A. Wolf (Plenum Press, New York, 1984).
2. P. W. Anderson, *Phys. Rev.* **109** (1958) 1492.
3. S. John, *Phys. Rev. Lett.* **53** (1984) 2169.
4. T. R. Kirkpatrick, *Phys. Rev.* **B31** (1985) 5746.
5. P. W. Anderson, *Phil. Mag.* **B52** (1985) 505.
6. P. Sheng and Z. Q. Zhang, *Phys. Rev. Lett.* **57** (1986) 1879.
7. S. M. Cohen and J. Machta, *Phys. Rev. Lett.* **54** (1985) 2242.
8. K. Arya, Z. B. Su and J. L. Birman, *Phys. Rev. Lett.* **54** (1985) 1559.
9. P. Sheng, Z. Q. Zhang, B. White and G. Papanicolaou, *Phys. Rev. Lett.* **57** (1986) 1000.
10. P. Sheng, B. White, Z. Q. Zhang and G. Papanicolaou, *Phys. Rev.* **B34** (1986) 4757.
11. D. Sornette, to be published in *Acoustica*.
12. E. N. Economou, *Green's Function in Quantum Physics*, 2nd edn. (Springer-Verlag, New York, 1983).
13. D. Vollhardt and P. Wölfle, *Phys. Rev.* **B22** (1980) 4666, and *Phys. Rev. Lett.* **45** (1980) 842.
14. A. Z. Genack, *Phys. Rev. Lett.* **58** (1987) 2043.
15. L. Tsang and A. Ishimaru, *J. Opt. Soc. Am.* **A1** (1984) 836.
16. M. P. Van Albada and A. Lagendijk, *Phys. Rev. Lett.* **55** (1985) 2692.
17. P. E. Wolf and G. Maret, *Phys. Rev. Lett.* **55** (1985) 2696.
18. M. J. Stephen, *Phys. Rev. Lett.* **56** (1986) 1809.

19. E. Abraham, P. W. Anderson, D. C. Licciardello and T. V. Ramakrishnan, *Phys. Rev. Lett.* **42** (1979) 673.
20. See, for example, N. F. Mott and G. A. Davis, *Electronic Processes in Non-Crystalline Materials*, 2nd edn. (Claredon Press, 1979).
21. A. D. Zdetsis *et al.*, *Phys. Rev.* **B32** (1985) 7811.

# NOVEL CORRELATIONS AND FLUCTUATIONS IN SPECKLE PATTERNS


SHECHAO FENG

*Department of Physics and the*
*Solid State Science Center*
*University of California, Los Angeles*
*Los Angeles, CA 90024*

# Contents


1. Introduction                                                      181

2. Definition of the Problem and Various Assumptions                 181

3. Conventional Theory of Speckle Patterns                           186

4. Novel Correlation Effects of Speckle Patterns in the
   Multiple Scattering Regime                                        189

5. Experimental Situations and Related Work                          200

6. Summary                                                           205

References                                                           206

# 1. INTRODUCTION

The study of wave propagation through disordered scattering media began at least as early as the days of Lord Rayleigh.[1] By now, work in this field has evolved into an industry of its own, with many different angles of emphasis and different languages of communication by many workers.[2] From a practical point of view, this is a subject of high importance. For instance, in the field of communications, it is of interest to study the effect of the scattering of radio waves propagating through atmospheric and oceanic turbulences. In geophysics, people are concerned with using wave transmission through planetary atmospheres to remotely determine their turbulence and other dynamic characteristics. As it is impossible to even make a sketch of the collective information gathered in this general field over the years, I shall confine my attention in this chapter to the special topic of the correlation and fluctuation effects of the so-called "speckle patterns", which are generated by coherent wave transmission through disordered multiple scattering media. The chapter shall be organized according to the following. In Sec. 2, the definition of the problem will be carefully given, with attention paid to all the various assumptions and different regimes of validity. In Sec. 3, the conventional theory of the speckle patterns will be reviewed briefly, mainly through physical discussions. In Sec. 4, new theoretical results on the novel correlations and fluctuations of speckle patterns shall be presented, using diagrammatic perturbation and numerical techniques. In Sec. 5, a comparison of this theory with recent optical experiments will be made. Relations of this theory to the related work shall also be discussed. Section 6 will contain some brief concluding remarks.

# 2. DEFINITION OF THE PROBLEM AND VARIOUS ASSUMPTIONS

The phenomenon of "speckle patterns" refers to the complex intensity patterns generated by propagation of a coherent wave either through the bulk of a disordered scattering medium, or by reflection of such a wave off of a disordered rough surface (the medium here can be opaque). It is essentially a complex interference pattern of the scattered wave, owing to the randomness of the scattering medium. For concreteness, let us first concentrate on the first situation, namely the case of a bulk scattering media. The situation can be made precise by providing the following generic geometry: A coherent wave is incident on the left side of a slab. The direction of this incident beam is characterized by the incident wave vector $\mathbf{k}_a = k\hat{k}_a$, where $k = \omega/c \equiv 2\pi/\lambda$, and $c$ is the wave speed. (See Fig. 1.) The incident beam is assumed to have a spot size (width) $W \gg \lambda$ (therefore it can be regarded as nearly a plane wave). The thickness of the slab is assumed to be $L$. The slab is assumed to contain random inhomogeneities so as to scatter the wave *elastically*. For definiteness, we can imagine that the elastic scattering is due to a certain concentration of point-like scatterers in an otherwise homogeneous and transparent medium, such as the case of air bubbles in a piece of glass. Another example of the type of scattering media that we are considering here is an inhomogeneous acoustical medium in which the speed of sound varies randomly from one region to another. We confine our attention to the case where only elastic scattering is present because in this case the various interesting fluctuation and correlation effects are most enhanced. The strength of the elastic scattering of the disordered medium is characterized by a transport mean free path (of the scattered wave) $l$, which is defined as the distance at which the coherent incident beam intensity (in the direction $\hat{k}_a$) has fallen to $1/e$ of its value before entering the sample, due to elastic scattering events which take the intensity of the incident beam into various other directions. If we place a screen to the right side of the slab, we would see an image of the intensity distribution of the scattered wave upon transmission. Experimentally one observes this pattern to have an extremely grainy look, therefore one usually refers to it as a *transmission speckle pattern*. (See Fig. 2.) Similarly, a screen which is placed on the left side of the slab will also exhibit a speckle pattern which we refer to as a reflection speckle pattern. The bulk of this chapter shall be devoted to the case of transmission speckle patterns, for the sake of concreteness and its clear connection to electron transport problems. We should at this point emphasize a few important physical considerations and assumptions with respect to the speckle pattern problem.

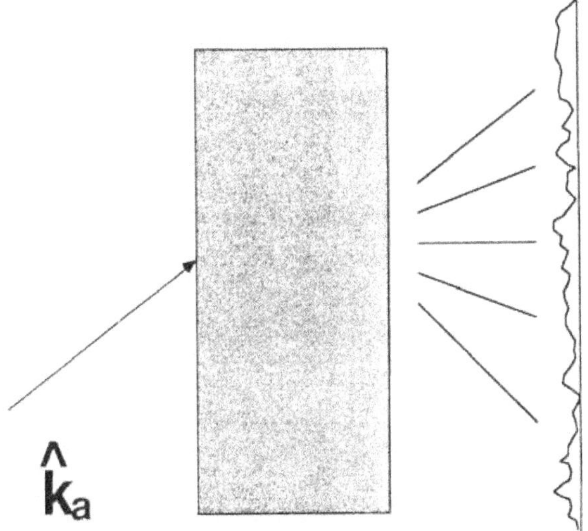

Fig. 1. Standard geometry to study transmission speckle patterns. The scattering medium is assumed to have only elastic scatterers. The incoming beam is assumed to be a coherent laser source, with a spot size $W$.

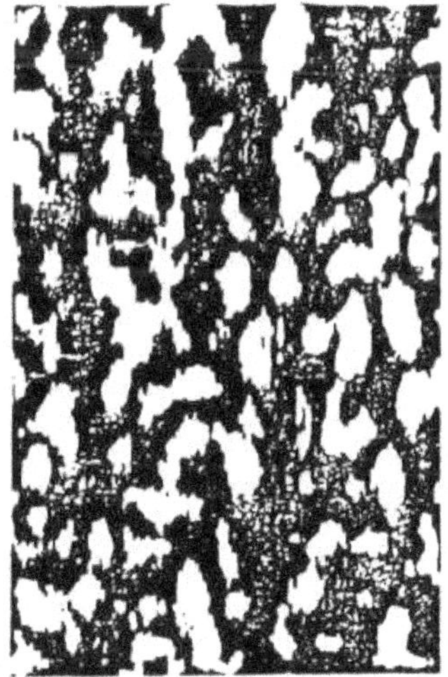

Fig. 2. A typical picture of (laser) transmission speckle pattern.

(1) Elastic scattering is qualitatively different from elastic scattering and absorption in that it is an entirely *reversible* process; i.e. during each elastic scattering event the scattered wave has a *definite*

phase relation with respect to the incoming wave. In contrast, during an inelastic scattering event, the phase information of the wave function is lost upon scattering. Thus elastic scatterings will give rise to complex interference patterns, i.e. speckle patterns, while inelastic scatterings tend to destroy such interference effects. To maximise the effect of interference in generating interesting correlation and fluctuation effects in speckle patterns, we choose to at first consider scattering media in which inelastic scattering and absorption effects are negligible. We will generalize our discussion to include these phase-breaking scatterings briefly towards the end of Sec. 4.

(2) We emphasize that the scattering medium under study is *static*, i.e. inhomogeneities which give rise to the speckle patterns do not fluctuate in time. This is to be contrasted to the situations one encounters in studying the multiple scattering properties of solutions containing suspended Brownian particles.

(3) From the last two points, one can conclude that speckle patterns are a *sample specific* property, namely they are a complicated interference "finger-prints" of all the positions of the scattering centers in the slab, despite the fact that the pattern may look quite random.

(4) I would like to discuss at this point the distinction between multiple versus single scattering. If the slab thickness $L$ is small compared to $l$, one sees from the definition of $l$ that, on average, the incident beam suffers only one or no elastic scattering events before getting out of the sample. We shall refer this case generically as the *Born regime*, since the well-known Born approximation techniques in wave mechanical scattering theory can be employed directly to study this regime.[3] What is more interesting is the opposite limit $L \gg l$. In this regime, the transmitted speckle pattern is produced as a result of scattered waves that suffer many elastic scattering events before reaching the screen to the right of the sample. We will refer to this regime as the *multiple scattering regime*, and our attention shall be concentrated on this case. It is to be noted that the speckle patterns generated by a reflectively rough surface can be formally mapped onto the Born regime of the bulk transmission speckle pattern problem. If one takes a very thin bulk scattering slab, and put a reflective wall on its back, it will mimic

rather closely the situation of the scattering off of a rough surface of an opaque material.

(5) We will make a simplifying assumption throughout this chapter to treat our wave as a *scalar* wave. This is a fine assumption when the wave refers to the one-electron wave function in the context of electron transport in mesoscopic conductors, but for light, which has two polarizations, it is not directly applicable. However, as indicated in recent work,[4] the effect of polarization is not really important insofar as the statistical properties of the interference patterns are concerned, since polarized incident wave gets quickly randomized after a few scattering events inside the random multiple scattering medium.

How does this problem of studying the speckle patterns fit into the framework of classical wave localization as elucidated in the other chapters of this volume? The answer lies again within the range of parameters that one chooses to concentrate on. We saw in Chapter 3 that a rough condition to achieve wave localization is the Ioffe-Regel conditions[5] $kl \approx 1$. To observe transmission speckle patterns, we need the wavefunction to be able to transmit through the sample, i.e. it should not be localized inside the sample. This means that we should concentrate on the extended regime $kl \gg 1$, (or $\omega\tau \gg 1$, where $\tau \equiv l/c$), which is sometimes referred to as the weak localization regime. In this regime, we have seen from previous chapters the interesting phenomenon of the coherent backscattering. Contrary to this effect, which becomes clearly visible only after one takes an average of the reflection coefficient near the backscattering direction over a large ensemble of similar but different samples, the speckle patterns, or aperiodic fluctuations of the scattered intensity exist in all directions and for a *single* given sample. The theoretical analysis of the complex intensity correlations and fluctuations behavior of these grainy and random-looking interference patterns, as far as we know, have not been made in the multiple scattering regime until our recent calculations. But before presenting our theoretical results, we shall take a look at the conventional wisdom on speckle patterns, and discuss the regime of validity of such a theory, to thus see both its merits and shortcomings.

To facilitate further discussions, we now need to define the relevant transmission coefficients. We define $T_{ab}$ to be the ratio of the scattered *intensity* in direction $b$ (characterized by the outgoing wave direction $\hat{k}_b$, measured in far field (i.e. at a distance $\gg L, W$), of a solid angle $\Delta\Omega \approx 1/(kW)^2$, on the right side of the slab, with the incoming beam in direction $a$. The reason for choosing this normalization solid angle $\Delta\Omega$ will become clear when we introduce an alternative waveguide geometry language.

## 3. CONVENTIONAL THEORY OF SPECKLE PATTERNS

The conventional theory of speckle patterns are best summarized in a series of papers by Goodman.[6,7] In this approach, one argues on physical ground that, given an incident beam in direction $a$, of unit amplitude, the complex scattered wave amplitude $(A_{ab})$ is a coherent superposition of a great many Huygens' wavelets, each coming from their last scattering events inside the sample. Because of the randomness of the disorder, one assumes the phases of each of these wavelets to vary greatly (compared to $2\pi$) and randomly. One can visualize this sum as a "random walk" of wavelet amplitudes on a complex plane, as illustrated in Fig. 3. The random nature of the phase of each wavelet is reflected by the random directions and magnitudes of the respective wavelet amplitudes, which are represented by the small "vectors". Mathematically, one writes for the total scattered wave amplitude in direction $b$ as:

$$A_{ab} = \sum_{k=1}^{N} a_k e^{i\phi_k} , \tag{1}$$

where $a_k$ and $\phi_k$ are assumed to be uncorrelated random numbers. From this description, it is clear then that the statistical distribution function for the intensity $T_{ab} \equiv |A_{ab}|^2$ can be obtained, through a complete analogy with the analysis of random walks (of step number $N \gg 1$) on a two dimensional complex plane. In particular, one can show that,[7]

$$P(T_{ab}) = \frac{1}{\overline{T}_{ab}} e^{-T_{ab}/\overline{T}_{ab}} \tag{2}$$

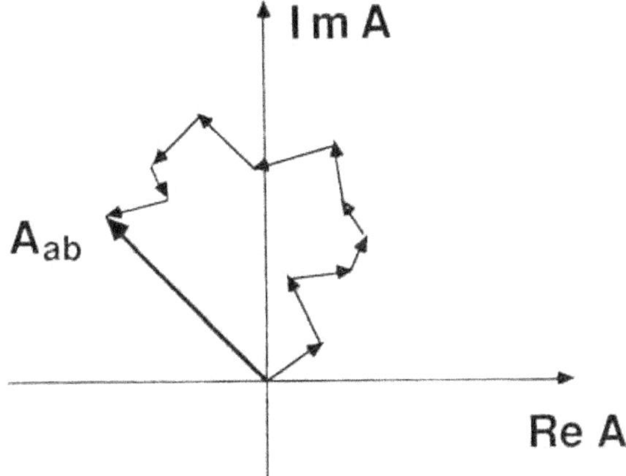


Fig. 3. Scattered amplitude $A_{ab}$ on the screen as a coherent superposition of Huygens' wavelets, which are presumed to have random amplitudes and phases.

describes the probability distribution function for the intensities. Here $\overline{T}_{ab}$ is the *ensemble averaged* intensity in direction $b$. An important consequence of this form (for an ensemble of samples) of the distribution function is that the standard deviation of the intensity, defined as,

$$\delta T_{ab}^2 \equiv \langle (T_{ab} - \overline{T}_{ab})^2 \rangle \tag{3}$$

has the value given by,

$$\delta T_{ab}^2 = \overline{T}_{ab}^2 \, , \tag{4}$$

i.e. the typical variations in intensity on a speckle pattern are of the same value as the average intensity itself. This is the origin of the extremely grainy look of a speckle pattern. This relation is also well verified by experiments. A schematic drawing of how the transmitted intensity varies as a function of direction $b$ is shown in Fig. 4.

This picture is obviously very attractive, since it provides a simple and physically reasonable statistical picture of the properties of the speckle patterns. In particular, it gives correctly the relation between the variance in intensity and the average intensity. What this approach has left out, of course, is any possible *correlations* in the intensity variations in a given speckle pattern, since all the $a_k$'s and $\phi_k$'s are assumed

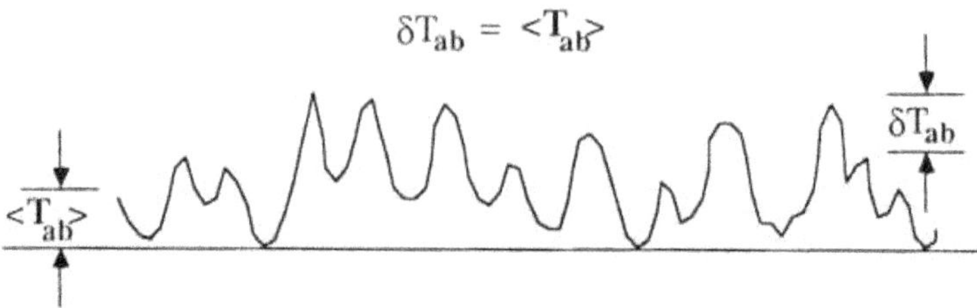



Fig. 4. Schematic illustration of the typical behavior of the intensity variations on the far field screen (on the transmitted side of the scattering medium). Notice the relation $\delta T_{ab}=(T_{ab})$, which is the reason behind the grainy look of the speckle patterns.

to be entirely random. How good is this assumption? Judging the randomness of the speckle pattern by the eye, this appears to be quite a reasonable assumption. But we can illustrate the fact that the $a_k$'s and $\phi_k$'s are not totally without correlations by considering the following simple Gedanken experiment.

Suppose our slab is thin such that $L \ll l$, thereby Born approximation can be employed to describe its scattering properties. For simplicity we will take the scatterers to be point-like particles. Let us now consider the transmitted speckle pattern generated by an incoming beam in direction $a$. Suppose we then rotate slightly the incident beam direction, by a small angle $\delta\theta$. From the rotational invariance of the scattering centers, we can easily see that in the limit of very small $\delta\theta$, the produced speckle pattern will simply have to shift down uniformly by the same angle. This situation is illustrated in Fig. 5. Thus we see that at least in this simple case, the $a_k$'s and $\phi_k$'s are not simple random quantities, but rather are (complex) deterministic functions of the incoming wavefront characterized by $\theta$ (or $\mathbf{k}_a$). This effect, which we coin the "memory effect", will be discussed further in the next section. One might argue intuitively then, that as $L$ increases to be much larger than $l$, so that we enter the multiple scattering regime, a typical path of a scattered wave is so tortuous and windy that the resulting transmission intensity pattern must have "forgotten" the incident direction $a$, so that the above random statistical description might become adequate. But as we shall see later, even in the multiple scattering regime $L \gg l$, due to the phase coherent nature of wave propagation through random media, the memory effect is still present (but weaker than in the Born regime).

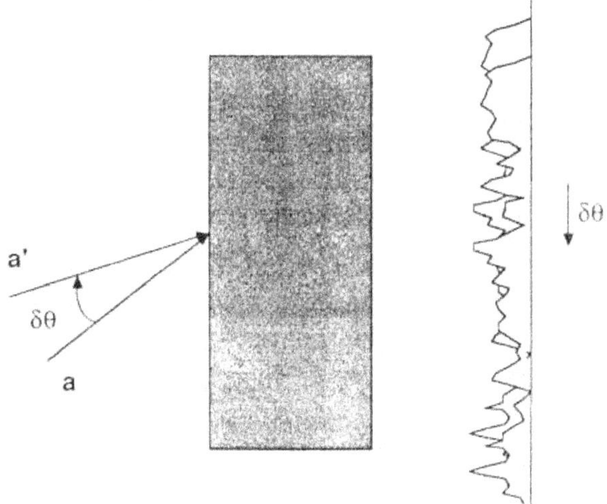



Fig. 5. Illustration of the "memory" effect. When the incoming beam is shifted by an angle $\delta\theta$, the transmitted speckle pattern will on average shift by the same angle, provided that $\delta\theta$ is less than $\sim 1/kL$. This is a consequence of the $C^{(1)}$ correlation term.

We are now ready to present results of our recent calculation on the novel correlation effects of the speckle patterns.

## 4. NOVEL CORRELATION EFFECTS OF SPECKLE PATTERNS IN THE MULTIPLE SCATTERING REGIME

This section mainly contains work performed by Kane, Lee, Stone, and myself[8] on the subject, which was recently published. In order to perform a careful study of the correlations in a speckle pattern generated by the propagation of a coherent scalar wave through a random elastic scattering medium, we first need to make a further simplification about the geometry under study, in order to save us some trouble in doing the calculations. Instead of the open geometry we have been describing, i.e. a slab on which a coherent beam of width $W$ is incident, we opt to put reflecting boundaries of width $W$ all the way through, making the geometry a wave-guide in the horizontal direction, as illustrated in

Fig. 6. The reason for imposing this wave-guide geometry is that in the open geometry, the incident wave is not exactly a plane wave, but rather a Gaussian wave packet of width $W$. While this poses no conceptual difficulties, it does introduce unnecessary complications to the various response functions that we must calculate. In the waveguide geometry of Fig. 6, an incident beam and all the scattered beams are quantized, with their transverse wave vectors given by $\mathbf{q} = \frac{\pi}{W}(n_x, n_y)$; while the $z$-component of the wave vector given by $k_z = \sqrt{k^2 - (\mathbf{q})^2}$. The transmission coefficients $T_{ab}$ can now be defined very precisely, namely it is the ratio of the transmitted *intensity* in waveguide mode (or "channel") $b$ to the incident intensity which is entirely in channel $a$. We assume the weak localization condition $kl \gg 1$, so that all properties of the wave propagation in such a scattering medium can be described by the diffusion

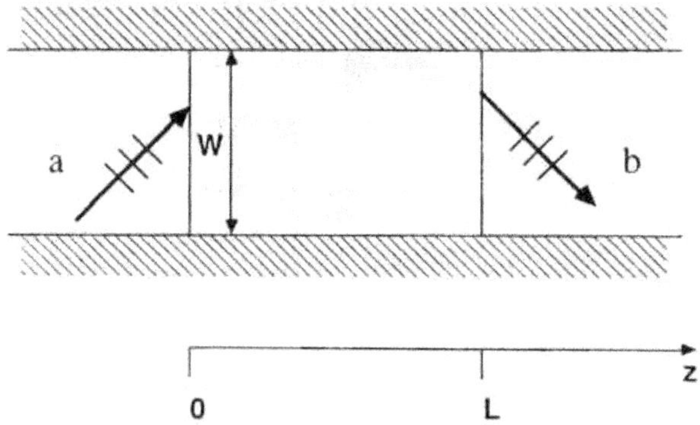


Fig. 6. The waveguide geometry for studying the transmission speckle patterns. The shaded region ($0 < z < l$) represents the disordered scattering medium. This geometry corresponds closely to the theoretical model of a disordered mesoscopic conductor. The side walls are assumed to be prefectly reflecting. This boundary condition then determines the waveguide modes (or channels) in the ordered regions ($z < 0$ and $z > L$).

diffusion approximation. We also concentrate only on the more interesting case of multiple-scattering regime, namely $L \gg l$. The scatterers in the medium are assumed to be point-like objects which only scatter an incoming wave by generating an $s$-wave (spherical) scattered beam. The correlation effects can be studied by considering the intensity-intensity correlation function

$$C_{aba'b'} = \langle \delta T_{ab} \delta T_{a'b'} \rangle \, , \tag{5}$$

where $\delta T_{ab} \equiv T_{ab} - \langle T_{ab} \rangle$, and $\langle \ \rangle$ stands for an average over the ensemble of different realizations of the positions of the scattering centers. We note that the ensemble average is *distinct* from an average over the direction ($b$) of a given speckle pattern generated by a single realization of the positions of the scatterers. Before ensemble averaging, one can easily show that $T_{ab}$ is given by a product of two Green's functions

$$T_{ab} = c_a c_b G_{ab}^+(z = 0, z = L) G_{ab}^-(z = 0, z' = L) \, , \tag{6}$$

where $c_a \equiv c\hat{k}_a^z$ and similarly for $c_b$. Also $G_{ab}^\pm(z, z')$ is defined as the Fourier transformation in the transverse directions of the Green's function in real space,

$$G_{ab}^\pm(z, z') = \frac{1}{A} \int d\boldsymbol{\rho} \int d\boldsymbol{\rho}' e^{-i\mathbf{q}_a \cdot \boldsymbol{\rho}} e^{i\mathbf{q}_b \cdot \boldsymbol{\rho}'} G^\pm(\mathbf{r}, \mathbf{r}') \, , \tag{7}$$

where $A = W^2$ is the cross-sectional area of the waveguide, $\mathbf{r} = (\boldsymbol{\rho}, z)$ and $G^\pm(\mathbf{r}, \mathbf{r}')$ is the standard retarded $(+)$ and advanced $(-)$ Green's function for propagating a coherent wave (of frequency $\omega$) generated by a point source of unit amplitude at $\mathbf{r}$, to another point $\mathbf{r}'$, taking into account *all* the scattering events for that given configuration of the scatterers. Symbolically, we have,

$$G^\pm(\mathbf{r}, \mathbf{r}') = \langle \mathbf{r} | [k_0^2 \pm i0^+ + \nabla^2 - V(\mathbf{r})]^{-1} | \mathbf{r}' \rangle \, , \tag{8}$$

where $k_0 = \omega/c$, and $V(\mathbf{r})$ denotes the complex scattering potential in the scattering medium. We re-emphasize that we are considering here a fixed frequency $\omega$, at which the mean free path for the wave is $l \equiv l(\omega)$. All sources of absorption and inelastic scattering are to be neglected.

As a first step, we compute the ensemble averaged intensity transmission coefficient $\langle T_{ab} \rangle \equiv \overline{T}_{ab}$. In the weak localized regime $(kl \gg 1)$, it can be calculated by standard diagrammatic perturbation techniques, by summing up the so-called "diffusion ladder diagram", as shown in

Fig. 7. Here the solid lines represent the ensemble averaged Green's function, which in $k$-space takes the form $G^{\pm}(\mathbf{k}) = 1/(k_0^2 - k^2 \pm i/2\tau)$, with $\tau = l/c$ being the inverse elastic scattering rate. The dotted lines represent ensemble averaged scattering potential $u^2 \delta_{\mathbf{k}_1 + \mathbf{k}_2 + \mathbf{k}_3 + \mathbf{k}_4, 0}$, where $u^2$ is the square of the Fourier transformation of the scattering potential of a single scatterer, multiplied by the density of scatterers. This is in turn related to $\tau$ through $1/\tau = 2\pi u^2 N(\omega)$, with $N(\omega)$ being the density of levels of plane wave states in a unit volume, at frequency $\omega$. (A good source to learn about the standard Green's function perturbation techniques is Doniach and Sondheimer's book on the subject.[9]) Evaluation of the diagram in Fig. 7 yields, up to a constant of order unity which we do not worry about,

$$\overline{T}_{ab} \approx \frac{c_a c_b}{c^2} \frac{l}{NL} , \tag{9}$$

where $N \approx k^2 A$ is the total number of propagating waveguide modes (or "channels"). In the next section, we will demonstrate that this result is physically reasonable by discussing the relation of our results for classical wave correlations to the problem of "universal conductance fluctuations" in mesoscopic metal systems.

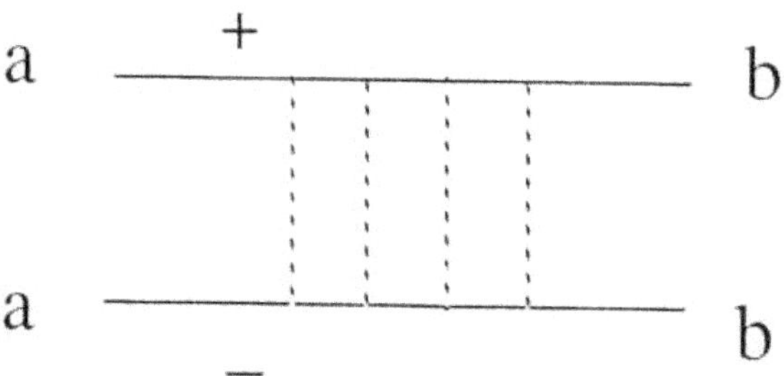


Fig. 7. Feynman diagram for the average transmitted intensity $\langle T_{ab} \rangle$. Here the solid lines denote the ensemble averaged Green's function, and the dotted lines represent the ensemble averaged scattering potential due to the elastic impurities in the medium.

We now proceed to compute the correlation function $C_{aba'b'}$. Following the standard diagrammatic practices, the largest contribution comes from the diagram drawn in Fig. 8. Its value can be evaluated to yield,

$$C^{(1)}_{aba'b'} = \overline{T}_{ab}\overline{T}_{a'b'}\,\delta_{\Delta q_a, \Delta q_b}\,F_1(\Delta q_a L)\,, \tag{10}$$

where the "form factor" $F_1(x)$ takes the form, within our diffusion approximations,

$$F_1(x) = \frac{x^2}{\sinh^2 x}\,. \tag{11}$$

and $\Delta q_a \equiv q_a, -q_a$, similarly for $\Delta q_b$. The physical meaning of this expression is quite interesting. First let us consider the case $a = a'$ and $b = b'$. Then the correlation function reduces to the mean-square variance of the intensity in direction $b$:

$$\langle \delta T^2_{ab}\rangle = (\overline{T}_{ab})^2\,. \tag{12}$$

This agrees with what conventional theory predicted (cf. last section), and also with numerous experimental observations. While conventional theory assumes there are no correlations for these rapidly fluctuating transmitted intensities, an examination of the Delta function in our result of Eq. (10) yields the interesting memory effect.

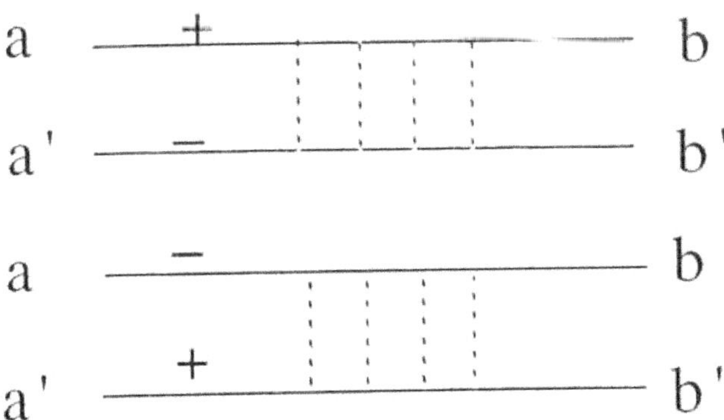


Fig. 8. Feynman diagram for the $C^{(1)}_{aba'b'}$ correlation function (the largest term in the intensity-intensity correlation function).

Since $C^{(1)}_{aba'b'}$ is identically equal to zero unless $\Delta q_a = \Delta q_b$, and taking into account that the $C^{(1)}_{aba'b'}$ term is the largest contribution to the overall intensity-intensity correlation function under study; this represents the physical fact that the speckle pattern generated by an incoming beam in direction $\mathbf{q}_a$ is strongly correlated to that generated by the incoming beam in direction $\mathbf{q}_a + \Delta\mathbf{q}_a$, *if and only if* one shifts the transmitted speckle pattern by the same angle characterized by $\Delta\mathbf{q}_a$. In particular, if one shifts the incoming beam, say, up by a small angle $\delta\theta$, the entire transmission speckle pattern will on average shift down by the same angle, provided $\delta\theta$ is not too large. Considering the fact that we are dealing with the propagation of waves in the multiple scattering regime $L \gg l$, a typical scattered path consists of $\sim (L/l)^2$ scattering events before leaving the sample and reaching the speckle pattern screen, this effect of the speckle pattern still somehow "remembering" the direction of an incident beam is quite remarkable! It physically represents the fact that the transmission speckle pattern, while looking innocuously random, still contains a great deal of information about the shape of the wavefront of the incident beam, i.e. it is a complicated coding function of the incoming wave's direction and the positions of all the scatterers in the sample. The maximum angle $\Delta\theta_{\max}$ for observing this effect is determined by requiring $\Delta q_a^{\max} \approx 1$, or $\Delta\theta_{\max} \approx 1/kL$. We shall see in the next section that recent optical experiments are in excellent agreement with this scaling of $\Delta\theta_{\max}$ with $L$. We also note that $\Delta\theta_{\max}$ is independent of $l$, so that this critical angle for observing the memory effect is quite universal.

As $C^{(1)}_{aba'b'}$ is identically zero for $\Delta\mathbf{q}_a \neq \Delta\mathbf{q}_b$, and decays exponentially even for $\Delta\mathbf{q}_a = \Delta\mathbf{q}_b$, as $\Delta\mathbf{q}_a L$ increases to be larger than unity, we need to search for the next order term in the correlation function $C_{aba'b'}$. This is accomplished through an evaluation of the diagram in Fig. 9, which contains two more diffusion ladder diagrams compared to Fig. 8. The vertex "$H$" denotes the so called "Hikami box", which is shown in Fig. 10. The result, which we name $C^{(1)}_{aba'b'}$, takes the form (up to a constant of order unity),

$$C^{(2)}_{aba'b'} = \frac{\overline{T}_{ab}\overline{T}_{a'b'}}{g}\left[F_2(\Delta q_a L) + F_2(\Delta q_b L)\right] , \qquad (13)$$

where $g = Nl/L$ is the so-called Thouless number which is much greater than unity in the weak localization regime $kL \gg 1$. $g$ is also the dimensionless average conductance in the electronic equivalent of our problem, which is to be described later. The form factor $F_2(x)$ is given by

$$F_2 = \frac{2}{x}\left[\coth x - \frac{x}{\sinh^2 x}\right].$$

Because of the form of the $C^{(2)}_{aba'b'}$, it makes the largest contribution to the intensity-intensity correlation function when *either* $a \approx a'$ or $b \approx b'$, but need not both (which was the case for $C^{(1)}_{aba'b'}$). Thus, taking the case $a = a'$, one concludes that the intensities at different spots of a transmission speckle pattern are more or less uniformly and positively correlated, by an amount roughly equal to

$$\frac{\overline{T}^2_{ab}}{g} \sim \frac{l}{N^3 L}.\tag{15}$$

This result again shows that speckle patterns contain novel correlations which have not been included properly by the conventional argument.

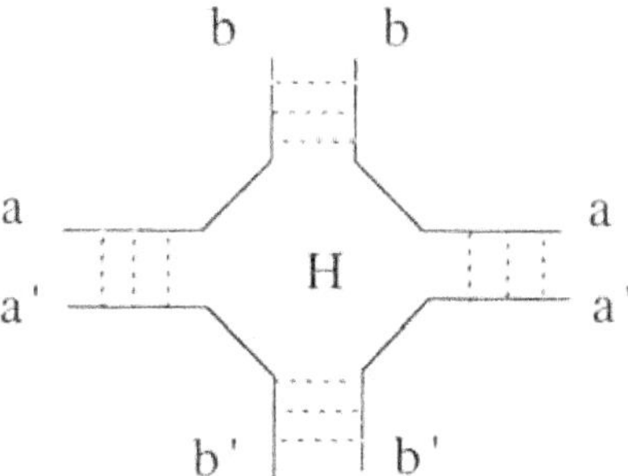


Fig. 9. Feynman diagram for the $C^{(2)}_{aba'b'}$ correlation function (the second largest term in the intensity-intensity correlation function).

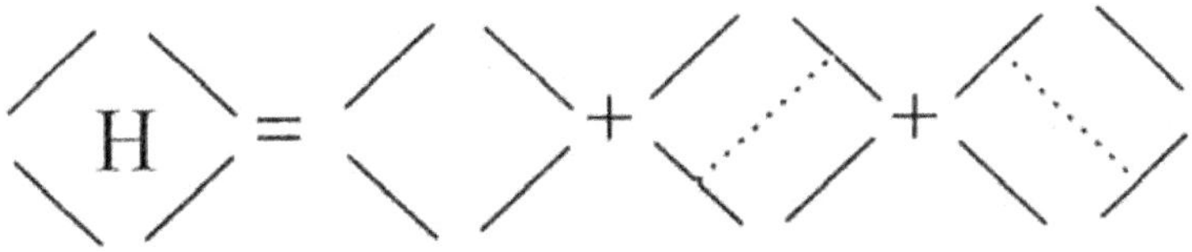

Fig. 10. The definition of the Hikami vertex (Ref. 16), in the Feynman diagrammatic representation.

In the most general situation $a \neq a'$ and $b \neq b'$, the contribution of $C^{(2)}_{aba'b'}$ decays to zero quickly. Thus in order to describe any remaining correlations in this case, we need to look for the next order term in our expansion of $C_{aba'b'}$, in the small parameter $1/g$. This is done by considering a diagram given in Fig. 11, which contains two more diffusion ladders than the previous term. Since all the external momentum lines are "dressed" by diffusion graphs for this diagram, this new term, $C^{(3)}_{aba'b'}$, contains no form factors such as $F_1$ and $F_2$. Explicitly, we have,

$$C^{(3)}_{aba'b'} \approx \frac{\overline{T}_{ab}\overline{T}_{a'b'}}{g^2} . \tag{16}$$

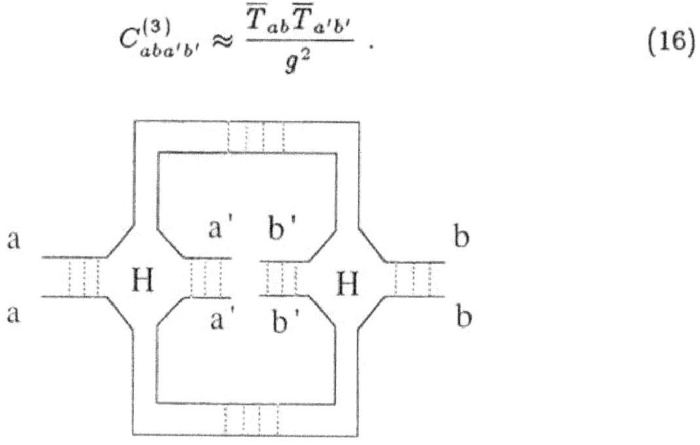


Fig. 11. Feynman diagram for the $C^{(3)}_{aba'b'}$ correlation function (the smallest term in the intensity-intensity correlation function).

As $g \gg 1$ in the regime we are considering, this contribution to the correlation function is tiny in magnitude indeed. But since it does not depend strongly on any of the incoming/outgoing directions, it represents a *long range* nearly uniform background correlation, which exists for all the channels. Physically, this term represents the tendency that the transmitted intensity, when integrated over all incoming and outgoing directions for a given realization of the scatterer positions, $\widetilde{G} \equiv \sum_{ab} T_{ab}$, is a fluctuating quantity from sample to sample. Thus a given realization of the random scattering medium is always a little more transparent or more opaque than the ensemble mean. The variance of the integrated intensity fluctuations is given by $\langle \delta \widetilde{G}^2 \rangle \approx 1$, while the ensemble averaged value of $\widetilde{G}$ is roughly given by $g \gg 1$. We shall see in the next section that this is exactly the statement of the so-called "universal conductance fluctuations" in the context of mesoscopic electron transport in disordered metals. Another way to see the physical significance of $C^{(3)}_{aba'b'}$ is that it indicates the distinction between ensemble averaging and spatial averaging in coherent wave transport; namely the ensemble averaged intensity $\langle T_{ab} \rangle$ is *not* the same as the directionally averaged intensity $\sum_{ab} T_{ab}/N^2$ for a given sample. This is an explicit statement about the non-self averaging nature of a multiple scattered wave in random media.

The above diagrammatic results

$$C_{aba'b'} = C^{(1)}_{aba'b'} + C^{(2)}_{aba'b'} + C^{(3)}_{aba'b'} \, , \tag{17}$$

have been checked numerically, by looking at the transmission of scalar waves through a two-dimensional disordered scattering medium described by the tight-binding Anderson model. In order for the $C^{(3)}_{aba'b'}$ term to be measured numerically with a reasonable number of realizations, we had to take a value of the disorder parameter $kl$ to be rather close to one, i.e. the simulations were done rather close to the localization transition. The Thouless parameter that we actually used was $g \approx 1.2$. This formally puts the simulations in a regime outside the regime of validity $(kl \gg 1)$ of the diagrammatic calculations. But one hopes that the qualitative features of the above diagrammatic predictions are still preserved in such simulations. This is indeed the case. Figure 12 shows the numerically computed $C_{aba'b'}$ for two special choices of the various channels. The square line represents a function $C_{k_0,k_1,k_0,k_0+\Delta}$ with $\Delta$ being plotted as the $x$-axis, so that for all values of $\Delta$, we have $a = a'$. The circle line plots the function $C_{k_0,k_1,k_1,k_0+\Delta}$, so that when $\Delta = k_1 - k_0$, we have $b = b'$. $k_1$ and $k_0$ are defined as the integers such that $q_1 = k_1 \frac{\pi}{L}$, etc. It is then clear that for $\Delta \neq k_1 - k_0$, the circle trace should be roughly given by $C^{(3)}$, — the constant background correlations. But when $\Delta = k_1 - k_0$, we have $b = b'$, and the correlation function should now contain an extra contribution from the $C^{(2)}$ term. As shown in Fig. 12, this is indeed true. For the square trace, we have $a = a'$ always; thus for $\Delta \neq k_1 - k_0$, it is roughly $C^{(2)} + C^{(3)}$ for all values of $\Delta$. When $\Delta = k_1 - k_0$, the $C^{(1)}$ term contributes, therefore we see a peak at this particular value of $\Delta$. Figure 12 shows these effects clearly. It is also satisfying to see that the peak of the circle trace corresponds very well with the background of the square trace, as it should per our diagrammatic results. This is a strong statement about the structures of the intensity-intensity correlations that we predict. Some other simulation parameters should be mentioned: $L = 20, W = 20$, and the number of realizations is 20,000. The Anderson disorder parameter is $W/V = 4$.

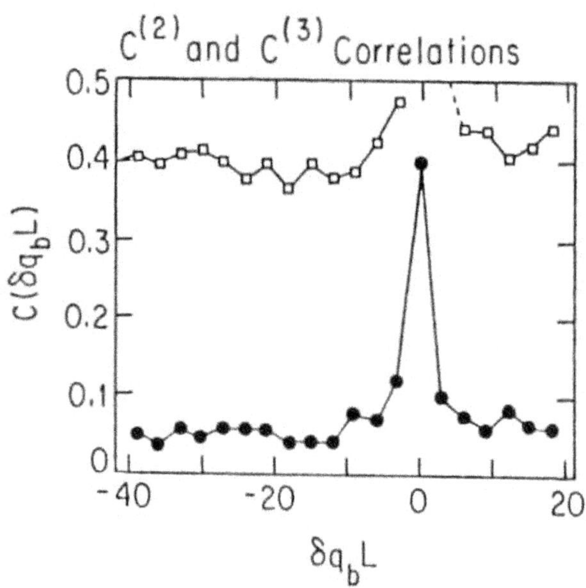


Fig. 12. Numerical simulation data for illustrating the $C^{(1)}_{aba'b'}$, $C^{(2)}_{aba'b'}$ and $C^{(3)}_{aba'b'}$ correlations. The circle trace represents the function $C_{k_0,k_1,k_1,k_0+\Delta}$ as a function of $\Delta$. Here $k_0 = 7$, and $k_1 = 14$, etc. are the "channel" indices for the various waveguide modes, with $\delta q_b L \equiv \pi \Delta$. Anderson model with $g \approx 1.2$ (corresponding to disorder parameter $w/v = 4$) are being used, with the number of averaged samples being over 20,000. The square line represents the function $C_{k_0,k_1,k_0,k_0+\Delta}$. The significance of these data are described in the text.

We can easily generalize our calculation to treat the intensity-intensity correlations of speckle patterns at two slightly different frequencies. Namely, we can compute

$$C_{aba'b'}^{\Delta\omega} \equiv \langle \delta T_{ab}(\omega)\delta T_{a'b'}(\omega + \Delta\omega)\rangle \ . \tag{18}$$

The $C^{(2)}$ and $C^{(3)}$ terms in this case should be modified from Eq. (13) and Eq. (16) by noting that whenever the parameter $\Delta q_a$ or $\Delta q_b$ is smaller than $(\Delta\omega/\omega_c)^{1/2}$, where $\omega_c \equiv D/L^2$ with $D = cl/3$, it should be replaced by the parameter $(\Delta\omega/\omega_c)^{1/2}$. Whereas for the $C^{(3)}$ term, it should be multiplied by a crossover function $G(\Delta\omega/\omega_c)$, where $G(x) \approx 1$ for $x \ll 1$, and $G(x) \sim x^{-1/2}$ for $x \gg 1$.

We can also similarly generalize our calculation to include the effect of inelastic scattering. Suppose we are dealing with a system where the wave inelastic scattering rate is $1/\tau_{in}$ at frequency $\omega$. This rate then enters the diffusion propagators as a cutoff, in the form of $\frac{1}{Dq^2} + \frac{1}{\tau_{in}}$. Qualitatively, we find that if the inelastic diffusion length $L_{in} \equiv (D\tau_{in})^{1/2}$ is long compared with the sample dimensions $L, W$, the intensity-intensity correlation function $C_{aba'b'}$ is unaffected. But when $L_{in} < L, W, C_{aba'b'}$ should be modified in the same way as in the case of finite frequency shift just described, with the parameter $(\Delta\omega/\omega_c)^{1/2}$ replaced by $L/L_{in}$ (we assume $L \sim W$ for simplicity).

Our study of the correlation functions $C_{aba'b'}$ can also be generalized to treat the dynamical situations, namely, when we allow the scatterers to move randomly inside a fluid. Assuming the speed of a typical scatterer $v$ is much smaller than the wave speed $C$, we can treat the scattering problem quasi statically. This problem has been the subject of Chapter 6 (by D. J. Pine *et al.* in this volume), with the name "diffusive wave spectroscopy" (DWS). In our language, this problem amounts to considering a quantity

$$C_{aba'b'}(t) \equiv \langle \delta T_{ab}(0)\delta T_{a'b'}(t)\rangle \tag{19}$$

where $T_{ab}(t)$ is the intensity transmission coefficient measured at a time $t$, which is a function of the configuration at this time. It can be easily shown that, for the case of Brownian motion of the scatterers, this dynamical correlation function (in the weak localized regime) can be obtained by replacing the arguments of the form factors $F_1$ and $F_2$ by

$X_t = L\sqrt{(\delta q_a)^2 + \left(\frac{t}{t_0 l^2}\right)}$, where $t_0 = 1/D_p k^2$ is a microscopic time scale, with $D_p$ being the diffusion constant of the Brownian scatterer particles. For the $C^{(3)}_{aba'b'}$ term, which does not have a form factor, it should now be multiplied by a function $G(x_t)$, with $G(x \ll l) \approx 1$, and $G(x \gg l) \approx x^{-1/2}$.

One important corollary of this description of dynamic correlations in multiple scattering systems is that if we measure the quantity $C_{abab}(t) = \langle \delta T_{ab}(0) \delta T_{ab}(t) \rangle$ which is usually done in the DWS experiments, it maps directly with the *static* correlation function $C^{(1)}_{a,b,a+\delta,b+\delta}$, as $C_{abab}(t) = \langle T_{ab} \rangle^2 F_1\left(\sqrt{\frac{t}{t_0}}\left(\frac{L}{l}\right)\right)$, and $C_{a,b,a+\delta,b+\delta} = \langle T_{ab} \rangle^2 F_1(\delta q_a L)$. This prediction is amenable to experimental verifications.

This concludes the presentation of our recent investigation of the intensity-intensity correlation function for the transmission speckle patterns for multiply scattered waves through elastic random media.

## 5. EXPERIMENTAL SITUATIONS AND RELATED WORK

Can the various novel correlation effects such as the memory effect and the background correlations be measured experimentally? The answer is that it depends on what kind of wave function one is considering. In the optical realization of the problem, where the speckle patterns were first observed, only the $C^{(1)}$ term can be detected. This is due to the fact, as elucidated in Chapter 1 of this book, that the elastic mean free path $l$ is generally much larger than the wavelength $\lambda$, i.e. $kl \gg 1$ is almost always true in a typical optical setup. This makes the $C^{(2)}$ and $C^{(3)}$ terms very small, due to the very large value of $g$. Thus with the limited number of samples of which one can realistically store the scattered intensity data, it is difficult to observe these more subtle correlation effects. The $C^{(1)}$ correlations, i.e. the memory effect, has just been observed and characterized experimentally, by the recent work of Freund *et al.*[10] The scattering medium in this work is a slab of opal glass. The coherent light source is a He-Ne laser with $\lambda = 0.63 \mu m$. The thickness of the slab is 400 $\mu m$ and 800 $\mu m$, and the transport mean free path at this frequency is $l \approx 100 \mu m$. This gives a value of $kl$ to be about $10^3$. This system is then the ideal setup for

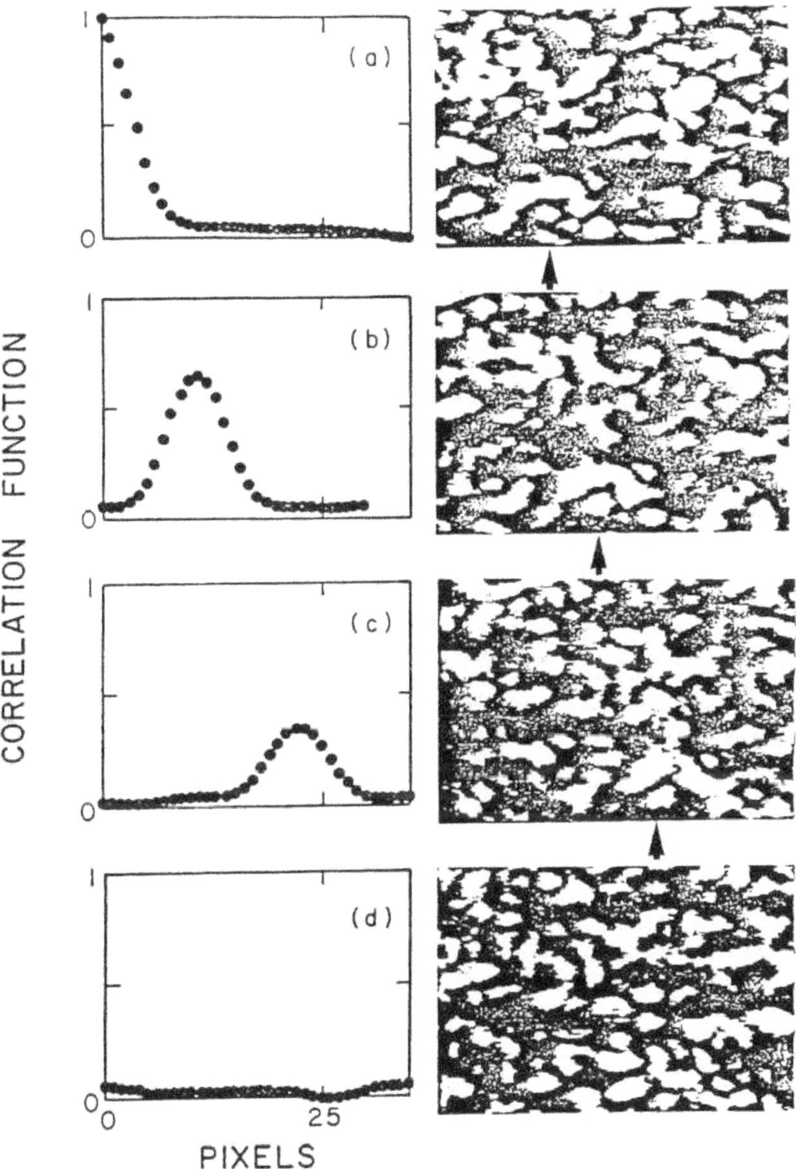

Fig. 13. Experimentally measured static correlation function for different values of $\delta\theta_a$ (or $\Delta\mathbf{q}_a$), as indicated by the arrows. The figures (a)–(d) are cross-correlations of the transmission speckle patterns before and after the angular shift in the incoming direction.

comparison with the diagrammatic results. Upon rotating the angle of incidence slowly, the video image of the transmitted speckle pattern (in the far field) can be visually observed to shift at the same rate as that of the incoming wave direction, amongst other fluctuating effects, just as we predicted theoretically. In fact, the visual effect of the speckle pattern generated by a moving source is quite similar to the image of a river in which the water is flowing with some speed, with all the random flickerings but an average motion in one direction. We can represent this situation by Fig. 13, where the cross correlation functions are computed experimentally, along with pictures of the speckle patterns upon changing the incoming beam direction. The experimentally determined form factor $F_1(\Delta q L)$ agrees well with the theoretical predictions for $\Delta q L > 1$; whereas for $\Delta q L < 1$, deviations from the theory are observed, in that the experimentally determined form factor is essentially given by $e^{-2(\Delta q)L}$ for all values of $(\Delta q)L$, whereas the diagrammatic prediction gives a function that flattens out for small values of $(\Delta q)L$. This discrepancy is reasonable considering the crude way the actual multiple scattering processes are approximated diagrammatically by a diffusion propagator $P(\mathbf{r},\mathbf{r}')$, which was assumed to have a simple boundary condition $P(\mathbf{r},\mathbf{r}') = 0$ at $z = 0$ and $z = L$.

We now discuss the relationship of our calculation of $C_{aba'b'}$ with the conductance fluctuations in mesoscopic metals. A mesoscopic conductor in this context refers to a disordered metal wire, whose dimensions are so small that the inelastic diffusion length $L_\phi = \sqrt{D\tau_\phi}$ is large compared to all the sample dimensions. Here $D = v_F^2\tau/3$ is the diffusion constant corresponding to multiple elastic scattering of the electronic wavefunction (at the Fermi surface) by static impurity atoms, and $\tau_\phi^{-1}$ is the inelastic scattering rate which, unlike elastic scatterers, randomizes the phase information of the electron wavefunction. Referring back to Fig. 6, we can now imagine the disordered regime to be our dirty metal, and the ordered regime to be the (two-terminal) leads made of clean metal pads. Recent experimental[11] and theoretical[12] studies have established a result coined as the "universal conductance fluctuations" (UCF), which states that for this two-probe geometry, the conductance varies from realization to realization of the impurity configurations by an amount roughly given by $e^2/h$. This result can be cast in terms of the intensity transmission coefficient for the electron wave function (at $\varepsilon_F$) $T_{ab}$ via the generalized Landauer formula[13]:

$$G = 2\frac{e^2}{h} \sum_{ab} T_{ab} \ . \tag{19}$$

Before going to the considerations of the fluctuations, we can show from the above equation that, upon using our earlier diagrammatic result for $\langle T_{ab}\rangle$, the average conductance $\langle G\rangle$ is given by

$$\langle G\rangle \approx \frac{e^2}{h}\frac{Nl}{L} \ . \tag{20}$$

This expression reduces to the simple Ohm's law $G = \sigma\frac{A}{L}$. From the Landauer formula Eq. (19), we can also see that the statement of UCF is equivalent to saying that

$$\sum_{aba'b'} \langle \delta T_{ab}\delta T_{a'b'}\rangle \approx 1 \ , \tag{21}$$

as long as $L, W < L_\phi$.

Using our results for $C_{aba'b'}$, as discussed in the last section, this result easily follows. It is interesting to note that the three different terms contribute to the UCF in a reverse order of importance. Namely, we can easily show that $\sum_{aba'b'} C^{(1)}_{aba'b'} \approx (l/L)^2 \ll 1$, $\sum_{aba'b'} C^{(2)}_{aba'b'} \approx (l/L) \ll 1$, but $\sum_{aba'b'} C^{(3)}_{aba'b'} \approx 1$. Therefore, the uniform background correlations, although smallest in magnitude for a given set of incoming and outgoing directions, make the dominant contribution to UCF, owing to its very long range of influence. In this sense, our result of $C^{(3)}_{aba'b'}$ has been verified experimentally, through the observation of the UCF. In fact, the UCF development was the primary impetus for our recent calculation of the intensity-intensity correlations, in that this calculation reveals the microscopic correlations and fluctuations of different transmission channels in giving rise to the universal conductance fluctuations.

We now discuss the relation of our calculation of $C_{aba'b'}$ to some recent related work. Shapiro[14] considered sometime ago the intensity correlations problem in an infinite disordered scattering medium, where a point source is placed at the origin, and the following correlation function was computed diagrammatically in the weak localization regime:

$$C(\mathbf{r},\mathbf{r}') = \langle \delta I(\mathbf{r})\delta I(\mathbf{r}')\rangle \ . \tag{22}$$

He used only the factorized diagram (our Fig. 8) to obtain the result:

$$C(\mathbf{r},\mathbf{r}') = \langle I(\mathbf{r})\rangle\langle I(\mathbf{r}')\rangle\left(\frac{\sin(k\Delta r)}{k\Delta r}\right)^2 e^{(-\Delta r/l)} \ . \tag{23}$$

Since he did not look at the case of transmission or reflection through a finite sample, the experimental ramifications of his calculations were not clear. It can be easily shown that in the geometry that he considered, the $C^{(1)}$ diagram is the only dominant one. But because of the infinite-space geometry he used, the interesting memory effect was not made evident.

Stephen and Cwilich[15] later concentrated on the correlation function for the slab geometry, for the case of normal incidence and *near field* transmission and reflection speckle pattern. In particular, they considered a slab geometry where the incident wave was in the normal direction on the left side of the slab, and they computed the correlation

$$C^{(t)}(\mathbf{R}) = \langle I^T(0)I^T(\mathbf{R})\rangle \ , \tag{24}$$

and

$$C^{(r)}(\mathbf{R}) = \langle I^R(0)I^R(\mathbf{R})\rangle \ , \tag{25}$$

where $I^T(\mathbf{r})$ refers to the intensity of the transmitted wave *on the surface* of the slab on the transmission side, and $I^R(\mathbf{r})$ is that on the reflection side. By using the diagram which is Fig. 9 in our language, they showed that

$$C^{(t)}(\mathbf{R}) \approx \frac{1}{(kl)^2}\left(\frac{l}{L}\right)^3\left(\frac{L}{R}\right) \ . \tag{26}$$

It can be shown that in this geometry, the most dominant diagram is that of Fig. 9, since we have, roughly speaking, the situation $a = a'$

and $b \neq b'$ in our language. In principle these near-field correlation effects of the speckle pattern can be measured experimentally. But as we mentioned before, in the usual optical setups we have $kl \gg 1$, therefore making the observation of these correlation effect difficult. Upon integrating $C^{(t)}(R)$ over $R$, we obtain the mean square fluctuations of the total transmitted intensity, which can be related to the quantity $\delta\widetilde{T}^2 \equiv \sum_{bb'} C_{0b0b'} \approx l/NL$ in our formulation. Calculations from these two different approaches agree, which is quite satisfactory. We should note that apart from this integrated result, the $C^{(t)}(R)$ function (for near field) and our $C_{aba'b'}$ (for far field) cannot be related directly. It is also worth noting that of the three terms $C^{(1)}, C^{(2)}$ and $C^{(3)}$, the $C^{(2)}$ term is the most important, as far as their contributions to $\delta\widetilde{T}^2$ are concerned. Thus we see again that even though the $C^{(1)}, C^{(2)}$ and $C^{(3)}$ terms in the intensity-intensity correlation function are arranged in descending order of magnitudes, the term which actually makes the dominant contribution depends on the actual quantity and the specific geometry that one is concerned with.

## 6. SUMMARY

The main purpose of this chapter has been to provide the reader with a general picture of the recent development in the study of novel correlation and fluctuation effects of coherent wave transmissions through disordered scattering media. We have concentrated on the purely elastic multiple scattering regime under the assumption of weak localization. We first described the conventional statistical theory of the speckle patterns in this regime, and then argued about its inadequacy in describing various correlation effects buried inside such random looking patterns. We then presented both diagrammatic and numerical calculations of the main quantity of our investigation, the intensity-intensity correlation function $C_{aba'b'}$, for describing the fluctuations and correlation properties of the transmission speckle patterns. We thus uncover interesting correlation effects such as the memory effect and the long-range background correlations. Relations of these effects to recent development in the universal conductance fluctuations in mesoscopic electron transport in disordered metals were discussed, along with a brief presentation of the recent optical experiment which verified the memory effect in multiple scattering wave transport.

## REFERENCES


1. J. W. Strutt (Lord Rayleigh), *Phil. Mag.* **10**, (1880) 73.
2. Various aspects of this field, particularly from an engineering perspective, are contained in a special issue of the *J. Optical Soc. Am.* **2** (1985) 2069.
3. See, for example, *Quantum Mechanics*, L. D. Landau and E. M. Lifshtz, Vol. 3 of course of theoretical physics, (Pergamon Press, Oxford, 1977) pp. 513–18.
4. M. J. Stephen and G. Cwilich, *Phys. Rev.* **34** (1986) 7564.
5. A. F. Ioffe and A. R. Regal, *Prog. Semicond.* **4** (1960) 237.
6. J. W. Goodman, *J. Optical Soc. Am.* **66** (1976) 1145.
7. J. W. Goodman, in *Laser speckle and related phenomena*, ed. J. C. Dainty (Springer-Verlag, Heidelberg, 1975), Vol. 9, pp. 9–15.
8. S. Feng, C. Kane, P. A. Lee and A. D. Stone, *Phys. Rev. Lett.* **61** (1988) 834.
9. S. Doniach and E. H. Sondheimer, *Green's functions for solid state physicists*, (Benjamin, London, 1974).
10. I. Freund, M. Rosenbluh and S. Feng, to appear in *Phys. Rev. Lett.*
11. J. C. Licini, D. J. Bishop, M. A. Kastner and J. Melngailis, *Phys. Rev. Lett.* **55** (1985) 2987; R. A. Webb, S. Washburn, C. P. Umbach and R. B. Laibowitz, *Phys. Rev. Lett.* **54** (1985) 2696; W. J. Skocpol, P. M. Mankiewich, R. E. Howard, L. E. Jackel, D. M. Tennant and A. D. Stone, *Phys. Rev. Lett.* **56** (1986) 2865.
12. B. L. Al'tshuler and D. E. Khmel'nitskii, *Pis'ma Zh. Eksp. Teor. Fiz.* — **42** (1985) 291; P. A. Lee, A. D. Stone and H. Fukuyama, *Phys. Rev.* **B35** (1987) 1039.
13. D. S. Fisher and P. A. Lee, *Phys. Rev.* **23** (1981) 6851.
14. B. Shapiro, *Phys. Rev. Lett.* **57** (1986) 2168.
15. M. Stephen and G. Cwilich, *Phys. Rev. Lett.* **59** (1987) 285.

# FLUCTUATIONS, CORRELATION AND AVERAGE TRANSPORT OF ELECTROMAGNETIC RADIATION IN RANDOM MEDIA


AZRIEL Z. GENACK

*Department of Physics, Queens College of CUNY*
*65-30 Kissena Blvd., Flushing, NY 11367*


In this chapter we present a universal model of wave propagation illustrated and stimulated by measurements of electromagnetic propagation in random media. These measurements exploit the spatial, spectral and temporal resolution of optical and microwave sources to reveal the statistical character and scale dependence of wave transport. Measurements of average transmission, of the correlation function of intensity fluctuations at a point as a function of input wave frequency and time and of pulse transmission in the diffusive regime and near the Anderson localization threshold are discussed. We show that the Fourier transform of the distribution of travel times to a point is the field-field correlation function as the frequency is varied. The half width of this correlation function corresponds to the average width of eigenstates of the sample. For diffusive transport it is given by, $\delta\nu = 0.96 D/L^2$. In the strong scattering regime, the scale dependent diffusion coefficient is defined by the relation $\delta\nu(L) = 0.96D(L)/L^2$. The relationship between the full variety of transport phenomena depends upon the correlation frequency $\delta\nu$ and the level spacing $d\nu/dN$. The ratio $\delta = \delta\nu/(d\nu/dN)$, is the universal dimensionless parameter from which a scaling theory of localization and correlation can be constructed. The mobility edge is reached when $\delta = 1$. This gives quantitative expression to the Thouless criterion for localization which relates the sensitivity of the eigenstate energy to the presence of a boundary and the energy spacing between eigenstates. The scattering strength needed to achieve localization depends upon the sample geometry and boundary conditions. For a cubic sample without reflecting walls, localization is predicted to occur when $kl = 1.1$, where $k$ is the wave number and $l$ is the transport mean free path. The number of eigenstates within the level width $\delta\nu$ is precisely $\delta$. This is the number of statistically independent parameters needed to represent a monochromatic wave in the medium, $N_{\text{ind}} = \delta$. $1/\delta$ is shown to be the probability of return of a photon to a coherence volume of the sample as well as the degree of intensity cross correlation between points in the medium. The influence of correlation is directly observed in microwave experiments in which $\delta$ can be made small by choice of sample geometry and scattering strength. Both the shape of the intensity correlation function and of the intensity probability distribution are observed to change as $\delta$ decreases and correlation builds up. From the point of view of wave coherence, the mobility edge is reached, when all transmitted speckle spots are correlated, independent of the length scale, i.e. $N_{\text{ind}}(L) = 1$. The magnitude of fluctuations of the total transmitted intensity or conductance is evaluated using these notions and is in agreement with previous theories of long range correlation and universal conductance fluctuations in the weak scattering limit.

# Contents


1. Introduction　　210

2. Universal Aspects of Wave Propagation in Random Media　　213

3. Comparison of Electron and Photon Propagation in Disordered Media　　234

4. Experimental Approach: Methods, Samples and Applications　　239

5. Photon Diffusion　　248

6. Local Fluctuations　　260

7. Weak Localization　　281

8. Long Range Correlation　　285

9. Scale Dependence of Diffusion Coefficient　　300

10. Conclusion　　305

Acknowledgements　　305

References　　306

## 1. INTRODUCTION

The aim of this article is to describe the statistical character of electromagnetic (EM) propagation in random media. On length scales greater than the transport mean free path $l$, EM propagation is not influenced by microscopic details of the interaction with the material, except insofar as these determine the velocity $v$, the inelastic scattering length, $l_i$ and $l$ itself. We, therefore, seek a universal description of wave propagation. In the absence of interaction between quanta of the field, this description should apply equally to classical and quantum mechanical waves. It should utilize only a small number of key parameters and provide the relationship between the full variety of transport phenomena.

In this article we discuss observations associated with five transport phenomena: (1) particle diffusion, (2) local fluctuations, (3) weak localization, (4) long range correlation, and (5) the transition from extended to localized waves. In order to make clear the plan of this article, we first present an overview of the relationship between these phenomena.

The essential character of wave propagation in a sample is contained in the path length distribution which is proportional to the time of flight distribution of particles, $T(t)$. Therefore, a fundamental parameter is the temporal width of the distribution, $\delta t$. The value of $\delta t$ for a wave of a given frequency $\nu$ is determined by the scattering parameters $l, l_i$ and $v = \nu\lambda$, as well as by the physical dimensions and reflectivity of the boundary.

A key question is the dependence of $\delta t$ upon the scale of the sample. In the weak scattering limit, wave interference does not significantly influence average transport and $\delta t$ is proportional to $L^2/D$, where $D$ is the classical diffusion coefficient and $L$ is the sample thickness. In general, however, constructive interference of time reversed partial waves scattered back to a point enhances the return to that point and reduces the rate of transport in the sample.[1,2] Since the number of paths returning to a point increases as the size of the sample increases, the role of such coherent backscattering can be represented by a scale dependent effective diffusion coefficient $D(L)$ (Ref. 3), which describes the pulse width, $\delta\tau \sim L^2/D(L)$. Even though coherent backscattering does not lead to a substantial renormalization of $D$ in the weak scattering limit its influence is observed as fluctuations in transmission or conductance.[2−6] In the strong scattering limit, however, the wave is exponentially localized[7−10] by coherent backscattering. $\delta t$ then increases exponentially with $L$. In three dimensions, the threshold for Anderson localization is given by the Ioffe-Regel criterion $kl \sim 1$ where $k = 2\pi/\lambda$ is the wavenumber.[11]

As a result of the random phase and amplitude of partial waves arriving at a point, the values of the field and intensity fluctuate in space, giving rise to the familiar speckle pattern for light.[12,13] Fluctuation also occurs at a point as the frequency of the wave is varied.[14−21] The field-field correlation function is the Fourier transform of $T(t)$.[15] Thus, by the uncertainty relationship, $\delta t$ is inversely proportional to the field-field correlation frequency $\delta\nu$. $\delta\nu$ can be identified with the width of the eigenstates of the sample.[22] The level width is thus proportional to $D(L)/L^2$.

The scaling of $\delta t, \Delta\nu$ or $D$ depends upon the density of states of the sample $dN/d\nu$, which is the inverse of level spacing $d\nu/dN$. The fundamental dimensionless ratio,

$$\delta(L) = \frac{\delta\nu(L)}{d\nu(L)/dN} , \tag{1}$$

is the variable which describes the scale dependence of propagation.[3,8,22] In accord with the scaling theory of localization, the scaling of $\delta$ for a given dimension is given by a universal scaling function which depends only upon $\delta$.[3,8,22−24] This parameter can be directly determined from measurements of field or intensity fluctuations with incident frequency. The dimensionless frequency or Thouless number $\delta$ describes the coupling between statistically equivalent blocks of a sample and determines the proximity to the Anderson localization threshold in any dimension.[3,8,25,26] The threshold is given by the Thouless criterion, $\delta = 1$ (Refs. 3,8). This criterion holds independent of sample geometry and dimension. For $\delta < 1$, the eigenstates within adjacent blocks of the sample are not likely to overlap. Transport is inhibited and the wave is exponentially localized.

The dimensionless frequency is also related to the degree of intensity correlation in the sample. The probability that a photon at a point in the output wavefront also passes through a phase coherence volume containing an arbitrarily selected point in the sample is $1/\delta$ (Ref. 22). Hence $1/\delta$ is the degree of photon or intensity correlation between points in the sample. It is proportional to the intensity cross correlation function when $\delta > 1$. The degree of long range intensity correlation in the weak scattering regime has been calculated by diagramatic Green's function techniques,[17−20,27−30] the method of random matrices,[30,31,32] and by a Langevin approach.[33] Here we emphasize another approach which focuses upon the number of independent parameters which are needed to represent a wave of frequency $\nu$ in a random medium .[22,34] This number, $N_{\text{ind}}$, is inversely proportional to the degree of correlation. $N_{\text{ind}}$ is the number of eigenstates of the sample within a frequency range $\delta\nu$ of $\nu$. Since the number of speckle spots in the transmitted wave, $N_s$, is greater than $N_{\text{ind}}, N_s > N_{\text{ind}}$, speckle spots in the transmitted wave are correlated, even in the weak scattering limit. Remarkably, we find that,

$$N_{\text{ind}} = \delta , \tag{2}$$

This equation is a succinct statement of the relationship between fluctuations, correlation and average transport (phenomena 1–5). The increased correlation as the scattering strength and sample size increase is compensated by an increasing number of speckle spots and reduced transmitted intensity to give a constant variance in the conductance or total transmission for an ensemble of samples. This is called universal conduction fluctuations.[5,6,27−29] Increased correlation between speckle spots as the scattering strength increases is associated with the Anderson transition. The mobility edge[35] separating extended from localized states is reached when $\delta(L) = N_{\text{ind}} = 1$, which corresponds to the point at which the intensity throughout the transmitted wave is correlated.

We will discuss the circumstances in which the abovementioned transport phenomena occur and their experimental observation for EM radiation. In this discussion an intuitive and often quantitative understanding of the universal character of wave phenomena is arrived at by consideration of the experimental observation of local wave characteristics within the speckle pattern. The spatial, spectral and temporal resolution of laser and microwave sources affords a unique opportunity to study universal aspects of wave propagation in random media. Measurements have been made of $cw$ (Ref. 14) and pulsed[14,15,36] transmission as well as of spectral[9,14-16] and temporal[37-43] intensity fluctuations of optical and microwave radiation in the diffusive regime. Quite recently there have been some measurements of transport in the critical regime.[44,45] These results are discussed within a unified framework.

The rest of this article is organized as follows: In Sec. 2, we discuss the scaling hypothesis and present a simple model of the buildup of correlation in a sample. The differences between electron and photon propagation are highlighted in Sec. 3. There, experimental opportunities in studies of EM propagation and difficulties which must be overcome to achieve photon localization are contrasted respectively with the experimental difficulties encountered in testing predictions of the theory of electron localization and the ease with which electron localization occurs in amorphous solids. The experimental approaches used in the study of EM propagation are outlined in Sec. 4. Disordered samples in which photon localization may be achieved and periodic structures in which EM propagation,[46,47] emission[46] and resonant dipolar interactions[48] are suppressed are also considered. The physical model sketched in Sec. 2 is developed more fully in Secs. 5–9 in connection with experimental results relating sequentially to the five propagation effects mentioned above. For the reader primarily interested in the experimental results, Secs. 5–9 can be read independently of the rest of the chapter. We conclude in Sec. 10.

## 2. UNIVERSAL ASPECTS OF WAVE PROPAGATION IN RANDOM MEDIA

Wave transport through a random medium may be described in terms of a distribution of Feynman paths, with transverse dimensions of the phase coherence length, $L_c \sim \lambda/4$, and with direction which is randomized in a distance $l$. In the limit $kl \gg 1$, in a three-dimensional sample the probability that a path passing through a phase coherence volume, $V_c = L_c^3$, will find its way back to that volume is $\sim 1/k^2 l^2 \ll 1$

(Sec. 7). Thus interference can be neglected in calculating the path length distribution in the medium. Transport may, therefore, be modeled by the random walk of photons[30,49]. In the limit that sample size, $L$, is much greater than $l$, photon propagation can be described by the diffusion equation for the photon density, $n(\mathbf{r}, t)$,

$$D\nabla^2 n(\mathbf{r}, t) - \frac{1}{\tau_a} n(\mathbf{r}, t) = \frac{\partial n(\mathbf{r}, t)}{\partial t} , \tag{3}$$

where $D = vl/d$ is the Boltzmann value of the diffusion coefficient $d$ in $d$ dimensions and $1/\tau_a$ is the photon absorption rate. This is analogous to the description of electron transport in metals in terms of the diffusion coefficient, which is related to the conductivity by the Einstein relation. In the strong scattering regime, transport can no longer be described in terms of an intensive diffusion coefficient. However, some of the consequences of wave interference can be described in terms of an anomalous extensive diffusion coefficient $D(L)$ (Refs. 9,10,50,51).

Though the diffusion coefficient is not appreciably modified from the Boltzmann value in the weak scattering regime, nonetheless, the interference of backscattered waves on electrical conductance or optical reflectivity has dramatic consequences corresponding to weak localization.[1-4,52] Measurements of magnetoresistance give the phase breaking time for the electrons.[2] In these experiments, the application of a magnetic field breaks time reversal invariance and reduces coherent backscattering.

Weak localization, which enhances the return of the wave to individual coherence volumes, can be distinguished from universal conductance fluctuations, or total transmission fluctuations, which reflects correlation between different coherence volumes. In the absence of inelastic scattering, the magnitude of conductance fluctuations in a sample with longitudinal dimensions, $L$, greater than or equal to its transverse dimensions, is independent of sample composition and size. Fluctuations in the dimensionless conductance,[28,29] $g = G/(e^2/h)$, are of order unity as long as $L$ is less than both the localization length and the inelastic length, $L_i = (D\tau_i)^{1/2}$, which is the decay length of transmission of particles that have not been inelastically scattered.

The difference between weak localization and universal conductance fluctuations is seen in magnetoresistance experiments in which both effects appear together[5,6]. Measurements by Webb, Washburn, Umbach and Laibowitz[5] of resistance fluctuations in a gold ring are shown in Fig. 1 as the flux threading the ring, shown in the insert, is varied. The phase difference between electron waves travelling in two halves of the ring is given by, $(2\pi e/h) \oint \mathbf{A} \cdot d\mathbf{l}$, where the integral is taken around the ring. This gives rise to a periodic modulation of conductance with a flux period $h/e$ due to the Aharanov-Bohm effect.[53] But the presence of the magnetic field also modifies interference of electron waves that travel along identical paths, but in opposite senses, completely around the ring. Conductance fluctuations due to such paths, have a magnetic flux period of $h/2e$ (Ref. 4) which is half that of the ordinary Aharonov-Bohm effect, in inverse proportion to the increased path length of the interfering electron waves. This is borne out in the power spectrum of magnetoresistance fluctuations shown in Fig. 1 which has a peak in flux period at $h/e$ corresponding to the ordinary Aharonov-Bohm effect and at $h/2e$ corresponding to weak localization. The nonperiodic oscillations in Fig. 1 are related to flux penetrating the wire of the ring which results in a distribution of flux values enclosed within the actual electron paths.[27]

Nonperiodic, but reproducible, spatial fluctuations within a speckle pattern as well as spectral fluctuations within a single speckle spot are observed in optical and microwave scattering. The fluctuation of transmission with frequency is analogous to fluctuations in electronic conductance as the chemical potential difference across the sample is varied. The phase of the partial waves in the sample, directly associated with the path length $s$, $2\pi s/\lambda$, changes with laser wavelength. Thus, for a random distribution of path lengths and additional random phases of partial waves, the intensity of transmitted light varies randomly with frequency. When the intensity is measured within a single coherence area of the speckle pattern in the weak scattering limit, the fluctuations in intensity are as large as the average intensity. An example of optical intensity fluctuations which we have measured within a single speckle spot as a function of sample thickness, $L$, is shown in Fig. 2. The sam-

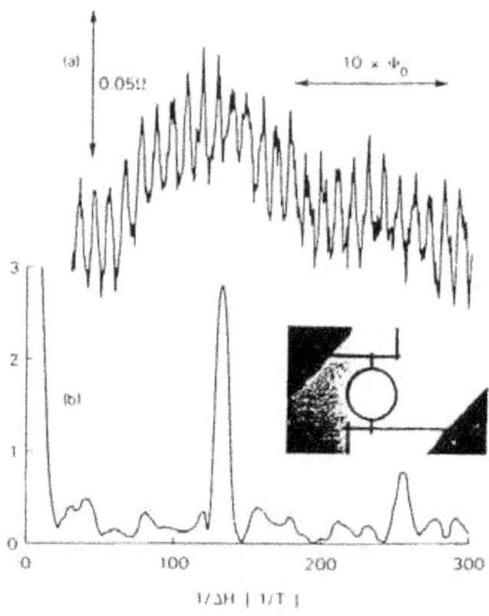



Fig. 1. (a) Measurements of magnetoresistance, $\Delta R/R$, in the gold ring with inside diameter of 784 nm shown in the insert. Measurements are made at $T =$ 0.01K. (b) The power spectrum of the magnetoresistance measurements. The peaks at $1/\Delta H \sim 125\ 1/T$ and $250\ 1/T$ correspond to magnetic flux periods of $h/e$ and $h/2e$ which are associated respectively with Aharanov-Bohm effect and weak localization. These results are from R. A. Webb, S. Washburn, C. P. Umbach and R. B. Laibowitz, *Phys. Rev. Lett.* 54 (1985) 2696.

ple is composed of equal volumes of 0.5 $\mu m$ and 0.8 $\mu m$ polystyrene spheres deposited in a wedged glass sample holder as the colloid of the spheres in water dries. The intensity-intensity correlation function with laser frequency in a single speckle spot in the weak scattering limit is directly related to the path length distribution represented in the pulse profile of transmitted light. In Sec. 6 it is shown to be associated with the Fourier transform of the pulse distribution. Thus the lengthened pulse in thicker regions of the sample correspond to smaller correlation frequencies as seen in Fig. 2.

Recently, coherent backscattering has been directly observed in an enhancement by a factor of two of reflected intensity in a cone of radial width $1/kl$ about the incident laser beam direction.[54-69] These observations are discussed in detail in this volume in the chapter by van

Albada and Lagendijk. Enhanced backscattering represents a reduction in the forward propagation of light, and hence of the optical diffusion coefficient, in proportion to the ratio of the solid angle of the coherent backscattering cone to the total reflected solid angle. For weak scattering in three dimensions the fractional reduction in $D$ is proportional to $1/k^2 l^2$. The observation of weak localization of light suggests the equivalence of quantum and classical wave transport and supports the conjecture by John[9] and others[10,50,70−74] that, for sufficiently strong scattering, photons may be localized.

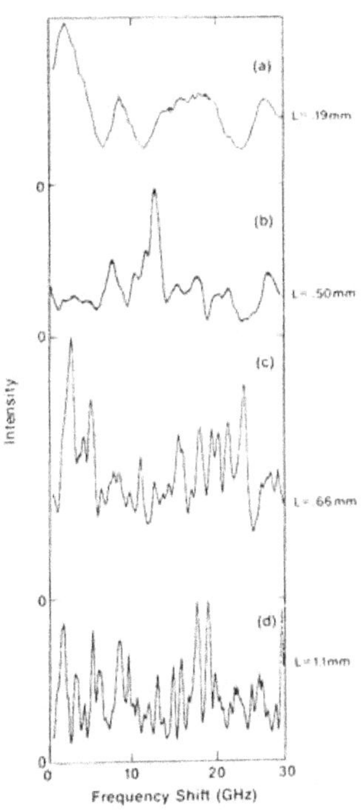


Fig. 2. Fluctuations in intensity in a single speckle spot of light ($\lambda \sim 589$ nm) transmitted through a random sample composed of polystyrene spheres of diameters 0.5 and 0.8 $\mu$m. The sample thickness $L$ increases from (a) to (d).

Anderson originally considered electronic transport in a random potential within the context of a tight binding model for a single electron moving in a periodic array of atoms with nearest neighbor coupling $V$

and random site energy, $\epsilon_i$, uniformly distributed over a range of values $W$ (Ref. 7). All states in the band are localized when $W/V$ is larger than a constant of order unity which is proportional to the number of nearest neighbors. The exact value of this constant is hard to establish and is model specific.

In order to develop a sharper and more general test for localization, Edwards and Thouless[25] proposed a criterion for localization based upon the sensitivity of electrons to the boundary of the sample. It can be expressed in terms of the ratio of the energy shift of a state which is caused by the presence of a boundary, $\Delta E$, to the average spacing between energy states of the sample, $dE/dN$. Within the Anderson model, this ratio can be written as $\Delta E/(W/N)$, where $N$ is the number of sites in the volume.

The energy shift, $\Delta E$, can be taken as the difference in the energy of states calculated with periodic and antiperiodic boundary conditions for a periodic array of internally disordered blocks. The energy shift is a measure of the coupling between states in adjacent blocks. For localized states, the ratio $\Delta E/(W/N)$ decreases as the block size increases due to the exponential fall off of the electron wave function at the boundary. On the other hand, for extended states, $\Delta E \sim V/N$, and $\Delta E/(W/N) = V/W$. Thus, we expect the states in the band will be localized when,

$$\frac{\Delta E(L)}{(W/N)} < \frac{V}{W} , \tag{4}$$

in which case the ratio $\Delta E(L)/(W/N)$ decreases as the sample scale $L$ increases.

More generally, $\Delta E(L)/(dE(L)/dN)$ is a scale dependent parameter which characterizes the change of energy levels as the sample size increases. Its fundamental importance is seen in that the condition for the onset of localization can be directly expressed through this parameter using a simple physical argument. Specifically, localization occurs when the level shift, $\Delta E$, is less than the level spacing since waves in neighboring regions of the sample are then not coupled. Furthermore, it was argued that the parameter, $\Delta E/(dE/dN)$, is directly related to electronic transport in the sample, since it can be identified with the dimensionless conductance, $g(L)$,

$$\frac{\Delta E(L)}{\frac{dE(L)}{dN}} = g(L) \; . \tag{5}$$

This can be seen in the weak scattering regime by identifying $\Delta E$ with $\hbar/t$, where $t \sim L^2/D$ is the time for the electron to diffuse across a block of length $L$. This gives, $D \sim \Delta E L^2/\hbar$. The conductance, $G = \frac{A}{L}\sigma = \sigma L^{d-2}$, where, $A = L^{d-1}$, is the generalized area of a $d$ dimensional hypercube, can be expressed using the Einstein relation,[3,8]

$$\sigma = e^2 D \frac{dn}{dE} \; , \tag{6}$$

where, $dn/dE = (dN/dE)/AL$, is the density of states per unit volume. This gives, $G = (e^2/h)(\Delta E/dN) = (e^2/h)g$, in accord with Eq. 5.

The ratio, $\Delta E(L)/(dE(L)/dN)$, was assumed by Abrahams *et al.*[3] to be the only statistically relevant feature of the energy levels. Consequently, they argued that the conductance at any length scale, $bL$, is determined by the value of the conductance at $L$ and the scale factor $b$, i.e. $g(bL) = f(b, g(L))$. By considering the differential form of this scaling relation, a one-parameter scaling function, $\beta_d(g(L))$, is constructed,

$$\frac{d \ln g(L)}{d \ln L} = \beta_d(g(L)) \; , \tag{7}$$

which depends upon the sample dimension $d$. For $g < 1$, the states are localized and $g$ decays exponentially, $g \sim g_0 e^{-L/\xi}$, where $\xi$ is the localization length. In this case,

$$\lim_{g \to 0} \beta_d(g) = \ln(g/g_0) < 0 \; . \tag{8}$$

For $g > 1$, $g$ is the usual macroscopic conductance, $g = \sigma L^{d-2}/(e^2/h)$. Consequently,

$$\lim_{g \to \infty} \beta_d(g) = d - 2 \; . \tag{9}$$

For $d < 2$, $\beta_d$ is always negative and $g$ decreases with $L$, falling off exponentially for large $L$. Thus $d = 2$ is the marginal dimension for localization. For $d > 2$, there is an unstable fixed point at $\beta = 0$, corresponding to the mobility edge. The value of $g$ at this point $g_c \sim 1$ is the minimum metallic conductance. Since the sensitivity to the boundary always decreases as $g$ decreases, $\beta$ is assumed to be continuous and to decrease monotonically as $L$ decreases. A schematic plot of $\beta_d$ is shown in Fig. 3.

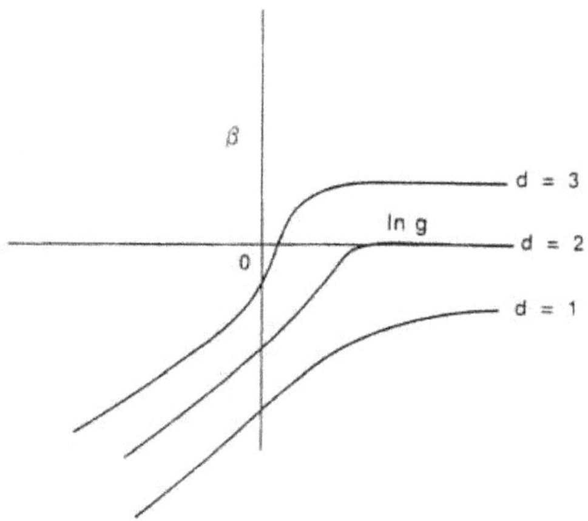


Fig. 3. Plot of $\beta_d(g)$ versus ln $g$ for $d = 3,2,1$ showing the scaling of conductance. The figure is adapted from E. Abrahams, P. W. Anderson, D. C. Licciardello and T. V. Ramakrishnan, *Phys. Rev. Lett.* **42** (1979) 673.

The scaling theory is in agreement with field theoretic results of Wegner[75] near the mobility edge and consistent with many experiments.[2,76] It is, however, contradicted by other experiments.[77,78] One source of the disagreement may be due to the role of electron-electron interactions. Further Anderson, Thouless, Abrahams and Fisher[79] pointed out that localized systems have a broad distribution of conductance[23,76–78] with an average conductance which may reflect the tail of the distribution rather than a representative value within the distribution. Indeed, for one dimensional samples, the average conductance may diverge due to transmission resonances in a small fraction of configurations in the ensemble of random samples.[80,81] Even in three dimensional metallic samples, departures from a Gaussian distribution of conductances in finite samples are predicted to result in the divergence of high moments of the conductance.[82] Therefore, the description of the scaling of conductance must include the scale dependence of the full distribution of $g$.[83-85] Large fluctuations in $g$ reflect long range density correlation within the sample and underline the complexity of the conductance and its inadequacy as a fundamental scaling parameter.

Anderson *et al.*[79] developed a scaling theory of localization for topologically one dimensional systems using quantum scattering theory. Reflection and transmission amplitudes are related to conductance by the method of Landauer.[86-88] Complete randomness is assumed between transmission matrix elements of different blocks of the samples. These authors identify a proper scaling variable in this system as one in which the probability distribution is not singular but is approximately Gaussian. Further they require that the overall resistance should not depend upon the order in which scaling is done, so that the scaling parameter should have an additive mean. A proper parameter is found to be $\ln(1 + \rho)$, where $\rho = 1/g$ has the property,

$$\langle \ln(1 + \rho) \rangle = \langle \ln(1 + \rho_1) \rangle + \langle \ln(1 + \rho_2) \rangle , \tag{10}$$

when the sample is composed of two lengths with resistances $\rho_1$ and $\rho_2$. For $\rho < 1$, the resistance is additive, whereas for $\rho > 1$ the log of the resistance is additive, corresponding to exponential localization. Since the sample is composed of a large number of blocks with random phasing, $\langle \ln(1+\rho) \rangle$ can be written as the sum of a large number of terms such as those in Eq. 10. A Gaussian distribution may, therefore, exist for the sum of these random terms. Recently Edrei *et al.* have found a crossover from a Gaussian distribution of transmission values in the weak scattering limit to a log-normal distribution for strong localization in two dimensions.[85]

We argue here that the fundamental character of the dimensionless ratio $\delta$ commends it as a universal scaling variable. The universality of $\delta$, however, appears paradoxical in view of its identification with $g$. This difficulty is resolved by noting that $\delta = \langle g \rangle$ only in the Boltzmann limit. Indeed $\delta$ and $g$ are entirely different sorts of entities. Unlike conductance, $\delta$ is not a distribution of values for each energy in each sample configuration but the ensemble average of a quantity which self averages. The width of the distribution of the ratio of $\delta\nu$ to $\delta\nu/dN$ for different sample configurations approaches zero as the energy range or sample size increases.

The nature of propagation in an ensemble of samples is expressed in the probability distribution of path lengths $P(s)$ of particles transmitted through the sample. Though the random phases associated with partial waves in a given sample lead to large fluctuations in intensity, $P(s)$ and the temporal profile $T(t)$, are well defined distributions. The shape of the pulse transmitted through a slab of thickness $L$ in the diffusive limit is given in Eq. 52 in Sec. 5.

The form of the time of flight distribution in the absence of absorption is shown in Fig. 4 plotted in the dimensionless units $t/(L^2/D(L))$. It is assumed that $D$ is constant over the range of frequencies contained in the pulse.

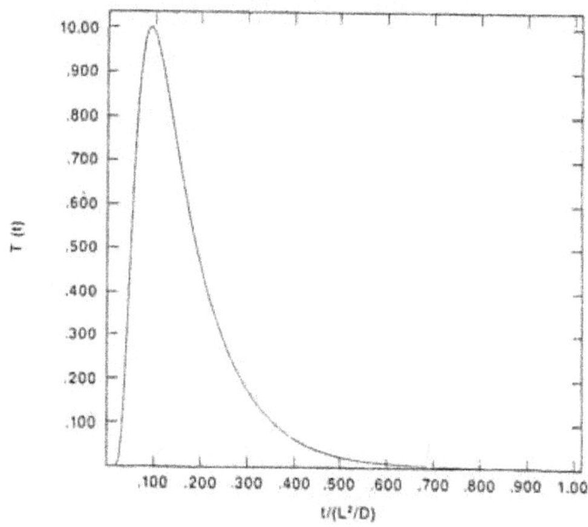



Fig. 4. The profile of the pulse intensity transmitted through a non-absorbing slab versus time delay in dimensionless units $t/(L^2/D)$. This form should hold for all extended waves as long as the value of $D$ is constant over all frequency components of the incident pulse.

In order to define the linewidth of a state, we consider the ensemble average of the field-field correlation function in the transmitted wave versus the frequency shift $\Delta\nu$ of the incident wave, $G^\psi(\Delta\nu) = \text{Re}\langle\psi(\nu)\psi^*(\nu + \Delta\nu)\rangle$. The field $\psi$ may be a classical field such as the electromagnetic field $E$ or a quantum mechanical field for which the frequency shift is $\Delta\nu = \Delta E/h$. The wave function can be represented as a linear combination of eigenstates,

$$\psi_\nu(\mathbf{r}) = \Sigma c_i(\nu) u_i(\mathbf{r}) . \tag{11}$$

The eigenstates, $u_i(\mathbf{r})$, of the electromagnetic wave equation for a random medium, in the absence of inelastic processes, satisfy the radiation boundary condition that the incident flux, assumed from the left, equals the sum of the transmitted and reflected flux. The eigenstates are normalized such that the integral of the flux for the eigenstate over the output face of the sample at $z = L$ is equal to the transmission $T_i$ for unit intensity in the direction of propagation, for the incoming part of the wave on the left. The frequency range over which $\psi_\nu$ or $c_i(\nu)$ are correlated is the half width of $G^\psi(\Delta\nu)$ which may be identified with the half width of the eigenstates, $\delta\nu$. For EM waves, we write the field correlation function as $G^E(\Delta\nu)$. In Sec. 6 (Eq. 59), $G^E(\Delta\nu)$ is shown to be the Fourier transform of the ensemble average of the intensity profile of the transmitted pulse due to an incident pulse which is a delta function in time. $G^E(\Delta\nu)$ can thus be determined either from pulsed experiments or from direct measurements of the frequency dependence of the transmitted field. Such measurements can be made with a mixer for microwave radiation and with homodyne mixing of the transmitted optical radiation with a plane wave generated from the excitation source. In the weak scattering limit, the field correlation function can be determined from the intensity correlation function. The normalized field correlation function for the wave transmitted through a slab in the absence of absorption in the weak scattering limit is plotted versus the dimensionless parameter $\Delta\nu/(D/L^2)$ in Fig. 5. The half-width of $G^E(\Delta\nu)$ is,

$$\delta\nu = 0.96 D/L^2 , \tag{12}$$

and corresponds to the frequency width of the wave function in the medium. The precise value of the level width is presumably not far from that given in Eq. 17 because $G^E(\Delta\nu)$ falls rapidly with $\Delta\nu$ at the half width of $G^E(\Delta\nu)$. Further consideration will be necessary to obtain the constant of proportionality between the level width and $D/L^2$, but for the present we will assume that the level width is the half width of $G^E(\Delta\nu)$ as indicated in Eq. 12.

In the presence of strong scattering, transport can no longer be described in terms of an intensive diffusion coefficient $D$. Equation 12 may then be generalized to define a scale dependent diffusion coefficient, $D(L)$,

$$\delta\nu(L) = 0.96D(L)/L^2 \tag{13}$$

The value of $D(L)$ may be expressed in terms of $\delta\nu$ while the scaling of $D$ depends upon the Thouless number $\delta$ which specifies the proximity to the localization threshold. $\delta$ obeys the universal scaling relation,

$$\frac{d\ln\delta(L)}{d\ln L} = \beta(\delta(L)) . \tag{14}$$

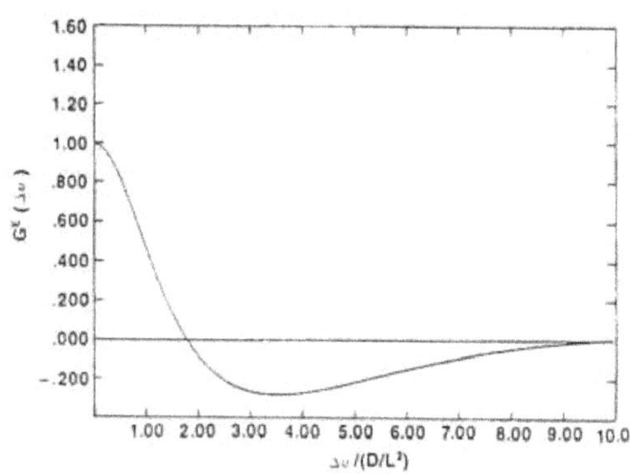

Fig. 5. The normalized field-field correlation function versus frequency shift in dimensionless units $\Delta\nu/(D/L^2)$. This is the Fourier transform of the pulse shape in Fig. 4 (Eq. 59).

A function closely related to $G^E(\Delta\nu)$ is the intensity-intensity correlation function with frequency shift $\Delta\nu, G^I(\Delta\nu)$. In general, the shape of $G^I(\Delta\nu)$ depends upon the degree of correlation in the sample. However it is shown in Sec. 6 that in the limit $N_{\text{ind}} \gg 1$, in which correlation between speckle spots is small, the cumulant of the intensity-intensity correlation function in transmission in a single coherence area is the absolute value squared of the complex field correlation function $G^E(\Delta\nu)$, i.e. $C_0^I(\Delta\nu) = |G^E(\Delta\nu)|^2$ (Refs. 15,18). This function is plotted in Fig. 6 in terms of the dimensionless variable $\Delta\nu/(D/L^2)$. The experimental verification of the connection between $C_0^I(\Delta\nu)$ and $T(t; L)$ via $G^E(\Delta\nu)$ in the limit $N_{\mathrm{ind}} \gg 1$ is presented in Sec. 6. In Sec. 8 we present experimental evidence of the breakdown of this relationship when $N_{\mathrm{ind}}$ is small.

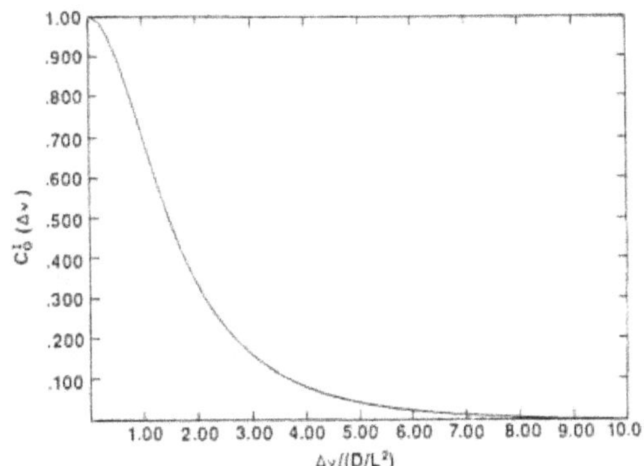

Fig. 6. The cumulant intensity-intensity correlation function at a point in the transmitted speckle pattern as a function of frequency shift in dimensionless units $\Delta\nu/(D/L^2)$. The sample is a nonabsorbing slab and $N_{\mathrm{ind}}$ is assumed to be large so that the effects of long range intensity correlation are negligible.

Since $G^E(\Delta\nu)$ is the Fourier transform of $T(t)$, it and, hence, $\delta\nu$ and $\delta$ are defined independently of intensity correlation. Hence, the value of $\delta(L)$ is not changed by the degree of correlation between speckle spots which influences the distribution of conductances. Rather, $\delta(L)$ is the underlying parameter in terms of which the degree of intensity correlation in the sample may be quantitatively expressed.

As long as $\delta\nu > d\nu/dN$, the wave functions in adjacent blocks of a sample can couple and the wave is extended. However, when $\delta\nu < d\nu/dN$, an energy mismatch can occur between a wave in a block and all states in the neighboring block. This mismatch becomes more severe as the size of the blocks increase. The ensemble average of the intensity of the wave will then decay exponentially. The sample is at the mobility edge, separating extended from localized states when,

$$\delta_c = 1 \; . \tag{15}$$

Since $\delta\nu \sim D(L)/L^2$, with $D(L) \lesssim vl/d$, and $d\nu/dN = L^{-d}/(dn/d\nu)$, it follows that $\delta = D(L)(dn/d\nu)L^{d-2} < (vl/d)(dn/d\nu)L^{d-2}$. Thus for

$d < 2, \delta(L) \to 0$ for large $L$. Thus, the Thouless condition[8] anticipates the result obtained from the scaling theory of localization that $d = 2$ is the marginal dimension for localization. Only for $d > 2$ can extended waves exist in the presence of disorder.

When the level width is less than the level spacing, the transmission coefficient will fluctuate as the input energy is varied. Azbel has shown that the transmission in a one dimensional sample can approach unity when the input energy is resonant with a localized state at the center of the sample.[80,81] When averaged over a frequency band greater than the level spacing, transmission is reduced exponentially because the level width itself is exponentially small, reflecting the exponentially long time for the incident energy to propagate through the sample.

Expressing Eq. 15 in terms of the diffusion coefficient at the mobility edge, and using Eqs. 1 and 12, gives,

$$D_c(L) = \frac{L^2}{0.96(dN(L)/d\nu)} \, . \tag{16}$$

The density of states depends upon the frequency of the wave, the dimensions of the sample and the reflectivity at the boundary. For the case of EM radiation, the density of states which enters a three dimensional medium only on a single input face is,

$$\frac{dn}{d\nu} = \frac{1}{2^p}(8k^2/\pi v), \quad p = 0, 1, 2, \tag{17}$$

where $p$ is the number of pairs of reflecting boundaries. These pairs refer to opposing surfaces of a parallelpided. Reflecting boundaries are mirrors for EM radiation and the interface with air for electronic systems. For the case of a cubic sample without reflecting surfaces, i.e. $p = 0$, Eq. 16 becomes,

$$D_c(L) = (\pi v/15.36k^2)/L \, . \tag{18}$$

The above result for $D_c(L)$ may be compared to the prediction of the scaling theory,[3,10,47] which we apply to $\delta$, rather than to $g$. The effective diffusion coefficient for extended waves for a three dimensional sample is reduced by a factor, $l/\xi$, as a result of coherent backscattering,

$$D(L) = (vl/3)(l/\xi), \tag{19}$$

where $\xi$ is the correlation length, given by,

$$\frac{1}{\xi} = \frac{1}{\xi_0} + \frac{1}{L} + \frac{1}{L_i} \tag{20}$$

where,

$$\xi_0 = l^2/(l - l_c) , \tag{21}$$

is the coherence length in an infinite sample in the absence of inelastic processes. The expression for $\xi_0$ has the limiting behavior that $\xi_0 \to l$ for $l \gg l_c$ and $\xi_0$ diverges as $l$ approaches $l_c$ as $[(\lambda - \lambda_c)^{-1}]$ (Ref. 3). In the absence of inelastic scattering or absorption, $\xi = L$ at the mobility edge. The diffusion coefficient may then be written as,

$$D_c(L) = (vl/3)(l_c/L) = vl_c^2/3L . \tag{22}$$

Equating the expressions for $D(L)$ in Eqs. 18 and 22 gives,

$$kl_c = 1.1 . \tag{23}$$

A recent calculation by Soukoulis *et al.*[74] using the coherent potential approximation for similar conditions gives $kl_c = 0.84$. Of course both these predictions for the value of $kl_c$ are quite close to the conjecture of Ioffe and Regel, $kl_c = 1$. Ioffe and Regel suggested that when, $kl = 1$, the uncertainty in $k, 1/l$, appears to be larger than $k$ itself so that an extended travelling wave description is not appropriate and the wave is localized.[11]

The value of $kl_c$ is modified in the case of reflecting surfaces in accord with Eqs. 16 and 17. This is quite reasonable since the presence of the reflecting surfaces confines the wave and increases the probability that the wave will scatter back to a point. In the case of four reflecting side walls, $p = 2$ and $kl_c = 2.2$.

For the case of a sample which is scaled only in one dimension, so that the cross sectional area, $A$, is constant, Thouless pointed out that the wave will always be localized for sufficiently large $L$ (Ref. 8). Setting $\delta \nu$ equal to $d\nu/dN$ and using Eq. 13 for $\delta \nu$ gives,

$$L_0 = 0.96 A \frac{dn}{d\nu} D(L_0) , \tag{24}$$

for the length at which localization occurs. In this relation $D(L_0) < vl/3$ and depends upon $A$. Consequently, using Eq. 17, with $p = 2$ gives,

$$L_0 < 0.21 A k^2 l \; . \tag{25}$$

Localization of microwave radiation in a long reflecting tube should be observable, since in this case $Ak^2$ is approximately the number of speckle spots $N_s$, which can be quite small for tubes with diameters comparable to the wavelength.

The scaling function, $\delta(L)$, also determines the magnitude of fluctuations in the total transmitted intensity and the degree of correlation between speckle spots in the sample. This can be seen by noting that the number of eigenstates in the sum in Eq. 11 which contribute appreciably to the field at a given input frequency for the wave is just $(dN/d\nu)\delta\nu = \delta$. Thus the number of independent variables, $c_i$, needed to describe the wave in Eq. 11 is $N_{\mathrm{ind}} = \delta$, as stated in Eq. 2. This number, however, is smaller than the number of distinct transverse momentum states in the input or output wave, which is equal to the number of speckle spots in the transmitted wave,

$$N_s = 4k^2 L^2 / \pi 2^p \; . \tag{26}$$

Since $N_s > N_{\mathrm{ind}}$, for $L > l$, the intensity in different coherence areas of the transmitted speckle pattern is correlated. The degree of correlation in the transmitted wave may be defined as,

$$\eta = N_s / N_{\mathrm{ind}} \; . \tag{27}$$

This may be expressed as,

$$\eta(L) = vL/3.84 D(L) \; . \tag{28}$$

In the Boltzmann limit this gives,

$$\eta(L) = 0.78 L/l \tag{29}$$

These results suggest that the degree of wave correlation is inversely proportional to $\delta$ and proportional to the width of the transmitted pulse $\delta t$.

Since intensity correlation occurs because the intensity at one point influences the intensity at another, it can be characterized by the probability that a path traversing the medium passes through a typical coherence volume $V_c = L_c^d$, where $L_c \sim \lambda/4$. The average time spent per coherence volume is $t/N_c$, where $t \sim L^2/D$ is the transit time through the sample and $N_c \sim L^d/V_c$ is the number of coherence volumes in the sample. The travel time through a single coherence volume is $L_c/v$. The probability that a path will pass through a coherence volume $V_c$ is the dimensionless ratio,

$$\tau = (t/N_c)/(L_c/v) \sim (L^2/DN_c)/(L_c/v) \ . \tag{30}$$

However since $N_c \sim (dN/d\nu)\nu$, we have,

$$\tau \sim 1/\delta \ . \tag{31}$$

This provides the physical basis for the relationship of $1/\delta$ with the degree of long range correlation. At the mobility edge, $\delta = N_{\mathrm{ind}} = 1 \sim \tau$ and $\eta = N_s$, showing that, on the average, the wave visits each coherence volume of the sample once and the entire transmitted wave is correlated. This shows explicitly that localization is related to long range correlation.

The condition, $N_{\mathrm{ind}} = 1$, at the mobility edge implies that the correlation length $\xi$ equals the transverse dimensions $A^{1/2}$ at the output. This allows us to evaluate the localization length $L_0$ in a topologically one dimensional sample. The diffusion coefficient in Eq. 22 can be written as,

$$D(L_0) = (vl^2/3)A^{1/2} \ . \tag{32}$$

Substituting this into Eq. 24 and using Eq. 17 gives,

$$L_0 = 4A^{1/2}k^2l^2/3\pi \ . \tag{33}$$

In the weak scattering regime, $N_{\mathrm{ind}} \gg 1$, and we expect that long range intensity correlation will not appreciably influence the local intensity correlation function. Therefore $C_0^I(\Delta\nu)$ can be expressed as the complex amplitude square of $G^E(\Delta\nu)$. This is confirmed in optical[14] and microwave[26] experiments discussed in Sec. 6. We expect that the effect of long range intensity correlation upon the local intensity correlation function can be included in the weak scattering regime as an expansion in the small parameter $1/N_{\mathrm{ind}}$. This is confirmed by experiments and by calculations of Feng, Kane, Lee and Stone[13] in which $C_0^I(\Delta\nu = 0)$ is expressed as an expansion in $1/g$. As we have seen, $g = \delta = N_{\mathrm{ind}}$, in the weak scattering regime. These calculations are discussed in the chapter by S. Feng and the influence of long range correlation upon measurements of the intensity autocorrelation function and cross correlation function is discussed in Sec. 8 of this chapter.

Correlation among coherence areas of the transmitted signal is of importance even in the weak scattering regime. It is the source of universal conductance and transmission fluctuations. The conductance may be expressed by the generalized Landauer formula,[86,89,90]

$$G \sim (e^2/h) \sum_{\alpha,\beta} T_{\alpha\beta} \; , \tag{34}$$

in terms of the coupling coefficients of intensity in an incoming channel $\alpha$ of unit intensity to the outgoing channel $\beta$, $T_{\alpha\beta}$. The total optical transmittance[91] for unit intensity in each of the incoming channels can similarly be expressed as,

$$T = \sum_{\alpha,\beta} T_{\alpha\beta} \; . \tag{35}$$

We will assume the sides of the sample of length $L$ are reflecting and that incoming and outgoing waves as well are confined by reflecting surfaces to a cross sectional area $A = L^2$. This corresponds to the waveguide geometry. For the electrical case it corresponds to a resistor of cross section $A = L^2$ and length $L$ sandwiched between perfect conductors with the same cross sectional area. The incoming and outgoing channels may be transverse momentum states at a given energy, $c^2(k_x^2 + k_y^2 + k_z^2) = c^2(\pi/L)^2(n_x^2 + n_y^2 + n_z^2) = (2\pi\nu)^2$, where $c$ is the velocity outside the medium and the integers $n_i$ are the mode numbers, $n_i = k_i L/\pi$, with $k_z$ or $n_z$ positive and with two possible polarization states. For the waveguide geometry, a standing transverse wave exists and only positive values of $n_x$ and $n_y$ need be included. For samples for which the outgoing wave is not confined the outgoing channels correspond to polarized optical spots within the far field speckle pattern. The channels could also be defined in a spatial representation as the intensity within the coherence area, $A_c = L_c^2$, at the input or output surface of the sample for a given polarization. This is the near field speckle pattern and can be observed for light by imaging the output face of the sample with a microscope. The number of speckle spots may be written as, $N_s = 2A/A_c$ where the factor of 2 in the expression for $N_s$ accounts for the two polarization states for EM radiation (or spin states for electron waves) with given momentum. For waves that couple to the sample from an infinite half space, which is typical in optical experiments, the ratio $A/A_c$, can be written as the area of the half shell in number space with radius $n = kL/\pi$. This gives $A/A_c = 2\pi n^2 = \frac{2}{\pi} k^2 L^2$, and $N_s = \frac{4}{\pi} k^2 L^2$. On the other hand, for incident waves with "waveguide" boundary conditions, $N_s = k^2 L^2/\pi$, since waves with opposite transverse momenta are not counted separately, because they must occur together with equal amplitude to give zero flux at the transverse boundaries. These results are in accord with Eq. 26.

The average of total transmission $T$ for a given incident channel $\alpha$ of unit intensity for diffusive transport is, $\langle T_\alpha \rangle = \overline{\langle \Sigma_\beta T_{\alpha\beta} \rangle} \sim l/L$. Therefore, the average of $T_{\alpha\beta}$ for the $N_s$ outgoing states is,

$$\overline{\langle T_{\alpha\beta} \rangle} \sim l/LN_s \sim \langle T \rangle/N_s \, , \tag{36}$$

where the average is over all configurations and over all incoming and outgoing channels. Since the intensity in each polarized speckle spot is the square of a superposition of fields due to a distribution of Feynman paths with random phase, the average field value is zero, and we expect $\mathrm{var}(T_{\alpha\beta}) = \langle T_{\alpha\beta} \rangle^2$ in the weak scattering regime when $N_{\mathrm{ind}} \gg 1$ (Sec. 6). If we assume the intensity in each of the incoming and transmitted channels is uncorrelated, the fluctuations of the dimensionless conductance $G/(e^2/h)$, to which all input and output modes contribute would be $\mathrm{var}(g) = \mathrm{var}(T) = N_s^2 \overline{\mathrm{var}(T_{\alpha\beta})} = N_s^2 \overline{\langle T_{\alpha\beta} \rangle^2}$. Using Eq. 36 gives $\mathrm{var}(g) = \mathrm{var}(T) \sim l^2/L^2$. This is smaller than the universal value predicted by Lee and Stone[28] and by Altshuler,[29] which is near unity, by a factor of $l^2/L^2$. Universal conductance fluctuations which do not self average as the scale of the sample increases have been observed,[5,6] though an exact comparison with theoretical predictions is complicated by the thermal width in the electron energy distribution, inelastic scattering and the nature of coupling to the probe wires. Nonetheless, these observations demonstrate that coherence areas of the electronic wave are not independent, but rather are highly correlated, even in the limit $kl \gg 1$. Intensity correlation in the transmitted speckle pattern occurs because the total transmission depends upon $N_{\mathrm{ind}}$ rather than $N_s$ independent parameters for a single input channel and upon $N_{\mathrm{ind}}^2$ rather than $N_s^2$ independent parameters when all input channels are present.

The magnitude of fluctuations can be qualitatively evaluated by grouping $\eta$ correlated spots together. We first compute the variance of fluctuations in total transmission for a single input mode. We need to consider $N_{\mathrm{ind}} = N_s/\eta$ groups of independent speckle spots. The integrated intensity in $\eta$ correlated speckle spots is $\eta\overline{\langle T_{\alpha\beta}\rangle} = \eta(l/LN_s) = l/LN_{\mathrm{ind}}$. The variance of transmission fluctuations is, therefore, $\mathrm{var}(T) = N_{\mathrm{ind}}(l/LN_{\mathrm{ind}})^2 = l^2/L^2 N_{\mathrm{ind}} = \eta(l^2/L^2 N_s)$. This is a factor $\eta$ greater than the prediction for independent speckle spots. This factor is in agreement with Green's function calculations of Stephen[19] and of Feng, Lee, Kane and Stone.[20] For the case of transmittance or conductance, in which all modes are present at the input, there are $N_{\mathrm{ind}}$ independent groups at the output and an equal number of groups at the input which affect the output independently. When the fluctuations for all input and output modes are calculated, $\mathrm{var}(T)$ is increased by a factor of $\eta^2$ over the value predicted for independent speckle spots. This can be seen by noting that the integrated transmission in $\eta$ correlated spots on the output surface is $\eta l/LN_s$ for a single input mode. But input modes are correlated so that different input modes produce approximately the same intensity at the output. Consequently, these input modes must be included together as well. Thus the total correlated intensity for $\eta$ correlated input modes and $\eta$ correlated output modes is, $\eta^2 l/LN_s$. The variance in transmission is, therefore, $\mathrm{var}(T) = N_{\mathrm{ind}}^2(\eta^2 l/LN_s)^2 = \eta^2 l^2/L^2$. But from Eq. 29, this gives,

$$\mathrm{var}(T) \sim 1 \ . \tag{37}$$

This analysis explains qualitatively the physical origin of universal conductance fluctuations. A more quantitative calculation and justification of the arguments given here is given in Sec. 8.

Though the typical path length of reflected photons is much shorter than for transmitted photons, a high degree of correlation exists even within the reflected speckle pattern. Long range correlation is imposed on diffusely reflected light by the requirement of conservation of energy, that, in the absence of absorption, all the energy incident upon the sample is either reflected or transmitted. In terms of normalized quantities, $1 = T + R$. Thus, fluctuations from the average values of $R$ and $T$ due to a change in sample configuration or wave frequency are related,

$$\delta T = -\delta R \ . \tag{38}$$

Thus spectral fluctuations of the entire transmitted and reflected signal have the same magnitude and correlation frequency. Similarly, temporal fluctuations of the total reflected and transmitted signal in a sample with internal motion have the same correlation time. This is a consequence of long range correlation. We note further that the correlation frequency of the integrated reflected light differs from that for the intensity in a single reflected speckle spot, which is proportional to the inverse of the reflected pulse width, and is independent of sample thickness for $L \gg l$. On the other hand, the correlation frequency of the total reflected signal $R$ equals that of $T$ and depends upon $L$. Thus the correlation frequency in reflection depends upon the number of speckle spots averaged, indicating that reflected speckle spots are not statistically independent.

The universal magnitude of conductance fluctuations, which is obtained here by a consideration of the number of eigenstates which are required to accurately represent an arbitrary wave in a random medium, was first evaluated by Lee and Stone[28] and by Altshuler[29] using a perturbative calculation in the weak disorder limit. Recently Imry[31] has related conduction fluctuations to the properties of random matrices which characterize the coupling of input and output channels of a disordered medium.[32,92-93] The transfer matrices have statistical characteristics similar to ensembles of random matrices introduced by Wigner and Dyson. Such ensembles have been useful in studying the distributions of nuclear energy levels and mechanics of classically chaotic systems. Imry has shown that the effective number of active transmission channels is associated with eigenvalues which are closest to unity. This number is equal to $N_{ind}$ (Ref. 31).

## 3. COMPARISON OF ELECTRON AND PHOTON PROPAGATION IN DISORDERED MEDIA

The concept of localization was first advanced by Anderson to explain the absence of electron diffusion in glasses. Localization probably occurs for phonons in amorphous solids as well. It is very likely the source of the plateau in thermal conductivity as a function of temperature of glasses[95,96] and can be discerned in computer simulations of the density of phonon states in amorphous semiconductors.[97] Phonon localization is particularly likely to occur in samples of reduced dimensions and in samples in which the random coupling between atoms is highly anisotropic such that the underlying vibrational excitations are fractals.[98,99] The prospect of photon localization has been proposed relatively recently. Interest in this field has been stimulated as much by the similarities between quantum and classical wave propagation, as demonstrated in the observation of weak localization in backscattered light, as by differences in length scale over which electronic and optical coherence is manifest and in experimental approach.

The scaling theory of localization describes wave propagation in random media in which the quanta of the field do not mutually interact. The theory thus applies properly to the propagation of photons while its application to electron transport requires substantial modification.

The nature of the interactions and statistics that determine $l$ as well as the experimental possibilities and limitations in electronically and optically disordered samples are very different. Electronic conductivity experiments represent an average over energy and momentum states in a Fermi distribution as opposed to optical experiments, in which a coherent boson field makes possible the measurement of coupling between specific incoming and outgoing modes. Also the coupling of transverse momentum states to a sample is well defined for EM waves but difficult to characterize in electron conductivity experiments. The dependence of conductivity fluctuation upon the configuration, geometry and resistance of the leads is not fully understood.

The tunability of lasers and microwave generators makes it possible to study the Anderson transition as a function of the injected photon energy. Critical behavior in the transmission and in the correlation length, $\xi \sim |\nu - \nu_c|^{-1}$, where $\nu_c$ is the frequency of the mobility edge, can thus be studied by tuning the probe frequency. The study of the scale and frequency dependence of EM transmission near the mobility edge should make it possible to test fundamental concepts and measure key parameters in the scaling theory of localization.

The scale dependence of conductivity in the localized regime has been studied using thin films.[76,77] Critical exponents of the conductivity near the electronic mobility edge have been measured in a series of experiments by Hesse, De Conde, Rosenbaum and Thomas in which the phosphorous concentration in silicon is varied.[100,101] The transition from localized to delocalized electronic states is charted by varying phosphorous concentration in a series of samples. The critical exponent observed in the conductivity as a function of phosphorous concentration is very likely influenced by electron-electron interactions, which are not included in the scaling theory. The critical exponent is 1/2 in these experiments but is 1 in studies using charge compensated dopants.

Inelastic scattering of electrons by phonons leads to inelastic diffusion lengths, $L_i$, that are on the order of $10^3$A even for weakly scattering samples at cryogenic temperatures. This makes possible the measurement of the electron dephasing time $\tau_i$, but often restricts studies to microscopic samples. In contrast, optical and microwave propagation can be studied in well characterized macroscopic samples of continuously variable length. The most important inelastic process for electromagnetic propagation in static samples is absorption. For diffusing colloidal or particulate samples, wave coherence is also destroyed by motion in the sample. The analysis of fluctuations in intensity of multiply scattered light or microwave radiation makes possible the measurement of the particle velocity and diffusion coefficient.

The diffusion of waves in macroscopic samples leads to long transit times which can be resolved with picosecond laser or pulsed microwave techniques. Time resolved measurements should display a dramatic lengthening of the transport time as the length increases. For diffusive transport, in the absence of absorbtion the typical transport time is $t \sim L^2/D$, whereas, for scale dependent transport, we expect $t \sim L^2/D(L)$. For $L < \xi$, $D(L) \sim vl^2/3L$ and the transit time should increase as $L^3$ (Ref. 47). For a localized wave the transit time should increase essentially exponentially with $L$.

That major differences between electron and photon propagation exist is clear from the fact that electron localization is ubiquitous in disordered solids while there are only a few recent reports of propagation in the strong scattering regime for photons in disordered three dimensional samples.[44,45] The electron is more readily localized than light because its energy eigenvalue can be negative leading to exponentially decaying wave function in parts of the sample, whereas the photon energy is always positive. For electrons, low energy states are more readily localized, whereas for photons the strength of scattering is related to $(\nu^2/c^2)\varepsilon_1(\mathbf{r})$, where $\varepsilon_1(\mathbf{r})$ is the fluctuation about the mean of the dielectric function, and the scattering strength consequently, decreases as the photon energy decreases. The low and high frequency limits correspond to Rayleigh scattering and geometric optics respectively in which states are extended. Thus, for photons in a highly disordered sample, localization is expected to appear in a narrow wavelength window between extended states.[9] Sheng and Zhang[72] have suggested that localization can occur near the second, and third higher Mie resonances of individual scatterers so that a series of localization windows may exist.[74] This situation may be altered in nearly periodic structures in which a photonic bandgap could appear.[47] In this case photons would be localized in the tails of the band in a manner reminiscent of electron localization.[35,102]

Strong scattering exists for electrons because $s$ wave scattering is important whereas $p$ wave scattering is the lowest order for photons.

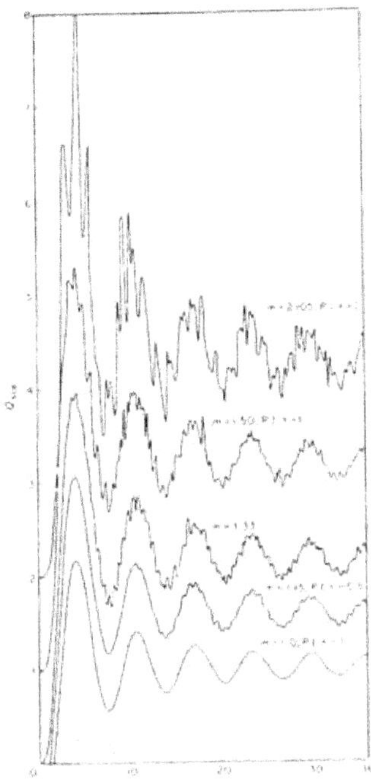

Fig. 7. Efficiency for scattering $Q_{sca}$ plotted against the parameter $\rho = 2\alpha (m-1)$ for various values of the index contrast $m$. *PI* denotes the plotting increments. This figure is taken from *The Scattering of Light and Other Electromagnetic Radiation* by M. Kirker (Academic Press, New York, 1969).

Consequently, the ratio of the cross section for scattering of photons out of the incident beam direction to the geometrical cross section, $Q_{sca}$, rises rapidly with radius, $a$, of scattering spheres. A plot of, $Q_{sca}$, versus additional phase shift, $\rho$, accumulated by the wave in passing through the center of the sphere as calculated by M. Kirker is shown in Fig. 7 (Ref. 103). It reaches a maximum at $\rho \sim 5$ corresponding to a radius, $a \sim 5\lambda/4\pi(m - 1)$, for isolated spheres, where $m$ is the ratio of the index of refraction of the spheres to that of the background. Since the maximum value of $Q_{sca}$ is $\sim 6$ for index contrast achievable with visible radiation, and maximum scattering is predicted to occur with filling fractions between 0.2 and 0.4, it is not immediately apparent whether or not the Ioffe-Regel condition for localization can be met for random scattering of visible light. This is particularly true since small angle scattering, which contributes to $Q_{sca}$, has little effect on $l$.

The strength of scattering in the visible is also reduced because peaks in the scattering cross section are likely to be washed out by polydispersity of the scattering particles and the random configuration of the spheres. This is supported by measurements of R. B. Stephens *et al.*[104] of scattering from samples composed of increasing numbers of random close-packed monolayers of monodisperse polystyrene spheres with $2\mu$m diameter. Measurements for a single random monolayer are shown in Fig. 8a. Direct transmission along the incident direction shows strong Mie resonances consistent with Fig. 7. Total diffuse reflection and transmission are measured with an integrating sphere placed on the incident and transmitting side of the sample respectively. Sharp resonances are not seen. The direct, unscattered, transmission is barely measurable for two monolayers as shown in Fig. 8b. Resonances are not observed in the total reflected and transmitted light and their sum which is denoted as total emitted light is equal to 1, the incident intensity, within experimental uncertainty. The direct transmission could not be measured for three monolayers but the flat transmission spectrum shown in Fig. 8c drops in intensity as the diffuse reflection rises in comparison to the results for two monolayers.

Recent experiments[44,45] and calculations[71-74] indicate that optical localization can be achieved in random three dimensional media. It is likely that larger EM scattering can be achieved than is predicted by extrapolating the Mie scattering approximation as a result of collective scattering related to sample anisotropy. The wave can be severely confined in either the low or high index regions of the sample or completely excluded from metallic parts of the sample. Strong scattering related to waveguide cut-off occurs when the wave is confined to regions of dimensions $\sim \lambda$. In addition, in nearly periodic structures, collective scattering leads to optical pseudogaps in which the density of states is low and localization is facilitated.[47] It should be noted that going beyond the Ioffe-Regel limit is more readily achieved in the microwave frequency range where the dielectric function, $\varepsilon$, can be much larger than in the visible.

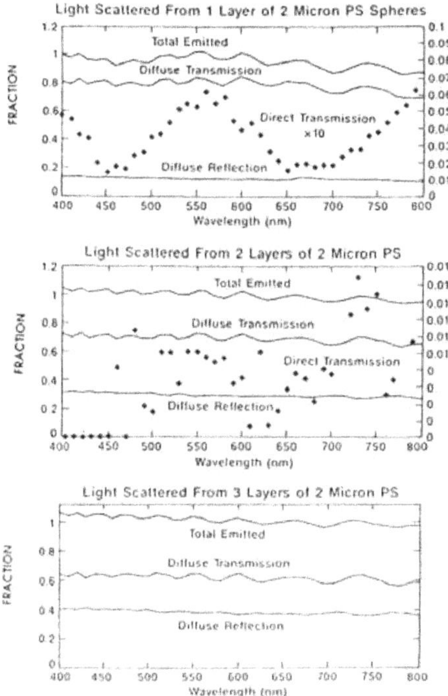


Fig. 8. Light scattered from one, two and three random monolayers of polystyrene spheres with diameter $d = 2\mu$m. The fraction of diffuse reflection and transmission are given by the scale on the left. The optical density of direct transmission along the incident beam direction is given by the scale on the right. The total reflected and transmitted radiation is denoted total emitted.

Studies of EM propagation will be guided by concepts developed to describe electron transport, and universal aspects of propagation may be critically tested. However, significant difference in the nature of electron and photon interactions may be expected to reveal new phenomena.

## 4. EXPERIMENTAL APPROACH: METHODS, SAMPLES AND APPLICATIONS

Multiple scattering of light in disordered media dramatically broadens the distribution of transit times, reduces the average transmitted intensity, and gives rise to long range spatial and spectral intensity correlation. The spatial, temporal and spectral resolution achievable with lasers and microwave sources makes them ideal tools with which to study universal aspects of wave propagation. The characterization of EM propagation in a variety of environments is of practical importance in itself and as a diagnostic of the time average and fluctuating properties of a medium such as the atmosphere.

In the diffusive regime, effects of scattering and absorption can be disentangled by measuring the scale dependence of EM transmission. Here, we will focus primarily on the measurements of the following quantities: (1) total transmission, which is normalized to the incident intensity, $T(L)$, (2) the temporal profile of the transmitted pulse, $T(t; L)$, (3) intensity fluctuations at points in the transmitted wave as the source frequency is tuned for a static sample, $I(\nu; L)$, and (4) intensity fluctuations in time in a single coherence area of a sample irradiated with monochromatic radiation, $I(t; L)$. The measurement of $T(L)$ gives $l$ and $l_a = v\tau_a$. For $L > L_a = (D\tau_a)^{1/2}$, we find $T(L)$ falls exponentially with decay length $L_a$, (Ref. 14), which may also be expressed as $L_a = (ll_a/3)^{1/2}$, in the Boltzmann limit. For $L < L_a, T(L) \sim l/L$ (Refs. 14,105). The quantity $I(t; L)$ or the corresponding intensity autocorrelation function with time, $G^I(\tau; L)$, are measures of the sensitivity of the transmitted intensity to the motion of the sample. If the motion of the sample is known in a statistical sense, then the spatial wave propagation parameters $l$ and $l_a$ can be determined from the measurement of $G^I(\tau; L)$. On the other hand, once $l$ and $l_a$, are known the path length distribution for photons is determined and the statistical properties of the internal motion can be determined from $G^{(2)}(\tau; L)$. Measurements of $T(t; L)$ and $I(\nu; L)$, in the time and frequency domain respectively, can be used to obtain $D$ and $\tau_a$ from a fit to a model of photon diffusion. The average velocity of propagation can be determined from the ratio of $D$ to $l$ or of $l_a$ to $\tau_a$. In addition the cross correlation function of intensity in the sample is a measure of the degree of correlation and the proximity to the Anderson localization threshold.

In the strong scattering limit, the effect of wave interference upon $\delta t$ or $\delta\nu$ can be described in terms of the scale-dependent diffusion coefficient, $D(L)$. A central issue in the study of photon localization is to find the form of $D(L)$ and its relationship to propagation in the space, time, and frequency domains. This can be determined from measurements of the scale dependence of transmission. John has predicted that optical transmission falls essentially exponentially with $L$ for $L > L_a$ with an absorption coefficient $\alpha = 1/L_a = (1/D(L)\tau_a)^{1/2}$ (Ref. 9). Anderson has made the conjecture that transmission over all length scales can be renormalized by substituting $D(L)$ for $D$ in the classical expression for transmission.[46] We have suggested that the substitution of $D(L)$ for $D$ in expressions for transmission in the frequency domain is the essence of scaling. Experimental support for the extension of scaling ideas of conductance to optical transmission is given in Sec. 9. Though the scaling of $\delta t$ or $\delta \nu$ would be known once the universal scaling function for $\delta$ and the density of states are known, the full temporal distribution is not known. Furthermore, enormous fluctuations occur in the temporal distribution between different samples in the strong scattering regime so that the ensemble average of the temporal distribution does not give an adequate understanding of tranport. Similarly, enormous fluctuations occur in the field or intensity spectrum at a point in the transmitted wave. Understanding the statistical character of the full distribution of the temporal and spectral distribution is the essence of the Anderson transition. Exploring these issues remains a challenge to both experimentalists and theorists.

We will discuss the use of the four types of measurements mentioned above to characterize optical and microwave propagation. Though most of the interest in and potential application of studies of photon propagation and localization is for light, the study of wave propagation is greatly facilitated by microwave studies because (1) the wavelength is long, (2) metallic reflectivity is high, and (3) samples with high dielectric constant and low loss are available. Since scattering is maximized when the scale of the structures comprising the sample are of the order of $\lambda$, typical structures in microwave studies are of the order of 1 cm. It is therefore easy to produce samples with a wide range of volume fraction of high index structures and to measure transmission even in samples in which the thickness is less than $l$. Of course, there is an associated disadvantage, that three dimensional samples for which all dimensions are much greater than $l$ are sometimes unpractically large. The high microwave reflectivity of metals makes it possible to study metallic samples and to confine the wave within a cavity. Long samples, which approach being one dimensional with respect to wave correlation in the transverse direction or flat two dimensional samples bounded by reflecting plates can be produced. Strong scattering may be produced in dielectric samples since the dielectric constant, $\varepsilon = n^2$ can be as large as 100 for titania or 9 for alumina, which is more readily obtainable as compact solid particles.

In microwave studies at moderate powers, the signal is measured in a single coherence area using a diode detector. It is, difficult, therefore, to measure the total transmitted intensity or the integrated intensity in an area greater than a speckle spot. In contrast, in optical studies, any portion of the total transmitted light can be detected by using an aperture and focussing the light on a photodiode or photomultiplier tube (PMT).

In this chapter, we will emphasize measurements of the scale dependence of properties of the transmitted intensity. In the case of rigid optical samples we use a wedge-shaped sample and select a specific value of $L$ by translating the point of focus of the beam on the sample by moving the sample. The temporal profile of the pulse averaged over many speckle spots is measured by time-correlated single photon counting. Local intensity fluctuations with frequency can be measured in a single speckle spot and averaging is achieved by analyzing data in many different speckle spots. The experimental setup used to measure optical transmission is shown in Fig. 9. An argon-ion laser pumps a Coherent Radiation, 599-21 single-frequency dye laser which can be scanned continuously over 30 GHz. The laser beam passes through a Coherent Associates electro-optic intensity stabilizer and approximately 20 mW of laser power is focused to a 12 $\mu$m diameter spot on the wedged sample. The transmitted intensity is a stable speckle pattern at a fixed laser frequency which fluctuates in a reproducible manner as the laser frequency is scanned under computer control. The speckle pattern also changes as temperatures in the room changes by several degrees. The sensitivity to temperature change increases as the sample thickness increases. To observe high contrast speckle spectra, $I(\nu, L)$, as the laser is scanned with $L$ fixed, a single coherence area is selected with an aperture, and a single polarization component is chosen with a polarizer. The spectra of random intensity fluctuations are detected in a PMT and displayed on a recorder as shown in Fig. 9. The signal is normalized to the incident intensity, which is monitored by a photodiode, and stored in a computer. To measure the transmission normalized to the incident intensity the transmitted light is passed through a large-area opal diffuser placed immediately after the sample and before the PMT so that the incident and transmitted light will have the same angular distribution when they are detected and can be compared. The sample thickness is varied by moving the sample with a computer controlled translation stage. For large values of $L$, the effect of light scattered around the edges of the sample is eliminated by imaging of the transmitted spot through an aperture before it is detected. Since the dependence of transmission on $L$ for the two measurement configurations is the same for intermediate values of $L$, the data sets can be spliced together to yield $T(L)$. In the absence of absorption, the transmitted beam has transverse dimensions $\sim L$, so the wedge angle must be small for $L$ to be well defined. In the presence of absorption, the dependence of the intensity of the transmitted light from the center of the spot falls even more rapidly. Thus, $l$ and $l_a$ can alternatively be determined from measurements of the transverse intensity profile of light transmitted through a slab.

Many of the results reported here are for the wedged sample shown in Fig. 10. It is a 40%–60% mixture, by volume, of rutile $TiO_2$ microparticles with $n \sim 2.7$ embedded in a polystyrene matrix with $n = 1.59$. The size distribution of the $TiO_2$ powder supplied by du Pont de Nemours and Company is shown in the electromicrograph of a cut surface in Fig. 10. The sample is produced by mixing the titania powder with a colloid of $0.091\mu$m polystyrene spheres, in water, provided by Dow Chemical. After being stirred for some time, the mixture undergoes a sudden exothermic transformation to a rich lather. This may correspond to the point at which a layer of polystyrene spheres coats the oppositely-charged $TiO_2$ particles. This produces a structure in which $TiO_2$ particles do not touch. The mixture is allowed to dry

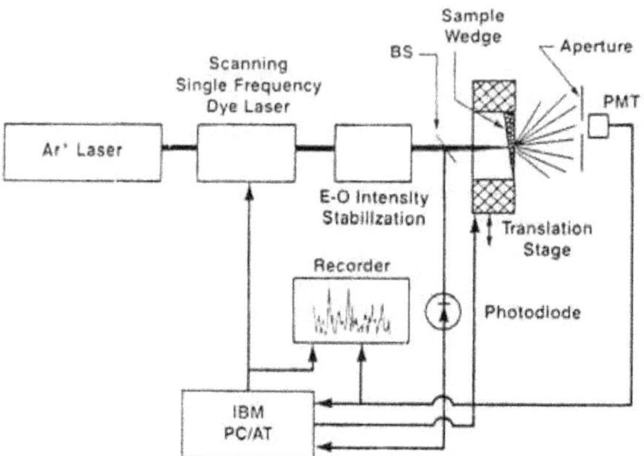


Fig. 9. Experimental setup for observing intensity fluctuations in a wedge shaped sample.

and is then pressed into a pellet. It is embedded at an angle into a Lucite block at a temperature which melts the polystyrene. The block is then polished and sawed in half to display the wedge of 0.075 rad, shown in the photograph of Fig. 10a. A photograph of the flat side of the wedge is shown in Fig. 10b. It shows that the edge is ragged with numerous pits in the surface. The electromicrograph of the tip of the wedge shown in Fig. 10c shows that the wedge starts at a thickness of approximately $5\mu$m.

For colloidal samples or other samples with moving constituents in which multiple scattering occurs, the fluctuations of reflected or transmitted intensity in time depends upon both the characteristics of light propagation and of the motion of the sample. Either can be determined once the other is known. Maret and Wolf[37] and others[38-43] have shown that the long photon pathlength in a randomly moving medium enhances the sensitivity of optical intensity to sample motion over that in samples in which the light suffers only a single scattering event.[106] Theoretical and experimental work show that photon correlation spectroscopy in the multiple scattering regime can be used to probe particle diffusion in colloidal samples. This is discussed in detail in the chapter by Pine *et al.* in this volume. Related fluctuations of microwave intensity for macroscopic particles can be used to probe

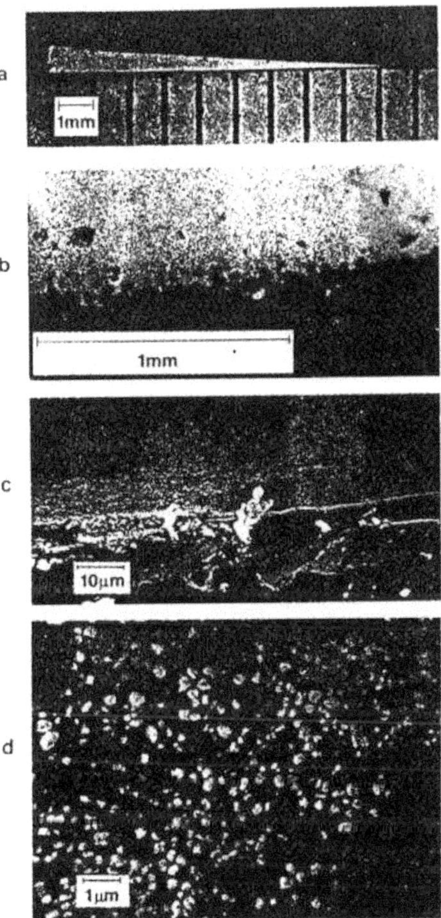

Fig. 10. Sample of random $TiO_2$ particles (40% by volume) embedded in polystyrene. (a) Photograph of wedge. (b) Photograph of face of sample. (c) Electromicrograph of tip of wedge which has been sliced with a diamond saw. Debris from cutting is seen. (d) Electromicrograph of wedge showing titania particles in the polystyrene matrix.

random motion on a time scale in which the particle motion is ballistic.

Microwave experiments are performed using the sample holder shown in Fig. 11. The sample is tumbled by rotating the holder about its axis and a stationary string may be strung through the sample to facilitate vigorous mixing so that a configuration average can be rapidly obtained by averaging over time. The mixing can be stopped and the properties of a particular sample can be studied by scanning the microwave frequency or moving the Schottky diode detector. The detector is mounted on a plunger at the output side of the cylinder. As the plunger moves out, the cylinder rotates while sample material falls from the hopper. The configuration average of transmission as a function of sample thickness, $T(L)$, is obtained by averaging the intensity detected in a single coherence area. Temporal fluctuations of the intensity can be analyzed to characterize the internal motion of the particles.

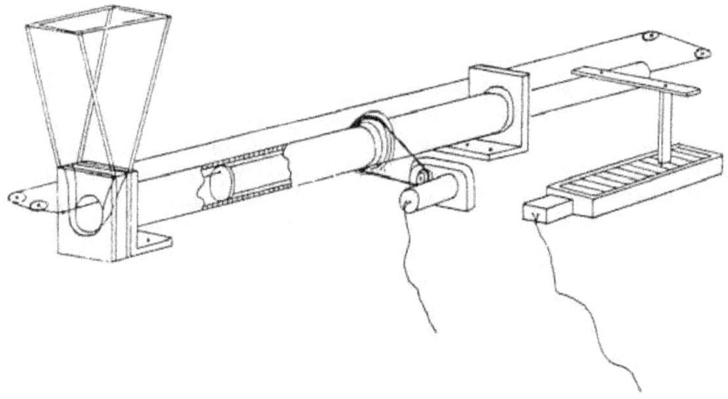

Fig. 11. Sample holder for microwave experiments. The cyclindrical copper sample holder is rotated by a computer controlled motor. The thickness of the sample is determined by the computer controlled translation stage. Material is fed into the sample chamber from the hopper. The string through the sample facilitates mixing.

Among samples in which photon localization is most likely to be observed are the following (1) Samples with large $n$ including $TiO_2$ with $n \sim 2.2$–$2.7$ in the visible and sub bandgap radiation in the infrared for Si and Ge with $n = 3.4$ and 4 respectively. In the microwave range $n = 3$ for alumina and $n \sim 10$ for titania. (2) Metallic samples with negative $\varepsilon$ and low loss at microwave frequencies. (3) Samples which are random in one[107–110] or two[64] dimensions. This can be achieved in one dimension, for example, with layered samples for visible or microwave radiation or by using metallic sample holders with dimensions less than $\lambda$. (4) Samples in which the radiation is confined by reflecting walls so as to increase the degree of coherence within the sample and hence facilitate localization. Such confinement can be achieved electronically in long wires[8] or thin films[76] and can also be accomplished for microwave radiation by containing long[16] or flat samples between metallic reflectors. And finally (5) samples consisting of nearly periodic dielectric structures.[47,111,112] The presence of short range order plays a key role in localizing electrons in amorphous semiconductors[113] in which localized states exist in gaps in the presence of disorder and are separated from extended states by a mobility edge. For sufficiently high dielectric contrast a photonic bandgap can exist in analogy with electronic bandgaps. For the face centered cubic crystal structure with $\sim 20\%$ solid fraction in air, a photonic bandgap can exist with dielectric constant $\varepsilon \sim 2.1$ for the solid.[112] Colloidal particles of titania may be suitable samples in which to observe such effects in the visible. Yablonovitch has observed a photonic bandgap exists at microwave frequencies with an index contrast of 3 (Ref. 114).

The density of states in a perfect crystal drops sharply at the band edge. Localization will then occur even in the case of weak disorder since the density of states becomes arbitrarily small. A tail of localized states will then extend into the gap which becomes a pseudogap separated from extended states by the mobility edge. In this case, the Ioffe-Regel criterion, $kl_c \sim 1$, in which $k$ corresponds to the Bloch wave vector and $l$ to the mean free path in which the Bloch wave vector is disrupted by Bragg scattering from disorder, is readily satisfied.[47]

Aside from providing a pathway to localization in the case of small departures from perfect order, macro-crystalline dielectric structures are potentially of practical importance. Yablonovitch has suggested that spontaneous emission can be quenched while lasing may exist in a single propagating mode inside the sample as a result of phase slip in the medium.[46] The suppression of spontaneous emission would enhance energy stored in the medium and could lead to more efficient semiconductor lasers and solar cells.[46] Since emission is from the band edge it is only necessary that the photonic band gap overlap the band edge. Thus even a narrow EM bandgap centered at the electronic band edge can have a significant effect on semiconductor optical devices. The suppression of the photonic density of states can also suppress resonant dipole-dipole interactions.[48] As a result donor-acceptor energy transfer, collisional dynamics, molecular spectra and dissociation energies in molecular systems having strong optical transitions within a photonic bandgap can be profoundly modified.

## 5. PHOTON DIFFUSION

In this section we will discuss measurements of the scale dependence of transmission, $T(L)$, and of the distribution of photon transit times, $T(t; L)$, in the weak scattering regime. The measurement of $T(L)$ over a range of $L$ or of $T(t; L)$ over a range of $t$ for fixed $L$ will be used to unscramble the effects of scattering and absorption. Measurements of $T(L)$ give $l$ and $L_a$, whereas measurements of $T(t; L)$ give $D$ and $\tau_a$. Measurements of the diffusion coefficient of optical and microwave radiation in random media allow us to assess the prospects for localizing photons in three dimensions.

In the weak scattering regime, the probability of a partial wave scattering back to a point in the medium is small. Therefore, to lowest order in the scattering strength, $1/kl$, consideration of the phasing of waves can be neglected in calculations of average transport. $T(L)$ and $T(t; L)$ can, therefore, be calculated using a statistical model for the random walk of photons[14,15,36,105,115] or by solving the transport equation.[116,117] In the bulk of the material, the velocity distribution is isotropic and the diffusion equation can be applied. Within a mean free path of the sample surface, however, the diffusion equation does not accurately represent propagation. As a consequence, for example, the angular distribution of light emerging from the sample is not accurately given by the diffusion model but depends upon details of the scattering near the surface.[42,64,105,116,117] Nevertheless, when $L \gg l, T(L)$ and $T(t; L)$ can be calculated fairly accurately up to a multiplicative constant by solving the diffusion equation. We will first consider the diffusion model for propagation and then show that, up to a multiplicative constant which can be determined from transport theory, it adequately describes optical and microwave transmission in the weak scattering limit.

In the interior of a slab, the photon current density, for uniform illumination at the input face, is given by

$$j(z,t) = -D\frac{\partial n(z,t)}{\partial z} \ , \tag{39}$$

where $n$ is the diffuse photon density with isotropic velocity distribution. For a thick slab with $L \gg l$, the maximum diffuse photon density, $n'$, due to a collimated beam of photon flux $j_{in}$ incident normal to the surface at $x = 0$ occurs within a few mean free paths of the surface and has a magnitude calculated from transport theory of $n' = 5j_{in}/v$. When $L \gg l$, the steady-state current density transmitted through the slab $j_{trans}(L)$ can be closely approximated by solving the diffusion equation, (Eq. 3) with the boundary conditions $n(0) = n'$ and $n(L) = 0$. This gives $n(x) = n' \sinh[\alpha(L - x)]/\sinh[\alpha L]$ with $\alpha = L_a^{-1} = (D\tau_a)^{-1/2}$. Evaluating the current in the limit $x \to L$ gives $j_{trans}(L) = n' \sinh(\alpha D)/\sinh(\alpha L)$. The normalized transmission, $T(L) = j_{trans}(L)/j_{in}$, can therefore be written as

$$T(L) = 5\sinh(\alpha D/v)/\sinh(\alpha L) = \begin{cases} 5D/vL\,, & L < L_a \\ (10\alpha D/v)\exp(-\alpha L)\,, & L > L_a \ . \end{cases} \tag{40}$$

Using the relation $D = vl/3$, $T(L)$ can be simply expressed when absorption is negligible as,

$$T(L) = 5l/3L\,, \quad L < L_a \ . \tag{41}$$

The distribution of photon transit time is proportional to the path length distribution, such that for pathlength $s = vt$

$$P(s) = T(t)/v \ . \tag{42}$$

$T(t)$ can be obtained by considering the first passage of a random walk through a sample of thickness $L$ without a zero crossing[115] or by solving the diffusion equation with the boundary conditions that the photon density is zero at the slab interfaces. This can be simulated by the method of images in which positive sources which are delta functions

in space and time are placed at $z' = a + 2Ln$ and equal and opposite
negative sources are placed at $z' = -a + 2Ln$, where $n$ ranges over
all the integers and $a \ll L$. This is illustrated in Fig. 12. Thus, we
compute the response to the incident pulse,

$$I_{\text{in}}(\mathbf{r}',t') = \delta(t')\delta(x')\delta(y') \sum_{n=-\infty}^{\infty} \delta(z'-a-2Ln) - \delta(z'+a-2Ln) \ . \quad (43)$$

We calculate the distribution at $L - a$, which is essentially the distribution of transmitted light. Since the detected light is in the far field, it
emerges from all points on the output face. The measured response is,
therefore obtained by integrating Eq. 43 over the transverse dimensions
in the plane at $z = L - a$. This gives,

$$T(t; L) = \int dx dy \int d^3 r' dt' G_0(\mathbf{r} - \mathbf{r}', t - t') I_{\text{in}}(\mathbf{r}',t') \ , \quad (44)$$

where,

$$G_0(\mathbf{r}, t) = \frac{1}{(4\pi Dt)^{3/2}} \exp(-\mathbf{r}^2/4Dt) \ , \quad (45)$$

is the Green's function in an infinite medium.
This gives,

$$T(t; L) = \frac{e^{-t/\tau_a}}{(4\pi Dt)^{1/2}} \sum_{n=-\infty}^{\infty} \left\{ e^{-[(2n-1)L-2a]^2/4Dt} \right.$$
$$\left. - e^{-[(2n-1)L]^2/4Dt} \right\} \ . \quad (46)$$

Beyond the peak in the transmitted pulse, the pulse decays nearly exponentially. The tail of the pulse exhibits a single exponential decay
with rate $1/\tau$ which is the lowest eigenvalue of the diffusion equation in
the presence of absorption,

$$1/\tau = 1/\tau_a + \pi^2 D/L^2 \ . \quad (47)$$

The total transmission versus thickness can be obtained by integrating Eq. 46 over time to give,

$$T(L) = \sinh(\alpha a)/\sinh(\alpha L) \ . \quad (48)$$

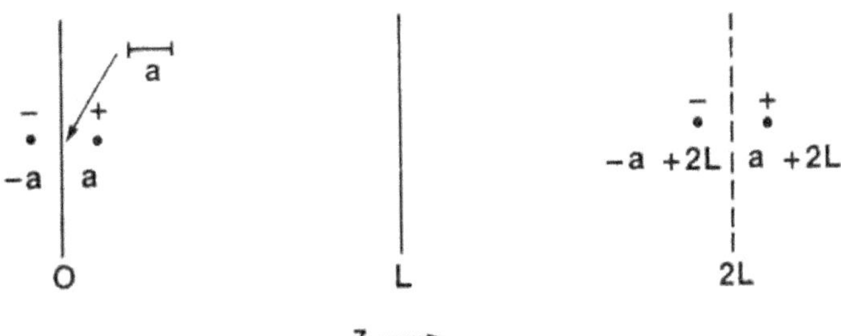



Fig. 12. The random sample is between the planes at $z = 0$ and $z = L$. It is excited by a pulse which is randomly scattered at $z' = a$. Some of the image charges introduced to satisfy the boundary conditions that the intensity at the sample boundary is zero are shown.

This result is the same as the solution of the steady state diffusion equation for diffuse incident radiation when we make the identification $a = 4l/3$. The solution of the transport equation for the case of normal incident radiation (Eq. 40) is obtained by taking $a = 5l/3$.

The measurement of $T(L)$ is performed as described in Sec. 4. Measurements in the sample shown in Fig. 10 are presented in Fig. 13 together with a fit of Eq. 40 to the data expected for a smooth wedge.[14] Jumps in transmission are caused by pits in an otherwise smooth surface. The lower envelope of the data, therefore, corresponds to transmission in a flat wedge which is found to start at $L = 6$ $\mu$m. The log-log plot of $T(L)$ in Fig. 13a displays an initial inverse dependence upon $L$ which is the signature of diffusion. The value of $T(L)$ in this region gives $l = 1.4$ $\mu$m. The semilog plot of the same data in Fig. 13b shows the eventual exponential decay of transmission due to absorption with decay length $L_a = 112 \pm 5$ $\mu$m. For $L > L_a$, the decay of $T(L)$ is essentially exponential. The actual path length of photon absorption in the sample, $l_a = v\tau_a$ in the Boltzmann diffusion limit is obtained from the relation $L_a^2 = D\tau_a = ll_a/3$,

$$l_a = 3L_a^2/l \; . \tag{49}$$

This gives $l_a = 2.6$ cm in this sample. This is shorter than the intrinsic absorption length of either polystyrene or titania and is likely caused by absorption by molecules adsorbed on the titania or polystyrene particles

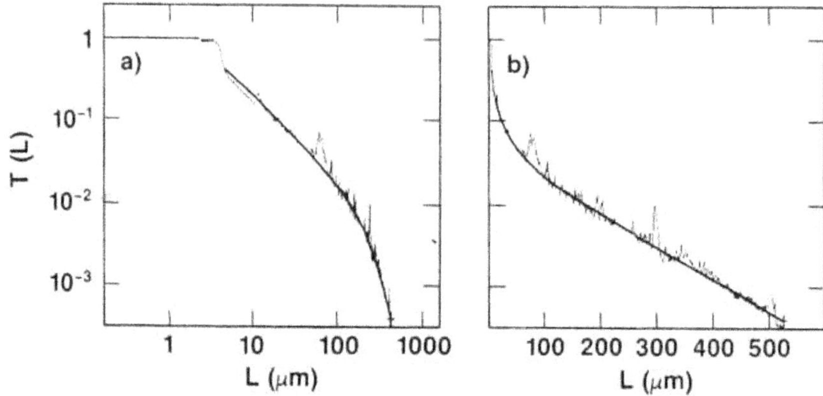

Fig. 13. (a) Log-log and (b) semilog plot of normalized transmission at $\lambda =$ 589 nm in the titania-polystyrene sample shown in Fig. 10. The solid line is a fit of the diffusion model (Eq. 40) to the lower envelope of the data.

from which the sample is produced or by surface oxidation states of titania.

In Fig. 14 we show a measurement of transmission versus thickness for a sample which is a mixture of equal volumes of 0.5 $\mu$m and 0.8 $\mu$m polystyrene spheres in a glass wedge. Two sphere sizes were mixed to ensure that the sample would be random. Measurements of intensity fluctuations in this sample were shown in Fig. 2. The sample thickness is only known reliably for $L > 60$ $\mu$m. From the initial fall off of $T(L)$ as $1/L$ in the regimes $L < L_a$, we find $l = 1.5$ $\mu$m.

Optical transmission has also been measured in a wedged titania sample. The sample is produced from a sintered 99.999% $TiO_2$ sputtering target supplied by International Advanced Materials. After additional sintering the sample has a porosity of 30% and an average particle size of 0.3 $\mu$m. The data is shown in Fig. 15. The data in the region in which $T(L)$ is inversely proportional to $L$ gives $l = 0.55 \pm 0.2$ $\mu$m.

Measurements of transmission at microwave frequencies are performed using the sample holder shown in Fig. 11 (Ref. 16). Approximately 20 mW of radiation is produced by an Alfred oscillator whose frequency can be controlled by an external voltage over several frequency ranges from 7 to 26.5 GHz. The wave is launched by a horn in each of the frequency ranges. The sample is set back sufficiently far

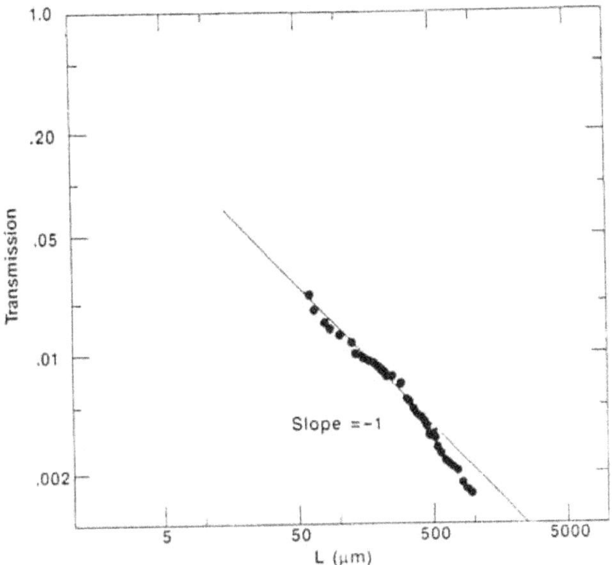


Fig. 14. Log-log plot of normalized transmission at $\lambda = 589$ nm in a wedged sample which is composed of equal volumes of 0.5 $\mu$m and 0.8 $\mu$m polystyrene spheres.

that the wave has uniform intensity within $\pm 10\%$ over the three inch diameter input face of the sample.

The signal is proportional to the intensity of radiation polarized along the direction of a probe wire of length $\lesssim \lambda/4$ attached to a Schottky diode. The signal detected is proportional to the intensity at the junction of the probe wire and Schottky diode. The signal across the diode fluctuates rapidly as the sample is tumbled. The integrated signal is an average over all configurations of the sample and is proportional to the ensemble average of the total transmitted intensity with the selected polarization. A signal proportional to the total radiation which can be compared to Eqs. 40 or 48 can be obtained by detecting radiation with field polarized at 45° to the direction of polarization of the incident wave. In this case, equal average intensity is detected along the incident polarization direction and perpendicular to it for all values of $L$.

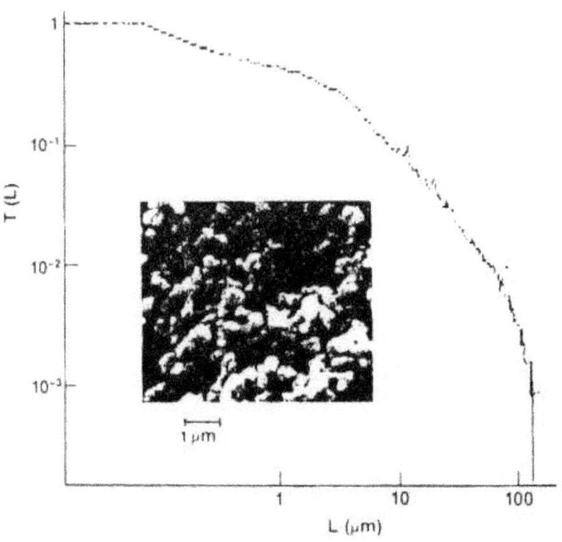


Fig. 15. Log-log plot of normalized transmission at $\lambda = 589$nm in a wedged sample of sintered titania. The sample structure is shown in the electromicrograph in the inset.

Results obtained by N. Garcia *et al.*[16] with the polarizer at 45° to the incident direction for a sample composed of 1/2 inch polystyrene balls with a volume filling fraction of $f = 0.56$ is shown in Fig. 16. For large $L$, the transmission falls exponentially with decay length $L_a = 25$ cm. In the next section we will show that $D$ can be accurately determined in this sample from measurements of the intensity-intensity correlation function with frequency. The mean free path can then be estimated using a calculated value for $v$. This gives $l = 4.3$ cm for this sample.

Measurements of transmission in the sample with the probe polarized parallel and perpendicular to the incident radiation direction are shown in Figs. 17 and 18 respectively. The signal rises to a peak for the perpendicular configuration as the incident wave is depolarized. At larger thicknesses the signal is similar in intensity to that for the parallel configuration. The signal is not zero at $L = 0$ for cross polarized detection because of scattering by the sample holder. The ability to study samples with $L < l$ makes it possible to study the transition from the ballistic to the diffusive propagation regimes.

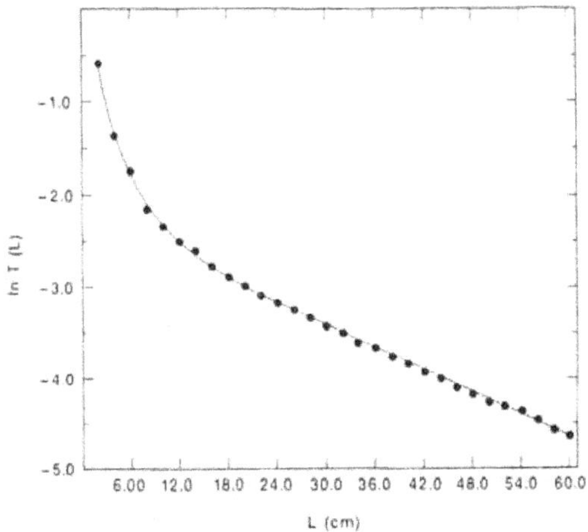

Fig. 16. Transmission versus thickness at a frequency of 22.25 GHz for a sample of 1/2 inch polystyrene spheres with a filling fraction of $f = 0.56$. The probe is polarized at $45°$ to the incident radiation.

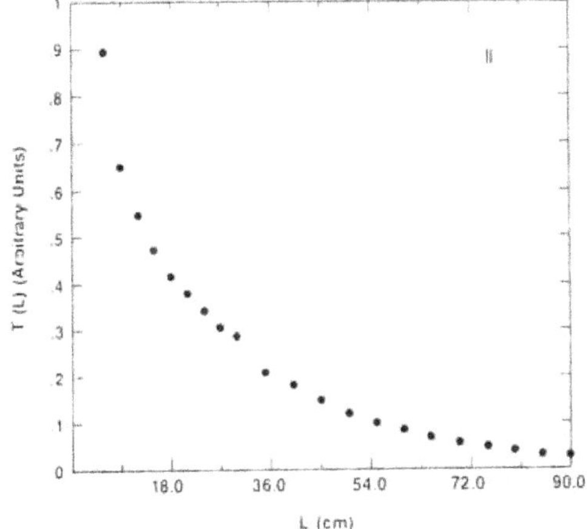

Fig. 17. Transmission versus thickness for the same conditions as in Fig. 16 except that the probe is polarized parallel to the incident ratiation.

Transmission of 19 GHz radiation through a sample composed of 1/8 inch aluminum spheres at a volume filling fraction of $f = 0.20$ is shown in Fig. 19. The remainder of the sample is composed of 1/8 inch polystyrene spheres that give a nearly close packed structure. The sample is, however, sufficiently loosely packed to facilitate configuration averaging as the cylindrical sample holder rotates about its axis.

256                     *Azriel Z. Genack*

The detector is oriented at 45° to the direction of incident polarization. Transmission is well fit by Eq. 40 for $L > 1$ cm. In this sample $L_a = 9$ cm.

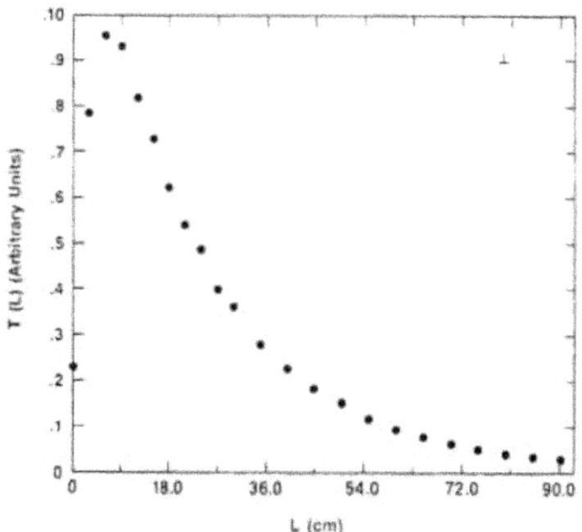

Fig. 18  Transmission versus thickness for the same conditions as in Fig. 16 except that the probe is polarized perpendicular to the incident radiation.

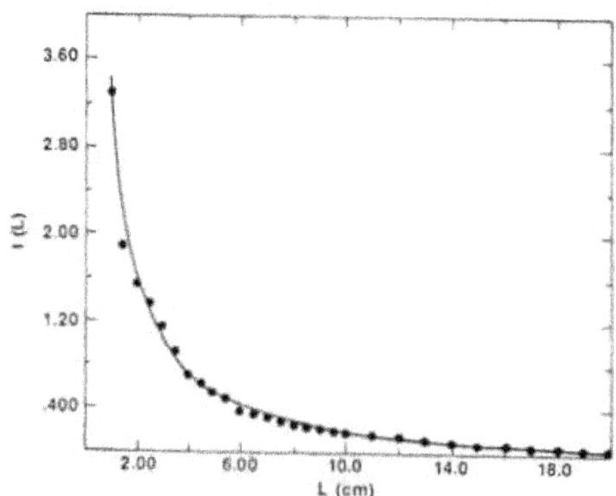

Fig. 19.  Transmission versus thickness at 19GHz through a sample composed of 1/8 inch aluminum spheres with a volume filling fraction $f = 0.20$. The solid line is a fit of the diffusion model (Eq. 40) to the data.

Scattering in metallic systems can become large when the wave is confined to regions of space outside the metal with channels which are of order $\lambda/\pi$. In this case, enormous scattering should be observed as a result of the analog of waveguide cutoff. Thus, just as in the case of a dielectric sample, a window of localization should exist between regimes of Rayleigh scattering and geometric optics. However, at high metal fraction in the long wavelength limit, the sample may be described by an effective dielectric function which is metallic in character. The wave would then be evanescent and would decay exponentially in the sample. The presence of localization could be distinguished from this situation by time or frequency domain measurements in which the scale dependence of the diffusion coefficient could be determined. Even without the effect of enhanced collective scattering by anistropic local configurations, localization has been predicted to occur for systems of metallic spheres whose influence upon total scattering is obtained by calculating the scattering from individual spheres embedded in an effective medium.[73] The results in Fig. 19 serve as a reference for comparison with results in this sample at higher frequencies and with results at larger sphere sizes, and higher concentrations.

Measurements of the temporal distribution of a transmitted optical pulse are made using 514 nm pulses derived from a mode locked cavity dumped $Ar^+$ laser as well as pulses derived from a mode locked $Ar^+$ laser which synchronously pumps a cavity dumped dye laser.[15] The laser pulses are 90 ps and 6 ps respectively. The signal is detected using a 12 $\mu$m two stage Hamamatsu microchannel plate photomultiplier. The time distribution is determined by time-correlated single photon counting. The photon count rate in the experiments is kept below 1 KHz by the use of neutral density filters. Single photon counting techniques are ideal for measuring small signals in pulsed transmission experiments. They require light levels which are far lower than the detection limits of real time detectors, which are generally not sensitive enough to observe the transmitted pulse.

Measurements of the instrumental response to the incident laser pulse at 581 nm and to the transmitted pulse for $L = 543$ $\mu$m made by J. M. Drake are shown in Fig. 20 (Ref. 15). The solid line drawn through

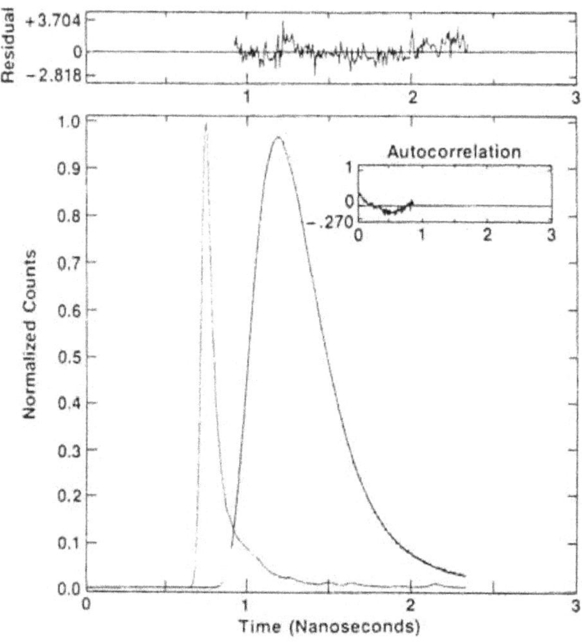


Fig. 20. Points are the distribution of time delay measured by time to amplitude conversion for a 581 nm pulse transmitted through the titania-polystyrene sample of Fig. 10 with $L = 543$ $\mu$m. The line drawn through the data is a reconvolution of the measured response to the incident pulse, shown in the figure, with Eq. 46 using $D$ and $\tau_a$ obtained from a least square minimization of the residual. The autocorrelation function of the residual is also shown.

the data is a nonlinear least squares fit of transmission calculated using the photon diffusion model (Eq. 46) reconvoluted with the instrument response. The time origin, diffusion coefficient $D$ and photon absorption time $\tau_a$ are determined independently from the fit. Only terms up to $n = 3$ in Eq. 46 contribute significantly to the fit of the transmitted pulse shown in Fig. 20. The shape of the pulse is insensitive to the value of $a$ for $a \ll L$. The fit of Eq. 46 to the pulse at 581 nm in Fig. 20 gives $D = 4.5 \pm 0.5 \times 10^5$ cm$^2$/s and $\tau_a = 243 \pm 40$ ps. A fit of the pulse profile at 514 nm excitation gives $D = 3.0 \pm 0.4 \times 10^5$ cm$^2$/s and $\tau_a = 126 \pm 20$ ps. A plot of the measured decay rate of the tail of the pulse, $1/\tau$, for various values of $L$ for 581 nm radiation is shown in Fig. 21. A fit of Eq. 47 to the data is shown in Fig. 21. The slope gives $D = 4.7 \pm 0.3 \times 10^5$ cm$^2$/s and the $y$ intercept gives $\tau_a = 293 \pm 40$ ps, in agreement with the values obtained from a fit of $T(t; L)$. The value of $L_a$ obtained from pulse transmission results at 581 nm is $(D\tau_a)^{1/2} = 105$ $\mu$m (Ref. 15) which is consistent with the value $L_a = 112$ $\mu$m obtained from a measurement of $T(L)$ at 589 nm (Fig. 13, Ref. 9). These results demonstrate that the temporal distribution and average transmitted intensity are well described by a model of photon diffusion.

Photon diffusion in the time domain has been observed by Watson *et al.* in samples of polystyrene spheres in water.[36] In these experiments, 50 ps pulses at 532 nm from a frequency-doubled mode-locked *YAG* laser are injected into a sample through an optical fiber placed at distance $s$ from the output window. Since, all the photons injected emerge from the sample, in the absence of absorption, this arrangement ensures a large signal. The evolution of the signal with increasing distance $s$ for a colloid of 0.19 $\mu$m polystyrene spheres with $f = 0.096$ is shown in Fig. 22. The solid lines drawn through the data are the best fits of the diffusion model convoluted with the input pulse shape using $D$ and $\tau_a$ as fitting parameters. The diffusion coefficient from these experiments is $D = 1.7 \pm 0.1 \times 10^7$ cm$^2$/s and is in good agreement with Mie scattering theory.

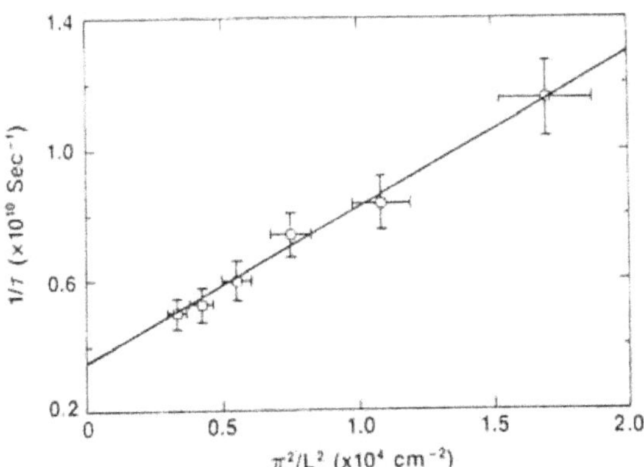


Fig. 21. The data are the exponential decay rates of the tails of $T(t; L)$ versus $\pi^2/L^2$. The line is a fit of Eq. 47 to the data.

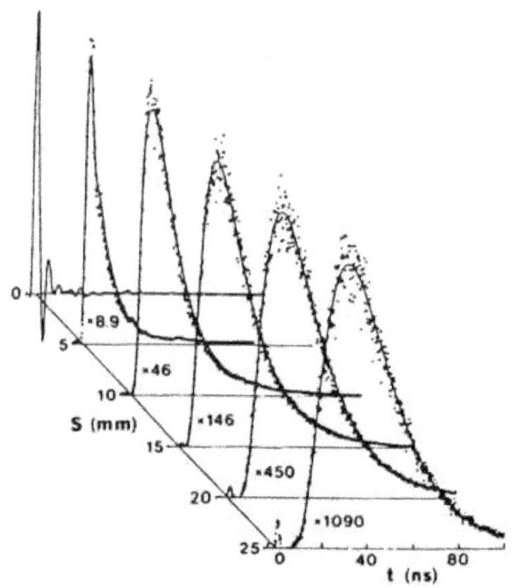



Fig. 22 Picosecond pulses from a mode-locked YAG laser are injected into a sample at a distance $s$ from the window. The evolution of the pulse emerging from the sample with increasing distance is shown for a 9.6% aqueous suspension of 0.198 $\mu$m polystyrene spheres. The solid lines are the best fits of the diffusion model convoluted with the measured pulse shape at $s = 0$. The vertical scales are adjusted for display, with gain indicated for each pulse. These results are from G. H. Watson Jr., P. A. Fleury and S. M. McCall, *Phys. Rev. Lett.* **58** (1987) 945.

## 6. LOCAL FLUCTUATIONS

In this section we discuss first and second order intensity statistics at a point in the weak scattering limit as well as field fluctuations in a medium at any scattering strength. We first consider the field and intensity autocorrelation functions with frequency, then the probability distribution of intensity and finally intensity fluctuations with time in a sample with random internal motion.

The field-field correlation function, $G^E(\Delta\nu)$ is of fundamental importance because its half width, $\delta\nu$, is the width of eigenstates in the sample. In the limit that $N_{\mathrm{ind}} = \delta \gg 1$, the intensity correlation function, $G^I(\Delta\nu)$ is not substantially influenced by intensity correlation and it can be expressed in terms of $G^E(\Delta\nu)$ (Ref. 15).

We have seen in the previous section that average transport in the weak scattering regime may be described using the language of particle diffusion. However, wave interference plays a crucial role in field and intensity fluctuations. The field produced at a point $r$ due to monochromatic excitation is the coherent sum of fields due to all Feynman paths which reach the point. The phase accumulated for a given path denoted by the index $\alpha$ of length $s_\alpha$ in traversing the medium is $\phi_\alpha = 2\pi s_\alpha/\lambda + \phi'_\alpha$ where $\phi'_\alpha$ is an additional random phase shift associated with scattering along the path $\alpha$. The phase $\phi'_\alpha$ is uniformly distributed in the interval $-\pi$ to $\pi$ radians when its value is taken modulus $2\pi$. For a single polarization component of the field at frequency $\nu$, the time independent field amplitude at a point can be represented as,

$$E(\nu) = \sum_\alpha p(\alpha)e^{i\phi_\alpha} \; , \tag{50}$$

where $p(\alpha)$ is real and,

$$\left\langle \sum_\alpha{}' p^2(\alpha) \right\rangle = P(s)ds \; , \tag{51}$$

is the ensemble average probability that an incident photon will reach the point along paths with length in the range $s$ to $s+ds$. The ensemble average, field-field correlation function at two frequencies $\nu$ and $\nu = \nu + \Delta\nu$ is,

$$\langle E(\nu)E(\nu')\rangle = \left\langle \sum_{\alpha,\beta} p(\alpha)p(\beta)\exp(i2\pi\nu s_\alpha/v) \right.$$
$$\left. \times \exp(-i2\pi\nu' s_\beta/v)e^{i(\phi'_\alpha - \phi'_\beta)} \right\rangle \; . \tag{52}$$

Paths with ensemble average other than zero are those for which $\phi'_\alpha - \phi'_\beta = 0$ for every configuration. This occurs only when $\alpha = \beta$. Thus, Eq. 52 gives,

$$\langle E(\nu)E(\nu + \Delta\nu)\rangle = \mathrm{Re}\left( \int P(s)\exp(i2\pi\Delta\nu s/v)ds \right) \; . \tag{53}$$

Using Eq. 42 the field correlation function can be shown to be equal to the Fourier Transform of the pulse profile at a point, $I(t)$,

$$G^E(\Delta\nu) = \int_0^\infty I(t)\cos(2\pi\Delta\nu t)dt \ . \tag{54}$$

Equations 53 and 54 hold regardless of the degree of long-range intensity correlation in the sample.

We have measured the intensity-intensity correlation function at a point, $G^I(\Delta\nu; L)$, rather than $G^E(\Delta\nu; L)$. This can be expressed as the ensemble average of four fields,

$$\langle I(\nu)I(\nu')\rangle = \Big\langle \sum_{\substack{\alpha,\beta \\ \gamma,\delta}} p(\alpha)p(\beta)p(\gamma)p(\delta)\exp(i2\pi\nu(s_\alpha - s_\beta)/v)$$

$$\times \exp(i2\pi\nu'(s_\gamma - s_\delta)/v)e^{i(\phi'_\alpha - \phi'_\beta + \phi'_\gamma - \phi'_\delta)}\Big\rangle \ . \tag{55}$$

Paths with ensemble average other than zero are those for which $(\phi'_\alpha - \phi'_\beta + \phi'_\gamma - \phi'_\delta) = 0$ for every configuration. This occurs only for pairs of paths (a) $\alpha = \beta, \gamma = \delta$, (b) $\alpha = \delta, \beta = \gamma$ and (c) for related pairs of paths which include closed loops where are identical except that the loops are traversed in opposite senses. Consequently, the summation can be performed in terms of pairs of paths. Equation 55 can, thus, be related to products of the square of the field amplitudes as follows,

$$\langle I(\nu)I(\nu + \Delta\nu)\rangle$$

$$= \Big\langle \sum_{\alpha,\beta} p^2(\alpha)p^2(\beta)\exp(i2\pi\Delta\nu s_\alpha/v)\exp(-i2\pi\Delta\nu s_\beta/v)\Big\rangle$$

$$+ \Big\langle \sum_{\alpha,\beta} p^2(\alpha)p^2(\beta)\Big\rangle \ . \tag{56}$$

When the effects of correlation are small, Eq. 56 can be written as,

$$\langle I(\nu)I(\nu + \Delta\nu)\rangle = \left| \int P(s)\exp(i2\pi\Delta\nu s/v)ds \right|^2 + \langle I(\nu)\rangle^2 \ . \tag{57}$$

Equation 57 may also be expressed as,

$$G^I(\Delta\nu) = \left| \int_0^\infty I(t) \exp(i2\pi\Delta\nu t) dt \right|^2 + \langle I(\nu) \rangle^2 . \tag{58}$$

The effect of correlation within a given sample is seen in Eq. 56 to be due to correlation of the intensities $p^2(\alpha)$ and $p^2(\beta)$, rather than of the phases. This suggests that the influence of long range correlation itself is not associated with long range phase coherence in a given sample but rather is the result of the diffusive propagation of intensity fluctuations. The contribution of spatial correlation within a coherence area of the far field speckle pattern has been evaluated recently by Feng *et al.* in the weak scattering limit.[20] They show that the contribution of long range correlation to the intensity correlation function at a point for $\Delta\nu = 0$ is smaller than the leading term, which does not include the effect of long range correlation, by a factor of $L/k^2$, which is much larger than unity when $kl \gg 1$ and $L \gg l$.

In the diffusive limit for transmission through a slab, Eqs. 54 and 58 can be evaluated using the time of flight distribution given in Eq. 46. Evaluating these integrals gives,

$$G^E(\Delta\nu; L) \sim \mathrm{Re}\left[ \sinh(q_0 a)/\sinh(q_0 L) \right] , \tag{59a}$$

for the field-field correlation function, and

$$C(\Delta\nu; L) \sim \left| \sinh(q_0 a)/\sinh(q_0 L) \right|^2 , \tag{60a}$$

for the intensity cumulant correlation function at a point, $C(\Delta\nu) = \langle I(\nu) - \langle I \rangle)(I(\nu + \Delta\nu) - \langle I \rangle) \rangle$, which correspond to the first term on the right hand side of Eqs. 57 and 58. Here $q_0$ is the root of, $q^2 = 1/L_a^2 + (i2\pi\Delta\nu)/D$, with negative imaginary part. These expressions are equivalent to previous results for transmission through a slab when those results are generalized to include the effects of absorption.[19,21] Note that for $\Delta\nu = 0, G^E(0; L) \sim T(L)$. In this case $q_0 = q = \alpha$ and Eq. 59 for $G^E(0; L)$ gives the result in Eq. 48. Equation 48 can be shown to be essentially identical to Eq. 40 for $T(L)$ since $5\alpha D/v = 5l/3 = a$.

Plots of $G^E(\Delta\nu; L)$ and $C^I(\Delta\nu; L)$ as functions of $\Delta\nu/(D/L^2)$ in the absence of absorption are shown in Figs. 5 and 6 respectively.

The half-width of $C^I(\Delta\nu; L)$ is $\Delta\nu_I = 1.46D/L^2$ for $L < L_a$ and $\Delta\nu_I = 0.265/(LL_a^3)^{1/2}$ for $L \gg L_a$. A plot of $(D/2\pi\Delta\nu_I L_a^2)$ versus $L/L_a$, determined from the half width of $C^I(\Delta\nu; L)$ obtained using Eq. 60 is given in Fig. 23.

In order to eliminate the influence of the instrumental response in the analysis of experimental results, each intensity value $I(\nu)$ is normalized to the average of intensity at frequency $\nu$ over all measured configurations $\langle I(\nu)\rangle$. The field correlation function and intensity cumulant correlation function for the normalized spectra $I(\nu)/\langle I(\nu)\rangle$ will henceforth be designated as $G^E$ and $C^I$. When long range correlation effects can be ignored, these correlation functions are given by,

$$G^E(\Delta\nu; L) = \mathrm{Re}[\sinh(q_0 a)/\sinh(q_0 L)]/[\sinh(\alpha a)/\sinh(\alpha L)], \quad (59b)$$

and

$$C^I(\Delta\nu; L) = |\sinh(q_0 a)/\sinh(q_0 L)|^2/[\sinh(\alpha a)/\sinh(\alpha L)]^2 \quad (60b)$$

Intensity fluctuations in the titania/polystyrene sample shown in Fig. 10 for various values of $L$ are presented in Fig. 24 (Ref. 14). The observations are made using a single frequency dye laser at a vacuum wavelength of 589 nm which is tuned over a range of 30 GHz. The average of eight intensity autocorrelation functions obtained by measurements in different speckle spots with $L = 452\ \mu$m is shown as the solid line in Fig. 25. The dotted line in Fig. 25 is a plot of Eq. 60 using values of $D$ and $\tau_a$ determined from pulsed transmission measurements which were discussed in Sec. 5 (Ref. 15). The results are in excellent agreement with experiment. In general, we find that the average of the autocorrelation function in a small number of spectra for $\Delta\nu l \lesssim 3\Delta\nu_I$ converges rapidly to the result in Eq. 60, though large differences occur at larger values of $\Delta\nu$. The results in Fig. 25 demonstrate that the character of local intensity fluctuations is not substantially influenced by long range intensity correlation in our sample. The value of $1/\Delta\nu_I(L)$ determined from measurements of $C^I(\Delta\nu; L)$ in the sample are plotted in Fig. 26. There is a large uncertainty in $\Delta \nu$ for small $L$ since the correlation frequency is large and the number of uncorrelated spectral regions in which the autocorrelation function is, calculated is correspondingly small. The solid line is a plot of $1/\Delta \nu_I(L)$ determined by using measured values of $D$ and $L_a$ in Eq. 60 and corresponds to the curve in Fig. 23 evaluated with these values for $D$ and $L_a$.

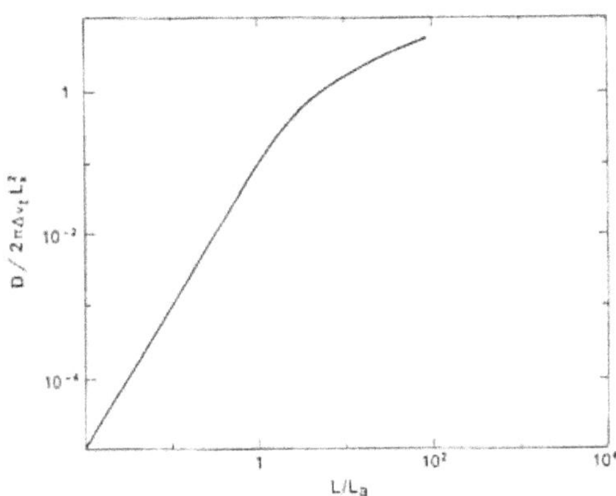

Fig. 23. Plot of the dimensionless inverse intensity correlation frequency $(D/2\pi \Delta \nu_I L_a^2)$ versus $L/L_a$, using the diffusion model (Eq. 60).

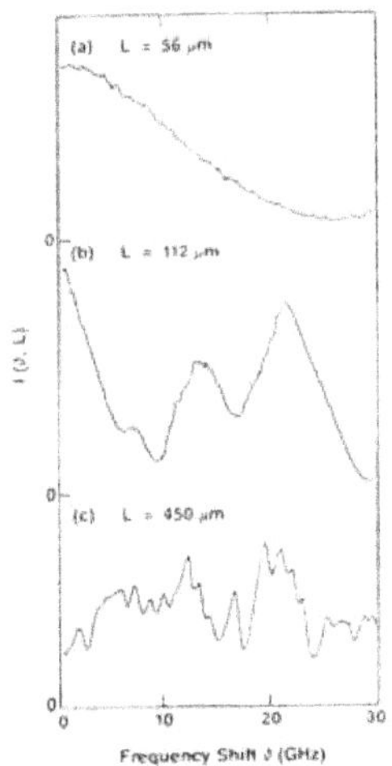


Fig. 24. Examples of intensity fluctuations with frequency within a coherence area of the transmitted speckle pattern for various thicknesses $L$ for the titania-polystyrene sample shown in Fig. 10.

                                   *Azriel Z. Genack*

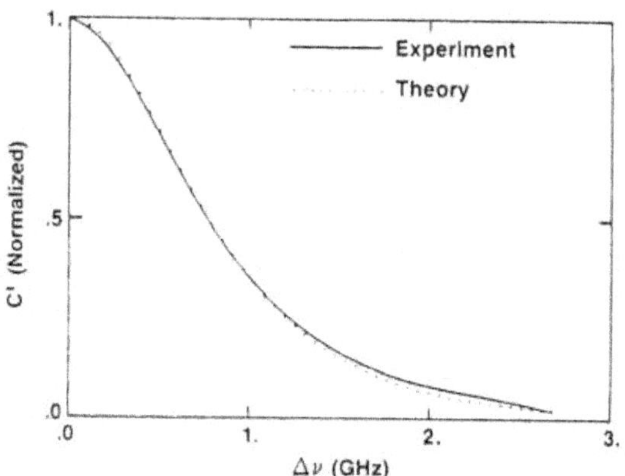

Fig. 25. The solid line is the measured normalized cumulant of the intensity correlation function with frequency shift for the sample shown in Fig. 10. The sample thickness is $L = 452$ $\mu$m. The dotted line is a plot of Eq. 60 using propagation parameters determined from the measurement of the time profile of the transmitted pulse shown in Fig. 20.

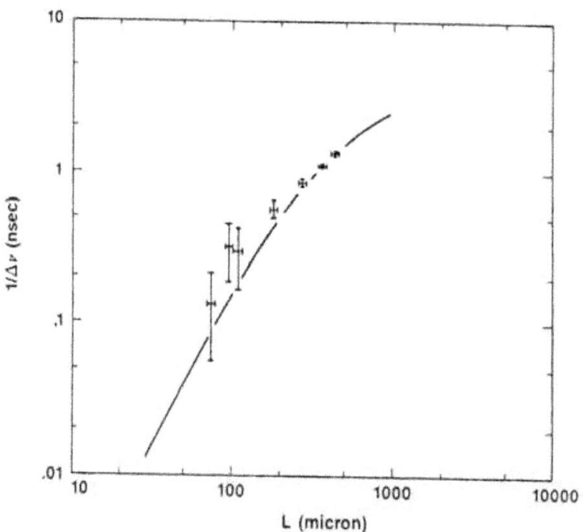

Fig. 26. The data are measurements of $1/\Delta\nu_I$ determined from measurements of the correlation function in the titania polystyrene sample. The solid line is obtained from the expression for the correlation function in Eq. 64 using the measured values of $D$ and $L_a$.

We now turn to a consideration of microwave intensity fluctuations and correlation. We present results in a sample composed of 1/2 inch polystyrene spheres with volume fraction $f = 0.56$ contained in the sample holder shown in Fig. 11. Measurements of the scale dependence of transmission in the sample are shown in Figs. 16–18. An accurate determination of the ensemble average of the intensity correlation function requires the average over a large number of configurations. This is accomplished under computer control. The microwave intensity spectrum $I(\nu)$ for a given random configuration is acquired by scanning the frequency of the oscillator with a voltage supplied by the computer. A new random configuration is then produced by activating a motor which turns the sample holder about its axis for a short time and mixes the sample before the next frequency scan is initiated. A typical spectrum $I(\nu)$ for $L = 60$ cm, is shown in Fig. 27. The microwave oscillator output is modulated at 10 KHz and the signal is detected using a lock-in amplifier. Note that the contrast in the detected signal is excellent and that the signal can approach zero. The intensity of microwave radiation is detected at the point of contact of a wire probe with the Schottky diode. The intensity at a point can thus be determined. The probe wire is 3.5 mm and the wavelength in air at the center frequency of the scan of 22.25 GHz is 1.35 cm.

In order to account for the frequency variation in the response of the experimental apparatus, the detected intensity, $I(\nu)$, is normalized to the configuration average of the intensity at a given frequency $\langle I(\nu) \rangle$. The autocorrelation function of $I(\nu)/\langle I(\nu) \rangle$ is then computed for each configuration and the average of the autocorrelation function for all configurations measured is then taken. The normalized intensity correlation function obtained by averaging the intensity correlation function of 380 transmission spectra $I(\nu)/\langle I(\nu) \rangle$ is shown as the points in Fig. 28. The correlation function in Fig. 28 is normalized to the value at $\Delta \nu = 0$. The solid line is a fit of Eq. 60 to the data obtained using $D$ as the fitting parameter and $L_a = 25$ cm as the absorption length as determined from measurements of $T(L)$. In the limit $|q_0 a|, |q_0 L| > 1$, the cumulant correlation function of transmission through a slab is seen from Eq. 60 to fall exponentially with the square root of the frequency

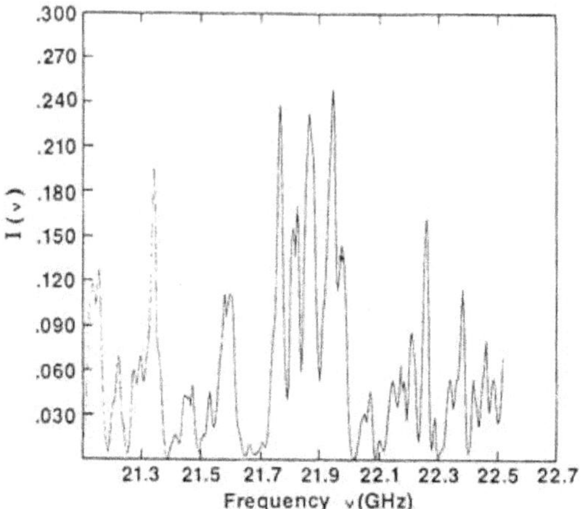

Fig. 27. Microwave intensity fluctuations as a function of frequency for a single random configuration of 1/2 inch polystyrene spheres at a thickness $L = 60$ cm.

shift,

$$C^I(\Delta\nu; L) \sim \exp -(4\pi\Delta\nu(L-a)^2/D)^{1/2} . \qquad (62)$$

In the present case for which the sphere diameter is comparable to the wavelength, the mean velocity in the medium can be estimated by the relation,[72]

$$\frac{1}{v} = \frac{nf}{c} + \frac{1-f}{c} . \qquad (63)$$

For polystyrene $n = 1.59$ is the microwave index of refraction and $v \sim 2.3 \times 10^{10}$ cm/s in the sample. The fit of Eq. 60 to the data gives $D = 3.3 \times 10^{10}$ cm$^2$/s. The mean free path is then $l = 3D/v = 4.3$ cm. This gives $kl = 2\pi\nu l/v \sim 27$, confirming that we are in the weak scattering limit.

In the absence of long range intensity correlation, $C^I(\Delta\nu; L) = |\int P(s)\exp(i2\pi\Delta\nu s/v)ds|^2/(\int P(s)ds)^2$. When $\Delta\nu = 0$, this expression equals unity, i.e. $C^I(0) = \mathrm{var}(I)/\langle I\rangle^2 = 1$. This is the well known result,[12,13]

$$\mathrm{var}(I) = \langle I\rangle^2 . \qquad (64)$$

For the data shown in Fig. 28, we find, $C^I(0) = 1.04$ which is quite close to the value of unity expected for the case of perfect spatial

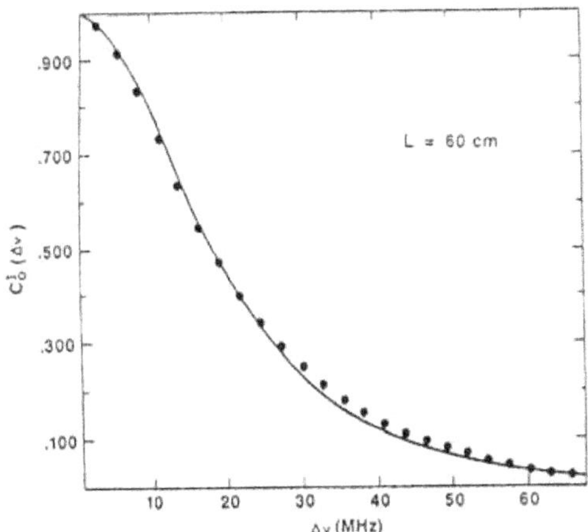



Fig. 28. The dots are the measured normalized cumulant intensity correlation function with frequency shift for the transmitted wave for a sample of 1/2 inch polystyrene spheres. The sample thickness is $L = 60$ cm. The solid line is a fit to the diffusion model in the absence of long range correlation (Eq. 60) using $D$ as a fitting parameter and the value $L_a = 25$ cm obtained from the measurements of $T(L)$.

resolution in the absence of long range correlation. The first order intensity statistics which underlie this result will be considered shortly.

A plot of $1/\Delta\nu_I$, which is corrected for the narrow width of the distribution of the time of arrival of partial waves from the oscillator to the input face of the sample, $(\Delta t)_{in}$, is shown in Fig. 29. $1/\Delta\nu_I$ is expressed in dimensionless form $D/2\pi L_a^2 \Delta\nu_I = 1/2\pi\Delta\nu_I \tau_a$. The spread in travel time through the sample itself is approximately given by $\Delta t = (\Delta t)' - (\Delta t)_{in}$ where $(\Delta t)'$ is the spread in time that would be measured in an experiment with delta function excitation at the oscillator output. This suggests the approximate relationship, $1/\Delta\nu_I = (1/\Delta\nu_I)' - (1/\Delta\nu_I)_{in}$, where $(\Delta\nu_I)'$ is the measured correlation frequency. $(\Delta\nu_I)_{in}$ is essentially the value of $\Delta\nu_I$ for $L \lesssim l$. For $(\Delta\nu_I)_{in}$ we use the value measured at $L = 2.5$ cm, $(\Delta\nu_I)_{in} = 320$ MHz. This gives $1/\Delta\nu_I = (1/\Delta\nu_I)' - (3.1 \times 10^{-9})$, which is plotted in Fig. 29. The solid line in Fig. 29 is a plot $1/\Delta\nu_I$ in dimensionless form using the value $D = 3.3 \times 10^{10}$ cm$^2$/s, which is determined at $L = 60$ cm using Eq. 60, and assuming $a \ll L$. This line corresponds to the curve in Fig. 23. The data is in good agreement with the diffusion model,

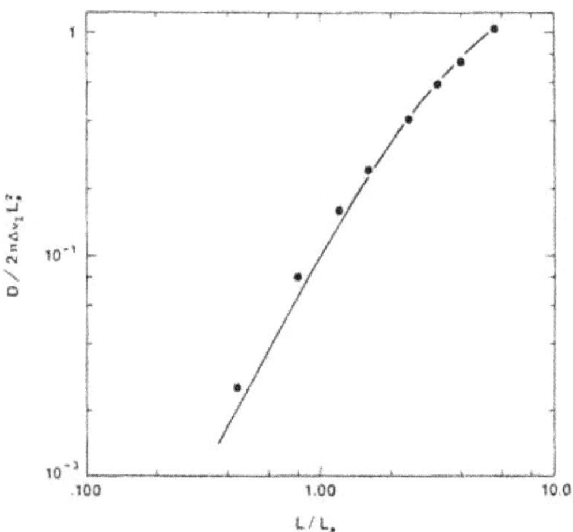

Fig. 29. Plot of the inverse of the normalized correlation frequency. The dots
are the measured correlation frequency corrected for the intensity fluctuations
present at the input face of the sample. The solid line is the half width
predicted by the diffusion model (Eq. 60) using the value $D$ obtained in the
fit of Eq. 60 to the measured intensity correlation function shown in Fig. 28.

particularly for values of $L$ much larger than $a$.

A plot of the intensity correlation function for fluctuations in the
reflected wave in an 80 cm thick sample of polystyrene sphere is shown
in Fig. 30. The results are the average of the correlation function of
400 spectra. The correlation function for the reflected intensity from
a sample of thickness $L$ can be evaluated by calculating the reflected
path length distribution and using this in Eq. 57. This gives,

$$C^I(\Delta\nu; L) \sim \left| e^{-q_0 a} + \frac{2\sin(q_0 a)}{\sin(q_0 L)} e^{-iq_0 L} \right|^2 , \qquad (65)$$

where $q_0$ is again the root of $q^2 = 1/L_a^2 + i2\pi\Delta\nu/D$ in the lower half
of the complex plane. For $L \gg a$, the second term in the sum on
the right hand side of Eq. 65 is much smaller than the first term. For
$\Delta\nu > D/2\pi L_a^2$, the intensity cumulant correlation function in reflection
may be approximated as,

$$C^I(\Delta\nu; L) \sim \exp -(4\pi\Delta\nu a^2/D)^{1/2} . \qquad (66)$$

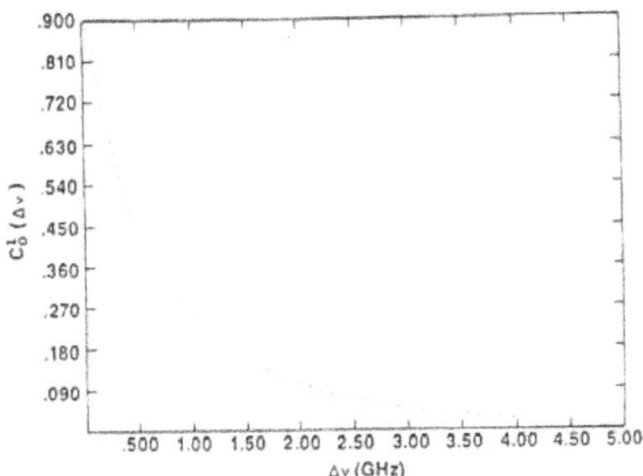

Fig. 30. Plot of the normalized cumulant of the intensity correlation function with frequency shift measured in reflection from the sample of 1/2 inch polystyrene spheres.

A plot of the natural logarithm of the data in Fig. 31 versus $\Delta\nu^{1/2}$ is shown in Fig. 31. Equating the slope of this curve of $2.2 \times 10^{-4}$ to $(-(4\pi a^2/D)^{1/2})$ and using the value of $D$ obtained from measurements of the intensity correlation function in transmission gives $a = 11$ cm, or $a = \gamma l$ with $\gamma = 2.6$. This is larger than the value $\gamma = 5/3$ expected for transmission measurements. This result may be compared to the measurements of the temporal intensity correlation function of backscattered light from a sample of diffusing particles by Pine $et\ al.$[31] In those experiments the fit of the data to the model of optical diffusion gives $\gamma = 2.0$. The model used to describe each of these experiments is only approximate since the statistics of backscattered waves depends crucially upon $a$ which appears in a simplified model as the specific depth into the sample at which a collimated beam is scattered isotropically. A more complete solution of the transport equation is given by MacKintosh and John[42] to explain statistics of backscattered light. This involves paths which are mostly within a few mean free paths of the input surface. They find that $\gamma$ is larger than 5/3 when scattering is not isotropic but stronger in the forward direction. This is the case in our sample in which the sphere diameter is comparable to $\lambda$.

In the present discussion, we have calculated the intensity correlation function of a scalar field. This gives the statistical properties of the intensity of a single polarization component of a vector field, which

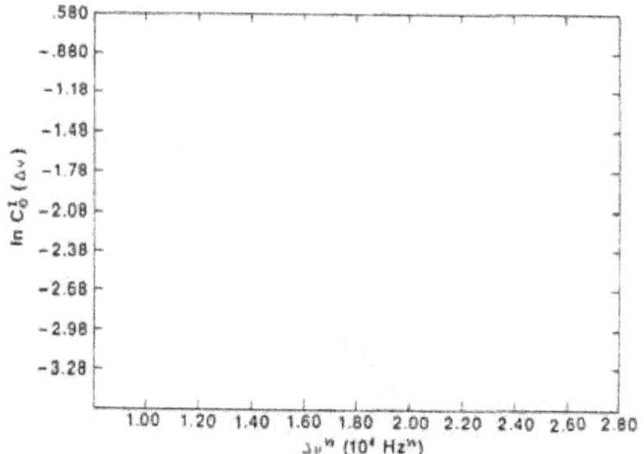

Fig. 31. Plot of logarithm of intensity correlation function shown in Fig. 30 versus the square root of the frequency shift.

corresponds to the optical and microwave experiments discussed here. We now calculate the probability distribution of intensity associated with a single component of the field in the case that the partial waves reaching a point are statistically independent for a fixed configuration.

We start by considering the random phasor sum in Eq. 50. In terms of its real and imaginary parts, it can be expressed as,

$$E(\nu) = E e^{i\theta} = \sum_\alpha p(\alpha) \cos \phi_\alpha + i \sum_\alpha p(\alpha) \sin \phi_\alpha$$
$$= E \cos \theta + i E \sin \theta = r + i \; . \tag{67}$$

This random phasor sum is shown schematically in Fig. 32. Since $\phi_\alpha$ and $p(\alpha)$ are uncorrelated, the real and imaginary parts are uncorrelated and, hence, $\langle ri \rangle = 0$. Further $\langle r \rangle = \langle i \rangle = 0$ and $\langle r^2 \rangle = \langle i^2 \rangle = \langle E^2 \rangle/2 = \langle I \rangle/2 \equiv \sigma^2$. The joint probability density for the real and imaginary parts, in the limit of a large number of partial waves in the sum in Eq. 67, is thus the product of a Gaussian probability distributions for $r$ and $i$, (Refs. 12, 13),

$$P(r,i) = \frac{1}{2\pi\sigma^2} \exp(-(r^2 + i^2)/2\sigma^2) \; . \tag{68}$$

The contours of constant probability in the $(r,i)$ plane are circles with radii $E = (r^2 + i^2)^{1/2}$. The Jacobian for the transformation from linear coordinates $(r,i)$ to polar coordinates $(E,\theta)$ is $E$. Hence we can write,

$$P(E,\theta) = \frac{E}{2\pi\sigma^2} \exp(-E^2/2\sigma^2) \; . \tag{69}$$

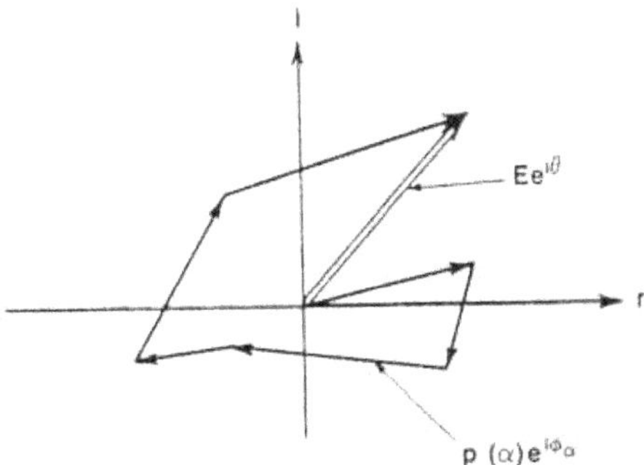



Fig. 32. Addition of random phasors.

Integrating over $\theta$ from $-\pi$ to $\pi$ gives the Rayleigh distribution,[13]

$$P(E) = \frac{E}{\sigma^2} \exp(-E^2/2\sigma^2) . \tag{70}$$

The probability density in terms of intensity is $P(I) = P(E)|\frac{dE}{dI}|$ $= P(E)/2I^{1/2}$. This gives $P(I) = \frac{1}{2\sigma^2} \exp(-I/2\sigma^2)$. Since $2\sigma^2 = \langle I \rangle$, this gives,

$$P(I) = \frac{1}{\langle I \rangle} \exp(-I/\langle I \rangle) . \tag{71}$$

In terms of the normalized intensity $(I/\langle I \rangle)$, the probability density is,

$$P(I/\langle I \rangle) = \exp(-I/\langle I \rangle) . \tag{72}$$

The value of $\langle I^2 \rangle$ is given by $\int_0^\infty I^2 P(I)dI = 2\langle I \rangle^2$. This gives var $(I) = \langle I \rangle^2$ in agreement with Eq. 64.

A departure of $\text{var}(I/\langle I \rangle)$ from unity or of $P(I/\langle I \rangle)$ from the form in Eq. 72 is an indication that either the detected wave is not fully polarized, the intensity is detected over an extended region rather than at a point or the square moduli of the partial waves are correlated. The presence of correlation can be demonstrated by showing that for similar detection procedure there is a different degree of deviation from the density in Eq. 72 in two samples.

Recently Kaveh *et al.*[60] have measured $P(I)$ for polarized light in backscattering from solid $BaSO_4$. They fit a histogram of $P(I)$ to the empirical gamma density distribution, $\langle I \rangle P(I/\langle I \rangle) = \mu^\mu/\Gamma(\mu)$

$\times (I/\langle I\rangle)^{\mu-1} \exp(-\mu I/\langle I\rangle)$, with the value $\mu = 2.5$. Equation 72 corresponds to $\mu = 1$. The deviation from the expected negative exponential statistics is attributed to correlation introduced as a result of strong scattering in the sample, in which $kl \sim 60$. Subsequently Wolf *et al.*[65] have reported measurements in reflection from a $BaSO_4$ sample in reflection in which $P(I/\langle I\rangle)$ is exponential for $I/\langle I\rangle > 1$ and which are inconsistent with the gamma density distribution. They find that $\langle I^2\rangle/\langle I\rangle^2$ is in the range 1.7–1.8 when $kl \sim 60$ as well as for much larger values of $kl$. These results are consistent with negative exponential distribution for the intensity which is measured with a finite experimental angular resolution.

Garcia *et al.*[16] have measured intensity statistics of a single polarization component of microwave radiation in the sample of polystyrene in which $kl \sim 27$ using the data obtained in the measurement of $I(\nu)/\langle I(\nu)\rangle$ which are used to determine the intensity correlation function.[26] The data for reflection with $L = 80$ cm and for transmission for $L = 11$ cm, $30$ cm, and $60$ cm are shown respectively in Figs. 33–36. The probe is perpendicular to the incident polarization for backscattered radiation and at $45°$ to the incident polarization for transmitted radiation. We find that the distribution $P(I/\langle I\rangle)$ falls exponentially with slope of $1 \pm .03$ in the distributions shown in Figs. 33–35 for $I/\langle I\rangle > 1$. For $L = 60$ cm a slight deviation from negative exponential statistics is observed. In backscattering $\langle I^2\rangle/\langle I\rangle^2 = 2.04$ which is very close to the value 2, characteristic of exponential statistics and $P(I/\langle I\rangle)$ agrees well with Eq. 72 except for $I/\langle I\rangle < 0.3$. This is true even though $kl$ is smaller than in previous measurements. We also note that $N_{\text{ind}}$, which can be no larger than $N$, is smaller than in the optical measurements of $P(I/\langle I\rangle)$. The extent of correlation between partial waves is, therefore, expected to be greater in this sample than in previous optical studies. Nevertheless, the backscattered light is very well described by negative exponential statistics. In transmission we find that the peak in the probability distribution moves out to larger values of $I/\langle I\rangle$ as the thickness increases. This can only be a result of increased correlation in samples with larger $L$ since the same detector is used in all these experiments. Thus correlation is seen to increase as $L$ increases. This is expected since $N_{\rm ind}$ decreases with $L$ in the restricted geometry of the cylindrical sample holder. The role of correlation in this sample will be considered further in Sec. 8 in connection with measurements in random polystyrene samples and in random metallic samples at $L = 140$ cm in which dramatic effects of correlation are observed.

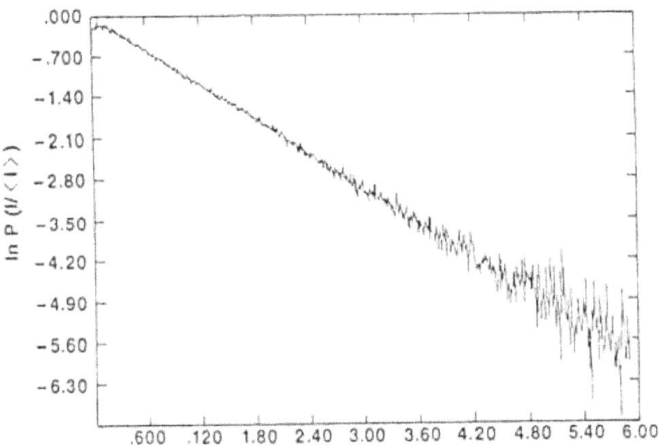

Fig. 33. Intensity probability density $P(I/\langle\, I\,\rangle)$ of the normalized intensity $I/\langle\, I\,\rangle$ of reflected microwave radiation from a sample of 1/2 inch polystyrene spheres.

We have seen that the scattered intensity from a sample fluctuates as the frequency is varied since the field at a point is a random sum of partial waves with phases which depend upon $\nu$. Similarly the intensity scattered from a sample fluctuates as the internal constituents of the sample move (Refs. 37–43). We have observed such intensity variation in the optical speckle pattern as a solid sample is heated. In the measurement of $T(L)$ for microwave radiation in the polystyrene sample, the sample holder is rotated at 1.6 Hz. This rapidly changes the relative positions of the polystyrene balls and allows us to measure the local intensity in a large number of configurations for the purpose of accurately measuring the average transmission. An example of the temporal intensity fluctuations of microwave radiation at 19 GHz measured in reflection from a 90 cm thick sample of 1/2 inch polystyrene

     *Azriel Z. Genack*

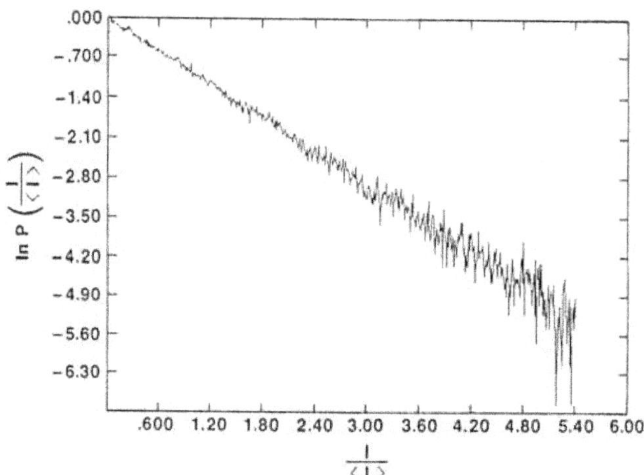

Fig. 34. Intensity probability density of the normalized intensity transmitted through a sample of 1/2 inch polystyrene spheres with $L = 11$ cm.

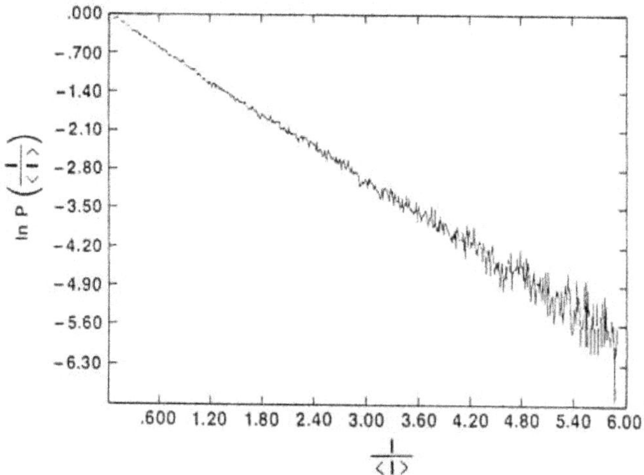

Fig. 35. Probability density for intensity transmitted through a sample of 1/2 inch polystyrene spheres with $L = 30$ cm.

spheres contained in a rotating cylinder is shown in Fig. 37 (Ref. 43). The filling fraction is $f = 0.49$.

The longer the pathlengths in the medium the more sensitive the

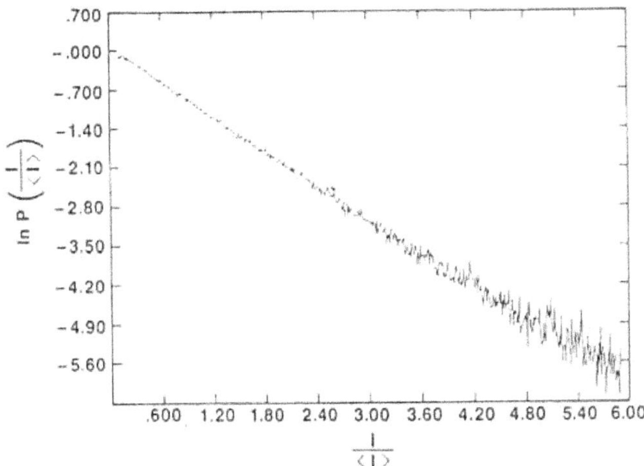

Fig. 36. Probability density for intensity transmitted through sample of 1/2 polystyrene spheres with $L = 60$ cm.

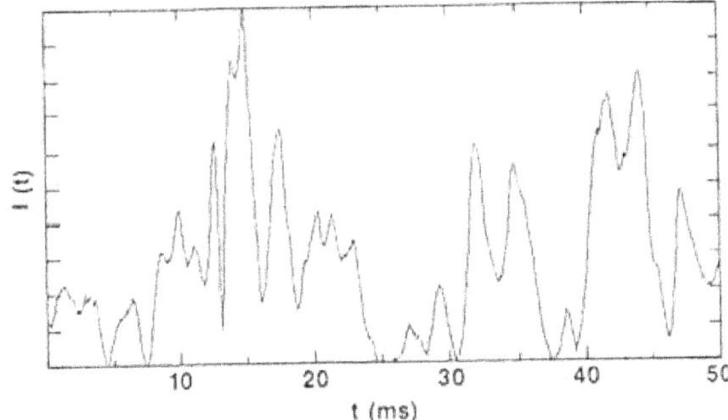

Fig. 37. Typical measurements of reflected intensity versus time at 19 GHz from a sample of 1/2 inch polystyrene spheres.

intensity to internal motion. Thus the measurement of the sensitivity of local intensity to configurational change can be used to determine the parameters of wave propagation. More importantly, if the character of wave propagation itself is known, the statistical description of the motion can then be determined. This makes possible the extension of powerful light scattering techniques to the multiple scattering regime. In this way the motion of dense collections of interacting particles can be studied.[37,40,41]

The first detailed experiments were carried out by Maret and Wolf.[37] They measured the temporal intensity autocorrelation function, $G^I(\tau)$, of light reflected from a colloid of polystyrene spheres in water. A rapid nonexponential decay of the cumulant intensity correlation function, $C^I(\tau) = G^I(\tau) - 1$ is observed. The loss of correlation is calculated by assuming that it is due solely to changes in the phase of partial waves along paths associated with a fixed sequence of scattering from the same particles. This leads to the result,

$$C^I(\tau) \sim \left| \int P(s) \exp\left[ - \left(\frac{s}{l}\right) \frac{t}{2\tau_\lambda} \right] ds \right|^2 , \tag{73}$$

where $\tau_\lambda = 1/4k^2 D_p$ is the time for a particle to diffuse a distance $1/k$ and $D_p$ is the particle diffusion coefficient. More generally, calculations by Stephen[39] of the cumulant intensity correlation function are equivalent to,

$$C^I(\tau) \sim \left| \int P(s) \exp\left[ - \left(\frac{5}{l}\right) \frac{1 - f_0}{f_0} \right] ds \right|^2 , \tag{74}$$

where $f_0 = \frac{1}{N} \sum_i J_0(k\delta R_i)$ and $\delta R_i = \langle R_i(\tau) - R_i(0) \rangle$ is the ensemble average of the displacement of the $i$th particle in the time $\tau$. When $k\delta R_i \ll 1$, we can write,

$$\frac{1 - f_0}{f_0} = \frac{k^2}{2N} \sum_i (\delta R_i)^2 / 2 = k^2 (\delta R)^2 / 2 . \tag{75}$$

In the limit of particle diffusion $(\delta R)^2 = 4D_p\tau$. This result together with Eqs. 74 and 75 gives Eq. 73. The validity of Eqs. 75 is supported by a number of light scattering experiments from polystyrene colloids performed in reflection by Rosenbluh *et al.*[40] and in both reflection and transmission by Pine *et al.*[41]

Equations 60 and 65 give the cumulant intensity correlation function for transmission and reflection respectively from a slab. The results can be readily generalized to include the case of intensity correlation between two configurations at different frequencies in the presence of absorption by noting that Eqs. 57 and 74 have similar forms. The role of absorption is included in $P(s)$, but it could have been included explicitly by substituting $P_0(s)e^{-s/l_a}$ for $P(s)$, where $P_0(s)$ is the pathlength distribution in the absence of absorption. In this case we can write $\langle I \rangle^2 = C^I(\Delta \nu = 0) = \left| \int P_0(s) \exp(-s/l_a)ds \right|^2$, which has the same form as Eq. 74. Thus, the intensity correlation function with frequency and configuration change in the presence of absorption has the same form (Eqs. 60 and 65) as for a static sample but $q_0$ is the root of

$$q^2 = \frac{1}{L_a^2} + \frac{3}{l^2} \frac{1 - f_0}{f_0} + \frac{i2\pi\Delta\nu}{D} , \tag{76}$$

with negative imaginary part.

The temporal correlation function in backscattering corresponding to the intensity fluctuations such as those shown in Fig. 37 is shown in Fig. 38. The results for 13.25 GHz and 15 GHz are shown.[43] The detection is polarized perpendicular to the incident wave. On the time scale shown, the polystyrene spheres are not likely to have suffered collisions which randomize the sphere velocity. We thus assume that we are in the ballistic regime for particle motion. The particles tend to be dragged along the rotating cylinder and have an average velocity $\boldsymbol{\omega} \times \boldsymbol{\rho}$, where $\boldsymbol{\omega}$ is the angular velocity of the sample holder and $\rho$ is the radial distance to the particle. In addition, the particles have a random velocity $v$ so that the instantaneous particle velocity is $\mathbf{v} + \boldsymbol{\omega} \times \boldsymbol{\rho}$.

The decay of the correlation function is dominated by the random velocity. The periodic motion associated with the average velocity of the particle would create a modulation of the signal at the turning frequency on a time scale which is longer than that of the decay observed. Using Eq. 65 in the limit $L \gg a$ gives, $C^I(\tau) \sim \exp(-2q_0 a)$. At a single frequency in the presence of weak absorption, we find $q_0 = \left(\frac{3}{l^2} \frac{1-f_0}{f_0}\right)^{1/2} \sim \left(\frac{3}{2}\right)^{1/2} \frac{k}{l} \left((\delta R)^2\right)^{1/2}$. For $\tau$ less than the velocity relaxation time, $(\delta R)^2 = \langle v^2 \rangle \tau^2$, and $\tau_\lambda = 1/\langle v^2 \rangle^{1/2} k$. Hence, the temporal correlation function can be expressed as,

$$C^I(\tau) \sim \exp(-2q_0 a) \sim \exp(-\sqrt{6}\gamma k \langle v^2 \rangle^{1/2}\tau) = \exp(-\sqrt{6}\gamma\tau/\tau_\lambda) . \tag{77}$$

This gives a single exponential relaxation in the time with a relaxation time which is proportional to the wavelength. This is consistent with

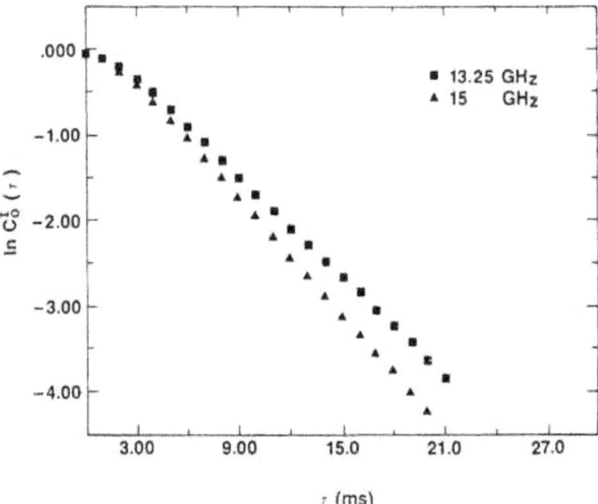



Fig. 38. Natural logarithm of the cumulant intensity correlation function versus time delay of reflected radiation at 13.25 GHz and 15 GHz from a sample
of 1/2 inch polystyrene spheres.

the data shown in Fig. 38. If we assume that $\gamma = 2.6$, as is found
in measurements of intensity fluctuations of backscattered light as the
frequency is varied in a polystyrene sample with $f = 0.56$, we find
$\tau_\lambda = 28.7$ ms at 15 GHz. From these results we obtain for the RMS
random velocity, $\langle v^2 \rangle^{1/2} = 8.5$ cm/s. The RMS velocity represents a
radial average as well as temporal average. These methods can be used
to characterize the motion of random granular mixtures.

In this section we have calculated the intensity correlation function at a point in the transmitted or reflected speckle pattern for fixed
direction of the incident beam. The intensity correlation function $\langle I_1 I_2 \rangle$
of intensity $I_1$ and $I_2$ at the specified point for different frequencies or
sample configurations was calculated by considering the phase shifts in
the fields induced by the change in the experimental parameters. It
was assumed that the distribution of path lengths in a given configuration is the same as the ensemble average and is given by the photon
diffusion model. This explicitly ignores the role of long range intensity
correlation and may be called the uncorrelated photon diffusion model.
The neglect of long range intensity correlation is equivalent to making

the field factorization approximation,

$$\langle I_1 I_2 \rangle \sim |\langle E_1 E_2^* \rangle|^2 \ . \tag{78}$$

The factorization approximation only holds in the limit $\delta \gg 1$.

Another case of local fluctuations that follows the paradigm described above is fluctuations in intensity as the sample is rotated by an angle $\Delta\theta$. This case was first discussed by Feng *et al.*[20] and called the "memory effect". The intensity-intensity cumulant correlation function as a function of the change in $k$ vector of the transmitted wave relative to the sample surface, $|\mathbf{k}_1 - \mathbf{k}_2| = \Delta k = 2k \sin \theta/2$ in the absence of absorption is,

$$\begin{aligned}
C(\Delta\theta) &= \langle I(\theta) I(\theta + \Delta\theta) \rangle - \langle I(\theta) \rangle \langle I(\theta + \Delta\theta) \rangle \\
&= D_1 \langle I(\theta) \rangle \langle I(\theta + \Delta\theta) \rangle F_1(\Delta k L) \ ,
\end{aligned} \tag{79}$$

where $D_1$ is a constant of order unity and

$$F_1(x) = x^2 / \sinh^2 x \ . \tag{80}$$

This result has been recently confirmed qualitatively in optical experiments.[120]

Higher order corrections to $C$ which arise as an expansion in the degree of long range correlation as well as the enchancement of total transmission fluctuations which occur even in the weak scattering limit will be discussed in Sec. 8.

## 7. WEAK LOCALIZATION

Optical weak localization is directly observed as enhanced reflection in a narrow cone about the backscattering direction. This was first observed by Kuga and Ishimaru.[54,55] Subsequently van Albada and Lagendijk,[57] Wolf and Maret[58] and others[59-65] performed experiments with sufficient angular resolution to accurately determine the magnitude and angular variation of the enhancement of backscattered light. These detailed measurements are in excellent accord with theoretical predictions. Here we discuss weak localization of scalar waves in order to place the subject in the context of the model of photon diffusion and local fluctuations presented in the two preceding sections.

The constructive interference of time reversed waves returning to a point doubles the probability of return of these waves as compared to the photon diffusion model in which such interference is not included. The fractional reduction of the diffusion coefficient in the weak scattering limit from the Boltzmann value is equal to the probability of return calculated using the photon diffusion model.[52] The probability of return of a particle which is at a point $\mathbf{r} = 0$ at $t = 0$ to that point integrated over time is given by,

$$\frac{V_c}{\tau_c} \int_{t_0}^{\infty} P(0,t)dt = \frac{V_c}{\tau_c} \int_{t_0}^{\infty} G_0(0,t)e^{-t/\tau_i}dt \ . \tag{81}$$

Here $P(0,t)V_c$ is the probability at a time $t$ that the particle is within a coherence volume $V_c = L_c^d$ centered on the origin, $\tau_c$ is the photon transit time through a coherence volume, $\tau_c \sim L_c/v \sim \lambda/4v$, and $t_0 \sim 2l/v$ is the time in which a photon is scattered back to the origin after a single collision. $G_0(\mathbf{r},t)$ is the Green's function which is given in Eq. 45 for $d = 3$. In $d$ dimensions, it is,

$$G_0(0,t) = \frac{1}{(4\pi D_B t)^{d/2}} \ , \tag{82}$$

where $D_B = vl/d$.

The average number of returns to the coherence volume centered about $\mathbf{r} = 0$ in $d$ dimensions, $N_{\text{return}}$, can be expressed within the framework of the photon diffusion model by using Eqs. 80 and 81 to give,

$$N_{\text{return}} \sim \frac{L_c^d}{\tau_c(4\pi D_B)^{d/2}} \int_{t_0}^{\infty} \frac{e^{-t/\tau_a}}{t^{d/2}}dt \sim \frac{L_c^d}{\tau_c(4\pi D_B)^{d/2}} \int_{t_0}^{\tau_a} \frac{1}{t^{d/2}}dt \ . \tag{83}$$

Equation 83 can be evaluated to give,

$$N_{\text{return}} = \begin{cases} \dfrac{L_c^d}{\tau_c(4\pi D_B)^{d/2}} \dfrac{1}{1-d/2} t^{1-d/2}\Big|_{t_0}^{\tau_a} \ , & d \neq 2 \ , & \text{(84a)} \\[2em] \dfrac{L_c^2}{\tau_c(4\pi D_B)} \ln(\tau_a/t_0) \ , & d = 2 \ . & \text{(84b)} \end{cases}$$

These results diverge when $\tau_a \to \infty$ for $d \leq 2$, indicating that the wave is localized. Only for $d > 2$ is $N_{\text{return}}$ finite. It becomes arbitrarily small as $kl$ becomes large. When $N_{\text{return}} \ll 1$ it equals the additional probability of return due to constructive interference of time reversed waves. In this limit it corresponds to the fractional reduction in the diffusion coefficient due to coherent backscattering, $(D_B - D)/D = N_{\text{return}}$. Evaluating Eq. 84 for $d = 3, \tau_a \gg t_0$ and $kl \gg 1$ gives the degree of renormalization of the diffusion coefficient in this limit,

$$\frac{D_B - D}{D_B} = (3\sqrt{3\pi}/32)/k^2 l^2 \sim 1/3.5 k^2 l^2 \ . \tag{85}$$

The reduced value of $D$ is associated with additional reflected light in the backscattered cone.

We now consider the angular dependence of enhance backscattering and stress the formal similarity with calculations of local fluctuations. Coherent backscattering of light can be understood by referring to Fig. 39. The trajectories of two partial waves for normally incident light which travel along time reversed paths in the random medium and emerge at an angle $\theta$ to the normal are shown. These wavelets interfere in the far field speckle pattern. Since the paths in the medium are essentially the same aside from the time order of scattering, the accumulated phase shifts in the medium are the same. The phase difference between the paths is,

$$\Delta\phi = \frac{2\pi}{\lambda}\Delta r = \mathbf{k} \cdot \boldsymbol{\rho} \ , \tag{86}$$

where $\Delta r$ and $\rho$ are defined in Fig. 39. The average intensity reflected at an angle $\theta$ is the ensemble average of the modulus squared of the field which is a sum of all partial waves,

$$\langle I(\theta)\rangle = \left\langle \left| \sum_\alpha \sum_{(\mathbf{r}_i,\rho)} P_{(\mathbf{r}_i,\rho,\mathbf{k})}(\alpha) e^{i\phi_\alpha} e^{i\mathbf{k}\cdot(\mathbf{r}-(\mathbf{r}_i+\rho))} \right|^2 \right\rangle \ . \tag{87}$$

In Eq. 87, $\mathbf{r}_i$ is the point at which the wave is incident upon the medium, $\mathbf{r}_i + \boldsymbol{\rho}$ is the point of exit, $P_{(\mathbf{r}_i,\rho,\mathbf{k})}(\alpha)$ is the real probability amplitude

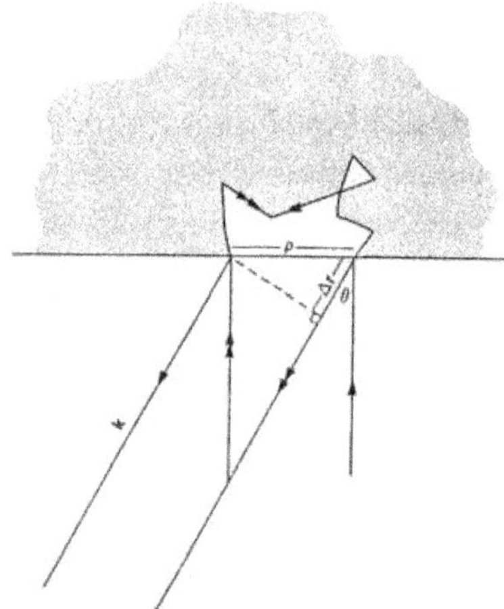

Fig. 39. Schematic diagram of two reflected partial waves from a random medium. The waves share the same path inside the medium but it is transversed in opposite senses. $\rho$ is the transverse displacement of the wave along the surface and $\Delta r$ is the additional path length in air of the wave emerging on the right $\rightarrow\rightarrow$ as compared to the wave emerging to the left $\rightarrow$.

for a wave to enter the medium at $\mathbf{r}_i$, to travel along path $\alpha$ in the medium, and leave at $\mathbf{r}_i + \boldsymbol{\rho}$ with wave vector $\mathbf{k}$. The sum in Eq. 87 may be rewritten by grouping together two partial waves in which the points of entry and exit are interchanged and the path $\alpha$ is the same except that the paths are traversed in opposite senses. These two partial waves have the same phase except for the additional phase $\Delta\phi$ in one of the paths. Thus Eq. 87 may be written as,

$$\langle I(\theta)\rangle = \left\langle \left| \sum_{\alpha}\sum_{(\mathbf{r}_i,\rho)}' P_{(\mathbf{r}_i,\boldsymbol{\rho},\mathbf{k})}(\alpha)e^{i\phi_\alpha}e^{i\mathbf{k}\cdot(\mathbf{r}-(\mathbf{r}_i+\boldsymbol{\rho}))}\left(1 + e^{i\mathbf{k}\cdot\boldsymbol{\rho}}\right)\right|^2\right\rangle, \quad (88)$$

where the prime on the sum indicates that paths in which the points of entry into and exit from the sample are reversed are not summed separately but are included together. Cross terms in the square of the field which do not correspond to the same path do not contribute to the ensemble average because of the random phasing of partial waves in different configurations.[54-65] This was discussed in detail in connection with the transition from the expression for the intensity correlation function in Eq. 55 to that in Eq. 56. This gives,

$$\langle I(\theta)\rangle = \left\langle \sideset{}{'}\sum_{\alpha}\sum_{(r_i,\rho)} P_{(r_i,\rho,k)}(\alpha)(2 + 2\mathrm{Re}(e^{ik\cdot\rho}))\right\rangle = I_{\text{in}} + I_c , \qquad (89)$$

where $I_{\text{inc}}$ is the value of the intensity that would be calculated neglecting interference of time reversed waves, corresponding to the incoherent background. The additional contribution of coherent backscattering to the reflected light is thus,

$$\langle I_c(\theta)\rangle = R_e \left\langle \sum_{\alpha}\sum_{(r_i,\rho)} P_{(r,\rho,k)}(\alpha)e^{ik\cdot\rho}\right\rangle . \qquad (90)$$

This can be expressed as,

$$I_c(\theta) = R_e \iint P_k(s,\rho)e^{ik\cdot\rho}dsd^2\rho , \qquad (91)$$

where $P_k(s,\rho)$ is the probability density along the surface that a photon incident upon the medium will exit the sample at a position displaced by $\rho$ with a path length of $s$ in the $k$ direction. After integration over $s$, the r.h.s. of Eq. 91 is $\int P_k(\rho)e^{ik\cdot\rho}d^2\rho$, which is the Fourier transform over the intensity at the surface. This is the field-field correlation function with angle whose square is given in Eq. 79 (Refs. 20,120) in the field factorization approximation.

A detailed comparison with weak localization experiments requires consideration of the vector nature of light and departures from the diffusion model which occur for pathlengths in the medium which are comparable to $l$ (Ref. 42,54–66,118,119).

## 8. LONG RANGE CORRELATION

In this section we consider the consequences of correlation in the square moduli of the amplitudes of different partial waves, $p^2(\alpha)$ in a given configuration which are ignored in the field factorization approximation (Eq. 78). Such correlation was neglected in going from the expression for the local intensity correlation function $G^I(\Delta\nu)$ in Eq. 56 to the expression in Eq. 57.

In Sec. 6, local intensity fluctuations are adequately described by the field factorization approximation in the limit $\delta \gg 1$. However, this approximation does not adequately describe local fluctuations for $\delta < 100$, nor can it describe long range intensity correlation even for $\delta \gg 1$. In the field factorization approximation, the normalized cumulant of the intensity cross correlation function between two points separated by a distance $R$ in an infinite medium is given by,[17]

$$C^I(R) = [\sin^2(kR)/(kR)^2]e^{-R/l} . \tag{92}$$

Stephen and Cwilich[18] have shown that as a result of long range intensity correlation there is an additional contribution to $C(R)$ which falls as a power law in $R$ for $R > l$. They find that the cumulant of the intensity cross correlation function between two points in a plane which is one mean free path from the transmitting surface, is given by,[18,19,30]

$$C(R) = (27/2k^2l^2)(l/L)^3[(L/R) + F(R/L)] \tag{93}$$

where $F(x) = \frac{1}{2}\int_0^\infty dq J_0(qx)\{[(\sinh(2q) - 2q)/\sinh^2 q] - 2\}$.

Feng, Kane, Lee and Stone have expressed $C$ as a function of sample rotation as an expansion using $1/g$ as the expansion parameter.[20] This is physically reasonable since $\delta = g$ in the weak scattering limit in the absence of absorption and $1/\delta$ is directly related to the degree of long range intensity correlation. The correlation function $C_{\alpha\beta\alpha'\beta'} = \langle\delta T_{\alpha\beta}\delta T_{\alpha'\beta'}\rangle$, where $\delta T_{\alpha\beta} = T_{\alpha\beta} - \langle T_{\alpha\beta}\rangle$, and $\alpha$ and $\alpha'$ are incoming and $\beta$ and $\beta'$ are outgoing transverse momentum states, is evaluated. It is expressed as $C_{\alpha\beta\alpha'\beta'} = C^{(1)}_{\alpha\beta\alpha'\beta'} + C^{(2)}_{\alpha\beta\alpha'\beta'} + C^{(3)}_{\alpha\beta\alpha'\beta'}$, where

$$C^{(1)}_{\alpha\beta\alpha'\beta'} = D_1\langle T_{\alpha\beta}\rangle\langle T_{\alpha'\beta'}\rangle\delta_{\Delta q_\alpha,\Delta q_\beta} F_1(\Delta q_\alpha L) , \tag{94a}$$

$$C^{(2)}_{\alpha\beta\alpha'\beta'} = D_2 g^{-1}\langle T_{\alpha\beta}\rangle\langle T_{\alpha'\beta'}\rangle[F_2(\Delta q_\alpha L) + F_2(\Delta q_\beta L)], \tag{94b}$$

$$C^{(3)}_{\alpha\beta\alpha'\beta'} = D_3 g^{-2}\langle T_{\alpha\beta}\rangle\langle T_{\alpha'\beta'}\rangle , \tag{94c}$$

where the $D$'s are constants of order unity $\Delta q_\alpha \equiv |q_\alpha - q_{\alpha'}|$ and $\Delta q_\beta = |q_\beta - q_{\beta'}|$ are wave vector differences and $F_1(x) = x^2/\sinh^2 x$ and

$F_2(x) = 2x^{-1}[\coth x - x/\sinh^2 x]$ (Ref. 20). This result is in qualitative agreement with recent experiments.[120] The variance of transmission for a single mode in and a single mode out is dominated by $C^{(1)}$, $\mathrm{var}(T_{\alpha\beta}) \sim C^{(1)}_{\alpha\beta\alpha\beta}$. The variance of transmission for one mode in and all modes out is dominated by $C^{(2)}$, $\mathrm{var}\sum_\beta T_{\alpha\beta} \sim \sum_{\beta\beta'} C^{(2)}_{\alpha\beta\alpha\beta'}$. This term gives a contribution $\eta$ times larger than the contribution from $C^{(1)}$. The variance of transmission for all modes in and all modes out is essentially given by the last term. It is $\mathrm{var}(\sum_{\alpha\beta} T_{\alpha\beta}) \sim \sum_{\alpha\beta\alpha'\beta'} C^{(3)}_{\alpha\beta\alpha'\beta'} \sim 1$, corresponding to universal conductance fluctuations.

Measurements were made of correlation between intensity at different points in the speckle pattern produced in transmission through the titania polystyrene sample shown in Fig. 10. They indicate that the correlation is extremely weak even when the two points are in neighboring speckle spots.[121] The measurements are made by recording speckle patterns with a video camera and using a frame grabber to digitize the pattern. Equation 94 shows that the correlation is reduced from the variance in intensity at a single point by a factor of $C^{(2)}_{\alpha\beta\alpha\beta'} \sim (Nl/L)^{-1} \sim (k^2 Ll/\pi)^{-1}$. In this sample $l = 1.5$ $\mu$m. At $L = 100$ $\mu$m, $g \sim 2 \times 10^4$ and the correlation is indeed very small. The correlation function in Eq. 60 for $\Delta\nu = 0$ does not contain any effects due to long range intensity correlation and corresponds to the term $C^{(1)}_{\alpha\beta\alpha\beta}$ in Eq. 94a. The influence of correlation for this sample is contained predominantly in the term $C^{(2)}_{\alpha\beta\alpha\beta}$ which we have seen is smaller by a factor of $2 \times 10^4$ than the leading term. This confirms that the neglect of amplitude correlation in the calculation of $C^I(\Delta\nu)$ for the titania-polystyrene sample shown in Fig. 25 was well justified.

It is of interest to compare the measured correlation frequency in the titania-polystyrene sample with the correlation frequency predicted for conductance fluctuations. The correlation function of conductance fluctuations with electron energy $E$ in a sample of thickness $L$ has been calculated by Lee and Stone in terms of the electron diffusion coefficient $D_e$ and phase breaking time $\tau_\phi$ (Ref. 28). When we identify $E$ with $h\nu$ and associate the measured values of $D$ and $\tau_a$ in our sample with $D_e$ and $\tau_\phi$, respectively, in the expression of Lee and Stone, we obtain the correlation function shown as the dashed curve in Fig. 40. The half-width of the correlation function is found to be $0.97\Delta\nu_I$, where $\Delta\nu_I$ is the measured value of the local intensity correlation function in our sample. We note that in general we do not expect the correlation frequency for local and total transmission to be identical. The dephasing time $\tau_\phi$ and the photon absorption time are not equal in general. Indeed for the case of lifetime limited dephasing $\tau_\phi = 2\tau_a$.[122] Further, the correlation function with frequency for total transmission should have a functional form related to that of $C^{(2)}$ rather than that of $C^{(1)}$ and the width is therefore expected to be somewhat larger than for $C^{(1)}$. In a recent experiment van Albada, de Boer and Lagendijk[123] showed that the frequency dependence of total optical transmission for a single input laser mode has the functional dependence of $C^{(2)}$ rather than $C^{(1)}$. The correlation function for conductance shown in Fig. 40 exhibits a long tail, falling as $1/\Delta\nu^{1/2}$ for large $\Delta\nu$. This is precisely the dependence of $C^{(2)}$ upon frequency in Eq. 94b when we make the identification $X = (2\pi\Delta\nu/D)^{1/2} L$ (Ref. 20).

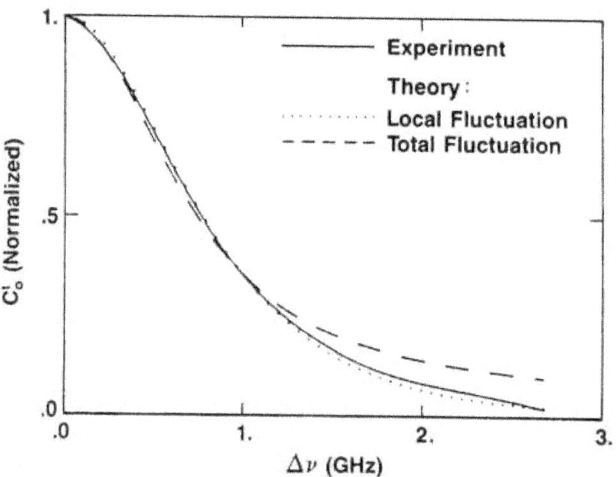


Fig. 40. The figure is to be compared with Fig. 25. The dashed line is obtained from the calculation of Lee and Stone for the conductance fluctuations using optical parameters as explained in the text.

Since $\Delta\nu_I$ for a single speckle spot is nearly the same as that calculated for the entire transmitted intensity, it appears that $\Delta\nu_I$ is nearly independent of the number of integrated spots, $N$. This is consistent with our observation shown in Fig. 41 in a sample composed of an equal volume of 0.5 $\mu$m and 0.8 $\mu$m polystyrene spheres. The value of the mean free path in this sample is $l \sim 1.5$ $\mu$m and the sample thickness is 570 $\mu$m. Within experimental error, $\Delta\nu_I$ does not vary as the aperture before the detector is opened to include from 1 to 300 speckle spots. The correlation functions shown in the figure are computed from the individual spectra shown. Only after substantial averaging does the correlation function approach the ensemble average. As the aperture is opened the contrast of the speckle pattern decreases. The intensity in each of the spectra shown are normalized to the average intensity. The $RMS$ value of transmitted intensity fluctuations with frequency for these normalized spectra falls roughly as $1/\sqrt{N}$, as expected for uncorrelated speckle spots. We expect that long range correlation in the speckle pattern should result in a slower decrease in the fluctuations in transmission only when $N > N_{\text{ind}}$. However, since $N_{\text{ind}} \sim g \sim 10^5$ for this sample at a thickness of 570 $\mu$m, correlation between speckle spots is not observed.

In the absence of absorption, the frequency correlation function for the total reflected and transmitted light must be the same (Eq. 38). Thus the correlation frequency of total reflected light is nearly the same as that observed at a point in the transmitted speckle pattern and depends upon the sample thickness. In contrast, the correlation frequency in a single speckle spot of reflected light is independent of sample thickness as seen in Eqs. 65 and 66. The correlation frequency is inversely proportional to the reflected pulse width which can be seen from Eq. 66 to be approximately $D/4\pi a^2$. For the titania-polystyrene sample, this is approximately 500 cm$^{-1}$. This is consistent with our observation that the intensity of the speckle pattern in reflection does not noticeably change when the laser is tuned by 1 cm$^{-1}$. We expect however, that, as the number of speckle spots detected in reflection increases, the correlation frequency will decrease, becoming equal to the correlation frequency in transmission when all the reflected light is detected.

                                   *Azriel Z. Genack*

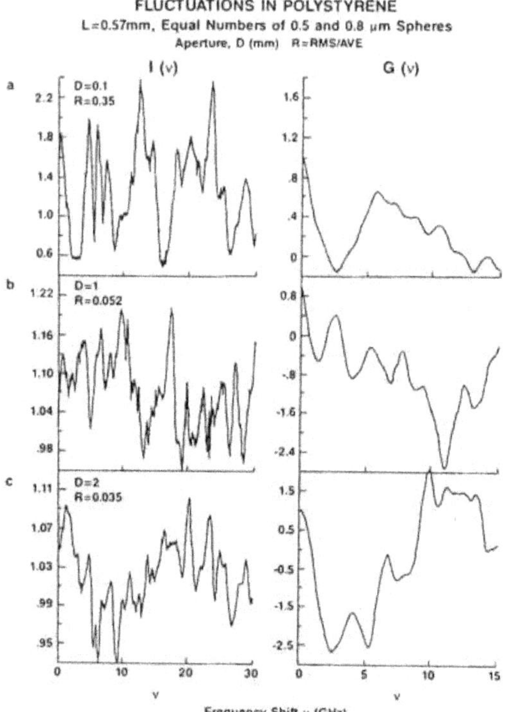



Fig. 41. Fluctuations in intensity transmitted through a sample of thickness $L = 570$ $\mu$m composed of a mixture of 0.5 and 0.8 $\mu$m polystyrene spheres. The total intensity transmitted through an aperture of increasing diameter $D$ given in milimeters is detected. The ratio of the RMS to the average intensity is $R$. The correlation function of each of the intensity spectra is shown.

We now consider fluctuations in transmission of local microwave intensity in a sample of 1/2 inch polystyrene spheres of length $L = 140$ cm[16,124]. Results for shorter lengths were discussed in Sec. 6. Since the cross section is constant, $N_{\text{ind}}$ for this sample decreases as $L$ increases. A typical intensity spectrum is shown in Fig. 42. The spectrum seems to float above the baseline in contrast to the spectrum in Fig. 27 obtained at $L = 60$ cm in which zeros in the spectrum are evident. The absence of small values of intensity is seen in the distribution of intensities shown in Fig. 43a. The peak in $P(I/\langle I \rangle)$ moves out to a higher value of $I/\langle I \rangle (\sim 0.25)$ continuing the trend towards a peak at higher values of $I/\langle I \rangle$ as $L$ increases, which was seen in Figs. 34–36. The reduced probability of low intensity values is a consequence of long range intensity correlation. In the presence of such correlation, the assumption of statistically independent partial waves, which led to negative exponential statistics (Sec. 6), is violated. The intensity at a point is correlated via incoherent photon diffusion to that at other points in the medium. Since the correlation with other points is positive, small values of intensity are unlikely. A departure from negative exponential statistics is also seen for $I/\langle I \rangle > 1$, giving a higher probability of large intensity fluctuations. The tail of $P(I/\langle I \rangle)$ is fit by a stretched exponential, $\exp -(I/\langle I \rangle)^\gamma$, with $\gamma = 0.75 \pm 0.05$. The plot of $\ln P(I/\langle I \rangle)$ versus $(I/\langle I \rangle)^{0.75}$ in Fig. 43b gives a straight line for $I/\langle I \rangle > 1$. $\gamma$ is found to decrease monotonically from the value $\gamma = 1$ for $L \leq 30$ cm as $L$ increases.[20]

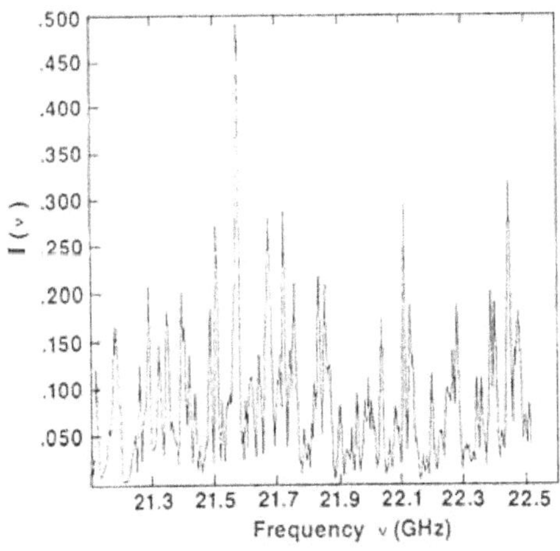

Fig. 42. Microwave intensity fluctuations as a function of frequency for a single random configuration of 1/2 inch polystyrene spheres at a thickness of $L = 140$ cm.

The measured correlation function with frequency is shown in Fig. 44 as the filled circles. The curve is a plot of Eq. 60 using the value $D$ obtained from the correlation function at shorter lengths. This value of $D$ gives a half width which is in accord with the uncorrelated diffusion model as seen in Fig. 29. The tail of the correlation function,

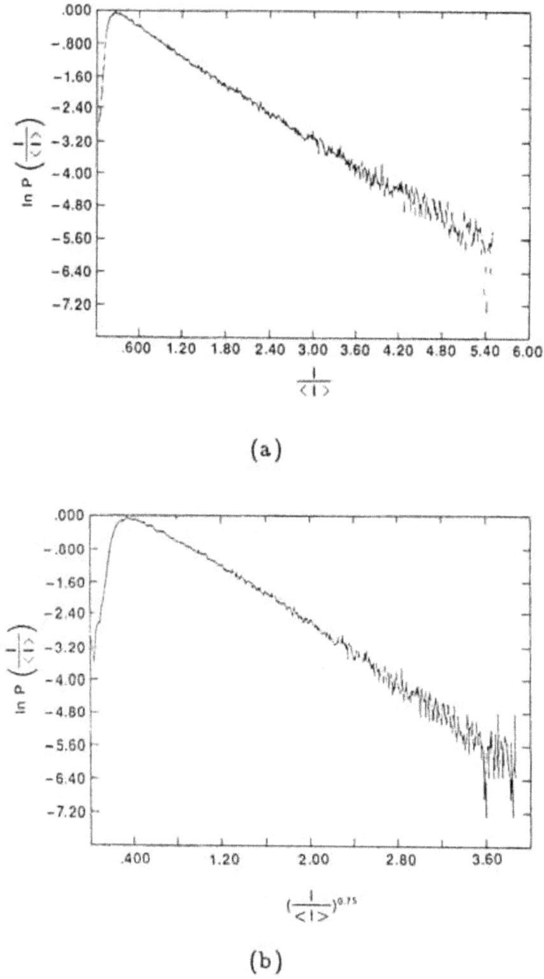

(a)

(b)

Fig. 43. Plot of ln $P(I/\langle I \rangle)$ for intensity transmitted through sample of 1/2 inch polystyrene spheres with $L = 140$ cm.

however, has a slow decay as opposed to the rapid fall off of Eq. 60. In Fig. 45 the correlation function is plotted versus $\Delta\nu^{-1/2}$. The straight line plot proves that the tail falls as $\Delta\nu^{-1/2}$ which is precisely the fall off of the correlation function of conductance fluctuations as seen in Fig. 40. Evidently this is the result of significant long range intensity

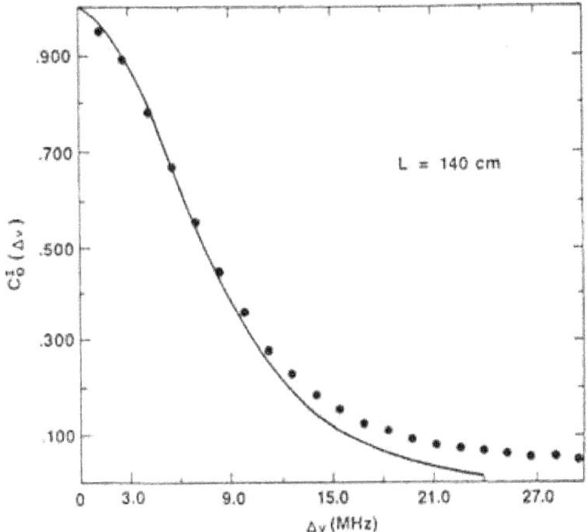



Fig. 44. The dots are the measured normalized cumulant intensity correlation function for the wave transmitted through a sample of 1/2 inch polystyrene spheres with $L = 140$ cm. The solid line is a plot of the diffusion model, in which long range correlation is neglected, (Eq. 60) using the values of $D$ obtained from measurements at $L = 60$ cm.

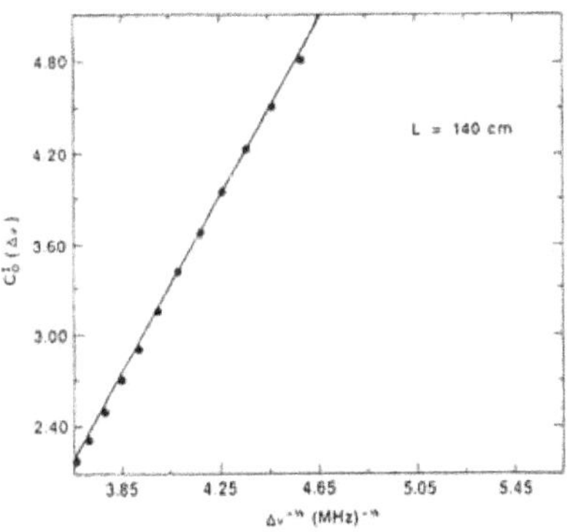



Fig. 45. Plot of the tail of the cumulant intensity correlation function shown in Fig. 44 versus $1/\Delta\nu^{1/2}$.

correlation in the sample.

To compare the results of intensity correlation in the strong correlation regime with theory we need to generalize Eq. 94 to include the effects of frequency shift and absorption. In the case that the time of flight distribution at a point is the same as that for the total transmission, it is given by Eq. 60 where $q_0$ is the negative imaginary root of

Eq. 76. This corresponds to the term $C^{(1)}$ in Eq. 94. The expression for $C^{(2)}$ in Eq. 94 may be generalized in a similar manner. Further, recognizing that in the limit $\delta \gg 1$, $1/\delta$ is the probability that a photon passing through a point near the output face of the medium will return to that point, we expect the effects of correlation to enter as an expansion in $1/\delta$. We note that, for a sample of fixed dimensions, $g^{-1}$ increases with absorption whereas the degree of long range intensity correlation should decrease as is the case for $\delta^{-1}$. Again $g$ is seen not to be a universal scaling parameter. Thus $C$ may be expressed as a universal function of the parameters $\delta\nu$ and $\delta$. $\delta\nu$ may be determined from measurements of either $T(t), G^{E}(\Delta\nu)$ or the leading term contributing to $C, C^{(1)}$. $\delta$ may be determined from the product of $\delta\nu$ and the density of states $dN/d\nu$ which can be estimated using Eq. 17. It may also be determined from the magnitude of the $C^{(2)}$ term.

Preliminary measurements of the intensity cross correlation function with frequency indicate that the dependence upon $R$ and $q_0$ can be factorized in each term contributing to $C^I$ for $R < L/2$. These considerations suggest that the cumulant intensity correlation function, normalized to $\langle I(\alpha L, r)\rangle \langle I(q_0 L, r+R)\rangle$, can be expressed as, $C^I(q_0 L, R) = C_1^I + C_2^I + C_3^I$, where

$$C^I = A_1 F_1(q_0 L) H_1(R), \tag{95a}$$

$$C^I = A_2 \delta^{-1} F_2(q_0 L) H_2(R), \tag{95b}$$

$$C^I = A_3 \delta^{-2}, \tag{95c}$$

Here, $A_1 = 1$ and $A_2$ and $A_3$ are of order unity for $\delta \gg 1$. Higher order corrections to $A$ should appear as an expansion in $\delta^{-1}$. We observe experimentally that $C_1^I$ decreases while $C_2^I$ and $C_3^I$ increase as $\delta^{-1}$ increases.[124] The $F$'s and $H$'s are normalized such that $F(\alpha L) = 1$ and $H(0) = 1$. $F_1$ is the normalized intensity correlation function in the field factorization approximation. For a point near the output face of the sample, $F_1$ is given by Eq. 60b. $H_1$ is the spatial intensity correlation calculated in the field factorization approximation. It falls rapidly with $R$ and is small for $R > \lambda$. $H_1$ was calculated by Shapiro[17] for points in an infinite medium in the absence of absorption

(Eq. 92). Following Eq. 94b, we write, $F_2(q_0 L) = |(\alpha/q_0)[\coth(q_0 L) - q_0 L/\sinh^2(q_0 L)]|/[\coth(\alpha L) - \alpha L/\sinh^2(\alpha L)]$ (Refs. 15,18,19,20,30). An expression for $H_2$ is given in Eq. 93 for two points separated by $R$ in a plane a distance $l$ from the output surface in the absence of absorption. We find experimentally that for points separated by $R$ lying on the axis of a cylinder with reflecting walls $H_2(R)$ has a correlation length equal to the sample length itself. In the absence of absorption, $H_3$ is constant in the sample.

The form of $C^I$ as given in Eq. 95 is compared to the experiment in Figs. 46 and 47. The points in Fig. 46 are the cumulant intensity autocorrelation function with frequency for $\frac{1}{2}$ inch polystyrene spheres at $L = 140$ cm calculated from $6 \times 10^3$ spectra. The least square fit of Eq. 95 to the data utilizing the value of $L_a$ obtained in the measurement of $T(L)$ and using $D, A_1, A_2\delta^{-1}$ and $A_3\delta^{-2}$ as adjustable parameters is shown in Fig. 46. The fit gives $D = 2.86 \times 10^{10}$ cm$^2$/s, $A_1 = .97, A_2\delta^{-1} = 0.065$ and $A_3\delta^{-2} = 0.0004$. The uncertainty in $C_3^I$ and $A_3$ is quite large. This is because there are fluctuations of $\sim \pm 15\%$ in the average intensity in individual spectra. Since $C_3^I$ is not frequency dependent, each spectrum contributes only a single number to the configuration average of $C_3^I$. The value of $D$ is slightly smaller than obtained for smaller $L$ and may reflect a renormalization of the diffusion coefficient by coherent backscattering. $C_1^I$ and $(C_2^I + C_3^I)$ as well as their sum is shown in Fig. 46. The residuals to the fit are also shown. Using this value of $D$ with the measured value of $L_a$ in Eq. 59 we are able to calculate $G^E(\Delta\nu)$. Its half width is $\delta\nu = 3.2$MHz. Using Eq. 17 to calculate the density of states $dN/d\nu = ALdn/d\nu$ with the value of $v$ from Eq. 63 gives $\delta = 20$. A different value of $\delta$ was given in Ref. 16 because the density of states used was twice the value given by Eq. 17 and because the half width of $C^I$ rather than of $C_1^I$ was used.

The relationship between the $C_2^I$ term and long range intensity correlation is confirmed by measurements of the intensity cross correlation function with frequency shift at two points in the plane of the output surface of the sample separated by $R = 6$ mm. The data shown in Fig. 47 is the average of over $3 \times 10^4$ configurations.[124] A fit of Eq. 95 to the data using the value of $D$ obtained from the fit to the autocorrela-

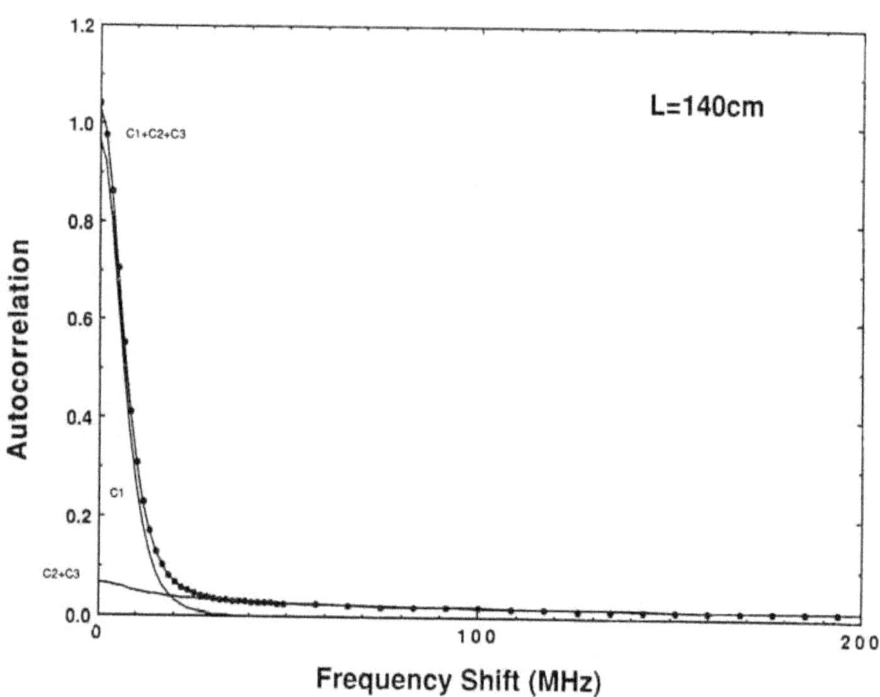

Fig. 46. Cumulant intensity autocorrelation function versus frequency shift normalized to $\langle I \rangle^2$. The sample is the same as in Figs. 42–45. The terms $C1$, $C2$ and $C3$ are the components of $C^I$ and are defined in Eq. 95.

tion function and using, $A_1 H_1 (R = 6$ mm$)$, $A_2 \delta^{-1} H_2 (R = 6$ mm$)$ and $A_3 \delta^{-2}$ as fitting parameters is also shown. $C_1^I$ and $C_2^I$ are smaller at $R = 6$ mm than at $R = 0$ because of the decrease of $H_1(R)$ and $H_2(R)$ with $R$. It is evident that $H_1(R)$ falls off more rapidly than $H_2(R)$. The results give $A_1 H_1 (R = 6$ mm$) = .035$, $A_2 \delta^{-1} H_2 (R = 6$ mm$) = .033$ and $A_3 \delta^{-2} = -0.002$. The negative value of $C_3^I$ reflects the high degree of noise in measurements of $C_3^I$, resulting from the fact that within a

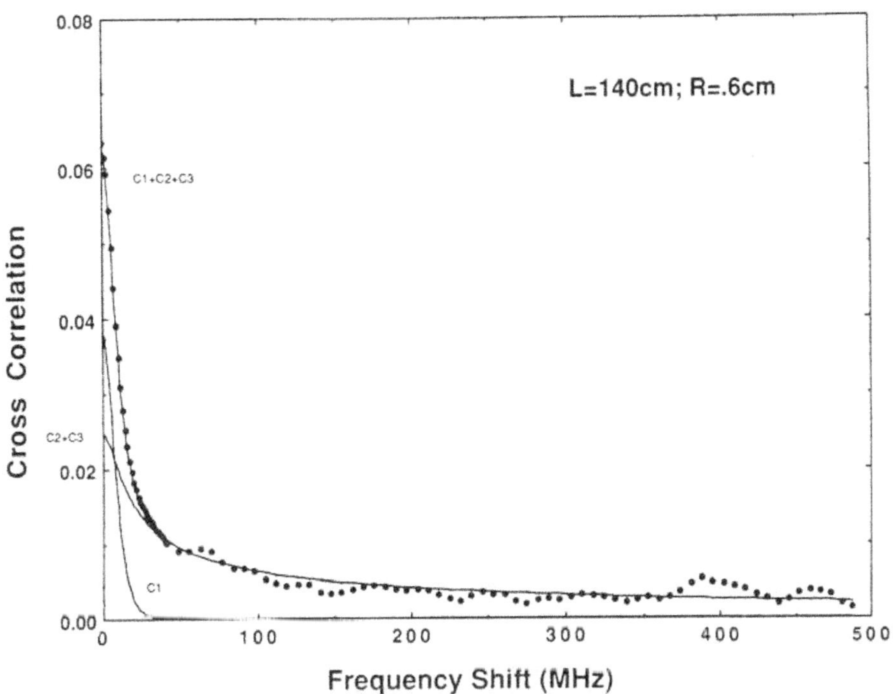



Fig. 47. Cumulant intensity cross correlation function versus frequency shift for probe separation of $R = 6$ mm.

single spectrum $C_3^I$ does not self average.

We have measured the intensity cross correlation function between a point in the center of the output surface and another point on the axis of the cylinder a distance $R$ from the output surface.[124] We find that for 20 cm $> R > 2$ cm, $C^I$ is essentially given by $C_2^I(q_0 L, R) = A_2 \delta^{-1} F_2(q_0 L) H_2(R)$. Further, we find that $H_2(R)$ does not fall as a power of $1/R$, as predicted by Eq. 93, but rather has a remarkably long range which is comparable to the sample length. In fact for 20 cm $< R <$ 2 cm, the magnitude of the cross correlation function is $A_2 \delta^{-1} H_2(R) = 0.042$. This behavior can be understood as follows. The intensity at any point is determined by the amplitudes of $N_{\text{ind}} = \delta$ eigenstates. The measured intensity at a point is a single number and leads to a fractional correlation of $1/N_{\text{ind}} = \delta^{-1}$ in the intensity at another point. For $R >$ 20 cm, the cross correlation function is smaller because only a fraction of the photon density emerges from the output face. Some photons reach the input face and contribute to the reflected signal. The effects of absorption are taken account of in the normalization of $C^I$. Thus, $\delta$ can be determined from the measurement of the cross correlation function, even without the knowledge of $A_2$. This gives $\delta = (0.042)^{-1} = 24$. This is nearly the same value as determined from the experimental determination of $C_1^I$ and the calculation of $dN/d\nu$. These measurements differ somewhat because the definition of $\delta\nu$ as the half width of $G^E(\Delta\nu)$ is somewhat arbitrary. In contrast, the determination of $\delta$ from the cross correlation measurement is independent of any assumption regarding the definition of $\delta\nu$.

Measurements of the intensity cross correlation function were also made in a sample of 3/16 inch aluminium spheres with a filling fraction $f = 0.2$ and 3/16 inch teflon spheres with $f = 0.4$ contained in the sample holder shown in Fig. 11 for a number of lengths. The amplitude of $C_2^I$ determined in measurements in both the polystyrene sample in which $kl = 27$ and the aluminium ball samples in which $kl = 2.5$ is found to be proportional to $1/\delta$.[124]

The total transmission through the sample can be written as[20,86,89] $T = \Sigma T_{\alpha\beta}$, where $T_{\alpha\beta}$ is the coupling coefficient of intensity in the incoming channel $\alpha$ to the outgoing channel $\beta$. These channels can be taken to be the $N_s$ transverse momentum states with the same energy. In the weak scattering limit, $T = g = \delta = N_{\text{ind}} \gg 1$ and var $(T_{\alpha\beta}) = \langle T_{\alpha\beta}\rangle^2$. If correlation between channels could be neglected, the variance of the total transmission in the case of a single input channel $\alpha$ would be $N_s \langle T_{\alpha\beta}\rangle^2 \sim \langle T\rangle^2/N_s$, since for the typical channels $\alpha$ and $\beta$, $\langle T_{\alpha\beta}\rangle \sim \langle T\rangle/N_s$. When all input modes are present there are $N_s^2$ terms contributing to $T$, with an average value $\langle T_{\alpha\beta}\rangle \sim \langle T\rangle/N_s$. In the absence of correlation between the terms, the variance of transmission would be $\langle T \rangle^2 / N_s$.

Intensity correlation in the transmitted speckle pattern occurs because the total transmission depends upon $N_{\mathrm{ind}}$ rather than $N_s$ independent parameters for a single input channel and upon $N_{\mathrm{ind}}^2$ rather than $N_s^2$ independent parameters when all input channels are present. The transmitted flux may be calculated by considering the wave with velocity $c$ on a hemispherical surface in the far field speckle pattern,

$$T_\nu = \int |\psi_\nu(\mathbf{r})|^2 c \cos\theta\, r^2 d\Omega \sim \sum_{i,j} c_i^* c_j \int u_i(\mathbf{r}) u_j(\mathbf{r}) c \cos\theta\, r^2 d\Omega. \quad (96)$$

The integral on the r.h.s. of Eq. 96 for $i = j$ is $T_i$, the transmission for unit incident flux associated with the incoming part of the wave on the left, $u_i$. For the case of a single input mode $\langle T_i \rangle = \langle T \rangle$ and $\langle |c_i|^2 \rangle \sim 1/N_{\mathrm{ind}}$. A typical value of a term in the sum in Eq. 10 with $i = j$ is, therefore, $\langle T \rangle / N_{\mathrm{ind}}$. Further, since $\mathrm{var}(|c_i|^2) = \langle |c_i|^2 \rangle^2$ in the weak scattering limit, the contribution to $\mathrm{var}(T)$ of these diagonal terms is $N_{\mathrm{ind}}(\langle T \rangle / N_{\mathrm{ind}})^2 = \langle T \rangle^2 / N_{\mathrm{ind}}$ for a single incident channel. This is a factor $N_s / N_{\mathrm{ind}} = \eta$ greater than the variance calculated under the assumption of uncorrelated speckle. For $i \neq j$, the phase of $u_i^* u_j$, in each of $N_s$ coherence area of the speckle pattern is random. Therefore, a typical value of an integral on the r.h.s. of Eq. 96 is $T/(N_s)^{\frac{1}{2}}$. A typical term in the sum on the r.h.s. of Eq. 96 for $i \neq j$ is thus $\langle T \rangle / (N_s)^{\frac{1}{2}} N_{\mathrm{ind}}$. Since there are $N_{\mathrm{ind}}^2$ terms in the off diagonal sum, the contribution of the diagonal sum to $\mathrm{var}(T)$ is $N_{\mathrm{ind}}^2(\langle T \rangle^2 / N_s N_{\mathrm{ind}}^2) = \langle T \rangle^2 / N_s$. This is the same as the variance found under the assumption of uncorrelated speckle spots. Thus, $\mathrm{var}(T)$ for a single input channel is dominated by the diagonal terms in Eq. 96. It is enhanced over the prediction of the uncorrelated speckle model by a factor $\eta$ which has the value $\eta = 1.56 L/l$ in the weak scattering limit. This result is in agreement with calculations of Stephen and Cwilich.[18]

When all the input channels are present there are only $N_{\mathrm{ind}}$ channels which couple effectively to the wave of frequency $\nu$ in the random medium. Thus, the total transmission can be written as a sum involving $N_{\mathrm{ind}}$ incoming channels and $N_{\mathrm{ind}}$ outgoing channels. A typical term in the sum, therefore, has a magnitude $\langle T \rangle / N_{\mathrm{ind}}^2$. Thus

$\mathrm{var}(T) \sim N_{\mathrm{ind}}^2 (\langle T \rangle / N_{\mathrm{ind}}^2)^2 = \langle T \rangle^2 / N_{\mathrm{ind}}^2$. Since $\langle T \rangle = g = \delta = N_{\mathrm{ind}}$ in the limit $\delta \gg 1$, this gives $\mathrm{var}(T) \sim 1$. This is a factor $\eta^2$ larger than obtained assuming $N_s$ statistically independent channels in both the incoming and outgoing wave. This is the well known result of universal conductance fluctuations obtained by diagrammatic Green's function calculations[20,28,29] as well as by the method of random matrices.[31]

## 9. SCALE DEPENDENCE OF DIFFUSION COEFFICIENT

In this section we discuss optical and microwave propagation in the critical regime in which the wave is extended but scattering is strong enough that $l < \lambda$. According to the scaling theory of localization, the diffusion coefficient is scale dependent in this regime and is renormalized by a factor $l/\xi$ for $L > \xi$ (Eqs. 19–21). It can be expressed as $D = v\tilde{l}/3$, where $\tilde{l}/l = l/\xi = D_B/D$. $\tilde{l}$ is essentially the diffusion coefficient expressed in dimensions of length and does not correspond to a physical length. For $L, L_a \gg \xi$, $\xi \simeq \xi_0$ and $\tilde{l} = l^2/\xi_0$. Using Eq. 21, this gives $\tilde{l} = l - l_c$. Thus when $l$ is close to $l_c$, $D$ is significantly renormalized.

At the mobility edge $\xi_0$ diverges and the diffusion coefficient $D_c$, approaches zero as $L \rightarrow \infty$ in the absence of absorption. In the presence of absorption, $\tilde{l}$ is essentially the diffusion coefficient expressed in dimensions of length and does not correspond to a physical length. $D_c$ can be written as

$$D_c(L) = (vl_c/3)\left(\frac{1}{L_a} + \frac{1}{L}\right) . \tag{97}$$

Using the value of $l_c$ given by the Ioffe-Regel criterion, $l_c = 1/k$, and taking the limit of large $L, L \gg L_a$ so that, $\xi_c \rightarrow L_a = (D_c \tau_a)^{1/2}$, we obtain,

$$D_c = (v^2/9k^4\tau_a)^{1/3} . \tag{98}$$

If we find that the measured value of $D$ is larger than $D_c$, we may conclude that the wave is delocalized. However, if at the same time $\tilde{l} < 1/k$, the wave is in the critical regime and $D$ is reduced by coherent backscattering.

The observation of optical propagation in the critical regime has recently been reported.[44] The sample studied is 129 $\mu$m thick and is composed of anatase titania spheres with index of refraction 2.3 and diameters $d = 0.6 \pm 0.2$ $\mu$m. The index contrast is nearly twice as large as in the titania polystyrene sample shown in Fig. 10. The sample is prepared by compacting a water colloid in a centrifuge. The volume fraction is expected to be between 0.5 and 0.7.

The values of $D$ and $\tau_a$ are determined from measurements of picosecond pulse transmission in the frequency range 514–640 nm. The measurement of the transmitted laser pulse are made by time-correlated single photon counting. The transmitted laser pulse at 590 nm and the instrumental response to the laser pulse are shown in Fig. 47. The solid line through the data is a nonlinear least squares fit of the photon diffusion model of Eq. 46 reconvoluted with the instrumental response to the transmitted laser pulse. This gives $D = 2.6 \times 10^4$ cm$^2$/s and $\tau_a = 390$ps. We expect the pulse profile is accurately given by Eq. 46 even in the strong scattering regime with $D(L)$ substituted for $D$ as long as $D(L)$ is the same for all the frequency components in the pulse and $L > L_a$.

The values of $D$, $\tau_a$ and $L_a$ obtained from measurements of pulsed transmission are shown in Fig. 49. At all wavelengths at which measurements were made, the sample is thicker than the absorption cutoff length $L > L_a$.

The value of $D$ at 514 nm is 20 times smaller than for the sample of polydisperse rutile titania particles in polystyrene, with transverse dimensions $d \sim 0.25 \pm 0.15$ $\mu$m which we have discussed previously. Because the distribution of sphere diameters is large and the local configuration of spheres is random, sharp scattering resonances are not expected[104] and $l, \tau_a$ and $v$ are expected to change gradually with wavelength.

To determine whether the wave is extended or localized we compare the measured value of $D$ to $D_c$ in Eq. 98 using the measured value of $\tau_a$. The wave velocity is obtained using Eq. 63. Using the value $f = 0.5$ gives $v = 1.7 \times 10^{10}$ cm/s. The largest value of $D_c$ occurs at 514 nm, $D_c \sim 4.5 \times 10^2$ cm$^2$/s. Since $D$ is larger than $D_c$ at all frequencies at which measurements were made, the wave is delocalized.

At 514 nm we find $\tilde{l} = 250$ A. This is smaller than $d$ by a factor of

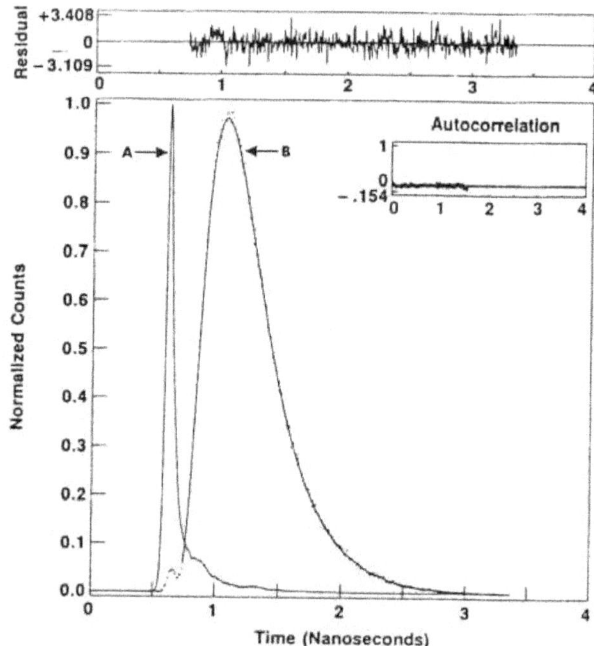



Fig. 48. The points are transmitted pulse profile through a 129 $\mu$m thick sample of titania particles. The solid line is a reconvolution of the measured response to the incident pulse, which is the narrow peak shown in the figure, with Eq. 46 using $D$ and $\tau_a$ as fitting parameters to minimize the residual. The autocorrelation function of the residual is also shown.

25 and does not correspond to a physically reasonable mean free path in the medium. Moreover, we find that $k\tilde{l} \sim 0.5$ at 514 nm and $k\tilde{l} < 1$ for all frequencies measured. This is consistent with conjecture that the wave is extended but the diffusion coefficient is renormalized. The diffusion coefficient is reduced to a fraction $D/D_B \simeq \tilde{l}/l \simeq (l - l_c)/l \simeq$ 0.35 of its classical value at 514 nm. This implies that $l$ is within $\sim 35\%$ of $l_c$.

Light localization is predicted to occur in a strong scattering window between regimes of weak scattering at long wavelengths, in which effective medium theory holds, and at short wavelengths, in which propagation is described by geometric optics. Using a self-consistent effective medium approach, 580 nm corresponds approximately to the third Mie resonance of the solid spheres embedded in a medium or the first

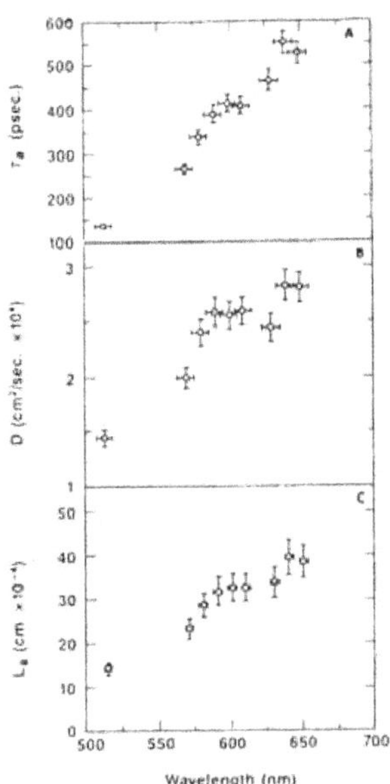

Fig. 49. Values of $\tau_a, D$ and $L_a = (D\tau_a)^{1/2}$ as a function of laser wavelength obtained from measurement of pulsed transmission.

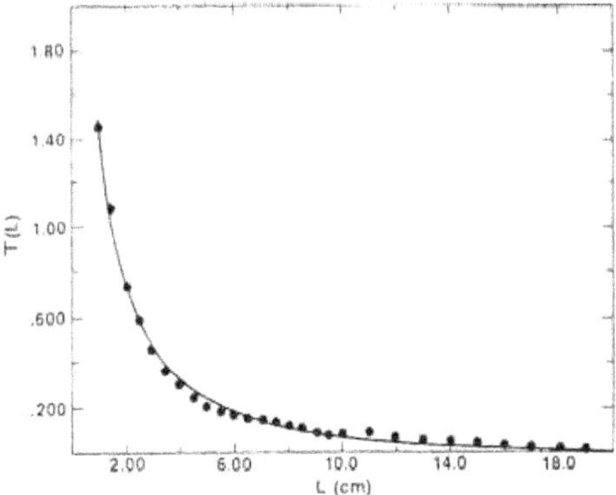

Fig. 50. Transmission versus thickness at 24.5 GHz through a sample composed of 1/8 inch aluminum spheres with a filling fraction $f = 0.20$ (see Fig. 19). The solid line is a fit to the data using the diffusion model (Eq. 40).

resonance from the air between the spheres. The short mean free paths observed may be associated with high order Mie resonances as predicted

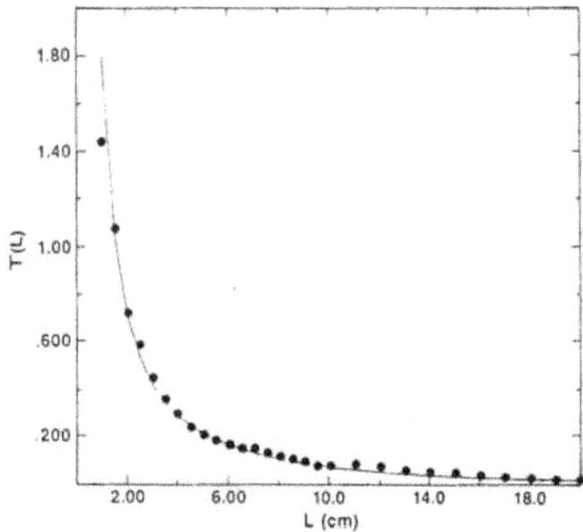

Fig. 51. The solid line is a fit to the same data as in Fig. 50 using a scale dependent diffusion coefficient with $\xi_0 = 0.8$ cm.

by Sheng and Zhang[72] but it would be necessary to measure transport over a broader frequency range to test this possibility. Soukoulis and Economu[125] have calculated the frequency dependence of $D$ for this sample using a coherent potential approximation. They are able to fit the measured frequency dependence of $D$ only if $f = 0.7$. They predict that even stronger scattering can occur in this sample outside the frequency range measured and that light could be localized in this sample for some frequencies. Stronger scattering is expected in samples with lower solid fraction, smaller particle dimensions, more uniform particle size, greater short range order and stronger index mismatch. It seems likely therefore that optical localization will be achieved in the near future.

Measurements of microwave propagation in the strong scattering regime have been carried out in random metallic samples. Measurements of diffusive propagation at 19 GHz in a sample of 1/8 inch dimension spheres mixed with polystyrene spheres with $f = 0.20$ were shown in Fig. 19 (Ref. 45). We find that at higher frequencies, the initial fall of $T(L)$ becomes faster than $1/L$. At 24.5 GHz, $T(L)$ is not adequately fit by the diffusion model as expressed in Eq. 40. The best fit to the diffusion model is shown in Fig. 50. An improved fit to the data is obtained by substituting $D(L)$ in Eqs. 19–21, for $D$ in Eq. 40

using the value $\xi_0 = 0.8$ and $L_a = 5.9$ cm in Eq. 20 for $\xi$. This fit of the expression for $T(L)$ to the data is shown in Fig. 51. These results suggest that the scattering strength is increasing with increasing frequency. The departure from diffusive behavior occurs near the upper end of our tuning range. To study the frequency dependence of propagation in the critical regime, we studied samples composed of 3/16 inch aluminum spheres. The use of larger spheres lowers the sphere resonance frequency and may bring critical behavior into the tuning range of our microwave oscillators.

## 10. CONCLUSION

A wide variety of measurements of EM propagation have been presented which support the conjecture that a universal description of wave propagation can be given in terms of the level width $\delta\nu$ and the level spacing $d\nu/dN$. $\delta\nu$ can be determined from the field-field correlation function with frequency or its Fourier transform, the time of flight distribution as well as from the intensity-intensity correlation function with frequency. Convenient combinations of the statistical parameters $\delta\nu$ and $d\nu/dN$ are the dimensionless level width $\delta$, which gives the proximity to the localization threshold and the density of states $dN/d\nu$, or $\delta$ and the diffusion coefficient $D$, or $\delta$ and $\delta\nu$.

We have shown that $1/\delta$ is the probability of return of a photon to a point in the medium. In the limit $\delta \gg 1$, $1/\delta$ is proportional to the degree of long range intensity correlation and is determined from measurements of intensity correlation. Nonetheless, $\delta$ is defined in a manner which is independent of spatial correlation in a sample.

In conclusion, we have shown that, aside from a parameter needed to describe inelastic processes, propagation in the regime $\delta > 1$ can be fully described using only two additional parameters which are determined from measurements of fluctuations, correlation and average transport of electromagnetic radiation in random media.

## ACKNOWLEDGMENTS

The experiments described in this work are the result of collaborations with many colleagues. I would especially like to acknowledge

the beautiful measurements of microwave propagation by N. Garcia and
W. Polkosnik of Queens College and of picosecond optical transmission
by J. M. Drake of the Exxon Research and Engineering Company. I
would also like to acknowledge the contribution of J. Zhu, L. A. Ferrari,
R. B. Stephens and H. Deckman and the expert technical assistance of
E. Kuhner and J. Varon. I thank S. John, G. Kurizki, B. Shapiro,
M. Stephen, M. Kaveh, I. Freund, S. Feng and M. H. Cohen for many
stimulating discussions. Finally, I thank S. Wasserman and S. Allen for
typing the manuscript. I would like to acknowledge the support of the
Exxon Research Foundation and of two PSC-CUNY Research Grants.

## REFERENCES

1. B. L. Altshuler, A. G. Aronov, D. E. Khmelnitskii and A. I. Larkin,
   in *Quantum Theory of Solids*, ed. I. M. Lifshitz (Mir Moscow, 1982) p.
   130; D. E. Khmelnitsii, *Physica* **126B** (1984) 235.
2. G. Bergmann, *Phys. Rep.* **107** (1984) 1.
3. E. Abrahams, P. W. Anderson, D. C. Licciardello and T. V. Ramakrishnan, *Phys. Rev. Lett.* **42** (1979) 673.
4. D. Yu Sharvin and Yu V. Sharvin, *Pisma Zh. Teor. Eksp. Fiz.* **34**
   (1981) 285 [*Sov. Phys. JETP Lett.* **34** (1981) 272].
5. R. A. Webb, S. Washburn, C. P. Umbach and R. B. Laibowitz, *Phys.
   Rev. Lett.* **54** (1985) 2696.
6. V. Chandrasekhar, M. J. Rooks, S. Wind and D. E. Prober, *Phys. Rev.
   Lett.* **55** (1985) 1610.
7. P. W. Anderson, *Phys. Rev.* **109** (1958) 1492.
8. D. J. Thouless, *Phys. Rev. Lett.* **39** (1977) 1167.
9. S. John, *Phys. Rev. Lett.* **53** (1984) 2169; *Phys. Rev.* **B31** (1985) 304.
10. P. W. Anderson, *Phil. Mag.* **52** (1985) 505.
11. A. F. Ioffe and A. R. Regel, *Prog. Semicond.* **4** (1960) 237.
12. J. W. Goodman, *J. Opt. Soc. Am.* **66** (1976) 1145.
13. J. W. Goodman, *Statistical Optics* (J. Wiley, New York, 1985).
14. A. Z. Genack, *Phys. Rev. Lett.* **58** (1987) 2043.
15. A. Z. Genack and J. M. Drake, to appear in Europhysics Lett.; A. Z.
    Genack and J. M. Drake, *Bull. Am. Phys. Soc.* **32** (1987) 792.
16. N. A. Garcia and A. Z. Genack, *Phys. Rev. Lett* **63** (1989) 1678.
17. B. Shapiro, *Phys. Rev. Lett.* **57** (1986) 2168.
18. M. J. Stephen and G. Cwilich, *Phys. Rev. Lett.* **59** (1987) 285.
19. M. J. Stephen, *Phys. Lett.* **A127** (1988) 371.
20. S. Feng, C. Kane, P. A. Lee and A. D. Stone, *Phys. Rev. Lett.* **61**
    (1988) 834.

21. R. Berkowits, I. Edrei and M. Kaveh, *Phys. Rev. B*, to be published.

22. A. Z. Genack, submitted for publication.

23. D. Volhardt and P. Wolfe, *Phys. Rev. Lett.* **48** (1982) 699.

24. P. A. Lee and T. V. Ramakrishnan, *Rev. Mod. Phys.* **57** (1985) 287.

25. J. T. Edwards and D. J. Thouless, *J. Phys.* C5 (1972) 807.

26. D. C. Licciardello and D. J. Thouless, *J. Phys.* C8 (1975) 4157.

27. A. D. Stone, *Phys. Rev. Lett.* **54** (1985) 2692.

28. P. A. Lee and A. D. Stone, *Phys. Rev. Lett.* **55** (1985) 1622; P. A. Lee, A. D. Stone and H. Fukuyama, *Phys. Rev.* B35 (1987) 1039.

29. B. L. Altshuler, *Pisma Zh. Eksp. Teor. Fiz.* **41** (1985); [*Sov. Phys. JETP Lett.* **41** (1985) 648]. B. L. Altshuler and D. E. Khmelnitskii, *Pisma Zh. Eksp. Teor. Fiz.* **42** (1985) 291; [*Sov. Phys. JETP Lett.* **42** (1985) 359].

30. R. Pnini and B. Shapiro, *Phys. Rev.* B39 (1989) 6986.

31. Y. Imry, *Europhys. Lett.* **1** (1986) 249.

32. P. A. Mello, E. Akkermans and B. Shapiro, *Phys. Rev. Lett.* **61** (1988) 459.

33. B. Z. Spivak and A. Yu Zyuzin, *Solid State Comm.* **65** (1988) 311; A. Yu Zyuzin and B. Z. Spivak, Zh. Eksp. Teor. Fiz. **93** (1987) 994 [*Sov. Phys. JETP* **66** (1987) 560].

34. A similar approach is taken independently by E. Reidel, unpublished.

35. N. F. Mott and J. H. Davies, *Electronic Processes in Non-Crystalline Materials* (Oxford University Press, Oxford, 1979); M. H. Cohen, H. Fritzche and S. R. Ovshinsky, *Phys. Rev. Lett.* **22** (1969) 1065.

36. G. H. Watson, Jr., P. A. Fleury and S. L. McCall, *Phys. Rev. Lett.* **58** (1987) 945.

37. G. Maret and P. E. Wolf, *Z. Phys.* B65 (1987) 409.

38. A. A. Golubentsev, *Zh. Eksp. Teor. Fiz.* **86** (1984) 47. [*Sov. Phys. JETP* **59** (1984) 26].

39. M. J. Stephen, *Phys. Rev.* B37 (1988) 1.

40. M. Rosenbluh, M. Hoshen, I. Freund and M. Kaveh, *Phys. Rev. Lett.* **58** (1987) 2754; I. Freund, M. Kaveh and M. Rosenbluh, *Phys. Rev. Lett.* **60** (1988) 1130.

41. D. J. Pine, D. A. Weitz, P. M. Chaikin and E. Herbolzheimer, *Phys. Rev. Lett.* **60** (1988) 1134.

42. F. C. MacKintosh and S. John, *Phys. Rev.* B40 (1989) 2383.

43. J. Zhu, W. Polkosnik and A. Z. Genack, to be submitted for publication.

44. J. M. Drake and A. Z. Genack, *Phys. Rev. Lett.* **63** (1989) 259.

45. N. Garcia, and A. Z. Genack, submitted for publication; A. Z. Genack, L. A. Ferrari, J. Zhu, N. Garcia and J. M. Drake, *Advances in Laser Science III*, AIP Conf. Proceedings (AIP, New York, 1988).

46. E. Yablonovitch, *Phys. Rev. Lett.* **58** (1987) 2059.

47. S. John, *Phys. Rev. Lett.* **58** (1987) 2486.

48. G. Kurizki and A. Z. Genack, *Phys. Rev. Lett.* **61** (1988) 2269.

49. U. Frisch in *Probabilistic Methods in Applied Mathematics*, Vol. 1, ed. A. T. Bharncha-Reid (Academic Press, NY, 1968).

50. S. John, H. Sompolinsky and M. J. Stephen, *Phys. Rev.* **B27** (1983) 5592; S. John and M. J. Stephen, *Phys. Rev.* **B28** (1983) 6358.

51. S. John, Comments on Cond. Mat. Phys. **14** (1988) 193.

52. *Localization, Interaction, and Transport Phenomena in Impure Metals*, eds. G. Bergmann, Y. Bruynserade and B. Kramer (Springer-Verlag, NY, 1985).

53. Y. Aharanov and D. Bohm, *Phys. Rev.* **115** (1959) 485.

54. Y. Kuga and A. Ishimaru, *J. Opt. Soc. Am.* **A1** (1984) 831.

55. A. Ishimaru, *Wave Propagation and Scattering in Random Media* (Academic Press, NY, 1978).

56. E. Akkermans and R. Maynard, *J. Phys. France Lett.* **46** (1985) L1045.

57. M. P. van Albada and A. Lagendijk, *Phys. Rev. Lett.* **55** (1985) 2692.

58. P. E. Wolf and G. Maret, *Phys. Rev. Lett.* **55** (1985) 2696.

59. S. Etemad, R. Thompson and M. J. Andrejco, *Phys. Rev. Lett.* **57** (1986) 575.

60. M. Kaveh, M. Rosenbluh, I. Edrei and I. Freund, *Phys. Rev. Lett.* **57** (1986) 2049.

61. E. Akkermans, P. E. Wolf and R. Maynard, *Phys. Rev. Lett.* **56** (1986) 1471.

62. M. P. van Alabada, M. B. van der Mark and A. Lagendijk, *Phys. Rev. Lett.* **58** (1987) 361; M. P. van Albada and A. Lagendijk, *Phys. Rev.* **36** (1987) 2357.

63. S. Etemad, R. Thompson, M. J. Andrejco, S. John and F. C. Mackintosh, *Phys. Rev. Lett.* **59** (1987) 1420; F. C. MacKintosh and S. John, *Phys. Rev.* **B37** (1988) 1884.

64. I. Freund, M. Rosenbluh, R. Berkovits and M. Kaveh, *Phys. Rev. Lett.* **61** (1988) 1214.

65. P. E. Wolf, G. Maret, E. Akkermans and R. Maynard, *J. Phys. France* **49** (1988) 63; E. Akkermans, P. E. Wolf, R. Maynard and G. Maret, *J. Phys. France* **49** (1988) 77.

66. M. J. Stephen and G. Cwilich, *Phys. Rev.* **B34** (1986) 7564.

67. A. Cohen, Y. Roth and B. Shapiro, *Phys. Rev.* **B38** (1988) 12125.

68. R. Vreeker, M. P. van Albada, R. Sprik and A. Lagendijk, *Phys. Lett.* **A132** (1988) 51.

69. K. M. Yoo, Y. Takiguchi and R. R. Alfano, *Appl. Opt.* **28** (1989) 2343.

70. K. Arya, Z. B. Su and J. L. Birman, *Phys. Rev. Lett.* **54** (1985) 1559.

71. K. Arya, Z. B. Su and J. L. Birman, *Phys. Rev. Lett.* **57** (1986) 2725.

72. P. Sheng and Z. Q. Zhang, *Phys. Rev. Lett.* **57** (1986) 1879.

73. T. R. Kirkpatrick, *Phys. Rev.* **B31** (1985) 5746; C. A. Condat and T. R. Kirkpatrick, *Phys. Rev. Lett.* **58** (1987) 226.

74. C. M. Soukoulis, E. N. Economu, G. S. Grest and M. H. Cohen, submitted for publication.

75. F. J. Wegner, *Z. Phys.* **25** (1976) 327.

76. D. J. Bishop, D. C. Tsui and R. C. Dynes, *Phys. Rev. Lett.* **44** (1980) 1153; *Localization, Interaction and Transport Phenomena*, eds. G. Bergmann, Y. Bruynserade and B. Kramer (Springer-Verlag, NY, 1985) p. 31.

77. R. A. Davies, M. Pepper and M. Kaveh, *J. Phys.* **C16** (1983) L285.

78. M. Kaveh and N. F. Mott, *Phil. Mag.* **B55** (1987) 19.

79. P. W. Anderson, D. J. Thouless, E. Abrahams and D. S. Fischer, *Phys. Rev.* **B22** (1980) 3519.

80. M. Ya. Azbel, *Solid State Comm.* **45** (1983) 527.

81. D. P. Vincenzo and M. Ya. Azbel, *Phys. Rev. Lett.* **50** (1983) 2102.

82. B. L. Altshuler, V. E. Kravtsov, and I. V. Lerner, *Pisma Zh. Eksp. Teor. Fiz.* **43** (1986) 342 [*JETP Lett.* **43** (1986) 441]; *Zh. Eksp. Teor. Fiz.* **91** (1986) 2276 [*Sov. Phys. JETP* **64** (1986) 1352]; N. Kumar and A. M. Jayannavar, *J. Phys.* **C19** (1986) L85.

83. B. Shapiro, *Phys. Rev.* **B34** (1986) 4394.

84. B. Shapiro, *Phil. Mag.* **B56** (1987) 1031.

85. I. Edrei, M. Kaveh and B. Shapiro, *Phys. Rev. Lett.* **62** (1989) 2120.

86. R. Landauer, *Phil. Mag.* **21** (1970) 863.

87. P. W. Anderson, *Phys. Rev.* **B23** (1981) 4828.

88. H. L. Enquist and P. W. Anderson, *Phys. Rev.* **B24** (1981) 1151.

89. D. S. Fisher and P. A. Lee, *Phys. Rev.* **B23** (1981) 6851.

90. P. A. Lee, *Physica* **140A** (1986) 169.

91. M. P. van Albada and A. Lagendijk, submitted for publication.

92. J. L. Pichard, *Europhys. Lett.* **2** (1986) 477.

93. K. A. Muttalib, J. L. Pichard and A. D. Stone, *Phys. Rev. Lett.* **59** (1987) 2475.

94. P. A. Mello, *Phys. Rev. Lett.* **60** (1988) 1089.

95. M. P. Zaitlin and A. C. Anderson, *Phys. Rev.* **B12** (1975) 4475.

96. J. Jackle, *Solid State Comm.* **39** (1981) 1261.

97. R. Biswas, A. M. Bouchard, W. A. Kamitakahara, G. S. Grest and C. M. Soukoulis, *Phys. Rev. Lett.* **60** (1988) 2280.

98. A. Aharony, S. Alexander, O. Entin-Wohlman, R. Orbach, *Phys. Rev. Lett.* **58** (1987) 132.

99. E. Courtens, J. Pelous, J. Phalippou, R. Vacher and T. Woignier, *Phys. Rev. Lett.* **58** (1987) 128.

100. T. F. Rosenbaum, K. Andres, G. A. Thomas and R. N. Bhatt, *Phys. Rev. Lett.* **45** (1980) 1723.

101. T. F. Rosenbaum, R. F. Mulligan, M. A. Paalanen, G. A. Thomas and R. N. Bhatt, *Phys. Rev.* **B27** (1983) 7509.

102. M. H. Cohen, *J. Non-Crystalline Solids* 4 (1970) 391.

103. M. Kirker, *The Scattering of Light and Other Electromagnetic Radiation* (Academic Press, NY, 1969).

104. R. B. Stephens, H. Deckman and A. Z. Genack, submitted for publication.

105. M. Lax, V. Nayaranamurti, and R. C. Fulton, Leningrad 1987 Conference on Light Scattering.

106. B. J. Berne and R. Pecora, *Dynamic Light Scattering with Applications to Chemistry, Biology and Physics*, (Wiley, New York, 1976).

107. M. Ya. Azbel, *Phys. Rev.* **B27** (1983) 3901.

108. P. Sheng, Z. Q. Zhang, B. White and G. Papanicolaou, *Phys. Rev. Lett.* **57** (1986) 1000.

109. B. S. White, P. Sheng and Z. Q. Zhang, *Phys. Rev. Lett.* **59** (1987) 1918.

110. J. E. Sipe, P. Sheng, B. S. White and M. H. Cohen, *Phys. Rev. Lett.* **60** (1988) 108.

111. P. Waterman and N. E. Pederson, *J. Appl. Phys.* **59** (1986) 2609.

112. S. John and R. Rangarajan, submitted for publication.

113. P. Ballone, W. Andreoni, R. Car and M. Parrinello, *Phys. Rev. Lett.* **60** (1988) 271.

114. E. Yablonovitch, *Phys. Rev. Lett.* **63** (1989) 1950.

115. W. Feller, *An Introduction to Probability Theory and Its Applications* (J. Wiley, New York, 1968).

116. P. M. Morse and H. Feshbach, *Methods of Theoretical Physics* (McGraw Hill, New York, 1953).

117. A. Ishimaru, *Wave Propagation and Scattering in Random Media* (Academic Press, Orlando, 1978).

118. D. Schmeltzer and M. Kaveh, *J. Phys.* **C20** (1987) L175.

119. M. Rosenbluh, I. Edrei, M. Kaveh and I. Freund, *Phys. Rev.* **A35** (1987).

120. I. Freund, M. Rosenbluh and S. Feng, *Phys. Rev. Lett.* **61** (1988) 2328; R. Berkovits, M. Kaveh and S. Feng, *Phys. Rev.* **B40** (1989) 737.

121. T. D. Cheung, H. W. Deckman, H. M. Lindsay, D. A. Weitz, S. John and A. Z. Genack, *Bull. Am. Phys. Soc.* **32** (1987) 793.

122. R. M. Macfarlane, A. Z. Genack and R. G. Brewer, *Phys. Rev.* **B17** (1978) 2821.
123. M. P. van Albada, J. F. de Boer and A. Lagendijk, submitted for publication.
124. W. Polkosnik, N. Garcia and A. Z. Genack, unpublished.
125. C. M. Soukoulis and E. N. Economu, submitted for publication.

# DYNAMICAL CORRELATIONS OF MULTIPLY SCATTERED LIGHT


D. J. PINE[ab], D. A. WEITZ[a], G. MARET[c], P. E. WOLF[d],
E. HERBOLZHEIMER[a], and P. M. CHAIKIN[ae]

[a] *Exxon Research and Engineering*
*Route 22 East, Annandale*
*New Jersey 08801, USA*

[b] *Department of Physics*
*Haverford College, Haverford*
*Pennsylvania 19041, USA*

[c] *Hochfeld Magnetlabor*
*Max Planck Institut für Festkörperforschung 166x*
*F-38042 Grenoble-Cedex, France*

[d] *Centre de Recherches sur les Très Basses Températures, CNRS*
*F-38042 Grenoble-Cedex, France*

[e] *Department of Physics*
*Princeton University, Princeton*
*New Jersey 07895, USA*


Motion of particles in optically dense media gives rise to temporal fluctuations in the intensity of multiply scattered light. We show that useful information about the dynamics of the scatterers can be obtained from measurements of the temporal autocorrelation functions of these fluctuations in the multiply scattered light. We develop a phenomenological theory, which models the transport of light as a random walk between scatterers, and obtain explicit expressions for the autocorrelation functions for several experimental geometries. These expressions are compared with experiments probing the dynamics of colloidal suspensions and are shown to be in excellent agreement with the data. The dependence of the autocorrelation functions on the experimental geometry provides a powerful means of exploring the particle dynamics over vastly different length and time scales. Thus, this technique extends the conventional single scattering technique of Dynamic Light Scattering to the multiple scattering regime. We call this new technique Diffusing Wave Spectroscopy (DWS). We illustrate the power of DWS by applying it to measure the particle size in concentrated suspensions and to study the diffusion of particles in porous media and the flow of particles under shear. In addition, we show that DWS can be extended to study the dynamics of interacting colloids by including the consequences of the correlations between the particle positions and velocities. DWS can also be used to study the nature of the transport of light in disordered systems and, in particular, the limitations of using a continuum diffusion approximation. To exploit this, we show that other quantities, such as the angular dependence of the coherent backscattering cone and the absorption dependence of the incoherent backscattering intensity, depend on the distribution of light paths through the sample in the same way as the temporal autocorrelation functions obtained in backscattering.

## Contents


| | |
|---|---|
| 1. Introduction | 315 |
| 2. Theory | 319 |
| 3. Transmission | 329 |
| 4. Backscattering | 335 |
| 5. Analogies with Static Measurements | 344 |
| 6. Applications | 348 |
| 6.1. Particle sizing | 349 |
| 6.2. Polydispersity | 351 |
| 6.3. Porous media | 355 |
| 6.4. Absorption | 358 |
| 6.5. Sheared suspensions | 360 |
| 6.6. Interacting particles | 364 |
| 7. Conclusions | 370 |
| Acknowledgements | 371 |
| References | 371 |

# 1. INTRODUCTION

A characteristic feature of the scattering of light from any random medium is the existence of a very strong spatial modulation of the scattered intensity called the speckle pattern. The speckles are due to the addition of the electric fields, with random phases, scattered from different illuminated points. This feature is characteristic not only of light, but also of any scattering phenomenon that can be described in terms of waves. It also results from multiply scattered light just as it does from singly scattered light. If the scattering medium is perfectly stationary, the speckle pattern will also be stationary. By contrast, if the positions of the scatterers in the medium change with time, the speckle pattern will also evolve with time. Therefore, the intensity of the scattered light at any point will fluctuate. Typically, the motion of the scatterers will occur on a time scale that is much longer than any propagation time of the light through the medium. Thus, the fluctuations in the scattered light intensity will reflect the dynamics of the scatterers.

In this chapter, we discuss the analysis of these temporal fluctuations of the intensity for the case of multiply scattered light, and determine the temporal autocorrelation function of the scattered intensity. We present a heuristic approach to derive the functional form of this autocorrelation function. This approach exploits the diffusion approximation to describe the transport of light in the strongly multiple scattering regime. The expressions derived are compared with experiment for several important geometries and are shown to be in remarkably good agreement. Furthermore, we show that useful information about the dynamics of the scattering medium can be recovered from our expressions for the autocorrelation functions.

The analysis of the temporal fluctuations of the scattered intensity in the strictly single scattering regime is a well developed and extremely powerful spectroscopy which is variously called quasielastic light scattering (QELS), dynamic light scattering, or photon correlation spectroscopy.[1,2] In this technique, knowledge of the scattering vector, **q**, is crucial as it sets the length scale over which particle motion is probed. Extensions of QELS to the multiple scattering regime have proved extremely difficult and have been limited essentially to only doubly scattered light.[3] In turn, this has severely limited the application of QELS to the singly scattering regime. By contrast, in the very strong multiple scattering regime, it is possible to exploit the diffusive nature of the transport of light and again obtain useful information about the particle dynamics.[4,5,6] We call this new technique diffusing wave spectroscopy (DWS).[7,8] In this chapter, we demonstrate the potential of DWS both in extending QELS to the multiple scattering regime and in studying the particle dynamics of interacting systems.

The analysis of the temporal intensity fluctuations also provides a unique probe of the nature of transport of light in strongly scattering media, complementing the many studies of the static properties of multiply scattered light (see the review by Lagendijk *et al.* in this volume).[9,10] To make this explicit, we show that there exists a direct analogy between the time dependence of the temporal autocorrelation function of the backscattered intensity and the angular dependence of the coherent enhancement of the static backscattered intensity.[11] Physically this arises from the fact that both measurements probe the same type of sum over light diffusion paths. In the case of the enhanced backscattering, these paths contribute to the increase of the scattered intensity, while in the case of the dynamics, they contribute to the decay of the autocorrelation function. We can, therefore, use the autocorrelation function as a sensitive experimental measure of the light diffusion paths. In particular, it provides a means of probing the limitations of the continuum nature of the diffusion approximation in describing the contributions of the very short scattering paths.

The temporal autocorrelation function of the intensity in one or more speckles is the most convenient quantitative measure of the dynamics of the scattered light. In the single scattering regime, the characteristic time dependence of the fluctuations is determined by the particle motion over the length scale set by $q^{-1}$, where the wavevector is $q = (4\pi n/\lambda)\sin(\theta/2)$, with $\lambda$ the wavelength, $n$ the index of refraction of the medium, and $\theta$ the scattering angle. Thus, particle motion is probed over length scales of $\lambda$ and larger. By contrast, in the multiple scattering regime, the characteristic time dependence is determined by the cumulative effect of many scattering events and, thus, the characteristic time scales are much faster. Consequently, particle motion is probed over length scales much smaller than $\lambda$.

This is illustrated in Fig. 1, where we show intensity autocorrelation functions from suspensions of 0.497-$\mu$m diameter spheres at vastly different concentrations. The data for Fig. 1(a) are obtained at a scattering angle of $\theta = 90°$ from an essentially transparent suspension of volume fraction $\phi = 10^{-5}$, which is in the single scattering regime. It is a single exponential, reflecting only one characteristic time scale $(Dq^2)^{-1}$, where $D$ is the particle diffusion coefficient. By contrast, the data in Fig. 1(b) are obtained in a backscattering geometry from a suspension with $\phi = 0.01$ which is in the strongly multiple scattering regime, and therefore appears white. It is both highly non-exponential and has a substantially faster initial decay than the single scattering data. However, the volume fraction of the suspension is still sufficiently low that the diffusion coefficient of the particles is unchanged from the dilute limit. Thus, the more rapid initial decay and non-exponential shape must arise from the much larger number of scattering events. The data for Fig. 1(c) are obtained for the same concentration, $\phi = 0.01$, but from light transmitted through a sample of thickness, $L = 2$mm. Its decay is even more rapid, although more nearly exponential. This reflects a larger average number of scattering events due to the more important contributions of longer paths to the transmitted light as compared to the backscattered light.

This chapter is organized as follows: In the next section, we discuss the extension of the traditional approach for QELS to the multiple scattering regime. We calculate the intensity autocorrelation function within the diffusion approximation for the transport of light and then

                                  *D. J. Pine et al.*

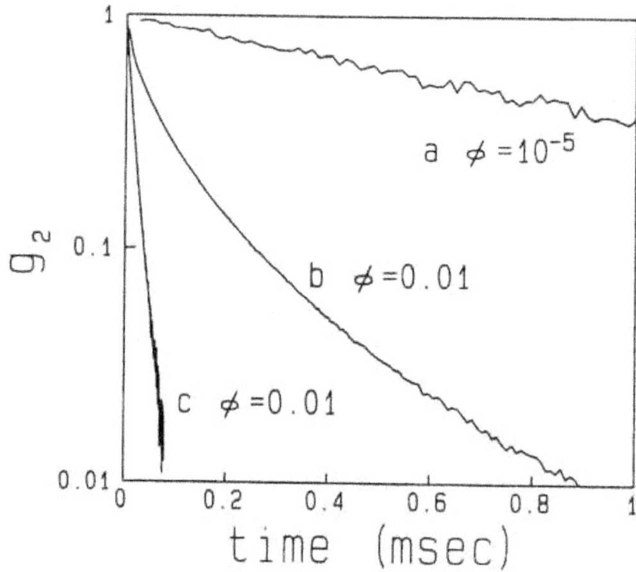


Fig. 1. Temporal intensity autocorrelation functions from aqueous suspensions of 0.497-$\mu$m-diameter polystyrene spheres at different volume fractions, $\phi$, and for different scattering geometries: (a) single scattering, $\phi = 10^{-5}$, $\theta = 90°$; (b) multiple scattering, $\phi = 0.01$, backscattering from a thick sample; (c) multiple scattering, $\phi = 0.01$, transmission through a 2-mm-thick sample.

obtain explicit functional forms of the complete intensity autocorrelation function for several important experimental geometries. The predictions are compared to the experimental results for transmission and backscattering geometries. In the next section, we demonstrate how DWS can provide a sensitive new probe of the nature of the transport of light in the multiple scattering regime by comparing the short-time behavior of the temporal autocorrelation functions to the angular dependence of the static coherent backscattering and the absorption dependence of the incoherent backscattered intensity. Finally, we illustrate the utility of DWS for studying the dynamics of particles in the multiple scattering regime.

## 2. THEORY

Two theoretical approaches have been used to calculate the temporal autocorrelation function of multiply scattered light. One approach is to start from the wave equation for the propagation of light and to use a diagrammatic description of the scattering from the random fluctuations in the dielectric constant. The consequences of motion of the scatterers was first considered by Golubentsev[12] and the actual temporal correlation functions were calculated by Stephen.[6] By constrast, the second approach considers the diffusive transport of individual photons directly, rather than starting from the wave equation.[4,5,7,8] The path of each photon is determined by random, multiple scattering from a sequence of particles. The loss of correlations due to the motion of scatterers is calculated for each light path. The contributions of all paths, appropriately weighted, are then summed to obtain the temporal autocorrelation function.

In this chapter, we adopt the second approach. It is mathematically simpler and physically more transparent than the diagrammatic methods. It can also be more readily generalized to larger particles which scatter light anisotropically and are typical of media which exhibit multiple scattering of light. The key to this approach is to model the transport of each photon through the sample as a random walk. This is done within the diffusion approximation, which allows the weighting of each path to be determined, and simplifies the calculation of the temporal autocorrelation.

We begin by considering a single light path and calculate its contribution to the decay of the autocorrelation function. For simplicity we consider independent spherical scatterers of uniform size undergoing Brownian motion. The fluctuating phase of the multiply scattered light caused by the motion of all the particles in the path leads to the decay of the autocorrelation function which, for a path of $n$ scattering events, is given by

$$G_1^{(n)}(\tau) \equiv \langle E^{(n)*}(0)E^{(n)}(\tau)\rangle = \langle |E^{(n)}(0)|^2\rangle e^{-i\Delta\phi^{(n)}}(\tau) \, . \qquad (1)$$

                                    *D. J. Pine et al.*

Here $E^{(n)}$ is the contribution to the scattered electric field from a path of order $n$ and $\Delta\phi^{(n)}(\tau)$ is the *total* change in its phase due to the motion of all $n$ scatterers,

$$\Delta\phi^{(n)}(\tau) = \sum_{i=1}^{n} \mathbf{q}_i \cdot \Delta\mathbf{r}_i(\tau) \, , \tag{2}$$

where $\mathbf{q}_i$ is the scattering wavevector (proportional to the momentum transfer) of the $i$th scattering event and $\Delta\mathbf{r}_i(\tau) = \mathbf{r}_i(\tau) - \mathbf{r}_i(0)$ is the change in position of the $i$th scattering particle in a time $\tau$. Assuming that the fields belonging to different paths add incoherently, the average contribution of all paths of order $n$ to the autocorrelation function becomes

$$G_1^{(n)}(\tau) = I_0 P(n) \left\langle \prod_{i=1}^{n} e^{-i\mathbf{q}_i \cdot \Delta\mathbf{r}_i(\tau)} \right\rangle \tag{3}$$

where $P(n)$ is the fraction of the total scattered intensity $I_0$ in the $n$th order paths. The total field autocorrelation function $G_1(\tau)$ is obtained by summing over $n$,

$$G_1(\tau) = I_0 \sum_{n=1}^{\infty} P(n) \left\langle \prod_{i=1}^{n} e^{-i\mathbf{q}_i \cdot \Delta\mathbf{r}_i(\tau)} \right\rangle . \tag{4}$$

This sum can be evaluated directly by computer simulation.[13] While this approach can yield accurate results, it is useful to have analytic expressions for $G_1(\tau)$. To this end, we assume that the dominant contribution to Eq. (4) comes from light paths with many scattering events, that is, large $n$. For large $n$, the transport of light through a sample can be accurately described within the diffusion approximation. This, in turn, allows $P(n)$ and the averages in Eq. (4) to be easily evaluated.

For large $n$, we can relax the condition that the sum of the intermediate scattering vectors must equal the difference between the incident and exiting wavevectors, $\Sigma_i \mathbf{q}_i = \mathbf{k}_n - \mathbf{k}_0$, and assume that successive scattering events are uncorrelated. This corresponds to a random distribution of the $q_i$. Then, for independent particles, the right hand side of Eq. (3) becomes $P(n)\langle\exp[-i\mathbf{q}\cdot\Delta\mathbf{r}(\tau)]\rangle^n$ where $\langle\rangle$ denotes averages over both the particle motion $\Delta\mathbf{r}(\tau)$ and the distribution of wavevectors $\mathbf{q}$. For Brownian motion, the distribution of $\Delta\mathbf{r}(\tau)$ is Gaussian and the average over particle motion is readily performed yielding

$$G_1^{(n)}(\tau) = I_0 P(n)\,\langle e^{-q_i^2\langle\Delta r^2(\tau)\rangle/6}\rangle_q^n \;, \tag{5}$$

where $\langle\Delta r^2(\tau)\rangle = 6D\tau$ and now $\langle\rangle_q$ denotes the average over $q$. Thus, Eq. (5) implies that the relaxation rate will be large for paths with large $n$, independent of the average over $q$.

The average over $q$ can be performed explicitly for point-like scatterers, which scatter light isotropically; this gives[6,12]

$$G_1^{(n)}(\tau) = I_0 P(n)\left[\frac{\tau_0}{4\tau}\left(1 - e^{-4\tau/\tau_0}\right)\right]^n \;, \tag{6}$$

where $\tau_0 \equiv (Dk_0^2)^{-1}$. However, for larger particles which scatter light anisotropically, the average over $\mathbf{q}$ is more difficult to perform. Thus, we restrict ourselves to small delay times $\tau \ll \tau_0$ and retain only the first term in a cumulant expansion,[4]

$$\left\langle e^{-Dq^2\tau}\right\rangle_q^n \simeq \langle 1 - Dq^2\tau\rangle_q^n = (1 - D\langle q^2\rangle\tau)^n \simeq e^{-D\langle q^2\rangle\tau n} \;. \tag{7}$$

We can compare this more general approximation with the exact expression, Eq. (6), for point-like scatterers by evaluating $\langle q^2\rangle$ for a flat distribution

$$\langle q^2\rangle = \int q^3\,dq \Big/ \int q\,dq = 2k_0^2$$

which gives

$$G_1^{(n)}(\tau) = I_0 P(n)e^{-(2\tau/\tau_0)n} \;. \tag{8}$$

Expanding the exact expression in Eq. 6 gives

$$G_1^{(n)}(\tau) = I_0 P(n)e^{-2n\tau/\tau_0[1-1/3(\tau/\tau_0)+O(\tau/\tau_0)^3]}$$

so that the cumulant expansion gives the correct results to leading order in $\tau/\tau_0$. Furthermore, the effective decay time is $\tau_0/2n$, so for large

$n$, $G_1^{(n)}(\tau)$ decays essentially to zero before the higher order terms make a significant contribution. Nevertheless, for large $\tau/\tau_0$ the cumulant expansion fails since the exact expression in Eq. (6) decays as a power law, $(\tau/\tau_0)^{-n}$, due to the contribution of scattering with small $q$. However, at these long times low order scattering events dominate. For small $n$, the distribution of $\mathbf{q}_i$ is not random but must satisfy the condition that $\Sigma_i \mathbf{q}_i = \mathbf{k}_n - \mathbf{k}_0$. This will modify the average in Eq. (6) and could have the effect of at least partially offsetting the error introduced by the cumulant expansion.[14]

The approximate expression can be easily generalized to treat the anisotropic scatterers typically used in experiments. Here, the single particle scattering intensity is peaked in the forward direction and $\langle q^2 \rangle$ is less than $2k_0^2$. Physically, this means that more scattering events will be required to randomize the direction of light propagation. The length scale over which this randomization occurs is the transport mean free path, $l^*$ (Refs. 15,16). For anisotropic scatterers, $l^*$ is generally greater than the scattering mean free path, $l$, the mean distance between scattering events. It is the transport mean free path which is related to the diffusion coefficient for multiply scattered light, $D_l = cl^*/3$, where $c$ is the speed of light in the medium.

We estimate $l^*$ by analogy to the calculation of the effective length of a statistical segment in semiflexible polymers.[17] Thus, we sum the projections of all steps of length $l$ in a long path onto the direction of the first step. Then

$$l^* = l \sum_{i=1}^{n} \langle \cos \theta \rangle^i$$

where $\langle \cos \theta \rangle$ is the average angle between successive steps. For large $n$, this becomes

$$l^* = l/(1 - \langle \cos \theta \rangle) \ .$$

Since $\langle 1 - \cos \theta \rangle = \langle 2 \sin^2(\theta/2) \rangle = 2\langle q^2 \rangle/(2k_0)^2$, we find

$$2\langle q^2 \rangle/(2k_0)^2 = l/l^* \ .$$

Therefore *both* $\langle q^2 \rangle$ and $l^*$ are simply obtained by averaging over the angular dependence of the form factor $F(\mathbf{q})$ of the scatterers

$$\langle q^2 \rangle = 2k_0^2 \frac{l}{l^*} = \int q^2 F(\mathbf{q})\, d\mathbf{q} \Big/ \int F(\mathbf{q})\, d\mathbf{q} \ . \tag{9}$$

It follows from Eqs. (5), (7), and (9) that for anisotropic scatterers, the contribution of an $n$th order scattering path is

$$G_1^{(n)}(\tau) = I_o P(n) e^{-(2\tau/\tau_0)(l/l^*)n} \ . \tag{10}$$

We note that the accuracy of Eq. (10) improves as the scattering becomes more anisotropic. The second cumulant of $\langle e^{-Dq^2\tau} \rangle$ is $e^{-\frac{1}{2}(D\tau)^2[\langle q^4 \rangle - \langle q^2 \rangle^2]}$. For typical form factors, $F(q)$, we expect $\langle q^4 \rangle$ to scale as $\langle q^2 \rangle^2$ so that the cumulant expansion in Eq. (7) should be valid to leading order in $D\langle q^2 \rangle \tau \simeq (\tau/\tau_0)(l/l^*)$. Thus, for large scatterers, where $l^* \gg l$, the range of validity of Eq. (10) may extend up to $\tau \simeq \tau_0$.

The total field autocorrelation function $G_1(\tau)$ is obtained by summing over scattering paths of all orders

$$G_1(\tau) = I_0 \sum_{n=1}^{\infty} P(n) e^{-(2\tau/\tau_0)(l/l^*)n} \ .$$

If we rescale $n$ by defining $n^* = (l/l^*)n$, the exponential term in this equation is identical to Eq. (8) for isotropic scatterers. Furthermore, in the diffusion approximation, $P(n)$, the fraction of photons which travel a path of length $s = nl$ through the scattering medium, is a function of $n^*$ rather than $n$, because the characteristic length scale of the light propagation is the transport mean free path, $l^*$, rather than the scattering mean free path, $l$. As a result, in this approximation,

$$G_1(\tau) = I_0 \sum_{n^*=1}^{\infty} P(n^*) e^{-(2\tau/\tau_0)n^*} \ . \tag{11}$$

This reflects a renormalization of the mean free path for anisotropic scatterers, so that the light may be viewed as undergoing an isotropic random walk with an average step length $l^*$. Finally, we note that the decay rate of a given path depends critically on its length. Long paths reflect the aggregate contribution of many scattering events. Thus, each particle need move only a small distance for the total path length to change by a wavelength. This occurs in a short time, leading to a rapid decay rate. By contrast, short paths reflect the contribution of a small number of scattering events. Thus, each particle must undergo substantial motion for the total path length to change by a wavelength. This takes a longer time, leading to a slower decay.

The key to the solution of Eq. (11) is the determination of $P(n^*)$. This task is greatly simplified if we restrict ourselves to the continuum limit. Thus, we approximate the summation over $n^*$ by an integral over the path lengths, $s = n^* l^*$, and obtain

$$G_1(\tau) = \langle E(0)E^*(\tau)\rangle = I_o \int P(s)e^{-(2\tau/\tau_0)s/l^*}\,ds \ . \tag{12}$$

In this case, $P(s)$ is the fraction of photons which travel a path of length $s$ through the scattering medium. Now we can use the diffusion approximation to describe the random walk of the light, and $P(s)$ can be obtained from the solution of the diffusion equation for the appropriate experimental geometry. We emphasize, however, that the continuum approximation to the discrete scattering events will break down whenever there are significant contributions from short light paths.

Physically, we can regard $P(s)$ as giving the distribution of path lengths of the light diffusing through the sample. In general, $P(s)$ depends on geometrical factors such as the size and shape of the sample cell and incident light beam. As we shall see, $P(s)$ can be determined for many different scattering geometries of experimental interest. First, however, it is instructive to consider the results for a case where the physics and mathematics are simple, but the experiment is quite difficult.

We consider a scattering medium of infinite extent and introduce the diffusing light at a point inside the medium. The light is detected at a point a distance $r$ away. If $r$ is much larger than $l^*$, then $P(s)$ is given by the usual Gaussian distribution which results from the central limit theorem,

$$P(s) = \left(\frac{3}{4\pi n^* l^{*2}}\right)^{3/2} e^{-3r^2/4n^* l^{*2}} = \left(\frac{3}{4\pi s l^*}\right)^{3/2} e^{-3r^2/4sl^*} , \quad (13)$$

where $n^* = s/l^*$ is the number of steps.[15] In Fig. 2 we show a plot of Eq. (13) as a function of path length $s$. As expected, $P(s)$ reaches a maximum when $s_{\max} \approx (r/l^*)^2 l^*/2$, which corresponds to a random walk with $n^* = (r/l^*)^2/2$ steps. For very short paths $P(s) \to 0$, since the light must diffuse a distance $r$ before it can be detected at all. Thus the relative probability of having random walks with path lengths $s \ll s_{\max}$ is small. For long paths, $s \gg s_{\max}$, $P(s)$ decays very slowly according to a power law, $s^{-3/2}$.

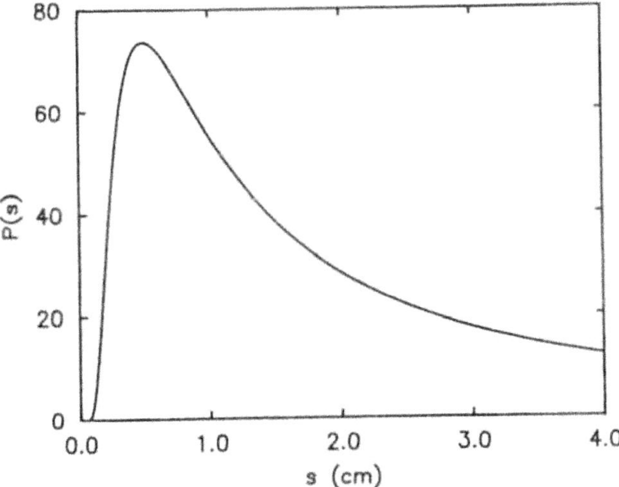

Fig. 2. $P(s)$ *vs* $s$: The fraction of photons which travel a distance $s$ between a point source and a point detector separated by a distance $r$ in an infinite random medium with strong multiple scattering. (Eq. (13), with $l^* = 100$ $\mu$m and $r = 1$ mm).

Using this simple form for the distribution of path lengths we can combine Eqs. (12) and (13) and obtain an expression for the temporal autocorrelation function

$$G_1(\tau) = I_0 \int_0^\infty \left(\frac{3}{4\pi s l^*}\right)^{3/2} e^{-3r^2/4sl^* - (2\tau/\tau_0)s/l^*} ds \ . \tag{14}$$

Some insight into the behavior of $G_1(\tau)$ as a function of delay time can be obtained using the method of steepest descents. Thus, the dominant contribution to the integral occurs when the argument of the exponential is at its maximum value,

$$\frac{\partial}{\partial s}\left(-\frac{3r^2}{4sl^*} - \frac{2\tau}{\tau_0}\frac{s}{l^*}\right) = \frac{3r^2}{4s^2l^*} - \frac{2\tau}{\tau_0 l^*} = 0$$

giving

$$s_d = r\sqrt{\frac{3\tau_0}{8\tau}} \ .$$

Thus, the length, $s_d$, of the paths which dominate the correlation function decreases with increasing delay time. Performing the integral in Eq. (14), we obtain an explicit expression for the autocorrelation function,

$$G_1(\tau) \propto e^{-(r/l^*)\sqrt{6\tau/\tau_0}} \ .$$

Since for $\tau \to 0$, $\partial G_1/\partial \tau \to -\infty$, at the shortest delay times, the autocorrelation function decays very quickly. This reflects the divergence of $s_d$ as $\tau \to 0$. The square root dependence arises from the form of the exponential cut-off of the short paths. However, the singularity at short times results from the fact that the contributions of $P(s)$ to the autocorrelation function are only a slowly decaying function of path length $(s^{-3/2})$. This singularity is a characteristic feature of any multiple scattering geometry which contains significant contributions from infinitely long paths, where $P(s) \sim s^{-3/2}$. As discussed below, this singularity is observed in backscattering from a semi-infinite sample, but not in transmission through a finite slab.

To describe experimental data, we must solve Eq. (12) to obtain the autocorrelation function for physically realizable geometries, where the mathematics are more complex but the experiment is simpler. Restricting ourselves to the continuum approximation, $P(s)$ can be obtained from the solution of the diffusion equation for the appropriate geometry. In fact we can further simplify our task by extending the limits of the integration in Eq. (12) from $s = 0$ to $s = \infty$, which makes $G_1(\tau)$ a Laplace transform of $P(s)$. Then we can solve the Laplace transform of the diffusion equation to obtain the desired autocorrelation function directly. However, we again emphasize the limitations in this approach. Extending the lower bound of the integral to $s = 0$ may allow unphysically short paths to contribute to the autocorrelation function, depending on the scattering geometry. Furthermore, the short paths cannot be properly treated within the continuum approximation inherent in the diffusion equation. Thus, this approach is strictly valid only for long paths, where the diffusion approximation is a good description of the transport of light. Nevertheless, the great simplicity afforded by this approach makes it worthwhile to try to extend its applicability to more general geometries.

Physical insight into the method for determining $P(s)$ by solving the diffusion equation can be obtained by considering a simple experiment. An instantaneous pulse of light is incident on the face of a sample which multiply scatters the light. The scattered photons will execute a random walk until they either escape the sample or arrive at the point $\mathbf{r}_d$ on the boundary where the light is detected. The intensity of light that reaches $\mathbf{r}_d$ is zero at $t = 0$, then increases to a maximum, and finally decreases back to zero at long times when all the photons have left the sample. At time $t$, the photons arriving at point $\mathbf{r}_d$ are those that have traveled a path length $s = ct$ through the sample. Thus, the time dependence of the intensity of light arriving at point $\mathbf{r}_d$ is directly proportional to $P(s)$. The details of this time dependence, and hence the shape of $P(s)$, will depend on the experimental geometry.

To obtain $P(s)$ for a given experimental geometry, we must use the diffusion equation to determine the dispersion induced in a delta function pulse as it traverses the scattering medium. To accomplish this, we denote the density of diffusing photons within the medium by $U(\mathbf{r})$. Then, the light detected at the point $\mathbf{r}_d$ at the boundary is the outward flux of diffusing light and is given by the normal derivative of $U$ evaluated at $\mathbf{r}_d$,

$$P(s) \propto -\hat{\mathbf{n}} \cdot \boldsymbol{\nabla} U|_{\mathbf{r}_d} , \tag{15}$$

where $\hat{\mathbf{n}}$ is the unit normal vector, directed outward. It is the intensity of diffusing photons that satisfies the diffusion equation,

$$\frac{\partial U}{\partial t} = D_l \nabla^2 U , \tag{16}$$

where $D_l = cl^*/3$ is the diffusion coefficient for light. The geometry for the solution of the diffusion equation will depend on the experiment. As initial conditions for the diffusion equation, we take an instantaneous pulse at $t = 0$, with a geometry suitable for the experiment. The boundary conditions must ensure that there is no flux of diffusing photons entering the sample from the boundaries. This is achieved with the boundary conditions[15]

$$U - \frac{2}{3}l^* \, \hat{\mathbf{n}} \cdot \boldsymbol{\nabla} U = 0 . \tag{17}$$

We use the transformation of variables, $s = ct$, to relate the time dependence to the required distribution in path lengths. We can then solve the Laplace transform of the diffusion equation, take its normal derivative at the surface and obtain the autocorrelation function directly.

Finally, to compare to experiment, we note that the electric field autocorrelation functions discussed above are usually not measured directly. Instead, one typically measures the intensity (or homodyne) autocorrelation function, $\langle I(\tau)I(0)\rangle/\langle I\rangle^2$, where $I$ is the intensity of the scattered light. For most systems of experimental interest, the intensity autocorrelation function is simply related to the electric field autocorrelation function by

$$\langle I(\tau)I(0)\rangle/\langle I\rangle^2 = 1 + f(A)g_2(\tau) ,$$

where $f(A)$ is a function determined by the collection optics,[1,2] $g_2(\tau)$ is related to $g_1(\tau)$ by the Siegert relation,

$$g_2(\tau) = |g_1(\tau)|^2$$

and $g_1(\tau)$ is the normalized field autocorrelation function, $g_1(\tau) \equiv \langle E(\tau)E^*(0)\rangle/\langle|E|^2\rangle$. The validity of the Siegert relation in the multiple scattering regime has been verified experimentally for $\tau = 0$, and no evidence for non-Gaussian statistics is found.[16] Experimental results are generally reported as normalized homodyne autocorrelation functions, $g_2(\tau)$.

## 3. TRANSMISSION

The first experimental geometry we consider is transmission through a slab of thickness $L$ and infinite extent.[6,7] For the transmission geometry, the detected light must diffuse a distance $L$ across the sample, as illustrated in the top of Fig. 3. This introduces a characteristic path length, $s_c = n_c l = n_c^* l^*$, where $n_c^* = (L/l^*)^2$ is the mean number of steps for a random walk of end-to-end distance $L$ and average step length $l^*$. The consequences of this characteristic length can be seen directly in the dispersion induced in a delta function pulse upon traversal of a sample with $L = 1$ mm and $l^* = 100$ $\mu$m, illustrated in the bottom of Fig. 3. The intensity is peaked sharply and then decays rapidly with increasing time. Since this time dependence is proportional to $P(s)$, this characteristic path length will be directly reflected in the autocorrelation function resulting in a characteristic decay time. Physically, this corresponds to the time taken for the characteristic path length to change by $\sim \lambda$, so that the total phase $\Delta\phi^{(n^*)} \approx 1$. Thus, we can estimate the typical distance an individual particle has moved from the condition

$$\langle (\Delta\phi^{(n^*)})^2 \rangle \approx n_c \langle q^2 \rangle \langle \Delta r^2 \rangle \approx \left( \frac{n_c^* l^*}{l} \right) \left( 2k_0^2 \frac{l}{l^*} \right) \langle \Delta r^2 \rangle \approx 1 \, .$$

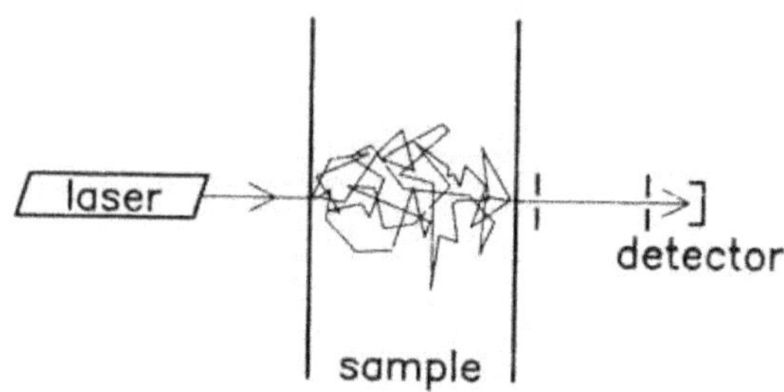


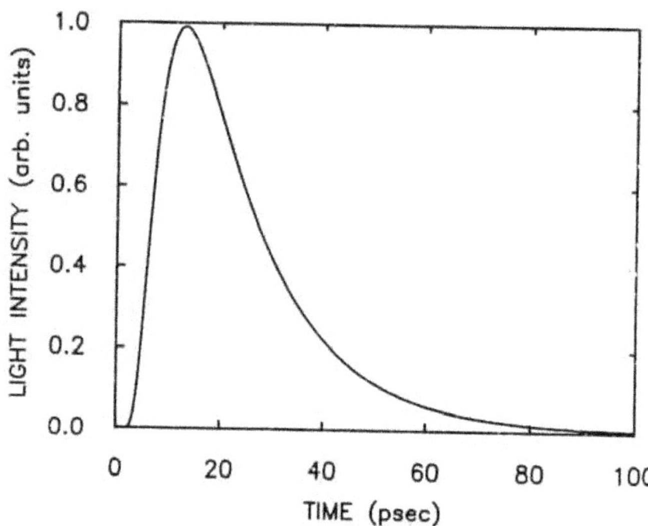

Fig. 3. Transmission of a short pulse of light through a multiply scattering medium: (top) Experimental geometry showing light paths which exit at a point directly opposite the input pulse. Other paths are present, but not shown; (bottom) Time-dependent output response to $\delta(t)$ input pulse ($L = 1$ mm, $l^* = 100$ $\mu$m, $c = 2 \times 10^{10}$ cm/s).

This gives $\Delta r_{\mathrm{rms}} \approx \lambda l^*/9L$. *Since $l^*/L \ll 1$, the typical length scale over which particle motion is probed by DWS in transmission is much smaller than the wavelength.* This reflects the fact that the decay of the autocorrelation function is due to the cumulative effect of many scattering events, so that the contribution of individual particles to the total decay is small. *This is in striking contrast to ordinary dynamic light scattering (QELS) where by varying q, length scales greater than or equal to the wavelength are probed.* Furthermore, an important feature of the transmission geometry in DWS is that the length scale over which particle motion is probed can be controlled experimentally by varying the sample thickness $L$. Similarly, the time scale over which motion is probed using DWS can be controlled by varying the sample thickness. This time scale is $\tau_0 (l^*/L)^2$, and is much shorter than the time scales probed using QELS, which are typically longer than $\tau_0$. Thus, DWS provides a useful and convenient means for extending the length and time scales over which particle motion can be measured using dynamic light scattering techniques.

Autocorrelation functions measured in transmission are shown in Fig. 4 (Ref. 7). The sample consisted of 0.497 $\mu$m diameter spheres at a volume fraction of $\phi = 0.01$ in a 2.0 mm thick cuvette. There is no unscattered light transmitted through the sample, insuring that the strong multiple scattering limit is achieved. The data in the lower curve were obtained when the sample was illuminated uniformly by a 1-cm diameter beam from a 488 nm argon ion laser. Imaging optics collected the transmitted light from a 50 $\mu$m spot on the opposite side of the sample from a point near the center of the illuminated area. For comparison, the upper curve shows data obtained when the incident beam was focused to a point on one side of the sample and transmitted light was collected from a point on axis with the incident spot. These data clearly decay somewhat more slowly than the data obtained with an extended source illumination. Physically, this difference reflects the fact that for the extended source there is a larger contribution from long paths, resulting in a somewhat faster decay than for the point source.

To quantitatively analyze these data, we obtain $g_1(\tau)$ by solving the Laplace transform of the diffusion equation as previously discussed. We take the source of diffusing intensity to be a distance $z = z_0$ inside the illuminated face where we expect that $z_0 \sim l^*$. We first consider the

                                  *D. J. Pine et al.*

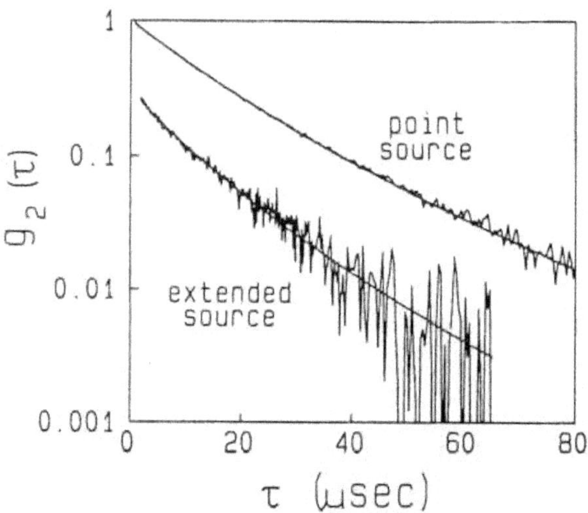

Fig. 4. Intensity autocorrelation functions *vs* time for transmission through 2-mm thick cells with 0.497-$\mu$m-diameter polystyrene spheres and $\phi = 0.01$. Smooth lines are fits to the data by Eqs. (15) and (16) with $l^* = 173\ \mu$m for the point source and $l^* = 175\ \mu$m for the extended source.

case of uniform illumination of one side by an extended source. From Eqs. (12) and (15), we obtain

$$g_1(\tau) = \frac{\frac{L+(4/3)l^*}{z_0+(2/3)l^*}\left\{\sinh\left[\frac{z_0}{l^*}\sqrt{\frac{6\tau}{\tau_0}}\right] + \frac{2}{3}\sqrt{\frac{6\tau}{\tau_0}}\cosh\left[\frac{z_0}{l^*}\sqrt{\frac{6\tau}{\tau_0}}\right]\right\}}{\left(1+\frac{8\tau}{3\tau_0}\right)\sinh\left[\frac{L}{l^*}\sqrt{\frac{6\tau}{\tau_0}}\right] + \frac{4}{3}\sqrt{\frac{6\tau}{\tau_0}}\cosh\left[\frac{L}{l^*}\sqrt{\frac{6\tau}{\tau_0}}\right]} \tag{18a}$$

$$\approx \frac{\left(\frac{L}{l^*}+\frac{4}{3}\right)\sqrt{\frac{6\tau}{\tau_0}}}{\left(1+\frac{8\tau}{3\tau_0}\right)\sinh\left[\frac{L}{l^*}\sqrt{\frac{6\tau}{\tau_0}}\right] + \frac{4}{3}\sqrt{\frac{6\tau}{\tau_0}}\cosh\left[\frac{L}{l^*}\sqrt{\frac{6\tau}{\tau_0}}\right]}, \tag{18b}$$

where the second expression holds for $\tau \ll \tau_0$. The characteristic time scale in these expressions is $\tau_0(l^*/L)^2$.

For the second geometry, we consider light incident from a point source on axis with the detector, and obtain

$$g_1(\tau) \propto \int_{(L/l^*)\sqrt{6\tau/\tau_0}}^{\infty}[A(s)\sinh s + e^{-s(1-z_0/L)}]ds, \tag{19a}$$

where

$$A(s) = \frac{(\varepsilon s - 1)\left[\varepsilon s e^{-s z_0/L} + (\sinh s + \varepsilon s \cosh s)e^{-s(1 - z_0/L)}\right]}{(\sinh s + \varepsilon s \cosh s)^2 - (\varepsilon s)^2} \qquad (19b)$$

and $\varepsilon = 2l^*/3L$. For convenience, we take $z_0 = (4/3)l^*$, but note that the solutions are insensitive to the exact value used since, in general, $z_0 \sim l^*$ and $l^*/L \ll 1$. This reflects the fact that the value chosen for $z_0$ affects only the first few steps of a random walk which typically consists of a great number of steps. Thus, the relative contribution of the first few steps is small.

These expressions can be compared directly to the experimental data. In Fig. 4, the solid lines through the data are fits to the appropriate equations above. The time constant $\tau_0 \equiv (Dk_0^2)^{-1}$ is set equal to 3.73 msec where $D$ was obtained from a QELS measurement in the single scattering limit at $\phi = 10^{-5}$. The diffusion coefficient $D$ remains independent of concentration for the particle concentrations used in these measurements. For both cases, the data are well-described by the predicted forms of $g_2(\tau)$ with $l^*$ the only fitting parameter. The fit gives $l^* = 175$ $\mu$m for the extended source and $l^* = 173$ $\mu$m for the point source. This excellent consistency for the fitted values of $l^*$ confirms that the somewhat different decay rates in Fig. 4 are due solely to geometric effects. The values of $l^*$ compare well with Mie theory which gives a value of $l^* = 195$ $\mu$m, about 10% higher than measured. This small difference may be due to the approximate nature of the boundary condition on the exit side. Indeed, a slightly smaller value for $l^*$ was obtained in an earlier analysis of this data, which used different boundary conditions, demonstrating the sensitivity of $g_1(\tau)$ to the boundary conditions.

While the autocorrelation functions shown in Fig. 4 are clearly not single exponentials, their curvature in the semi-logarithmic plot is not large. This reflects the existence of a dominant characteristic path length and time scale for these data, as expected. This also suggests the suitability of analyzing the data by means of the first cumulant $\Gamma_1$, or the logarithmic derivative at zero time delay. For Eq. (18), the first cumulant is given by

$$\Gamma_1 = \frac{\partial \ln g_1(\tau)}{\partial \tau}\bigg|_{\tau=0} \approx \frac{(L/l^*)^2 + 4(L/l^*) + 8/3}{\tau_0(1 + 4l^*/3L)} . \tag{20}$$

For Eq. (19), the first cumulant must be evaluated numerically. Measurements of $\sqrt{\Gamma_1}$ as a function sample thickness are shown in Fig. 5 for several volume fractions. The plots show the expected linear dependence of $\sqrt{\Gamma_1}$ over a broad range of concentrations confirming the diffusive nature of the transport of light. It is apparent that the data do not extrapolate through the origin but that the $x$-intercepts are a monotonically decreasing function of particle concentration and, to within experimental uncertainty, consistent with Eq. (20). The different slopes are due primarily to the concentration dependence of $l^*$, which is expected to scale inversely with particle density.

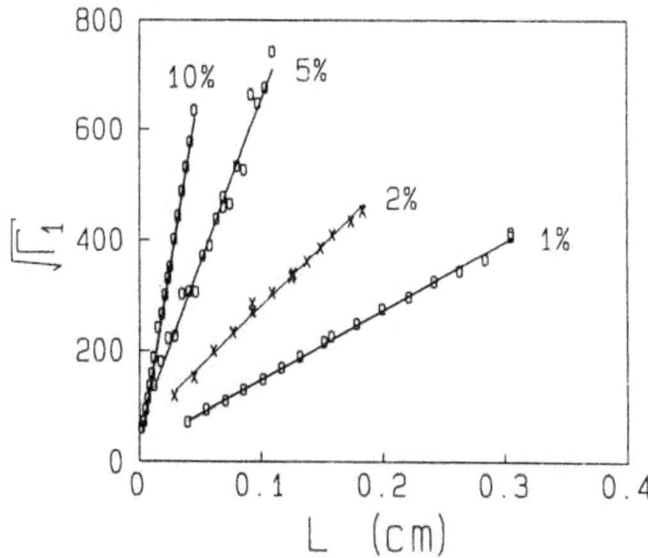


Fig. 5. Square root of the first cumulant $\Gamma_1$ *vs* sample thickness $L$ for 0.497-$\mu$m-diameter spheres for different volume fractions $\phi$.

# 4. BACKSCATTERING

Another interesting and important geometry is that of back-scattering.[4,5,6,7] Here, the light is incident uniformly on one face of a slab of thickness, $L$, and the scattered light is collected from the same face, as illustrated at the top of Fig. 6. In contrast to transmission, in backscattering there is no well-defined characteristic path length set by the sample thickness. This can be seen from the time dependence of the detected intensity of an incident delta function pulse, shown at the bottom of Fig. 6. The intensity is peaked at early times, corresponding to the fact that most of the light is scattered back after only a few scattering events. However, there is still considerable light detected at larger times, as reflected by the very slow decay of the intensity as $t$ increases. To emphasize this point, it is instructive to contrast the behavior of backscattered light with that of transmitted light. We do this in Fig. 7, which shows the dispersion induced in delta function pulses for transmission (dashed line) and backscattering (solid line), plotted logarithmically. The transmission pulse is sharply peaked at the time corresponding to the characteristic path length, and decays rapidly at later times. By contrast, the backscattered pulse is peaked at very early times, corresponding to the light immediately backscattered, but then has a very long, power-law decay. In Fig. 7, we also show on the upper axis the number of scattering events, $n^*$, corresponding to the transit time, $t$.

This power-law decay in the dispersion of a delta function pulse is directly reflected in $P(s)$ for the backscattering geometry. As a result, the autocorrelation function no longer depends sensitively on $l^*$, and, in principle, it becomes possible to determine $\tau_0$ without prior knowledge of $l^*$. In fact, since paths of all lengths contribute in backscattering, the autocorrelation function consists of contributions from all orders of multiple scattering. As a consequence, there is a much broader distribution of time scales in the decay. The longer paths consist of a larger number of scattering events and thus decay more rapidly, probing the

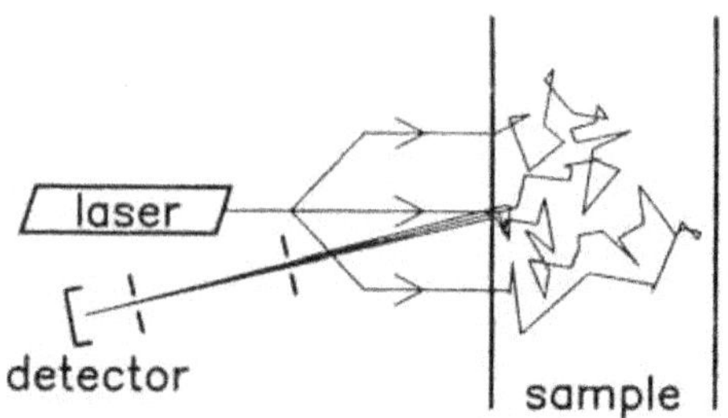


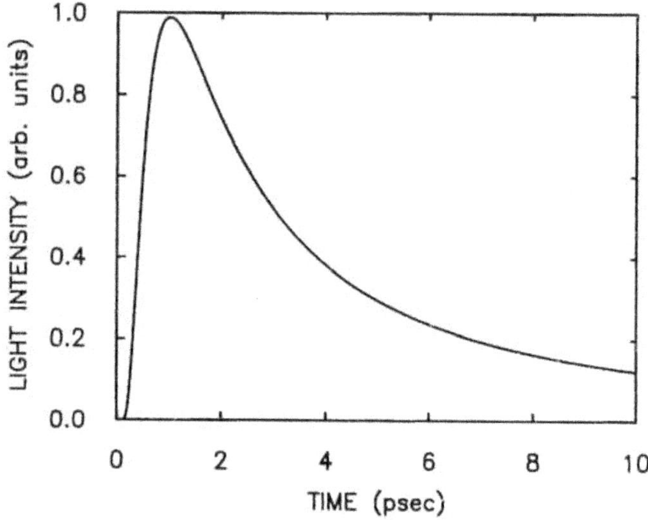


Fig. 6. Backscattering of a short pulse of light from a multiply scattering medium. Input beam is expanded and columated; light is detected near the center of the area illuminated by the input pulse. (top) Experimental geometry showing light paths which exit at a point directly opposite the input pulse. Other paths are present, but not shown. (bottom) Time-dependent output response to $\delta(t)$ input pulse ($L = 1$ cm, $l^* = 100$ $\mu$m, $c = 2 \times 10^{10}$ cm/s, $z_0 = 2l^*$).

motion of individual particles over shorter length and time scales. By contrast, the shorter paths consist of a smaller number of scattering events thus decaying more slowly and probing the motion of individual particles over longer length and time scales. This feature is particularly advantageous for exploring the dynamics of interacting systems, which can have a broad distribution of relaxation rates associated with motion over different lengths scales.

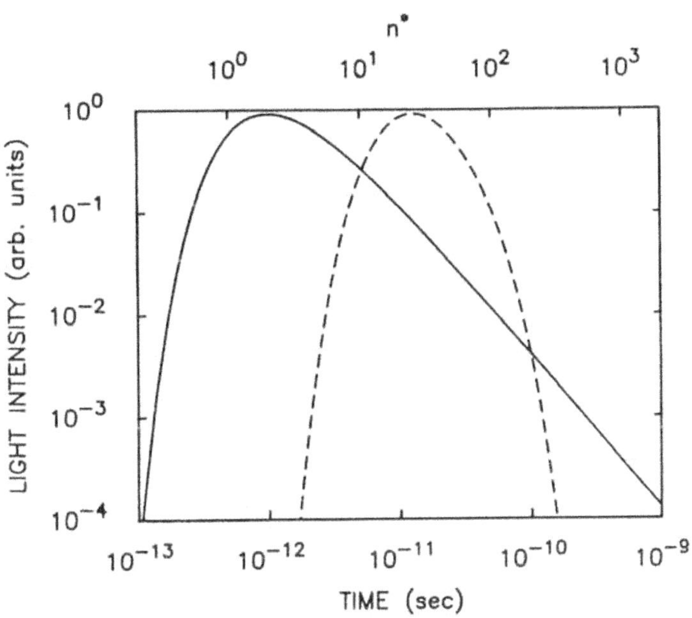


Fig. 7. Comparison of $I(t) \propto P(s)$ for backscattering (solid line) and transmission (dashed line) geometries.

To derive a functional form for the autocorrelation function for backscattered light, we consider a slab of thickness, $L$, and of infinite extent, uniformly illuminated from one side (Fig. 6). While we would like to maintain the simplicity and elegance of the Laplace transform approach used in the case of transmission, we must be extremely cautious in using this for backscattering. In obtaining the Laplace transform in Eq. (12), we have used a continuum approximation to change the summation paths in Eq. (11) to an integral. However, since $s = nl$ and we must have at least one scattering event, we require $n \geq 1$; thus, the lower bound on $s$ must be $l$ rather than 0. Equivalently, for isotropic scatterers, where $l = l^*$, the decay time for a path of length $s$ is

$\exp[-(2\tau/\tau_0)s/l]$. Allowing $s < l$ leads to unphysically long decay times, since $\tau_0/4$ represents the shortest decay time possible, corresponding to light singly scattered through $180°$. Similarly, for anisotropic scatterers, where $l^* > l$, we require $s \geq l^*$ to obtain physically meaningful decay times. However, in order to maintain the simplicity of the Laplace transform, the lower bound on the integral must be zero. For transmission, this does not present a problem as the shortest possible paths are $s = L$, so that $n \gg 1$. By contrast, for backscattering, short paths do contribute and thus, in using the diffusion approximation, we must ensure that the contribution of the unphysically short paths is suppressed.

A simple way of achieving this is to solve the diffusion equation using an initial condition of a source at a fixed distance, $z_0$, in from the illuminated face at $z = 0$. This ensures that there are no contributions from paths shorter than $z_0$. Physically we can regard this as the source of the diffusing intensity, which we expect to be peaked at $z_0 \sim l^*$. Thus, for the initial condition we take $U(\mathbf{r}, t = 0) = \delta(x, y, z - z_0, t)$ and for the boundary conditions again we use Eq. (17). The solution for this geometry is,

$$G_1(\tau) = \frac{\sinh\left[\sqrt{\frac{6\tau}{\tau_0}}\left(\frac{l}{l^*} - \frac{z_0}{l^*}\right)\right] + \frac{2}{3}\sqrt{\frac{6\tau}{\tau_0}}\cosh\left\{\sqrt{\frac{6\tau}{\tau_0}}\left(\frac{L}{l^*} - \frac{z_0}{l^*}\right)\right\}}{\left(1 + \frac{4}{9}\frac{6\tau}{\tau_0}\right)\sinh\left[\frac{L}{l^*}\sqrt{\frac{6\tau}{\tau_0}}\right] + \frac{4}{3}\sqrt{\frac{6\tau}{\tau_0}}\cosh\left[\frac{L}{l^*}\sqrt{\frac{6\tau}{\tau_0}}\right]} . \tag{21}$$

For a sample of infinite thickness, Eq. (21) simplifies,

$$G_1(\tau) = \frac{e^{-(z_0/l^*)\sqrt{6\tau/\tau_0}}}{1 + \frac{2}{3}\sqrt{\frac{6\tau}{\tau_0}}} . \tag{22}$$

We note that the initial condition of a source at $z_0$ appears explicitly in the solution, reflecting the importance of the contribution of the short paths. In fact, we expect there to be a distribution in the position of the apparent source, $z_0$. Thus we must integrate Eq. (22) over a distribution of sources, $f(z_0)$. While the exact form of $f(z_0)$ is unknown, we do know that $f(z_0)$ goes to zero as $z_0 \gg l^*$ and for $z_0 \ll l^*$. Furthermore, we expect $f(z_0)$ to be peaked near $z_0 \sim l^*$. Thus, to leading order in $\sqrt{\tau/\tau_0}$, $g_1(\tau)$ is given by Eq. (22), with $z_0/l^*$ replaced by $\langle z_0 \rangle/l^*$, the average over the source distribution $f(z_0)$. The first order correction is $[\langle z_0^2 \rangle/\langle z_0 \rangle^2 - 1]\tau/\tau_0$ which is small for a narrow distribution $f(z_0)$. Physically, we can think of $\langle z_0 \rangle$ as the average position of the source of diffusing intensity, and we expect that the exact form of the distribution $f(z_0)$, and therefore $\langle z_0 \rangle$, will depend on the anisotropy of the scattering as reflected by $l^*/l$.

An autocorrelation function measured in backscattering is shown in Fig. 8. The sample consisted of $0.412$ $\mu$m diameter spheres at a volume fraction of $\phi = 0.05$ in a 5 mm thick cuvette. It was illuminated by a uniform beam, 1 cm in diameter; light from a 50 $\mu$m diameter spot near the center of the illuminated area was imaged onto the detector. The scattering angle was $\sim 175°$, although we found virtually no dependence of the results on scattering angle when it was varied $\sim 20°$ from that used here. This is expected for diffusing light. After subtracting the baseline, the logarithm of the autocorrelation function, normalized by the baseline is plotted as a function of the square root of time, in units of $\tau_0$. We use $\tau_0 = 3.01$ msec, as measured experimentally in the single scattering limit, and consistent with the value calculated from the Stokes-Einstein relation. The solid line is a fit to the functional form given by Eq. (22), and is in reasonably good agreement with the data. However, the theoretical form exhibits somewhat more curvature than the data. In fact, to within the precision of experiment, the data shown in Fig. 8 is linear over three decades of decay when plotted logarithmically as a function of the square root of time. This suggests that the data can be more simply described using

$$G_1(\tau) = e^{-\gamma \sqrt{6\tau/\tau_0}}, \tag{23}$$

where $\gamma = \langle z_0 \rangle/l^* + 2/3$. This result was obtained previously using the simpler, but less rigorous, boundary condition, $U(z = 0) = 0$ (Refs. 7,8).

The behavior of the autocorrelation function shown in Fig. 8 is in fact quite general. Virtually all the autocorrelation functions that

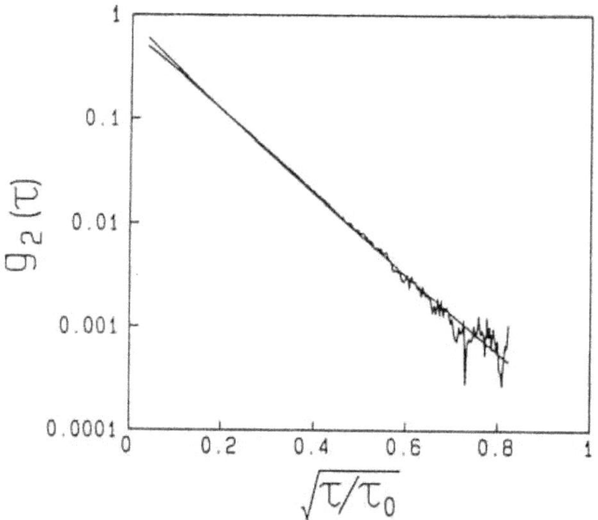

Fig. 8. Intensity autocorrelation function *vs* sqaure root of reduced time for backscattering from a 5-mm thick cell with 0.412-$\mu$m-diameter polystyrene spheres and $\phi = 0.05$. The line through the data is a fit to the theory in Eq. (22), and has a more rapid initial decay and more curvature than the data.

we have measured in backscattering for freely diffusing particles exhibit a similar behavior in that they decay exponentially in the square root of time. Therefore, it is most convenient to use the very simple form in Eq. (23) to describe their shape. The slope of the autocorrelation function, when plotted in this fashion, is determined solely by $\tau_0$ and by $\gamma$. While we have derived this functional form using rather heuristic arguments, similar results are also obtained using more rigorous diagrammatic techniques.[18] However, it is again essential to properly account for the contributions of the short paths. This can be done by ensuring that the difference between initial and final wavevectors is strictly $2k_0$, as required for backscattering, and by limiting the momentum transfer in any single scattering event to properly reflect the form factor of the scatterers. By contrast, if this is not done, the contributions of the short paths are over estimated in the diagrammatic approach, leading to a prediction[6] for the autocorrelation function that describes the initial decay reasonably well, but fails at longer times by predicting a power-law decay, in sharp disagreement with the data.

Using samples of known $\tau_0$, we find experimentally that $\gamma$ depends on both polarization and on the anisotropy of the scattering, as characterized by $l^*/l$. The dependence of $\gamma$ on polarization is strongest for isotropic scatterers, when $l^*/l \to 1$, as illustrated by the data in Fig. 9a. These data are obtained from a $\phi = 0.02$ sample of 0.091 $\mu$m diameter spheres, for which $l^*/l \approx 1.1$. The sample is illuminated by linearly polarized light, and an analyzer is used to detect scattered light whose polarization is either parallel or perpendicular to the incident light. The autocorrelation function for the perpendicular polarization decays more rapidly, because the analyzer discriminates against the low order paths which retain a high degree of their incident polarization. This reflects the contribution of multiple scattering paths for which the diffusion approximation is more appropriate. The value measured is $\gamma = 2.5$. By contrast, the autocorrelation function for the parallel polarization decays more slowly, because of the additional contribution of the low order scattering paths which have a longer decay time. We note, however, that the form of the autocorrelation function is still the same, remaining linear when plotted exponentially as a function of the square root of time. Here, the value measured is $\gamma = 1.6$. By contrast, for very anisotropic scatterers, the dependence of $\gamma$ on polarization is very weak. This is illustrated in Fig. 9b, which shows data obtained using a $\phi = 0.02$ sample of 0.605 $\mu$m diameter spheres, for which $l^*/l \approx 10$. The autocorrelation function for the perpendicular polarization again decays faster, with $\gamma_\perp = 2.1$; for the parallel polarization, $\gamma_\| = 2.0$.

We summarize the dependence of $\gamma$ on $l^*/l$ for both polarizations in Fig. 10. The data are obtained by varying the size of the spheres to vary $l^*/l$, since for the particle sizes used here this ratio is a monotonic function of sphere diameter. In all cases, $\phi = 0.02$. The data suggests that, for each polarization, $\gamma$ asymptotically approaches a constant as $l^*/l$ becomes large. Physically, this reflects the diminishing contribu-

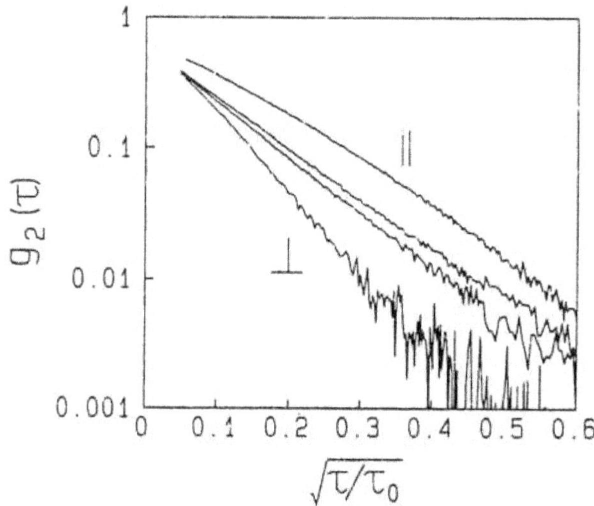


Fig. 9. Intensity autocorrelation functions *vs* square root of reduced time for backscattering for parallel and perpendicular polarizations. The upper and lower curves are for 0.091-$\mu$m-diameter spheres, with $\gamma_{\parallel} = 1.6$ and $\gamma_{\perp} = 2.9$; the two middle curves are for 0.605-$\mu$m-diameter spheres, with $\gamma_{\parallel} = 2.0$ and $\gamma_{\perp} = 2.1$.

tions of the short paths as the scattering becomes more strongly peaked in the forward direction. In this case, the diffusion approximation, which is scalar in nature, should apply. Therefore, $g_1(\tau)$ becomes a function of $\tau/\tau_0$ only, and thus, in this limit, $\gamma$ does not depend on $l^*/l$. In fact, an asymptotic value of $\gamma = 2.1$ is predicted theoretically by properly considering the contribution of the short paths to the autocorrelation function at long times.[18,19] Furthermore, the trends observed in Fig. 10 for the dependence of $\gamma$ on both $l^*/l$ and polarization are consistent with the predictions made using diagrammatic techniques.

The variation of $\gamma$ with particle size arises from the contributions of short paths, which are different for different polarizations and values of $l^*/l$. By contrast, the *absolute* decay of the unnormalized autocorrelation function, $G_1(\tau)$, at short times is due solely to the contributions of *long paths* which are well described within the diffusion

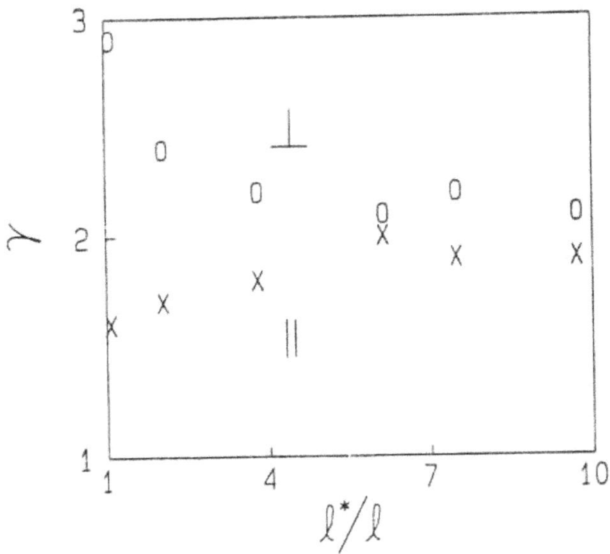

Fig. 10. Experimental values of $\gamma$ *vs* $l^*/l$ and particle size for parallel (crosses) and perpendicular polarizations (circles).

approximation. These contributions should be independent of polarization and $l^*/l$ due to the large number of scattering events. This leads to an important relation between the value of $\gamma$ and the static intensity, $G_1(0) \equiv \langle |E|^2 \rangle = \langle I \rangle$. To see this, we examine the short time expansion for $G_1(\tau)$ for a semi-infinite sample,

$$G_1(\tau) = G_1(0) - \gamma G_1(0)\sqrt{\frac{6\tau}{\tau_0}} + \cdots .$$

The characteristic square root of time dependence arises from the contributions of long paths and is independent of the boundary and initial conditions assumed. The rate of the *absolute* decay, $\partial G_1(\tau)/\partial\sqrt{\tau/\tau_0}$, is also determined solely by the long paths. Thus, the coefficient, $\gamma G_1(0)$, should be a constant independent of polarization and $l^*/l$. Hence, since $G_1(0) = \langle I \rangle$ we expect

$$\gamma \propto \frac{1}{\langle I \rangle} .$$

This implies that both $\gamma$ and $\langle I \rangle$ depend on the contributions of low order scattering and on the sample used. Physically, we expect $\langle I \rangle$ to decrease for large $l^*/l$ or for perpendicular polarization, both of which reduce the number of short paths contributing to the scattered intensity. Similarly, we expect $\gamma$ to increase when the contributions of short paths are reduced since it is the short paths which have the longest decay times.

To test these ideas, we again consider the behavior of the experimental autocorrelation function $g_2(\tau) = |g_1(\tau)|^2$. Figure 11 shows a plot of $g_2(\tau)$ *vs* $\sqrt{\tau/\tau_0}$ obtained using an extended light source with parallel polarizer and analyzer.[11] The measurements were taken using four different diameters of polystyrene spheres: 0.102 $\mu$m, 0.305 $\mu$m, 0.46 $\mu$m, and 0.797 $\mu$m. In each case, the sample size was 1cm $\times$ 1cm $\times$ 1cm and the volume fraction of spheres was $\phi = 0.10$. The autocorrelation function, $g_1(\tau)$, appears to have the same form as a function of $\tau/\tau_0$ for the three larger size spheres studied. Since the static average intensity $\langle I \rangle$ turns out to be the same for these samples, it is also consistent with the prediction that the coefficient $\gamma G_1(0)$ is constant. The autocorrelation function $g_2(\tau)$ decays slightly more slowly for the smallest beads, again consistent with the slightly larger static intensity observed in this case. Finally, the inset in Fig. 11 shows that the initial decay of the autocorrelation function is linear in $\sqrt{\tau/\tau_0}$, as expected from Eq. (23).

## 5. ANALOGIES WITH STATIC MEASUREMENTS

The functional form of the temporal autocorrelation function depends crucially on the distribution of paths explored by the diffusing light. In backscattering, $g_1(\tau)$ is particularly sensitive to the contributions of short paths as evidenced by the dependence of $\gamma$ on polarization and $l^*/l$. Thus, measurements of $g_1(\tau)$ become a sensitive probe of the nature of the transport of light when there is significant multiple scattering. Other optical properties of strongly scattering media also depend on $P(s)$: the angle ($q$) dependence of the coherent enhancement of the backscattering cone and the absorption dependence of the incoherent backscattered intensity. In fact, within the diffusion approximation, the description of all three of these quantities depends on $P(s)$ in exactly the same way. The analogy that exists between these three quantities dramatically illustrates the similarity of the underlying physics and the utility of the diffusion approximation in describing a wide variety of multiple scattering processes. The analogy also has practical consequences: it leads to a way of experimentally determining $\gamma$, and therefore also $\tau_0$, without recourse to any specific theory about the distribution of the low order multiple scattering paths which are not rigorously described within the diffusion approximation.

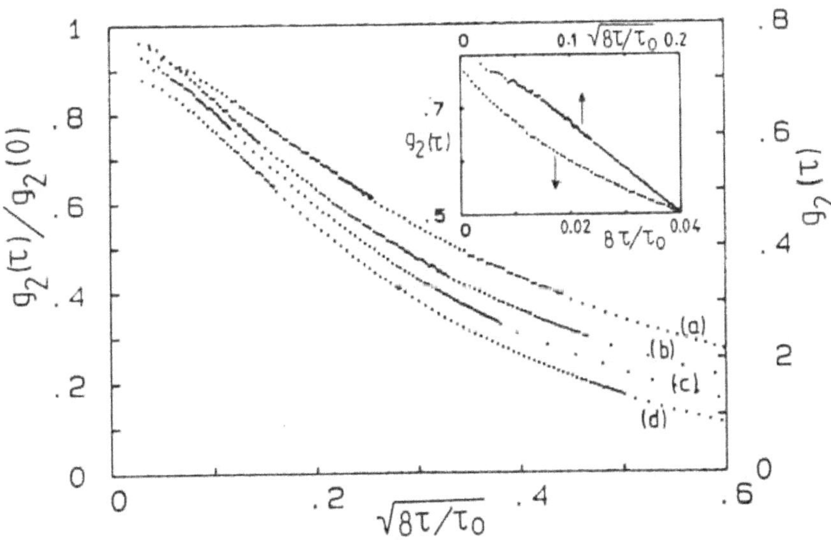

Fig. 11. Intensity autocorrelation functions *vs* reduced time for backscattering for polystyrene spheres of various diameters and $\phi = 0.10$. The inset shows that the initial decay is linear in the square root of time.

We have already considered the temporal autocorrelation function in the previous section where we found that for $\tau/\tau_0 \ll 1$,

$$\frac{G_1(\tau)}{G_1(0)} = 1 - \gamma\sqrt{\frac{6\tau}{\tau_0}} + \dots , \qquad (24)$$

where the coefficient $\gamma G_1(0)$ does not depend on the contributions of short paths.

Next, we consider the coherent backscattering cone. As outlined elsewhere in this volume, the average scattered intensity is higher in the backward direction than at wide angles. This is because for each scattering path there exists a time-reversed path of identical length which contains the same fraction of the total intensity and has the same phase shifts. Light emerging from both of these paths interferes constructively in the backward direction. The angular dependence of this coherent enhancement above the incoherent wide angle intensity is controlled by the spatial separation of the endpoints of the path, $\rho$. The enhancement for a given angle and a given path length $s$ is simply the Fourier transform of the probability distribution $p(\rho, s)$

$$\alpha(s, q_0) = I_0 \int p(\rho, s) e^{-q_0 \cdot \rho} d^2\rho \propto P(s) e^{-q_0^2 l^* s/3} \ ,$$

where $\mathbf{q}_0$ is the scattering vector with respect to the backscattering angle ($\theta_0 = 0$) and its magnitude is $q_0 = 4\pi n/\lambda \sin(\theta_0/2)$. This means that the paths of length $s$, which have a mean square end-to-end distance of $\sim \sqrt{sl^*}$, contribute a gaussian of angular width $\lambda/\sqrt{sl^*}$ and of amplitude $P(s)$ to the total coherent backscattering enhancement $\alpha(q_0)$ [albedo]. Thus, $\alpha(q_0)$ is obtained by integrating over these Gaussians,

$$\alpha(q_0) \propto \int P(s) e^{-q_0^2 l^* s/3} ds \ . \tag{25}$$

Equation (25) has the same form as Eq. (12) when the identification $2\tau/\tau_0 \rightarrow (ql^*)^2/3$ is made. Therefore, by analogy the angular variation of $\alpha(q_0)$ at small $q_0 \, (q_0 l^* \ll 1)$ is given by

$$\frac{\alpha(q_0)}{\alpha(0)} = 1 - \gamma q l^* + \dots \ , \tag{26}$$

where the coefficient $\gamma$ is the same as in Eq. (24) for the same sample, provided that single scattering is negligible.[11] The square root singularity of $G_1(\tau)$ with time maps into a linear singularity of $\alpha(q_0)$ with angle.

We can also compare $G_1(\tau)$ and $\alpha(q_0)$ with the dependence on absorption of the incoherent intensity near backscattering. If the absorption in the scattering medium is characterized by an absorption length, $l_a$, then the fraction $P(s)$ of the intensity scattered on average along a path of length $s$ is attenuated by $\exp(-s/l_a)$. The total scattered intensity can then be written as an integral over the distribution of path lengths,

$$\alpha_I(l_a) \propto \int P(s)e^{-s/l_a}\,ds \ . \tag{27}$$

This form also maps into $G_1(\tau)$ and $\alpha(q_0)$ in Eqs. (12) and (25) when the identifications $2\tau/\tau_0 \to (ql^*)^2/3 \to l^*/l_a$ are made. Therefore, the variation of the normalized backscattered intensity with absorption for weak absorption ($l^* \ll l_a$) bceomes

$$\frac{\alpha_I(l_a)}{\alpha_I(\infty)} = 1 - \gamma\sqrt{\frac{3l^*}{l_a}} + \cdots \ . \tag{28}$$

Equations (24), (26), (28) are three equivalent relations for the initial slope of the normalized temporal autocorrelation function, the normalized angular dependence of the coherent backscattering cone, and the normalized dependence of the incoherent scattering intensity on absorption. These equations contain essentially three potentially unknown parameters: $\gamma, \tau_0$, and $l^*$. Thus in principle, an absolute measure of a particle's diffusion coefficient $D$ can be obtained from $\tau_0$ by experimentally determining $g_1(\tau), \alpha(q_0)/\alpha(0)$, and $\alpha_I(l_a)/\alpha_I(\infty)$ on the same sample in the short time, small angle, and weak absorption regimes, respectively. Physically, this correspondence reflects the fact that all these processes probe the same long diffusion paths of the light. This analogy demonstrates the simplicity and power of the diffusion approximation in treating the consequences of multiple scattering.

We demonstrate this correspondence in Fig. 12. We compare the temporal decay of $g_2(\tau)$ plotted as a function of $\sqrt{\tau/\tau_0}$ with the variation of the static incoherent intensity $\langle I \rangle$ as a function of added absorbing dye for a $\phi = 0.10$ suspension of 0.46 $\mu$m diameter spheres.[11] We find good agreement with Eqs. (24) and (28). Even beyond the linear regime the two quantities show a very similar functional dependence when plotted as a function of the appropriate variables, $\sqrt{2\tau/\tau_0}$, and $\sqrt{l^*/l_a}$. This demonstrates the validity of Eq. 12. In fact, the value of $\tau_0$ used in order to achieve good matching between the two data set is 10% larger than the free particle diffusion time in backscattering. This could be due to a slight reduction of $D$ which is expected because of the hydrodynamic interactions between particles.

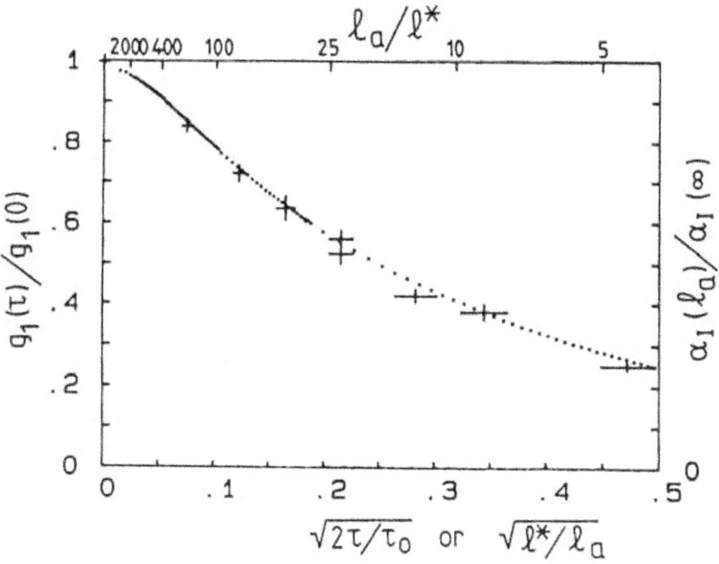

Fig. 12. Comparison of the temporal autocorrelation function *vs* $\sqrt{2\tau/\tau_0}$ with the incoherent intensity *vs* $\sqrt{l/l_a}$ for backscattering from 0.46-$\mu$m-diameter spheres with $\phi = 0.10$.

## 6. APPLICATIONS

In the previous sections we limited our theoretical treatment of the autocorrelation function of multiply scattered light to non-interacting, monodisperse particles whose dynamics were determined solely from their Brownian motion. In this section we generalize our treatment to investigate the behavior of the correlation function for other situations, including polydispersity in the particle size, absorption of light, and convective motion of the suspension. We also discuss the application of DWS to suspensions where interparticle interactions become important.

## 6.1. Particle Sizing

One of the most useful applications of DWS is for sizing particles.[4,8] In principle, we can combine DWS with measurements of the back-scattering cone or dependence on absorption to uniquely measure $\tau_0$ for an unknown sample. In practice, it is more straightforward to exploit the features of DWS alone for particle sizing. To demonstrate the utility of DWS, we measured autocorrelation functions in backscattering with perpendicular polarization for monodisperse suspensions of polystyrene spheres of different diameters. In Fig. 13, we have plotted $\log g_2(\tau)$ *vs* $\sqrt{\tau/\tau_0}$ for six different samples with $\phi = 0.05$ and with sphere diameters, $d$, ranging from 0.091 $\mu$m to 0.605 $\mu$m. For $d \geq 0.3 \mu$m, the scaled autocorrelation functions all fall on nearly the same curve, suggesting that the parameter $\gamma_\perp$ approaches a single value for large particle sizes. This is consistent with the data in Fig. 9 which show $\gamma_\perp$ saturating at a value of approximately 2.1 for large $l^*/l$. Thus, in the limit of large particle diameters, DWS can be used to unambiguously determine $\tau_0 = (Dk_0^2)^{-1}$. If the relationship between $D$ and $d$ is known, as it is for small $\phi$, DWS can be used to determine particle diameter to better than 10%.

For $d < 0.3$, $\gamma_\perp$ is no longer independent of particle size but steadily increases with decreasing particle size. Since $\gamma_\perp$ and $\tau_0$ enter the expression for the autocorrelation function multiplicatively, $\tau_0$ cannot be obtained in a single measurement without prior knowledge of $\gamma_\perp$. However the data shown in Fig. 9 suggest that $\gamma_\perp/\gamma_\parallel$ is a monotonic function of $l^*/l$. Therefore, it is possible to unambiguously determine $\tau_0$ using DWS for a given sample by measuring the autocorrelation functions for the different polarizations. In this case, the ratio of the decay rates gives the value of $l^*/l$ allowing the absolute values of $\gamma_\perp$ and $\gamma_\parallel$

                           *D. J. Pine et al.*

to be determined. Once $\gamma_\perp$ and $\gamma_\parallel$ are known, $\tau_0$ can be obtained from the rate of decay of the autocorrelation function.

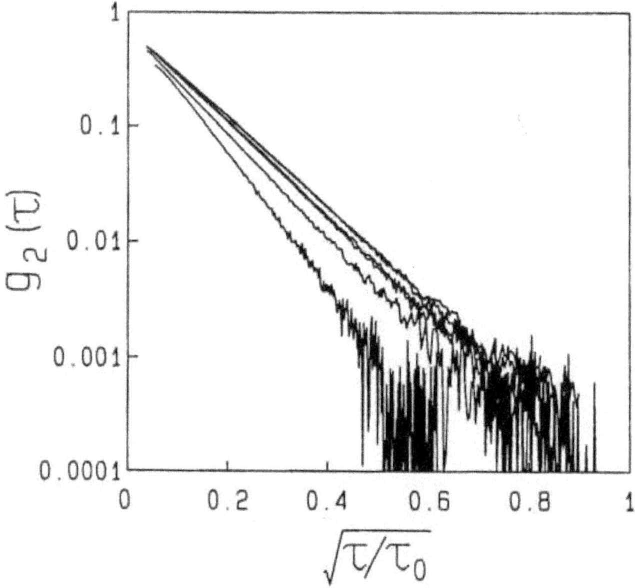

Fig. 13.  Intensity autocorrelation functions *vs* square root time for back-scattering for polystyrene spheres for six different samples with $\phi = 0.05$. From the lower left to the upper right corners, the curves correspond to data from spheres with the following diameters: 0.091 $\mu$m, 0.198 $\mu$m, 0.305 $\mu$m, 0.497 $\mu$m, 0.412 $\mu$m, 0.605 $\mu$m.

However, care must be exercised in using this scheme since the theory accounts only for the variations of $\gamma$ arising from the single particle form factor, $F(\mathbf{q})$, and ignores the effects of longer range correlations as measured by the static structure factor, $S(\mathbf{q})$. As outlined below, estimates of the influence of $S(\mathbf{q})$ suggest that these effects may be significant at particle volume fractions as low as 10%. Fortunately, there are alternative methods for determining $\gamma$ which do not depend on specific assumptions made about the details of the scattering properties of the medium. One of these methods depends on the analogy between DWS, coherent backscattering, and incoherent absorption discussed in the previous section. However, this method requires that three independent measurements be made. An alternative, which is significantly simpler to implement, is to exploit the result that $\gamma\langle I\rangle$ should be a constant, independent of particle size. Thus, the diffusion coefficient $D$ could be measured by comparing the decay of $g_2(\tau)$ to that of a calibration sample made up of polystyrene spheres with a known $D$. The ratio of the slope $\partial \ln g_2(\tau)/\partial\tau = \gamma\sqrt{6/\tau_0}$ times the ratio of the static intensities would provide the inverse ratio of $\sqrt{\tau_0}$ for the two different systems. In this way, the dynamics of concentrated suspensions could be probed and particle size could be determined provided, once again, that the relationship between $D$ and $d$ is known.

## 6.2. Polydispersity

We now consider the effects of polydispersity on the correlation function. In this case, the scattering medium possesses a distribution of species with different optical properties and different diffusion coefficients. In the case of conventional quasi-elastic light scattering, this leads to a non-exponential relaxation of the autocorrelation function. A considerable amount of research has gone into analysing the QELS data, to invert the shape of the non-exponential autocorrelation function and obtain information about the distribution of scattering species.[20] In the case of strong multiple scattering, the decay of the autocorrelation function is already non-exponential and geometry dependent. Should we then expect additional changes in the decay of the autocorrelation function due to the effects of polydispersity?

In the previous sections we have seen that the correlation function could be calculated if we knew the temporal dependence of the dephasing of a statistical path of $n$ steps. The dephasing occurs by a random walk of phase shifts caused by the motion of the individual particles along the scattering path. For a polydisperse system the scattering from different size particles leads to phase shifts of different average magnitude resulting from the dependence of $\langle q^2\rangle$ and $D$ on the particle diameter. This results in a random walk with different step sizes: In

   *D. J. Pine et al.*

the continuum limit, or diffusion approximation, the statistics of the random walks reduce to simple Gaussian functions. Thus, if the step size of a random walk varies for each step, then for large $n$, the statistics still remain the same as for a walk with a single average step size. This is the result of the central limit theorem. Thus, the time dependence of the autocorrelation function should be the same for a polydisperse system as for a monodisperse system. In contrast to QELS, there should be no additional information about polydispersity in these measurements. We will only measure average properties. The question which remains is how to calculate the appropriate average diffusion constant for a polydisperse sample.

Intuitively the answer is clear: We would expect the average diffusion constant to be weighted by the relative concentration of a species and by its scattering strength as expressed by its scattering cross section. As we detail below the intuitive answer is correct. We limit our discussion here to the case of non-interacting particles. For simplicity and illustration we first consider the case of a bimodal distribution, a density $\rho_j$ of particles with scattering cross section $\sigma_j$ and diffusion constant $D_j$, where $j = a, b$. We then replace Eq. (2) for the total phase shift for an $n$th order scattering process by

$$\Delta\phi^{(n)}(\tau) = \sum_{i=1}^{n} \mathbf{q}_i \cdot \Delta\mathbf{r}_i(\tau) = \sum_{i=1}^{n_a} \mathbf{q}_{a_i} \cdot \Delta\mathbf{r}_{a_i}(\tau) + \sum_{i=1}^{n_b} \mathbf{q}_{b_i} \cdot \Delta\mathbf{r}_{b_i}(\tau) \; , \quad (29)$$

where $n_a + n_b = n$. Since the average scattering wavevectors and rms displacements differ for the two species it is convenient to sum separately over the phase shifts for each species. The ensemble averages are calculated as in the monodisperse case but now keeping track of the species. Equation (5) becomes

$$G_1^{(n)}(\tau) = I_0 P(n) \left\langle e^{-q^2 \langle \Delta r_{a_i}^2(\tau) \rangle / 6} \right\rangle_q^{n_a} \left\langle e^{-q^2 \langle \Delta r_{b_i}^2(\tau) \rangle / 6} \right\rangle_q^{n_b} .$$

Performing the individual averages as before, we have

$$G_1^{(n)}(\tau) = I_0 P(n) e^{-n_a 2 k_0^2 (l_a / l_a^*) D_a \tau - n_b 2 k_0^2 (l_b / l_b^*) D_b \tau} \; , \quad (30)$$

where we have left the characteristic decay times $\tau_{0a}$ and $\tau_{0b}$ in terms of the diffusion coefficients for each species and $k_0$. Thus, $\tau_{0j} = (D_j k_0^2)^{-1}$.

We must now determine the average number of scattering events from particles $a$ and $b$ in a path of total length $s$. The relative probabilities of scattering by $a$ or $b$ are proportional to their number densities and total scattering cross sections or inversely proportional to their individual mean free paths

$$n_a/n_b = (\rho_a \sigma_a/\rho_b \sigma_b) = l_b/l_a$$

or $n_a l_a = n_b l_b$. Substituting this into Eq. (30) gives

$$\Delta\phi^{(n)}(\tau) = n_a l_a 2k_0^2 (D_a/l_a^* + D_b/l_b^*)\tau$$

where $\Delta\phi^{(n)}(\tau)$ is the exponent of Eq. (30). The total path length is given by $n$ steps with the actual mean free path of the system $l'$ which includes scattering from both species.

$$1/l' = \rho_a \sigma_a + \rho_b \sigma_b = 1/l_a + 1/l_b$$

$$s = nl' = n_a l_a$$

since $n = n_a + n_b = n_a(1 + l_a/l_b)$. For use below we also note:

$$1/l_{\text{eff}}^* \equiv (1/l_a^* + 1/l_b^*) \ .$$

The final result for the time dependent dephasing from the two species is therefore:

$$\Delta\phi^{(n)}(\tau) = s2k_0^2 (D_a/l_a^* + D_b/l_b^*)\tau$$

which is simply proportional to $s$ and $\tau$ as expected, and has the same form as the previously derived expression for a single species. Thus, the time dependence of the correlation functions will not change from the results for a single species provided we substitute the appropriate averages for $l^*$ and $\tau_0$. Writing the phase shift as:

$$\Delta\phi^{(n)}(\tau) = (s/l_{\text{eff}}^*)(\tau/\tau_{\text{eff}})$$

and generalizing to the case of many different species we have directly:

$$1/l^*_{\text{eff}} = \sum_j 1/l^*_j \tag{31a}$$

$$1/\tau_{\text{eff}} = k_0^2 D_{\text{eff}} \tag{31b}$$

$$D_{\text{eff}} = \left(\sum_j D_j/l^*_j\right)\Big/\left(\sum_j 1/l_j\right) . \tag{31c}$$

The effective diffusion constant is the weighted average of the diffusion coefficients. Since $1/l^*_j = \rho_j\sigma_j$, the weighting factor is the number density times the cross section for transport scattering or alternatively the inverse mean free path. The expressions for $\tau_{\text{eff}}$ and $l^*_{\text{eff}}$ in Eq. (31) can be directly substituted for $\tau_0$ and $l^*$ in the correlation functions previously described. To illustrate this behavior, we measured the autocorrelation function in both transmission and backscattering from a series of binary mixtures of 0.198 $\mu$m and 0.605 $\mu$m diameter polystyrene spheres. In all cases, the functional form of the data obtained from mixtures is identical to that obtained from a single species. As an example, in Fig. 14, we show the autocorrelation function obtained in backscattering from a mixture of $\phi = 0.02$ of 0.198 $\mu$m spheres and $\phi = 0.02$ of 0.605 $\mu$m spheres. The expected exponential decay in the square root of time is apparent. We also calculate $1/l^*_i = \rho_i\sigma_i\langle 1 - \cos\theta_i\rangle$ for each species using Mie theory, and $D_i$ from the Stokes-Einstein relation. The solid line through the data is the calculated decay using the calculated values of $D_{\text{eff}}$ and $l^*_{\text{eff}}$ and assuming that $\gamma = 2.1$. The agreement is excellent. Similar agreement is obtained for mixtures with different ratios. These results are summarized in Fig. 15 where we compare the value of $\tau_{\text{eff}}$ determined experimentally from backscattering measurements with that calculated theoretically from $D_{\text{eff}}$ and $l^*_{\text{eff}}$.

We conclude by again emphasizing that, unlike QELS, DWS cannot yield independent information about polydispersity. Instead, only average quantities can be determined. However, we can calculate the average quantities if the distribution is known which, in principle, can allow the effects of different possible distributions to be tested. One

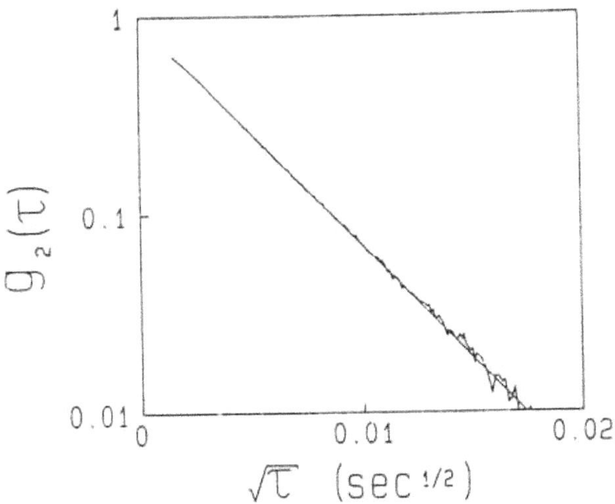

Fig. 14. Intensity autocorrelation functions *vs* square root reduced time for backscattering from a mixture with 0.198-$\mu$m-diameter spheres at $\phi_a = 0.02$ and 0.605-$\mu$m-diameter spheres at $\phi_b = 0.02$. The solid line through the data is determined without any fitting parameters using Eq. (31) as discussed in the text.

important effect that has not as yet been determined is the dependence of $\gamma$ on polydispersity.

## 6.3. Porous Media

One application of the calculation for a polydisperse system is the limit where part of the sample consists of strong scatterers which are static while another component scatters and is diffusing. Such would be the case for particles diffusing in a porous medium. The result is straightforward, for the static components we simply set $D_j = 0$ in Eq. (31) and use the correlation functions previously described. For example if we have a porous structure with a transport mean free path $l^*$ and we introduce a colloid which has the same $l^*$ and a diffusion constant $D$ (in the presence of the pores), then the correlation function would decay as for a homogeneous medium with effective diffusion constant $D/2$. The fact that the analysis of the data is so straightforward if we know the transport mean free paths of the porous structure

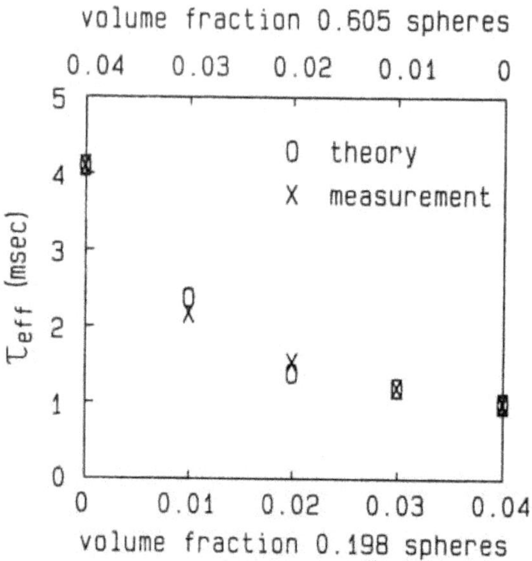


Fig. 15. Effective decay time $\tau_{\text{eff}}$ *vs* volume fraction $\phi$ for mixtures of 0.198-$\mu$m-diameter spheres and 0.605-$\mu$m-diameter spheres.

and the particles separately, implies that this should be a very useful way of seeing how the particle diffusion is affected by the presence of a confining geometry. To illustrate these ideas, in Fig. 16 we show an autocorrelation function obtained in backscattering from a glass frit with 4–8 $\mu$m diameter pores which were saturated with a $\phi = 0.01$ suspension of 0.497 $\mu$m diameter polystyrene spheres. The autocorrelation function exhibits the exponential decay in the square root of time. It is important to note that, in this case, the rigid glass frit provides a reference signal so that the autocorrelation function was measured in the *heterodyne* mode; hence, $g_1(\tau)$ is obtained.

There is a complication which may arise in studying porous media. We have implicitly assumed in the derivation of Eq. (31) that the geometry and density of particles of each species is such that there is a statistical sampling of each species in each path, that is, if we double the path length there are twice as many particles of each species con-

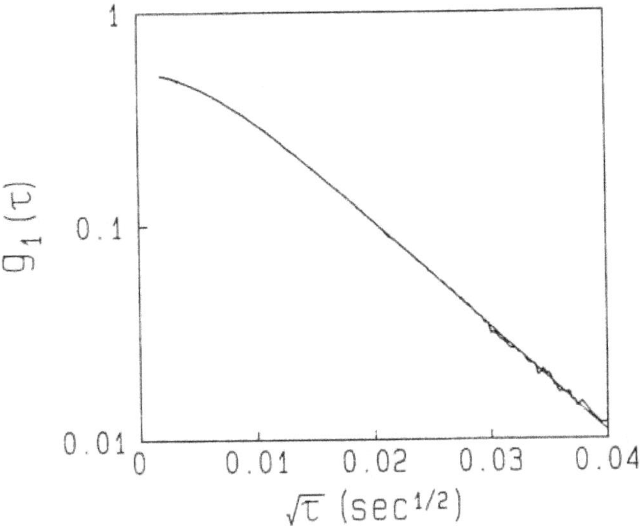

Fig. 16. Heterodyne autocorrelation function *vs* square root time for backscattering from a glass frit with 4-8 $\mu$m diameter pores saturated with a $\phi = 0.01$ suspension of 0.497-$\mu$m-diameter polystyrene spheres.

tributing to the scattering. If on the other hand we have a very dilute concentration of moving particles in a porous medium, such that all paths sampled by the correlation function involve at most one scattering from a mobile particle, then the description given above is invalid. Instead the correlation function will be like that for the single scattering limit, i.e. like that for QELS. The effect of the scattering from the porous medium will simply be to randomize the direction of incident and scattered wavevectors. Thus the correlation function will have the form of a QELS relaxation averaged over scattering angles and weighted by the form factor for scattering from a single sphere. For spheres much smaller than the wavelength of light, the scattering from a single sphere is isotropic and the average is easily performed:

$$g_1(\tau) = \frac{4\tau_0}{\tau} \left(1 - e^{-\tau/4\tau_0}\right) . \tag{32}$$

For $\tau \ll \tau_0$, $g_1(\tau) \sim \exp(-\tau/2\tau_0)$ and for $\tau \gg \tau_0$, $g_1(\tau) \sim \tau^{-1}$. Figure 17 shows an autocorrelation function obtained from a porous glass sample with $\sim 5\mu$m-diameter pores which were filled with an aqueous suspension of mobile $0.091\mu$m-diameter polyballs at $\phi = 0.01$. The data were found to be in excellent agreement with Eq. (32) where the value of $\tau_0 = (Dk_0^2)^{-1}$ is given by the free particle diffusion coefficient for the polyballs. This is consistent with the expectation that diffusion should be unimpeded when the pore size is much larger than the mean particle diameter.

## 6.4. Absorption

In any physical system there will always be absorption of light, either by the scatterers themselves or by the solvent in which they are suspended. Furthermore, the effects of absorption will be enhanced in the multiple scattering limit because the path lengths of the light and the number of scattering events are greatly increased. However, the consequences of absorption for the autocorrelation function can be readily determined within our approach. For any scattering geometry, $P(s)$ is the fraction of the scattered intensity associated with paths of length $s = nl$. With absorption, the intensity of light travelling a path of length $s$ is attenuated exponentially by a factor $\exp(-s/l_a)$, where $l_a$ is the absorption length. Thus, we write $P'(s) = P(s)\exp(-s/l_a)$, where $P(s)$ is determined solely by geometry, as before. Then Eq. (12) has the form:

$$G_1(\tau) = I_0 \int P(s) e^{-s/l_a - (2\tau/\tau_0)s/l^*} \, ds \; .$$

Since both terms in the exponent are linear in $s$, the effect of the absorption is mathematically the same as shifting the time scale. Thus all of our previous analysis for different geometries is directly applicable if we simply make the substitution:

$$\tau/\tau_0 \to l^*/2l_a + \tau/\tau_0 \; .$$

In this case, the form of the autocorrelation function for backscattering becomes

$$g_1(\tau) = \exp\left(-\gamma \sqrt{\frac{6\tau}{\tau_0} + \frac{3l^*}{l_a}}\right) \; . \tag{33}$$

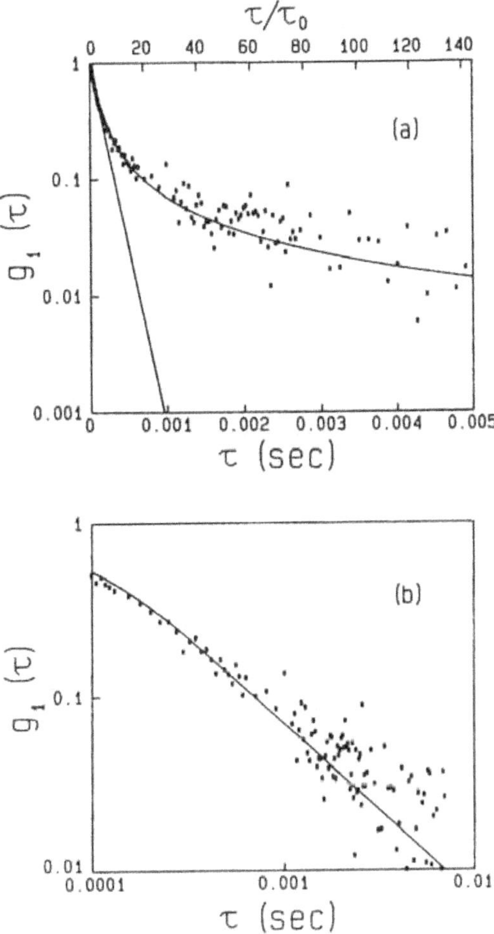

Fig. 17. Heterodyne autocorrelation function *vs* time for scattering from a porous glass sample filled with an aqueous suspension of mobile 0.091-$\mu$m-diameter polystyrene spheres at $\phi \simeq 0.01$. Light is multiply scattered by the immobile porous glass but only singly scattered from the mobile polystyrene spheres. (a) Straight line shows first cumulant approximation; (b) $\tau^{-1}$ decay at long times.

We illustrate this behavior in Fig. 18, where we show the autocorrelation function obtained in backscattering from 0.497 $\mu$m diameter polystyrene spheres with $\phi = 0.01$. The lower curves were obtained upon addition of varying amounts of methyl red, which absorbs the 488 nm laser light used. The absorption lengths are $l_a = 4.87$ mm for the upper curve and $l_a = 2.53$ mm for the lower curve. Physically, the effect of the absorption is to reduce the contribution of the longer paths to the decay of the autocorrelation function. These paths would otherwise contribute a rapid, initial decay of the correlation function. This effect is clearly evident in Fig. 18 by the rounding of the correlation function at early times.

A comparison of the prediction of Eq. (33) with the data is shown by the solid line in Fig. 18. The theoretical curves do not have any free parameters other than the overall normalization, which is determined by the collection optics, as reflected in $f(A)$. The value of $\gamma$ used is determined from a measurement without the dye, and the values of $l_a$ for each sample are determined from independent measurements of the transmitted intensity through the sample. The agreement between the theoretical prediction and the data is excellent, confirming the validity of our expression in Eq. (33).

The reduction in the contribution of long paths will also have profound effects in transmission. It will dramatically alter the functional dependence of $\Gamma_1$ on $L$, with $\Gamma_1$ becoming linearly dependent on $L$ rather than $L^2$. This reflects the fact that only the shortest paths can contribute to $G_1(\tau)$ in transmission with absorption, since the longer paths are attenuated. Thus the typical path contribution to the decay will have a length of $L$ rather than $(L/l^*)^2 l^*$.

## 6.5. Sheared Suspensions

The loss of correlations in the scattered light can be caused by almost any motion of the scattering particles, not just the diffusive motion that we have considered to this point. Convective motion can also cause a loss of correlations. However, uniform motion will not lead to dephasing of multiply scattered light, just as it will not contribute for singly scattered light. For uniform motion, the speckle pattern simply translates with the object. Thus the correlation function will decay only

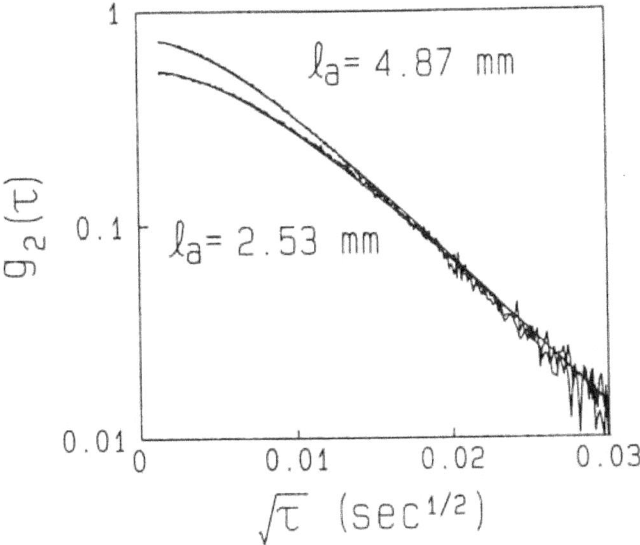



Fig. 18. Intensity autocorrelation functions *vs* square root time for back-scattering from samples with absorbing dye in aqueous solution.

as new parts of the sample are introduced into the incident beam or the viewing aperture. By contrast, for relative motion of the particles, the path lengths change and dephasing occurs. The essential difference for convective motion by comparison to diffusive motion is the time dependence of the relative mean square displacement. For convection we have $\langle \Delta r_i^2(\tau) \rangle \propto \tau^2$ whereas for diffusion we have $\langle \Delta r_i^2(\tau) \rangle \propto \tau$. Therefore, we expect the $\sqrt{\tau}$-dependence found up to now to be replaced by a $\tau$-dependence.

The simplest relative motion to treat is that for simple shear such as planar Couette flow.[21] Let us take a velocity profile given by $\mathbf{v} = \Gamma x\,\hat{\mathbf{e}}_x$, where $\Gamma \equiv \partial v/\partial z$. The relative displacement of a particle consists of a Brownian diffusive term, $\Delta r_i^B(\tau)$, and a convective shear term, $\Delta r_i^S(\tau)$. Then, if the Brownian motion is not affected by the laminar shear flow, the total change in phase along a given path of $n$

362                          *D. J. Pine et al.*

scatterers is:

$$\Delta\phi^{(n)}(\tau) = \sum_{i=1}^{n} \Delta\phi_i(\tau) = \sum_{i=1}^{n} \mathbf{q}_i \cdot \Delta\mathbf{r}_i^B(\tau) + \mathbf{q}_i \cdot \Delta\mathbf{r}_i^S(\tau) \ ,$$

where $\mathbf{q}_i = \mathbf{k}_i - \mathbf{k}_{i-1}$. Since the phase is changed only by relative motion of the particles, it is convenient to rewrite the contribution of convection to $\Delta\phi^{(n)}(\tau)$ as

$$\mathbf{q}_i \cdot \Delta\mathbf{r}_i^S(\tau) = \mathbf{k}_i \cdot [\Delta\mathbf{r}_{i+1}^S(\tau) - \Delta\mathbf{r}_i^S(\tau)] = \Gamma\tau k_0 \Lambda_i (\widehat{\boldsymbol{\kappa}}_i \cdot \widehat{\mathbf{e}}_x)(\widehat{\boldsymbol{\kappa}}_i \cdot \widehat{\mathbf{e}}_z)$$

where $\Lambda_i \equiv |\Delta\mathbf{r}_{i+1}^S(0) - \Delta\mathbf{r}_i^S(0)|$ is the distance between successive scattering events, $\widehat{\boldsymbol{\kappa}}_i$ is a unit vector in the direction of the scattered light, and $\widehat{\mathbf{e}}_x$ and $\widehat{\mathbf{e}}_z$ are unit vectors in the $x$ and $z$ directions, respectively. This can be rewritten in terms of the polar and azimuthal angles, $\theta_i$ and $\phi_i$, shown in Fig. 19:

$$\mathbf{q}_i \cdot \Delta\mathbf{r}_i^S(\tau) = \Gamma\tau k_0 \Lambda_i \cos\theta_i \sin\theta_i \cos\phi_i \ .$$

For small particles which scatter light isotropically, the successive $\Delta\phi_i(\tau)$ are uncorrelated and the average of the product in Eq. (3) is once again the product of the average of $n$ independent terms. Since we have assumed that $\Delta\mathbf{r}_i^B(\tau)$ and $\Delta\mathbf{r}_i^S(\tau)$ represent two independent motions, the average over them can be performed independently. We have previously considered the Brownian motion. For the contribution from shear, we perform a moment expansion. Since the leading non-vanishing term in such an expansion is the second moment, we average $\sin^2 2\theta \cos^2 \phi$ over the unit sphere and obtain

$$\left\langle \prod_{i=1}^{n} e^{-i\,\mathbf{q}\cdot\mathbf{r}_i(\tau)} \right\rangle = e^{-2[\tau/\tau_0 + (\tau/\tau_s)^2]} \ ,$$

where $\tau_s^{-1} = \Gamma l k_0/\sqrt{30}$ and $l = \langle\Lambda\rangle$ is the scattering mean free path. The total autocorrelation function is once again obtained by summing over all paths with $s = nl$

$$G_1(\tau) = I_0 \int P(s) e^{-2[\tau/\tau_0 + (\tau/\tau_s)^2]s/l}\,ds \ .$$

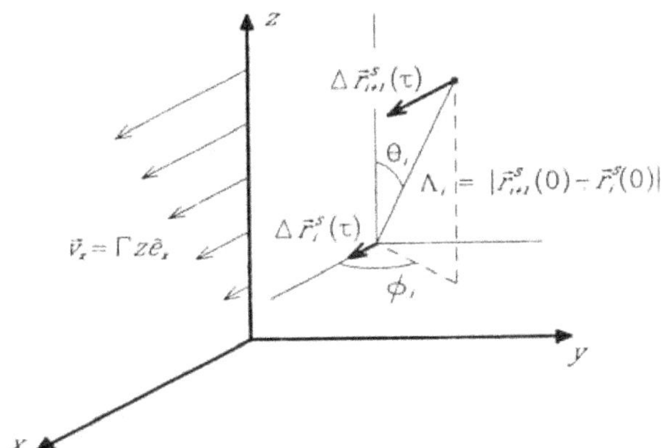


Fig. 19. Scattering geometry for calculation of the autocorrelation functions for shear flow.

For larger particles which scatter light anisotropically, the same results are obtained provided $l$ is replaced by $l^*$ everywhere.[21] Thus, in general

$$\tau_s^{-1} = \Gamma l^* k_0 / \sqrt{30} \; . \tag{34}$$

The form of our results are identical to those previously obtained for diffusion in the absence of shear provided we make the substitution

$$\tau/\tau_0 \rightarrow [\tau/\tau_0 + (\tau/\tau_s)^2] \; . \tag{35}$$

Thus, we can simply adapt our previous results for $G_1(\tau)$ in transmission and backscattering to the case of shear.

In Fig. 20, we show data obtained in backscattering and transmission for a $\phi = 0.02$ suspension of 0.415-$\mu$m-diameter polystyrene spheres subject to Poiseuille flow through a 5-cm-long rectangular glass cell. The transverse dimensions of the flow cell were 1mm $\times$ 12mm with light incident on the 12-mm face. Velocity gradients were probed along the 1mm dimension in both scattering geometries. The solid curves through the data in Fig. 20 are obtained without any fitting parameters. The shear rate $\Gamma$ was calculated from measurements of the flow rate and taken to be the rms average over the 1mm thickness of the cell. The other parameters, $\gamma, l^*$, and $\tau_0$, were obtained from measurements without flow. The agreement between theory and experiment is excellent.

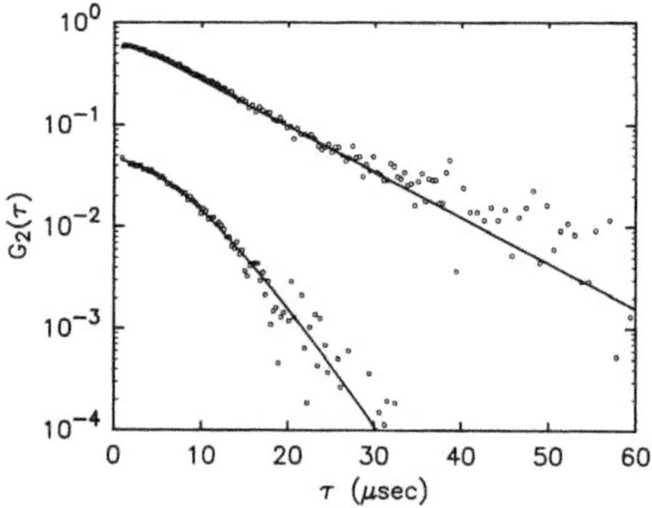

Fig. 20. Intensity autocorrelation functions for backscattering (top) and transmission (bottom) from a $\phi = 0.02$ suspension of $0.415$-$\mu$m-diameter polystyrene spheres subject to Poiseuille flow.

## 6.6. Interacting Particles

In the preceding discussion, we have implicitly assumed that the particles which scatter light are completely uncorrelated. Thus, we were able to separately average over the particle positions and the scattering wavevector. However, this is strictly true only in the limit of a dilute, non-interacting suspension of particles. More generally, both the positions and the velocities of the particles are correlated due to their interactions. These effects are particularly important in the dense suspensions for which DWS is ideally suited. The most important interactions are: (a) hard-core repulsion, which is short-range and important at high volume fractions, (b) hydrodynamic interactions, which are long range but generally become important as the volume fraction increases, and (c) Coulomb repulsion, which can be important even at low volume fractions provided the counterions in solution do not screen the interactions too severely.[22] To fully exploit the potential of DWS, we must consider the consequences of correlations between the particles on the measured autocorrelation functions.

To appreciate the problems of treating interacting systems, it is useful to recall the limiting case of single scattering. Quasielastic light scattering from interacting systems measures the dynamic structure factor, $S(\mathbf{q}, \tau)$, defined as,

$$\frac{\langle E(0) E^*(\tau)\rangle}{F(\mathbf{q})} = N S(\mathbf{q}, \tau) = \left\langle \sum_{ij=1}^{N} e^{i\,\mathbf{q}\cdot[\mathbf{r}_i(0) - \mathbf{r}_j(\tau)]} \right\rangle,$$

where the sum is over the $N$ scatterers, $\mathbf{q}$ is the scattering wavevector, $F(q)$ is the form factor of the scatterers and $\langle\rangle$ denotes the ensemble average. The dynamic structure factor is related to the Fourier transform of the pair correlation function of scatterers and depends on the interactions.[22] For independent scatterers undergoing Brownian motion, $S(q,0) = 1$ and $S(q,\tau) = \exp(-Dq^2\tau)$; interactions introduce correlations between particle positions and velocities and $S(q,\tau)$ no longer decays exponentially.

For short times, $S(q,\tau)$ can be expressed in terms of $S(q,0)$ and the diffusion coefficient, $D_0 = k_B T/\varsigma$, since the particles have moved a distance small compared to the average separation. Here, $\varsigma$ is the particle friction coefficient corrected for hydrodynamic interactions. Hence, a particle's motion is not yet affected by interaction with its neighbors and the initial decay of $S(q,\tau)$ is determined by the free particle motion.[22] Thus, we have,

$$S(q,\tau) = S(q,0) - q^2 \langle r^2(\tau)\rangle/6$$

or, for Brownian motion,

$$S(q,\tau) = S(q,0) - D_s q^2 \tau = S(q,0) e^{-D_s q^2 \tau / S(q)}$$

which amounts to considering the effects of correlations only on the average positions of the scatterers, and not on their velocities. Thus, the collective diffusion coefficient is

$$D_c(q) = D_s/S(q) , \qquad (36)$$

where $S(q) = S(q,0)$ is the static structure factor.

Considering now the multiple scattering regime, proper incorporation of the correlations requires that the basic expression for the scattering from particles in the medium be reformulated. Rather than considering the scattering from the individual particles as we have done, we must consider the scattering from the density fluctuations of the correlated particles which are described by $S(q,\tau)$. In general, the range of correlations, $\xi$, is smaller than the scattering mean free path, so that the multiple scattering is from an ensemble of *independent* cells of size $\xi$, which are described by the *local* structure factor, $S(q,\tau)$ (Refs. 11,18). One effect of these correlations will be to modify $l^*$ as can be seen directly from its definition, Eq. (9). To include the effects of interactions, we must replace $F(q)$ by the full scattering function, $S(q)F(q)$, giving

$$\frac{l^*}{l} = \frac{2k_0^2 \langle S(q)F(q)\rangle}{\langle q^2 S(q)F(q)\rangle} . \qquad (37)$$

With these modifications, we can adapt our former derivation for the autocorrelation function to the case of interacting particles. The time dependence of the average correlation function for a given path with $n$ cells is

$$\langle \prod_{i=1}^{n} S(q_i,\tau)\rangle / \langle \prod_{i=1}^{n} S(q_i,0)\rangle .$$

The normalization insures that this expression goes to 1 as $\tau \to 0$, so that it contains only the time dependence of the autocorrelation function, as modified by the interactions. This allows us to again use the diffusion approximation to determine the intensity of $n$th order paths,

$$G_1^{(n)}(\tau) = I_0 P(n)\frac{\langle \prod_{i=1}^{n} S(q_i,\tau)\rangle}{\langle \prod_{i=1}^{n} S(q_i,0)\rangle} ,$$

where $\langle\rangle$ is the average over $q$ weighted by the form factor $F(q)$ and $P(n)$ is the fraction of intensity in $n$th order paths. Within the diffusion approximation, $P(n) = P(s/l^*) = P(nl/l^*)$, with both $l$ and $l^*$ modified by the static structure factor, $S(q)$. For long paths, the successive $q_i$ are uncorrelated, and the sum over the paths samples all possible $q_i$, so that we can replace the average of the product by the product of the average,

$$G_1^{(n)}(\tau) = I_0 P(n) \frac{\langle S(q,\tau)\rangle^n}{\langle S(q,0)\rangle^n} .$$

For short times, $\langle S(q,\tau)\rangle / \langle S(q,0)\rangle \simeq 1$, and we express $G_1^{(n)}(\tau)$ as

$$G_1^{(n)}(\tau) \; I_0 P(n) \exp\left[ -\frac{nl}{l^*}\frac{l^*}{l}\left(1 - \frac{\langle S(q,\tau)\rangle}{\langle S(q,0)\rangle}\right)\right] ,$$

where we make explicit the fact that $nl/l^*$ is the relevant quantity in the diffusion approximation, since $s = nl$. Comparing this equation to Eq. (8), we find, as shown by MacKintosh and John,[18] it is possible to make the substitution,

$$\frac{2\tau}{\tau_0} \rightarrow \frac{l^*}{l}\left(1 - \frac{\langle S(q,\tau)\rangle}{\langle S(q,0)\rangle}\right)$$

in all the expressions for the autocorrelation function $G_1(\tau)$ to correctly account for particle correlations.

Let us now apply this result to the case considered above where $D_c(q) = D_s/S(q)$, or $S(q,\tau) = S(q,0) - D_s q^2 \tau$. Using Eq. (37), we find that the substitution gives[11]

$$\frac{\tau}{\tau_0} \rightarrow \frac{\tau}{\tau_0'}$$

with

$$\frac{\tau_0'}{\tau_0} = \frac{\langle q^2 S(q)\rangle}{\langle q^2\rangle} .$$

Thus, interactions do not modify the functional form of $G_1(\tau)$ for short times, and in particular, the square root singularity in backscattering is preserved. Only the time scale of the decay is modified according to the above equation. We can evaluate its modification for hard core interactions between the scatterers, which are probably dominant for the polystyrene latex suspensions discussed above, where the Coloumb interactions are effectively screened. In this case, the static structure factor is known.[23] For a suspension with $\phi = 0.10$, $S(q)$ increases from 0.45 at $q = 0$ to 1 at $q \sim \pi/R$, where $R$ is the particle radius. We then have to consider the average of $S(q)$, weighted by $q^2 F(q)$. For most of the scatterers we have used, this average emphasizes only the high $q$ region of $S(q)$, where $S(q) \sim 1$, because the diameter is comparable to the wavelength. Thus, the effect of the spatial correlations, which are reflected by the behavior at low $q$, will tend to be averaged out. Then we expect the time scale of the decay to be only slightly modified with respect to the non-interacting case. This probably explains why the data of Fig. 10, although obtained with suspensions of $\phi = 0.1$, fall on the same curve when the time scale is scaled by the free particle diffusion time, $\tau_0$.

By contrast, the long time behavior of $G_1(\tau)$ can be dramatically affected by the interactions. To illustrate this, we consider the autocorrelation functions obtained from a colloidal crystal. This is achieved by reducing the concentration of counterions in a suspension of polystyrene latex to the point where the Coulombic repulsion between the particles is sufficiently long range that the balls crystallize. In Fig. 21, we contrast the autocorrelation functions measured in backscattering from a non-interacting colloidal suspension and a colloidal crystal. We again use a logarithmic plot of $g_2(\tau)$ as a function of the square root of delay time. Both samples were comprised of 0.215 $\mu$m diameter polystyrene spheres at the same volume fraction, $\phi = 0.10$. The only difference between the samples is the range of the screened Coulomb interactions, which is experimentally controlled by the concentration of counterions which have been reduced in one sample, causing it to crystallize. However, the autocorrelation functions are markedly different. The autocorrelation function of the colloidal liquid decreases exponentially with the square root of delay time, as expected. By contrast, $g_2(\tau)$ for the colloidal crystal exhibits an initial rapid decay at the very early times, but then saturates, and does not decay to the baseline even at long time scales. To ensure that a correct measure of the baseline is obtained for these samples, data were taken at several different positions on the sample and the separate results were averaged.[24]

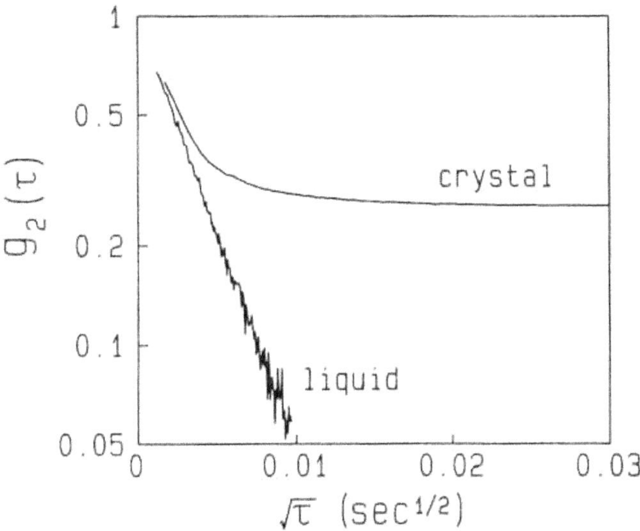


Fig. 21. Intensity autocorrelation functions from a weakly-interacting colloid (liquid) and from a strongly interacting colloid (crystal). Both suspensions consist of 0.215-$\mu$m-diameter spheres with $\phi = 0.10$.

This behavior is consistent with our expectations for the average displacement of the particles in a colloidal crystal. At early times they will move relatively freely, with the same diffusion coefficient as for the non-interacting particles. However, as their mean displacement approaches some fraction of the interparticle separation, they will feel the Coulombic repulsion of their neighbors and will be unable to move further. This is reflected in a decrease in their diffusion coefficient as time increases. The backscattering autocorrelation function reflects this directly. It probes a wide range of path lengths, and consequently a wide range of decay times. At early times, the very rapid decay reflects the contribution of the long paths, which consist of many scattering events, so that each ball need move only a small amount. This occurs very rapidly, since the particles are diffusing relatively quickly. By contrast, at later times, the slow decay reflects the contribution of the very short paths, which consist of a small number of scattering events, so that each ball must move a greater distance to cause the decay. This occurs on a much longer time scale due to the repulsive interactions.

These results illustrate the potential power of DWS in studying systems with interactions between the particles. There are clearly many other very interesting interacting systems that can be studied by means of DWS. However, caution must be exercised in the interpretation of the results in determining the exact consequences of the correlations of both the positions and motions of the particles. Indeed, considerable work remains to be done to fully determine the extent to which the relatively simple interpretations used here are valid.

## 7. CONCLUSIONS

In this chapter, we have presented a discussion of the temporal correlations of multiply scattered light. A phenomenological derivation of the autocorrelation functions of the scattered intensity is used to obtain expressions for several experimentally important geometries. This derivation specifically exploits the diffusion approximation for the transport of light in a multiple scattering medium. The geometries considered are transmission through a slab and backscattering from a slab. The expressions obtained are compared to experimental data and excellent agreement is found.

The temporal autocorrelation functions directly reflect the motion of the scatterers in the medium. By using our expressions, we are able to relate the temporal fluctuations to this motion and thereby obtain useful information about the dynamics of the scattering medium. This provides a new method for studying the dynamics of dense suspensions, which we have called Diffusing Wave Spectroscopy. We have demonstrated the utility of DWS by applying it to measure the particle size in concentrated suspensions, as well as to study the dynamics of particles in porous media and under shear. We have also considered the consequences of particle interactions, and the resulting correlations in the particle positions and velocities, on the measured autocorrelation functions.

Finally, we have shown that the temporal autocorrelation functions probe the same combination of diffusing light paths as does the enhancement of the static backscattering. This allows us to draw an elegant analogy between these measurements, and thereby use the measured autocorrelation functions as a sensitive probe of the physics of the multiple scattering of light.

## ACKNOWLEDGEMENTS

We have benefited greatly from valuable discussions with many of our colleagues. In particular, we wish to thank Michael Stephen who performed many of the important first calculations of the autocorrelation functions, and Fred MacKintosh and Sajeev John who were instrumental in determining the correct form of the autocorrelation function in backscattering. The measurements for the particles in a porous medium were obtained in collaboration with Penger Tong and Walter Goldberg; those for particles under shear with Xiao-Lun Wu; those for interacting particles with Xia Qiu and Dan Ou-Yang; and those for the dependence of $\gamma$ on $l^*/l$ with Jixiang Zhu. We also acknowledge many fruitful discussions with Fred MacKintosh and Rudolf Klein.

## REFERENCES

1. B. J. Berne and R. Pecora, *Dynamic Light Scattering: With Applications to Chemistry, Biology, and Physics* (Wiley, New York, 1976).

2. N. A. Clark, J. H. Lunacek and G. B. Benedek, *Am. J. Phys.* **38** (1970) 575.

3. J. K. G. Dhont, in *Photon Correlation Techniques in Fluid Mechanics*, edited by E. O. Schulz-DuBois, (Springer-Verlag, Berlin, 1983).

4. G. Maret and P. E. Wolf, *Z. Phys.* **B65** (1987) 409.

5. M. Rosenbluh, M. Hoshen, I. Freund and M. Kaveh, *Phys. Rev. Lett.* **58** (1987) 2754.

6.  M. J. Stephen, *Phys. Rev.* **B37** (1988) 1.

7.  D. J. Pine, D. A. Weitz, P. M. Chaikin and E. Herbolzheimer, *Phys. Rev. Lett.* **60** (1988) 1134.

8.  D. J. Pine, D. A. Weitz, P. M. Chaikin and E. Herbolzheimer, in *Proceedings of the Topical Meeting on Photon Correlation Techniques and Applications*, edited by A. Smart and J. Abbiss. (Optical Society of America, Washington, 1988).

9.  M. P. van Albada and A. Legendijk, *Phys. Rev. Lett.* **55** (1985) 2692.

10. P. E. Wolf and G. Maret, *Phys. Rev. Lett.* **55** (1985) 2696.

11. G. Maret and P. E. Wolf, (*to be published*).

12. A. A. Golubentsev, *Zh. Eksp. Theor. Fiz.* **86** (1984) 47 [*Sov. Phys. JETP* **59** (1984) 26].

13. D. Pine and E. Herbolzheimer, *private communication*.

14. Short paths can only occur in a backscattering geometry. In this case the distribution of $q_i$ can have a significant contribution from single scattering through $180°$, $q \sim 2k_0$.

15. A. Ishimaru, *Wave Propagation and Scattering in Random Media, Vol. I* (Academic, New York, 1978).

16. P. E. Wolf, G. Maret, E. Akkermans and R. Maynard, *J. Phys.* (France) **49** (1988) 63.

17. L. Landau and E. M. Lifshitz, *Statistical Physics* (Pergamon, Oxford, 1980), p. 396.

18. F. MacKintosh and S. John, *Phys. Rev.* **B40** 2383 (1989).

19. S. Milner, *private communication*.

20. See, for example, articles by N. Ostrowsky and D. Sornette or M. Bertero and E. R. Pike in *Photon Correlation Techniques in Fluid Mechanics*, edited by E. O. Schulz-DuBois (Springer-Verlag, Berlin, 1983).

21. X. L. Wu, D. J. Pine, P. M. Chaikin, J. S. Huang and D. A. Weitz, *J. Opt. Soc. Am. B* in press.

22. P. N. Pusey and R. J. A. Tough, in *Dynamic Light Scattering*, edited by R. Pecora (Plenum, New York, 1985) p. 85.

23. M. S. Wertheim, *Phys. Rev. Lett.* **10** (1965) 321.

24. P. N. Pusey and W. van Megan, *Phys. Rev. Lett.* **59** (1983) 2083.

# ANDERSON LOCALIZATION OF THE CLASSICAL ELECTROMAGNETIC WAVES IN A DISORDERED DIELECTRIC MEDIUM


KARAMJEET ARYA

*Physics Department*
*The City College of the City University of New York*
*New York, N. Y. 10031*

and

*Physics Department*
*San Jose State University*
*San Jose, CA 95192*


ZHAO-BIN SU

*Institute of Theoretical Physics*
*Academia Sinica*
*Beijing, China*


JOSEPH L. BIRMAN

*Physics Department*
*The City College of the City University of New York*
*New York, N. Y. 10031*

## Contents


1. Introduction     375

2. Average Electric Field and Intensity     378

3. Average One-Photon Green's Function     380

4. Average Two-Photon Green's Function     388

5. Results     394

Acknowledgements     398

Appendix     398

References     401

# 1. INTRODUCTION

Discussions of Anderson localization have mostly been related to electron transport in solids with random impurities or defects.[1,2] Recently, there has been growing interest in the studies of localization of classical waves such as electromagnetic (em) waves in a disordered dielectric[3-23] or elastic waves in solids with a random ionic potential.[24-25] The physical basis of the localization in both cases is essentially the same, i.e., the diffusion coefficient vanishes because of the coherent interference between scattered waves from random scatterers. However, there are some important differences which make the localization studies of em waves very interesting. For example,

(i) In case of electron transport, electrons with wavelength close to the Fermi wavelength take part in multiple scattering. This value is usually small and very difficult to control and measure. On the other hand the wavelength of an em wave, e.g., of visible light is orders of magnitude larger, which makes such controls very easy. In fact one can now study directly the coherent interference of multiple scattered em waves in detail.

(ii) the relevant spatial scale on which randomness is important is of the order of wavelength. This makes the control and characterization of disorder (or roughness) in case of an em wave much easier, due to its very large wavelength compared with the case of electron.

(iii) the theoretical treatment of electron localization is very involved due to the presence of strong electron-electron and electron-phonon interactions.[2,26] There are still some open questions regarding the nature of the transition from extended to localized electron state in 3-dimensions. Even if the transition is continuous, the value of the critical exponent remains in dispute. On the other hand, photons offer a great advantage for improved quantitative studies of localization due to almost no self (photon-photon) interaction.

In earlier studies, localization of em waves was considered in 1- and 2-dimensions.[4,5] For instance, on a semi-infinite metal surface, we showed that the surface plasmon polariton (SPP) which is an extended 2-dimensional em mode parallel to the metal surface, becomes localized in the presence of random roughness.[4] This is analogous to electron localization in 2-dimensions. We have also discussed the effects of this localization on surface enhanced Raman scattering which is due to the excitation of the SPP by the incident photon. Although localization of SPP on a rough metal surface has important contribution to the enhancement it has not yet been confirmed experimentally because of the large uncertainties in the experimental results.

Furthermore, as in the case of the electron, an em wave is always localized in 1- and 2-dimensions for arbitrarily small disorder. Therefore, localization studies become more important in 3-dimensions due to the existence of a mobility edge which separates the extended and localized state. In case of electrons, the Anderson transition or the mobility edge has been observed in many experiments through a sharp decrease in the electron conductivity due to the vanishing of the diffusion coefficient.[2] However, there is as yet no such successful experiment in the case of em waves in which the Anderson transition or the vanishing of the transmission coefficient has been reported.

Instead, in recent experiments, weak coherent interference effects have been directly observed in the case of scattering of em waves from a concentrated solution of polystyrene particles.[8,9] These show an increase in the scattered intensity in the backward direction by a factor of about 2 compared to the diffuse scattered intensity. This can be explained due to the presence of time reversal symmetry in the scattering process of em waves. Corresponding to any propagating wave in the medium, there is always another wave propagating in time reversed direction. These two waves have some phase correlation even in random medium. For example, one can easily show that these two waves have exactly the same phase when travelling on the same path in space.[12-13] Therefore, in reflection in the backward direction, their amplitudes add coherently to give increased intensity by a factor of two compared to scattered intensity in other directions (diffuse scattering). The angular width of this peak or the increase in intensity at other angles close to the backward direction depends on the phase correlation between waves travelling on other paths in space.

These results also agree with the many body theory of weak localization that includes multiple scattering by summing over both ladder and maximal cross diagrams.[12-15] The latter diagrams correspond to interference effects of a scattered wave with another scattered wave in the time reversed order. The angular width of the peak is also reported to be in agreement with the theory, i.e., width $\sim \lambda/2\pi\ell$, where $\lambda$ and $\ell$, respectively, are the wavelength and scattering length of the em wave. However, the small angular width of $\sim 1°$ (i.e., $\ell \gg \lambda$) in these experiments indicates that interference effects are important only for small number of scattered waves. Therefore, even for a 10% concentration of polystyrene solution, one is far from enough coherent interference leading to the Anderson transition that occurs for $\ell \sim \lambda$.

Furthermore, for the studies of strong localization, transmission experiments rather than scattering experiments are more suitable because only the former can give information about the diffusion of the em energy through the medium. Recently there have been a few experiments along these lines[10,11] and a decrease in diffusion constant due to coherent interference has also been reported.[11] However, the Anderson transition has not yet been observed in case of an em wave. This is because the scattering cross-section in these materials is rather small and the criterion for an Anderson transition is not satisfied. Studies of other dielectrics with larger scattering cross section is, therefore, necessary for strong localization.

In this article, we discuss strong localization of an em wave propagating through a random distribution of dielectric particles. To find the scattering length, $t$ matrix approach is used which calculates the scattering cross-section exactly from one sphere. Total scattering length is then calculated within the framework of multiple scattering theory. Earlier discussions are based on Maxwell-Garnett theory[3] which is valid only for a weakly disordered dielectric. By using the $t$-matrix and the diagrammatic Green's function method[4,7,27-29] we also derive the criterion for Anderson localization of an em wave which is similar to that for an electron in a random potential. We then discuss various possibilities where this condition of localization can be satisfied. In particular, the case of random suspension of metallic particles seems to be encouraging, e.g., those of Ag, Au etc.[7,17] This is because of the large scattering cross-section due to the Mie resonance present in case of a metal sphere with negative dielectric function.

The organization of the article is as follows. In Sec. 2 we have given expressions for the observable quantities, i.e., the average electric field and average electric field intensity at any point in terms of average one and two photon Green's functions. For example, the average intensity has been expressed in terms of the response function of the random medium to the external intensity source. Section 3 deals with the scattering $t$-matrix approach and its use to calculate scattering length of the em wave in the disordered medium. In Sec. 4, solution of the integral equation for the response function (averaged 2-photon Green's function) is discussed that gives the criterion for the Anderson localization. In Sec. 5, we have discussed these results, in particular, with respect to the possibilities of observing Anderson localization.

## 2. AVERAGE ELECTRIC FIELD AND INTENSITY

We consider a medium consisting of a random dispersion of spherical particles each of radius $a$ and local dielectric function $\varepsilon(\Omega)$; $\Omega$ is the frequency. The electric field $\mathbf{E}(\mathbf{r},\Omega)$ at any point then satisfies the Maxwell equation

$$-\vec{\nabla} \times \vec{\nabla} \times \mathbf{E}(\mathbf{r},\Omega) + \frac{\Omega^2}{c^2}\varepsilon(\mathbf{r},\Omega)\mathbf{E}(\mathbf{r},\Omega) = 4\pi\mathbf{j}^0(\mathbf{r},\Omega) , \qquad (2.1)$$

where $\mathbf{j}^0(\mathbf{r},\Omega)$ is some external current source and

$$\varepsilon(\mathbf{r},\Omega) = \varepsilon_0 + \big(\varepsilon(\Omega) - \varepsilon_0\big) \sum_i \theta(a - |\mathbf{r} - \mathbf{R}_i|) . \qquad (2.2)$$

$\mathbf{R}_i$ denotes the position of the $i$th dielectric sphere corresponding to particular configuration in the random suspension and $\varepsilon_0$ is the background dielectric constant. We define 1-photon Green's function as the solution of Maxwell equation (2.1):

$$-\vec{\nabla} \times \vec{\nabla} \times \overleftrightarrow{d}(\mathbf{r},\mathbf{r}',\Omega) + \frac{\Omega^2}{c^2}\varepsilon(\mathbf{r},\Omega)\overleftrightarrow{d}(\mathbf{r},\mathbf{r}',\Omega) = 4\pi\delta(\mathbf{r}-\mathbf{r}')\overleftrightarrow{I} , \quad (2.3)$$

and from Eqs. (2.1) and (2.3) one can then write

$$\mathbf{E}(\mathbf{r},\Omega) = \int d^3r'\, \overleftrightarrow{d}(\mathbf{r},\mathbf{r}',\Omega)\cdot\mathbf{j}^0(\mathbf{r}',\Omega) . \qquad (2.4)$$

In Eq. (2.3), $I_{ij}$ denotes unit dyadic. This result is valid for a particular given configuration. Any physical observable quantity, however, needs an ensemble average over all possible configurations given by the random distribution of $\mathbf{R}_i$. We therefore, have

$$\langle \mathbf{E}(\mathbf{r},\Omega) \rangle = \int d^3r' \langle \overleftrightarrow{\widehat{d}}(\mathbf{r},\mathbf{r}',\Omega) \rangle \cdot \mathbf{j}^0(\mathbf{r}',\Omega) \,, \qquad (2.5)$$

where $\langle \ \rangle$ denotes the ensemble average. Thus the average electric field requires the calculations of average one photon Green's function.

Similarly, the average electric field intensity can be expressed in terms of ensemble average of 2-photon Green's function:

$$\langle \mathbf{E}_i(\mathbf{r},\Omega)\mathbf{E}_j^*(\mathbf{r},\Omega') \rangle$$
$$= \sum_{k,l} \int d^3r' \int d^3r'' \langle \widehat{d}_{ik}(\mathbf{r},\mathbf{r}',\Omega)\widehat{d}_{jl}^*(\mathbf{r},\mathbf{r}'',\Omega') \rangle j_k^0(\mathbf{r}',\Omega) j_l^{0^*}(\mathbf{r}'',\Omega') \,. \tag{2.6}$$

For later use, we also give here the definitions of Fourier transformation as:

$$\mathbf{E}(\mathbf{r},\Omega) = \int \frac{d^3k}{(2\pi)^3} e^{i\mathbf{k}\cdot\mathbf{r}} \mathbf{E}(\mathbf{k},\Omega) \,, \qquad (2.7)$$

$$\widehat{d}_{ij}(\mathbf{r},\mathbf{r}',\Omega) = \int \frac{d^3k}{(2\pi)^3} \int \frac{d^3k'}{(2\pi)^3} e^{i\mathbf{k}\cdot\mathbf{r}-i\mathbf{k}'\cdot\mathbf{r}'} \widehat{d}_{ij}(\mathbf{k},\mathbf{k}',\Omega) \,. \tag{2.8}$$

Using Eqs. (2.7) and (2.8) in Eq. (2.6), one can write for the intensity tensor in a simplified form:

$$I_{ij}(\mathbf{r},\Omega,\omega) = \langle \mathbf{E}_i(\mathbf{r},\Omega+\omega)\mathbf{E}_j^*(\mathbf{r},\Omega) \rangle$$
$$= \sum_{k,l} \int \frac{d^3q}{(2\pi)^3} e^{i\mathbf{q}\cdot\mathbf{r}} L_{ikjl}(\Omega,\mathbf{q},\omega) J_{kl}^0(\mathbf{q},\Omega,\omega) \,, \tag{2.9}$$

where

$$L_{ikjl}(\Omega,\mathbf{q},\omega) = \int \frac{d^3k}{(2\pi)^3} \int \frac{d^3k'}{(2\pi)^3} L_{ikjl}(\mathbf{k},\mathbf{k}',\Omega,\mathbf{q},\omega) \,. \qquad (2.10)$$

In the above derivation, we have used

$$j_k^0(\mathbf{r},\Omega+\omega)j_l^{0*}(\mathbf{r}',\Omega) = J_{kl}^0(\mathbf{r},\Omega,\omega)\delta(\mathbf{r}-\mathbf{r}') , \qquad (2.11)$$

neglecting any fluctuations in the incident intensity source.[30] Also we have used the fact that after averaging, there is overall translational invariance in the system. Therefore,

$$\langle \widehat{d}_{ik}(\mathbf{k},\mathbf{k}',\Omega)\rangle = (2\pi)^3\delta(\mathbf{k}-\mathbf{k}')d_{ik}(\mathbf{k},\Omega) , \qquad (2.12)$$

$$\left\langle \widehat{d}_{ik}\left(\mathbf{k}+\frac{\mathbf{q}}{2},\mathbf{k}'+\frac{\mathbf{q}}{2},\Omega+\omega\right)\widehat{d}_{jl}^*\left(\mathbf{k}''-\frac{\mathbf{q}}{2},\mathbf{k}'-\frac{\mathbf{q}}{2},\Omega\right)\right\rangle$$
$$= (2\pi)^3\delta(\mathbf{k}-\mathbf{k}'')L_{ikjl}(\mathbf{k},\mathbf{k}',\Omega,\mathbf{q},\omega) . \qquad (2.13)$$

Equations (2.9) and (2.10) thus define the response of the random medium to an external intensity source $J^0$ and $L$ is the response function which we will discuss later.

## 3. AVERAGE ONE-PHOTON GREEN'S FUNCTION

We define the host Green's function $d_{ij}^0(\mathbf{r}-\mathbf{r}',\Omega)$ with background dielectric function $\varepsilon_0$ replacing $\varepsilon(\mathbf{r},\Omega)$ in Eq. (2.3). With this the differential Eq. (2.3) can be changed to an integral equation

$$\overleftrightarrow{d}(\mathbf{r},\mathbf{r}',\Omega) = \overleftrightarrow{d^0}(\mathbf{r},\mathbf{r}',\Omega)+\int d^3r''\,\overleftrightarrow{d^0}(\mathbf{r},\mathbf{r}'',\Omega)\Delta V(\mathbf{r}'',\Omega)\cdot\overleftrightarrow{d}(\mathbf{r}'',\mathbf{r}',\Omega),$$
$$(3.1)$$

where

$$\Delta V(\mathbf{r},\Omega) = -\frac{\Omega^2}{4\pi c^2}\left(\varepsilon(\mathbf{r},\Omega)-\varepsilon_0\right) . \qquad (3.2)$$

The average Green's function $\langle \widehat{d}_{ij}(\mathbf{r}-\mathbf{r}',\Omega)\rangle$ then satisfies the Dyson equation[29]

$$\langle \overleftrightarrow{d}(\mathbf{r},\mathbf{r}',\Omega)\rangle = \overleftrightarrow{d^0}(\mathbf{r}-\mathbf{r}',\Omega)$$
$$+ \int d^3r'' d^3r'''\,\overleftrightarrow{d^0}(\mathbf{r}-\mathbf{r}'',\Omega)\cdot\overleftrightarrow{\Sigma}(\mathbf{r}''-\mathbf{r}''',\Omega)$$
$$\cdot\langle \overleftrightarrow{d}(\mathbf{r}'''-\mathbf{r}',\Omega)\rangle , \qquad (3.3)$$

where from Eq. (3.1), we have

$$\int d^3 r'' \overleftrightarrow{\Sigma}(\mathbf{r} - \mathbf{r}'', \Omega) \cdot \langle \overleftrightarrow{\widehat{d}}(\mathbf{r}'', \mathbf{r}', \Omega) \rangle = \langle \Delta V(\mathbf{r}, \Omega) \overleftrightarrow{\widehat{d}}(\mathbf{r}, \mathbf{r}', \Omega) \rangle . \quad (3.4)$$

The self-energy $\Sigma_{ij}$ can be calculated by using a series expansion for $\widehat{d}_{ij}$ (Eq. (3.1)) and then taking the ensemble average term by term. This approximation is suitable for point scatterers. However, for finite size particles, we use scattering matrix approach in which the self-energy $\Sigma_{ij}$ is expressed in terms of average scattering matrix $T_{ij}$. Since scattering problem from one sphere can be solved exactly using Maxwell boundary conditions at the surface of the sphere,[31-33] one can therefore, obtain total scattering matrix $T_{ij}$ within the framework of multiple scattering theory.

The total average scattering operator $T_{ij}$ is defined as

$$\begin{aligned}
\langle \overleftrightarrow{\widehat{d}}(\mathbf{r}, \mathbf{r}', \Omega) \rangle = {}& \overleftrightarrow{d^0}(\mathbf{r} - \mathbf{r}', \Omega) \\
& + \int d^3 r'' d^3 r''' \, \overleftrightarrow{d^0}(\mathbf{r} - \mathbf{r}'', \Omega) \cdot \overleftrightarrow{T}(\mathbf{r}'' - \mathbf{r}''', \Omega) \\
& \cdot \overleftrightarrow{d^0}(\mathbf{r}''' - \mathbf{r}', \Omega) .
\end{aligned} \quad (3.5)$$

Comparing Eq. (3.3) with (3.5), we have

$$\overleftrightarrow{\Sigma} \cdot \langle \overleftrightarrow{\widehat{d}} \rangle = \overleftrightarrow{T} \cdot \overleftrightarrow{d^0} . \quad (3.6)$$

For simplicity, in Eq. (3.6) and in some of the lengthy expressions in the following we have suppressed the $\mathbf{r}, \mathbf{r}'$ dependence of the variables. We will write this dependence in the final results. Using Eq. (3.5) for $\langle \widehat{d}_{ij} \rangle$ in Eq. (3.6), one can solve for $\Sigma_{ij}$ in terms of $T_{ij}$ and we have

$$\overleftrightarrow{\Sigma} = \overleftrightarrow{T} \cdot [\overleftrightarrow{\mathbf{I}} + \overleftrightarrow{d^0} \cdot \overleftrightarrow{T}]^{-1} . \quad (3.7)$$

An expression for $T_{ij}$ can be obtained in expansion form by comparing Eqs. (3.1) and (3.5);

$$\overleftrightarrow{T} \cdot \overleftrightarrow{d^0} = \langle \Delta V \overleftrightarrow{\widehat{d}} \rangle , \quad (3.8)$$

and using expansion for $\hat{d}_{ij}$ from Eq. (3.1), we have

$$\overleftrightarrow{T} = \langle \Delta V \rangle \overleftrightarrow{\mathbf{I}} + \langle \Delta V \overleftrightarrow{d^0} \Delta V \rangle + \langle \Delta V \overleftrightarrow{d^0} \Delta V \cdot \overleftrightarrow{d^0} \Delta V \rangle + \ldots \quad (3.9)$$

Equation (3.9) gives $T_{ij}$ in terms of scattering potential $\Delta V(\mathbf{r}, \Omega)$. One can instead express it in terms of scattering matrix from one sphere. For this, we write

$$\Delta V(\mathbf{r}, \Omega) = \sum_i \Delta V_i(\mathbf{r}, \Omega) \, , \quad (3.10)$$

where from Eqs. (3.2) and (2.2),

$$\Delta V_i(\mathbf{r}, \Omega) = \frac{-\Omega^2}{4\pi c^2} \big( \varepsilon(\Omega) - \varepsilon_0 \big) \theta(a - |\mathbf{r} - \mathbf{R}_i|) \, . \quad (3.11)$$

Thus $\Delta V_i(\mathbf{r}, \Omega)$ denotes scattering potential from a single sphere centered at $\mathbf{R}_i$. Using Eq. (3.10) in Eq. (3.9) and rearranging terms, one can write

$$\overleftrightarrow{T} = \Big\langle \sum_i \overleftrightarrow{t_i} \Big\rangle + \Big\langle \sum_{i \neq j} \overleftrightarrow{t_i} \cdot \overleftrightarrow{d^0} \cdot \overleftrightarrow{t_j} \Big\rangle$$
$$+ \Big\langle \sum_{i \neq j, j \neq k} \overleftrightarrow{t_i} \cdot \overleftrightarrow{d^0} \cdot \overleftrightarrow{t_j} \cdot \overleftrightarrow{d^0} \cdot \overleftrightarrow{t_k} \Big\rangle + \ldots \, , \quad (3.12)$$

where

$$\overleftrightarrow{t_i} = \Delta V_i \overleftrightarrow{\mathbf{I}} + \Delta V_i \overleftrightarrow{d^0} \Delta V_i + \Delta V_i \overleftrightarrow{d^0} \cdot \Delta V_i \overleftrightarrow{d^0} \Delta V_i + \ldots \, . \quad (3.13)$$

To get physical meaning of $\overleftrightarrow{t_i}$ let $\overleftrightarrow{d^{(i)}}$ denotes the photon Green's function in the presence of single dielectric sphere centered at $R_i$. It is given by (see Eq. (3.1))

$$\overleftrightarrow{d^{(i)}} = \overleftrightarrow{d^0} + \overleftrightarrow{d^0} \Delta V_i \cdot \overleftrightarrow{d^0} \, . \quad (3.14)$$

Using Eq. (3.13), one can write

$$\overleftrightarrow{d^{(i)}} = \overleftrightarrow{d^0} + \overleftrightarrow{d^0} \cdot \overleftrightarrow{t_i} \cdot \overleftrightarrow{d^0} \, . \quad (3.15)$$

Thus $\overleftrightarrow{t_i}(\mathbf{r},\mathbf{r}',\Omega) = \overleftrightarrow{t}(\mathbf{r}-\mathbf{R}_i,\mathbf{r}'-\mathbf{R}_i,\Omega)$ describes scattering by a single sphere centered at $\mathbf{R}_i$ which can be calculated exactly as solutions of Maxwell equations for $\widehat{d}^{(i)}$ in the presence of one sphere is known.

In Eq. (3.12), the first term denotes single scattering of the em wave, second term denotes double scattering, i.e., the wave scatters from one sphere at $\mathbf{R}_i$ and then from another at $\mathbf{R}_j$, and similarly for other terms. The ensemble average over the distribution of spheres in Eq. (3.12) can be written as

$$\left\langle \sum_i \overleftrightarrow{t_i} \right\rangle = \sum_i \int d^3 R_i \, \overleftrightarrow{t_i} \, P_1(\mathbf{R}_i) \,, \tag{3.16}$$

$$\left\langle \sum_{i \neq j} \overleftrightarrow{t_i} \cdot \overleftrightarrow{d^0} \cdot \overleftrightarrow{t_j} \right\rangle = \sum_{i \neq j} \int d^3 R_i \int d^3 R_j \, \overleftrightarrow{t_i} \cdot \overleftrightarrow{d^0} \cdot \overleftrightarrow{t_j} \, P_2(\mathbf{R}_i,\mathbf{R}_j) \,, \tag{3.17}$$

$$\left\langle \sum_{\substack{i \neq j \\ j \neq k}} \overleftrightarrow{t_i} \cdot \overleftrightarrow{d^0} \cdot \overleftrightarrow{t_j} \cdot \overleftrightarrow{d^0} \cdot t_k \right\rangle$$

$$= \sum_{\substack{i \neq j \\ j \neq k}} \int d^3 R_i \int d^3 R_j \int d^3 R_k \, \overleftrightarrow{t_i} \cdot \overleftrightarrow{d^0} \cdot \overleftrightarrow{t_j} \cdot \overleftrightarrow{d^0} \cdot \overleftrightarrow{t_k}$$

$$\times \, P_3(\mathbf{R}_i,\mathbf{R}_j,\mathbf{R}_k) \,, \tag{3.18}$$

and so on. $P_1(\mathbf{R}_i)$, $P_2(\mathbf{R}_i,\mathbf{R}_j)$, $P_3(\mathbf{R}_i,\mathbf{R}_j,\mathbf{R}_k)$ etc. are, respectively, the one, two, three, particle probability distribution functions.

For uniform distribution of particles, $P_1(R) = 1/V$, where $V$ is the volume of the system and

$$P_2(\mathbf{R}_i,\mathbf{R}_j) = P_1(\mathbf{R}_i)P_1(\mathbf{R}_j) + P_1(\mathbf{R}_i)g(|\mathbf{R}_i - \mathbf{R}_j|)P_1(\mathbf{R}_j) \,. \tag{3.19}$$

Second term on the right hand side of (3.19) is due to the pair correlation and $g(R)$ is the pair correlation function. For instance, $g(R) \to 0$ as $R \to \infty$, i.e., correlation between two particles decreases as separation between them increases. For higher order distribution functions, we write them in terms of one and two particles distribution functions:

$$P_3(\mathbf{R}_i,\mathbf{R}_j,\mathbf{R}_k) = P_1(\mathbf{R}_i)P_1(\mathbf{R}_j)P_1(\mathbf{R}_k)$$

$$\times \, [1 + g(|\mathbf{R}_i - \mathbf{R}_j|) + g(|\mathbf{R}_j - \mathbf{R}_k|) + g(|\mathbf{R}_i - \mathbf{R}_k|)] \tag{3.20}$$

and so on. This approximation can be justified for Gaussian distribution of the dielectric particles, which is mostly the case in experiments.

One can thus carry out ensemble average term by term in Eq. (3.12) using these distribution functions. These calculations are easy to carry out in Fourier **k**-space. One can show that $T_{ij}(\mathbf{r},\mathbf{r}'\Omega)$ is a function of $(\mathbf{r}-\mathbf{r}')$ which in fact is a result of overall translational invariance of the averaged system. Therefore, using Fourier transform

$$T_{ij}(\mathbf{r}-\mathbf{r}',\Omega) = \int \frac{d^3k}{(2\pi)^3} e^{i\mathbf{k}\cdot(\mathbf{r}-\mathbf{r}')} T_{ij}(\mathbf{k},\Omega) \,, \qquad (3.21)$$

one can carry out ensemble average in Eq. (3.12). We give here results for $T_{ij}$ writing explicitly up to two terms on right hand side of Eq. (3.12) only:

$$\overleftrightarrow{T}(\mathbf{k},\Omega) = n\overleftrightarrow{t}(\mathbf{k},\mathbf{k}) + n^2\overleftrightarrow{t}(\mathbf{k},\mathbf{k},\Omega)\cdot\overleftrightarrow{d}(\mathbf{k},\Omega)\cdot\overleftrightarrow{t}(\mathbf{k},\mathbf{k},\Omega)$$
$$+ n^2\int \frac{d^3k}{(2\pi)^3}\overleftrightarrow{t}(\mathbf{k},\mathbf{k}',\Omega)\cdot\overleftrightarrow{d}(\mathbf{k}',\Omega)$$
$$\cdot\overleftrightarrow{t}(\mathbf{k}',\mathbf{k},\Omega)g(|\mathbf{k}-\mathbf{k}'|) + \dots \,, \qquad (3.22)$$

where

$$t_{ij}(\mathbf{k},\mathbf{k}',\Omega) = \int d^3r \int d^3r' e^{-i\mathbf{k}\cdot\mathbf{r}+i\mathbf{k}'\cdot\mathbf{r}'} t_{ij}(\mathbf{r},\mathbf{r}',\Omega) \,, \qquad (3.23)$$

and $n$ is the particle density. In Eq. (3.22) $g(k)$ is the Fourier transformation of the correlation function $g(|\mathbf{R}_i-\mathbf{R}_j|)$:

$$g(k) = \int d^3R\, e^{-i\mathbf{k}\cdot\mathbf{R}} g(R) \,. \qquad (3.24)$$

Because of the condition $i \neq j$ in Eq. (3.17), $g(|\mathbf{R}_i-\mathbf{R}_j|)$ is zero for $|\mathbf{R}_i-\mathbf{R}_j| < a$. For $|\mathbf{R}_i-\mathbf{R}_j| > a$, it depends upon the experiments. For example, for the Gaussian distribution, one can approximate it by a Gaussian function.

Writing Eq. (3.7) in expansion form:

$$\overleftrightarrow{\Sigma} = \overleftrightarrow{T} - \overleftrightarrow{T}\cdot\overleftrightarrow{d^0}\cdot\overleftrightarrow{T} + \overleftrightarrow{T}\cdot\overleftrightarrow{d^0}\cdot\overleftrightarrow{T}\cdot\overleftrightarrow{d^0}\cdot\overleftrightarrow{T}$$
$$- \overleftrightarrow{T}\cdot\overleftrightarrow{d^0}\cdot\overleftrightarrow{T}\cdot\overleftrightarrow{d^0}\cdot\overleftrightarrow{T}\cdot\overleftrightarrow{d^0}\cdot\overleftrightarrow{T} + \dots \,, \qquad (3.25)$$

and using (3.22) one can obtain expression for $\Sigma_{ij}(\mathbf{k},\Omega)$. Instead of writing explicit expression, we have expressed it in Fig. 1b through Feynman diagrams. Figure 1a shows the Dyson equation for the average $d_{ij}$ (thick line). Thin line in Fig. 1a is the background Green's function $d_{ij}^0$ and dashed line in Fig. 1b corresponds to pair correlation function $g(|\mathbf{k}-\mathbf{k}'|)$. Thick dot corresponds to $n$ times scattering matrix $t_{ij}(\mathbf{k},\mathbf{k}'\Omega)$. In Fig. 1c, we have also given the diagrammatic expansion of single particle scattering matrix $t_{ij}(\mathbf{k},\mathbf{k}'\Omega)$ in terms of the scattering potential $\Delta V_i$ (see Eq. (3.13)). The small dot in Fig. 1c corresponds to $\Delta V(\mathbf{k}-\mathbf{k}')$. The dotted line in this case which denotes the pair correlation $(g(0))$ of a particle with itself is equal to one. We have, however, shown it for completeness.

As mentioned earlier, the scattering matrix $t_{ij}(\mathbf{k},\mathbf{k}',\Omega)$ can be calculated exactly since the solutions of Maxwell equations for a single dielectric sphere are known from Mie theory.[34] Using a Maxwell Green's function for a single dielectric sphere, the expression for $t_{ij}$ has been derived in the Appendix (A). For the localization treatment, we need only calculate the imaginary part of the self-energy and that needs $t_{ij}(\mathbf{k},\mathbf{k}',\Omega)$ only for $k = k' = \Omega/c$ which is given as:

$$\overleftrightarrow{t}(\mathbf{k},\mathbf{k}',\Omega) = \frac{4\pi}{k} \sum_{l,m} \left[ \hat{\mathbf{k}} \times \mathbf{X}_{lm}(\Omega_k) \frac{c_{kl}^E}{1 + ic_{kl}^E} \hat{\mathbf{k}}' \times \mathbf{X}_{lm}^*(\Omega_{k'}) \right.$$

$$\left. + \mathbf{X}_{lm}(\Omega_k) \frac{c_{kl}^M}{1 + ic_{kl}^M} \mathbf{X}_{lm}^*(\Omega_k) \right] , \qquad (3.26)$$

where $\mathbf{X}_{lm}(\Omega) = [l(l+1)]^{-1/2} \mathbf{L} Y_{lm}(\Omega)$ is the vector spherical harmonic function. Coefficients $c_{kl}^\sigma$ ($\sigma = E$ and $M$ which correspond to electrical and magnetic modes in Mie theory) are obtained by matching the Maxwell boundary conditions at the surface of the dielectric sphere and are

$$c_{kl}^E = \frac{-\varepsilon(\Omega) j_l(k_i a)[ka j_l(ka)]' + j_l(ka)[k_i a j_l(k_i a)]'}{\varepsilon(\Omega) j_l(k_i a)[ka \eta_l(ka)]' - \eta_l(ka)[k_i a j_l(k_i a)]'} , \qquad (3.27)$$

$$c_{kl}^M = \frac{j_l(k_i a)[ka j_l(ka)]' + j_l(ka)[k_i a j_l(k_i a)]'}{-j_l(k_i a)[ka \eta_l(ka)]' - \eta_l(ka)[k_i a j_l(k_i a)]'} , \qquad (3.28)$$

(a)

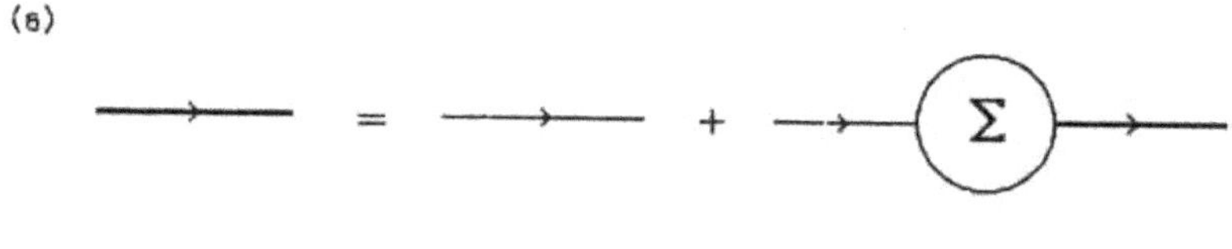

(b)

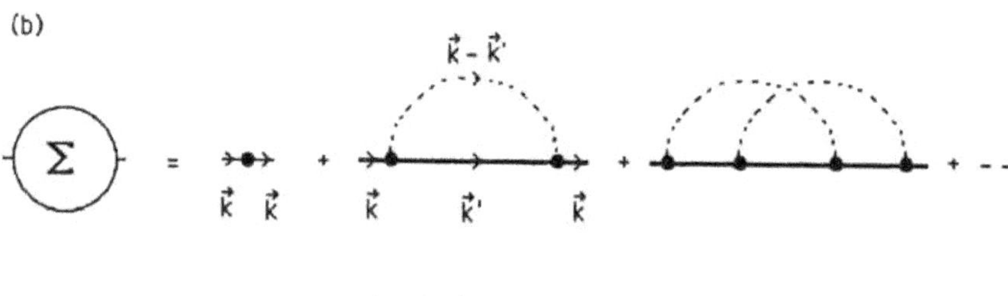

(c)

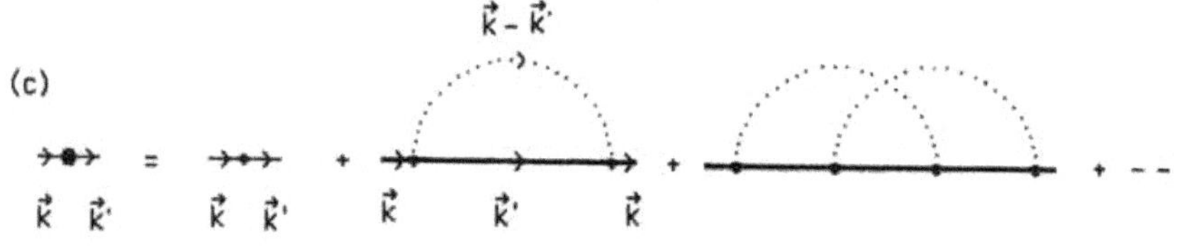

Fig. 1. (a) Dyson equation for the average one photon Green's function $d_{ij}$ (thick line). Thin line denotes background Green's function $d_{ij}^0$. (b) Diagrammatic expansion for the self-energy $\Sigma$. Dashed line corresponds to pair correlation function $g(\mathbf{k}-\mathbf{k}')$ and thick dot is $n$-fold scattering matrix $t$. (c) Diagrammatic expansion for single particle scattering matrix $t$ in terms of scattering potential $\Delta V$ (small dot). The dotted line denotes the pair correlation function $g(0)$ of a particle with itself and thus is equal to 1.

where $k_i = \sqrt{\epsilon(\Omega)}\, k$. $j_l(x)$ and $\eta_l(x)$ are spherical Bessel functions and the prime on the brackets in Eqs. (3.27) and (3.28) denotes differentiation with respect to the argument of the Bessel function.

Due to the vector nature of Maxwell fields, photon Green's function is a second rank tensor. For transverse em field which we consider here, one can write

$$\overleftrightarrow{d^0}(\mathbf{k},\Omega) = d^0(k,\Omega)(\mathbf{I} - \hat{\mathbf{k}}\hat{\mathbf{k}}); \quad d^0(k,\Omega) = \frac{4\pi c^2}{\Omega^2 - \Omega_k^2 - i\eta}, \quad (3.29)$$

where $\Omega_k = ck$ and $\eta \to 0$. $c$ is the velocity of em wave in the background dielectric medium. This simplification is also possible for average one photon Green's function $d_{ij}(k,\Omega)$ and its self energy $\Sigma_{ij}(k,\Omega)$. This is because random medium after averaging still looks like as though it has uniform dielectric constant. We will check this later by calculating the first order term in $\Sigma_{ij}$. We thus have

$$d_{ij}(\mathbf{k},\Omega) = d(k,\Omega)(I - \widehat{\mathbf{k}}\widehat{\mathbf{k}})_{ij} \; ; \quad \Sigma_{ij}(\mathbf{k},\Omega) = \Sigma(k,\Omega)(I - \widehat{\mathbf{k}}\widehat{\mathbf{k}})_{ij} \; . \quad (3.30)$$

Equations (3.3), (3.29) and (3.30) give a scalar Dyson equation for $d(k,\Omega)$:

$$d(k,\Omega) = d^0(k,\Omega) + d^0(k,\Omega)\Sigma(k,\Omega)d(k,\Omega) \; . \quad (3.31)$$

From Eqs. (3.29) and (3.31), we have

$$d(k,\Omega) = \frac{4\pi c^2}{\Omega^2 - \Omega_k^2 - 4\pi c^2 \Sigma(\mathbf{k},\Omega)} \; , \quad (3.32)$$

$\Sigma$ has both real and imaginary parts. In scattering theory, we are interested in near frequencies pole, i.e., $\Omega = \Omega_k$. Also the important quantity is the imaginary part of $\Sigma$ which is related to the scattering cross section. Therefore, neglecting small real part of $\Sigma$, we have

$$d(k,\Omega) \approx \frac{2\pi c^2}{\Omega(\Omega - \Omega_k + i\gamma)} \; , \quad (3.33)$$

where

$$\gamma = \frac{2\pi c^2}{\Omega}\,\mathrm{Im}\,\Sigma(\Omega/c,\Omega/c) \; , \quad (3.34)$$

corresponds to the scattering width due to the elastic scattering from the dielectric particles. We also give the definition of scattering length

$$\ell = c/2\gamma \; , \quad (3.35)$$

which will be used later.

We estimate here $\gamma$ in the lowest order approximation for $\Sigma$ (first term in Fig. 1b). This can be obtained by using identity for vector spherical functions

$$\sum_m \mathbf{X}_{lm}(\Omega_k)\mathbf{X}^*_{lm}(\Omega_k) = \sum_m \left(\widehat{\mathbf{k}} \times \mathbf{X}_{lm}(\Omega_k)\right)\left(\widehat{\mathbf{k}} \times \mathbf{X}^*_{lm}(\Omega_k)\right)$$

$$= \frac{3}{8\pi}(\mathbf{I} - \widehat{\mathbf{k}}\widehat{\mathbf{k}}) \; , \quad (3.36)$$

in Eq. (3.26). We thus have ($k = k' = \Omega/c$):

$$\text{Im}\,\Sigma_{ij}(\mathbf{k},\Omega) = \frac{\Omega}{c}U_0(\mathbf{I} - \widehat{\mathbf{k}}\widehat{\mathbf{k}})_{ij}\ , \tag{3.37}$$

and from Eq. (3.34)

$$\gamma = 2\pi c U_0\ , \tag{3.38}$$

where

$$U_0 = \frac{3}{2}\left(\frac{c}{\Omega}\right)^2 n \sum_{\sigma,l}\left|\frac{c_{kl}^\sigma}{1 + ic_{kl}^\sigma}\right|^2\ . \tag{3.39}$$

The contribution to $\Sigma_{ij}$ due to other terms in Fig. 1b involves correlation function $g(|\mathbf{R}_i - \mathbf{R}_j|)$ between two different spheres. These terms can be important, specially when particle density $n$ is large.[31]

## 4. AVERAGE TWO-PHOTON GREEN'S FUNCTION

To determine localization one needs to consider the transport of em energy density in a random medium, i.e., to investigate its transport diffusion coefficient. As discussed in Sec. 2, intensity or energy density of em wave at any point in a random medium is given by the average 2-photon Green's function (See Eqs. (2.9), (2.10), (2.13)). We calculate this using a diagrammatic Green's function method developed earlier for the case of the electron motion in a random impurity potential.[27] Following Ref. (29), $L_{ikjl}$ satisfies Bethe-Salpeter equation

$$
\begin{aligned}
L_{ikjl}&(\mathbf{k},\mathbf{k}',\Omega,\mathbf{q},\omega)\\
&= \sum_{mm'nn'} d_{im}(\mathbf{k}_+,\Omega+\omega)d_{jn}^*(\mathbf{k}_-,\Omega)\Big[(2\pi)^3\delta(\mathbf{k}-\mathbf{k}')\delta_{km}\delta_{nl}\\
&\quad + \int\frac{d^3k''}{(2\pi)^3}U_{mm'nn'}(\mathbf{k},\mathbf{k}'',\Omega,\mathbf{q},\omega)L_{m'kn'l}(\mathbf{k}'',\mathbf{k}',\Omega,\mathbf{q},\omega)\Big]\ ,
\end{aligned}
\tag{4.1}
$$

where $\mathbf{k}_\pm = \mathbf{k}\pm\mathbf{q}/2$. The kernal $U$ can be obtained following functional derivative method of Baym and Kadanoff. As discussed in Ref. (29) it is given by the functional derivative of $\Sigma_{mn}$ taken with respect to $d_{m'n'}$. This is equivalent to the diagrammatic method of Ref. (27) where $U$ is given by the sum of all irreducible diagrams in the four point vertex. For example, in the lowest order approximation for $\Sigma_{mn}(k,\Omega)$ (i.e., Fig. 1c (ii)), we have

$$U_{mm'nn'}(\mathbf{k},\mathbf{k}'',\Omega,\mathbf{q},\omega) = n\Delta V(\mathbf{k}-\mathbf{k}'',\Omega+\omega)\Delta V^*(\mathbf{k}-\mathbf{k}'',\Omega)\delta_{mm'}\delta_{nn'} \ . \tag{4.2}$$

Equation (4.1) is very complicated because of the tensor nature of the photon Green's function. As discussed in Sec. 3, for transverse em field one can factorize $d^0_{ij}$ and $d_{ij}$ into scalar and tensor parts (see Eqs. (3.29) and (3.30)). We write similar factorization for configuration dependent Green's functions also:

$$\widehat{d}_{ij}(\mathbf{k},\mathbf{k}',\Omega) = \widehat{d}(\mathbf{k},\mathbf{k}',\Omega)((\mathbf{I} - \widehat{\mathbf{k}}\widehat{\mathbf{k}}) \cdot (\mathbf{I} - \widehat{\mathbf{k}}'\widehat{\mathbf{k}}'))_{ij} \ . \tag{4.3}$$

Equation (4.3) is not exact. It is equivalent to an approximation in which the polarization of incident em wave is averaged over after each scattering. For example, an incident em wave with a given polarization can have different polarization after first scattering. However, before its next scattering, we average over the possible polarizations. Thus in each successive scattering, one has an unpolarized em wave. Therefore, in Eq. (4.3), polarization components corresponding to only initial and final wave vector appear. In this approximation one can write

$$\begin{aligned}
L_{ikjl}&(\mathbf{k},\mathbf{k}',\Omega,\mathbf{q},\omega)\\
&= ((\mathbf{I} - \widehat{k}_+\widehat{k}_+) \cdot (\mathbf{I} - \widehat{k}'_+\widehat{k}'_+))_{ik}((\mathbf{I} - \widehat{k}_-\widehat{k}_-) \cdot (\mathbf{I} - \widehat{k}'_-\widehat{k}'_-))_{jl}\\
&\quad \times L(\mathbf{k},\mathbf{k}'\Omega,\mathbf{q},\omega) \ ,
\end{aligned} \tag{4.4}$$

where $L$ now satisfies a scalar Bethe-Salpeter equation

$$\begin{aligned}
L(\mathbf{k},\mathbf{k}',\Omega,\mathbf{q},\omega) = d(\mathbf{k}_+,\Omega + \omega)d^*(\mathbf{k}_-,\Omega)\Big[&(2\pi)^3\delta(\mathbf{k} - \mathbf{k}')\\
&+ \int \frac{d^3 k''}{(2\pi)^3} U(\mathbf{k},\mathbf{k}'',\Omega,\mathbf{q},\omega)L(\mathbf{k}'',\mathbf{k}',\Omega,\mathbf{q},\omega)\Big] \ .
\end{aligned} \tag{4.5}$$

This problem is similar to the scalar problem of electron in a random potential. One can therefore, follow the diagrammatic method of Ref. (27)

to solve it. In Fig. 2a we have expressed Eq. (4.5) diagrammatically. Figure 2b contains the sum of all irreducible diagrams in the four point vertex, $U$. For example, Fig. 2b(i) when used alone for the vertex $U$ in Fig. 2a, gives the well known ladder series (see Fig. 2c). This leads to the conventional diffusion of the em energy in the random medium. Figures 2b(ii), 2b(iii), etc. are known as maximal cross diagrams in second and third order, respectively. These diagrams correspond to the time reverse path discussed in Sec. 1. This is because one can obtain them from the ladder diagrams by reversing the time direction of the bottom photon line in Fig. 2c. These diagrams give rise to coherent interference leading to the localization. Figures 2b(iv), 2b(v) are some of the other typical diagrams.

Before we discuss the complete solution of Eq. (4.5) we give its solution first in diffusion approximation, i.e., considering only Fig. 2b(i) for $U$ (Fig. 2c). One can write

$$
\begin{aligned}
L(\mathbf{k}, \mathbf{k}', \Omega, \mathbf{q}, \omega) = {} & d(\mathbf{k}_+, \Omega + \omega) d^*(\mathbf{k}_-, \Omega) \\
& + d(\mathbf{k}_+, \Omega + \omega) d^*(\mathbf{k}_-, \Omega) \Gamma(\mathbf{k}, \mathbf{k}', \Omega, \mathbf{q}, \omega) \\
& \times d(\mathbf{k}'_+, \Omega + \omega) d^*(\mathbf{k}'_-, \Omega) ,
\end{aligned} \tag{4.6}
$$

where following Ref. (27), we have for the vertex $(\omega \to 0, q \to 0)$:

$$
\Gamma(\mathbf{k}, \mathbf{k}', \Omega, \mathbf{q}, \omega) = \frac{2\gamma U_0}{-i\omega + D_0 q^2} . \tag{4.7}
$$

$D_0 = c^2/6\gamma = c\ell/3$ is the diffusion coefficient where $\ell = c/2\gamma$ is the diffusion length. Also $U_0 = \gamma/2\pi c$ is given by Eqs. (3.38) and (3.39).

We now solve Eq. (4.5) beyond diffusion pole approximation, i.e., including all diagrams in Fig. 2b. Since these diagrams are the same as in the case of electron in a random potential, we follow the diagrammatic treatment of Ref. (27). It can be easily shown that for small disorder all the approximations of Ref. (27) are also applicable to our case. One, however, should be careful in these calculations that photon dispersion is linear whereas for electron it is quadratic. For example, by using Ward identities, we obtain the continuity equation $(\omega \to 0, q \to 0)$

$$
\omega L(\Omega, \mathbf{q}, \omega) - \mathbf{q} \cdot \int \frac{d^3 k}{(2\pi)^3} \int \frac{d^3 k'}{(2\pi)^3} \frac{\partial \Omega_k}{\partial \mathbf{k}} L(\mathbf{k}, \mathbf{k}', \Omega, \mathbf{q}, \omega) = \frac{2\pi^2 c^2}{\Omega} i N(\Omega) , \tag{4.8}
$$

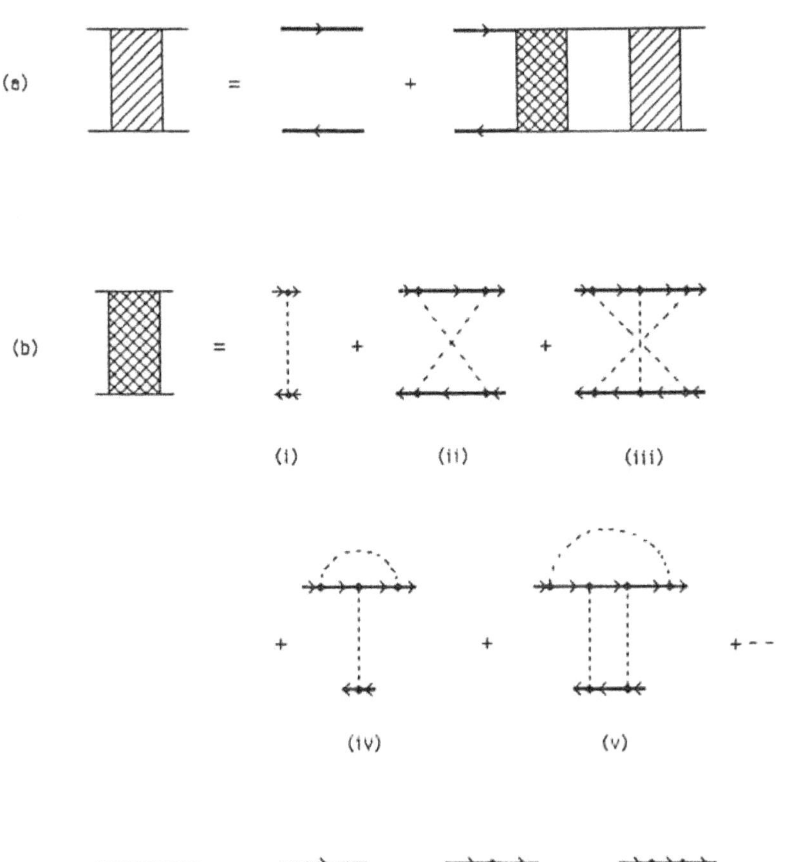

Fig. 2. (a) Diagrammatic representation of the integral equation (4.5) for the average 2-photon Green's function. (b) Irreducible diagrams for the four point vertex, $U$. (c) Ladder diagrams for the average 2-photon Green's function.

where

$$L(\Omega, \mathbf{q}, \omega) = \int \frac{d^3 k}{(2\pi)^3} \int \frac{d^3 k'}{(2\pi)^3} L(\mathbf{k}, \mathbf{k}', \Omega, \mathbf{q}, \omega) , \qquad (4.9)$$

is related to the em energy density in the medium. The second term on the left hand side in Eq. (4.8) is the photon current density and $N(\Omega) = \Omega^2/2\pi^2 c^3$ is the photon density of states. Similarly for weak disorder, following Ref. (27), one can derive another expression

$$[\omega + M(\mathbf{q},\omega)] \int \frac{d^3 k}{(2\pi)^3} \int \frac{d^3 k'}{(2\pi)^3} \widehat{\mathbf{q}} \cdot \frac{\partial \Omega_k}{\partial \mathbf{k}} L(\mathbf{k},\mathbf{k}',\Omega,\mathbf{q},\omega)$$
$$= \frac{q}{3} \left( \frac{\partial \Omega_k}{\partial k} \right)^2 L(\Omega,\mathbf{q},\omega) , \tag{4.10}$$

where

$$M(\mathbf{q},\omega) = 2i\gamma - \frac{3}{2\pi i N(\Omega)} \int \frac{d^3 k}{(2\pi)^3} \int \frac{d^3 k'}{(2\pi)^3}$$
$$\times \Delta D(k) U(\mathbf{k},\mathbf{k}',\Omega,\mathbf{q},\omega) \Delta D(k') . \tag{4.11}$$

In Eq. (4.11), $\Delta D(k) = \widehat{q} \cdot \mathbf{k}[d(\mathbf{k}_+,\Omega+\omega) - d^*(\mathbf{k}_-,\Omega)]$. From Eqs. (4.8) and (4.10) we thus obtain

$$L(\Omega,\mathbf{q},\omega) = \frac{(2\pi^2 c^2/\Omega) N(\Omega)}{-i\omega + D(q,\omega)q^2} , \tag{4.12}$$

where,

$$D(q,\omega) = \frac{2i\gamma D_0}{M(q,\omega)} , \tag{4.13}$$

is the renormalized diffusion coefficient. Note that in the ladder approximation, i.e., retaining only the lowest order irreducible diagram in $U$ (Fig. 2b(i)), the second term on right hand side of Eq. (4.11) can be shown to be zero, i.e., $M = 2i\gamma$. Therefore, we obtain $D(q,\omega) = D_0$ as the bare diffusion coefficient.

Equation (4.12) gives response function discussed in Sec. 2. Writing for external current source $J^0_{kl}$

$$J^0_{kl}(\mathbf{q},\Omega,\omega) = \widehat{\mathbf{e}}_k(\lambda,\mathbf{k}'_+) \widehat{\mathbf{e}}^*_1(\lambda,\mathbf{k}'_-) J^0(\Omega) , \tag{4.14}$$

and

$$I_{ij}(\mathbf{q},\Omega,\omega) = \widehat{\mathbf{e}}_i(\lambda,\mathbf{k}_+) \widehat{\mathbf{e}}^*_j(\lambda \mathbf{k}_-) I(\mathbf{q},\Omega,\omega) , \tag{4.15}$$

in Eq. (2.9), we find from Eqs. (4.4) and (4.12)

$$I(\mathbf{q},\Omega,\omega) = \frac{(2\pi^2 c^2/\Omega)N(\Omega)}{-i\omega + D(q,\omega)q^2}\, J^0(\Omega) \ . \tag{4.16}$$

In Eqs. (4.14) and (4.15), $\widehat{\mathbf{e}}(\lambda\mathbf{k})$ is the polarization vector for the transverse field. Also note that

$$\sum_\lambda \widehat{\mathbf{e}}(\lambda,\mathbf{k})\widehat{\mathbf{e}}^*(\lambda,\mathbf{k}) = (\mathbf{I} - \widehat{\mathbf{k}}\widehat{\mathbf{k}}) \ . \tag{4.17}$$

Equation (4.16) gives the response of the random medium to the external intensity source. In diffusion approximation, i.e., $D(q,\omega) = D_0$, Eq. (4.16) reduces to the diffusion equation:

$$\left(\frac{\partial}{\partial t} - D_0\nabla^2\right) I(\mathbf{r},t) = I^0\delta(\mathbf{r})\delta(t) \ , \tag{4.18}$$

which describes the diffusion of the intensity of a point source $I^0$ in the random medium.

We now calculate renormalized diffusion coefficient $D(q,\omega)$ by solving $M(q,\omega)$ from Eq. (4.11). Following Ref. (27), one can show that for $\gamma \ll \Omega$, the main contribution to $U$ comes from the maximal cross diagrams and the contribution of all other diagrams goes to zero in the limit $\omega \to 0, q \to 0$. The sum of maximal cross diagrams is calculated from the ladder summation by using the fact that photon being a Bose field, has time reversal symmetry and we thus have[27]

$$U(\mathbf{k},\mathbf{k}',\Omega,\mathbf{q},\omega) = \Gamma\left[\frac{1}{2}(\mathbf{k} - \mathbf{k}' + \mathbf{q}), \frac{1}{2}(\mathbf{k}' - \mathbf{k} + \mathbf{q}), \Omega, \mathbf{k} + \mathbf{k}', \omega\right]$$

$$\approx \frac{2\gamma U_0}{-i\omega + D(q,\omega)(\mathbf{k} + \mathbf{k}')^2} \ . \tag{4.19}$$

In writing this we have used Eq. (4.7) for $\Gamma$ with $D_0$ being replaced by renormalized diffusion coefficient $D(q,\omega)$ for self consistency. Inserting Eq. (4.19) into (4.11) we find after some simplification

$$M(q,\omega) = 2i\gamma + \frac{2i\gamma}{\pi N(\Omega)}\int \frac{d^3k}{(2\pi)^3}\frac{1}{-i\omega + D(q,\omega)k^2} \ . \tag{4.20}$$

Using this result in Eq. (4.13) we find an equation for $D(q,\omega)$

$$D(q,\omega) = D_0 \left[ 1 - \frac{\ell_c^2}{\ell} \int_0^{\ell/\ell_c} dk \frac{k^2}{k^2 - i\omega/D(q,\omega)} \right] , \qquad (4.21)$$

where $\ell_c = (3/\pi)^{1/2} c/\Omega$. Equation (4.21) determines renormalized diffusion coefficient. One can now find the condition for localization, i.e., the condition for $D(q,\omega) \to 0$ in the limit $\omega \to 0$. We find this condition from Eq. (4.21) to be satisfied only for $\ell < \ell_c \sim \lambda/2\pi$. This agrees with the localization condition derived in case of electron with $\lambda$ corresponding to Fermi wavelength.[2,27]

Thus in a disorder medium, an em wave undergoes a transition from an extended state to a localized state when scattering length $\ell \leq \ell_c \sim \lambda/2\pi$. One can also define the localization length

$$\xi = \lim_{\omega \to 0} (iD(q,\omega)/\omega)^{1/2} , \qquad (4.22)$$

which determines the distance over which the em intensity in the localized state decays exponently. From Eq. (4.21), $\xi$ is given by

$$(1 - \ell^2/\ell_c^2) = (1/\xi) \tan^{-1}(\xi/\ell) , \qquad (4.23)$$

which diverges as $(\ell - \ell_c)^{-1}$ for $\ell < \ell_c$.

We would like to mention that so far in our discussions we have assumed a lossless random dielectric medium, i.e., dielectric constant $\varepsilon(\Omega)$ is real. In general, along with elastic scattering which we considered here there is always some inelastic scattering usually due to the interband transitions in the dielectric. In case, the imaginary part of $\varepsilon(\Omega)$ is small, i.e., the inelastic scattering width $\gamma_i$ is much smaller than elastic scattering width $\gamma$, one can include the losses in the preceding treatment by replacing $i\omega$ by $i\omega + \gamma_i$.

## 5. RESULTS

We now discuss the possibility of whether in realistic experiments the criterion $(\ell/\ell_c < 1)$ for localization can be satisfied for the optical wave. For this we calculate scattering length $\ell = c/2\gamma$ using Eq. (3.38) for $\gamma$. We refer to experiments of Refs. (8)–(9) where propagation of em wave through a solution of polystyrene particles in water is studied. By definition, one can write for the particle density $n = f/v$ where $f$ is the volume fraction of the dielectric particles to the host material and $v$ is the volume of one particle. The scattering length $\ell$ thus calculated for polystyrene particles for $f = 0.1, a = 500$ Å, $\varepsilon(\Omega) = 2.53, \lambda = 6330$ Å is $\sim 35000$ Å. This is almost 35 times $\ell_c$. It can be reduced somewhat by suitably choosing the particle radius and the optical wavelength. But it seems difficult to attain the localization condition in case of polystyrene due to the small scattering cross-section from a single sphere. To see how this scattering can be increased (or decrease $\ell$) we write $c_{kl}^{\sigma}$ for $ka < 1$ for which the dominant contribution comes from $l = 1$ and $\sigma = E$. Thus from Eq. (3.22)

$$c_{kl}^{E} = \frac{2}{3}(ka)^3 \frac{1 - \varepsilon}{2 + \varepsilon} \, , \tag{5.1}$$

which can be increased by (note $\ell \propto 1/c_{kl}^{\sigma}$)
a) increasing the particle radius within the approximation $ka < 1$ for which Eq. (5.1) is valid. For $ka > 1$, the scattering cross-section reaches the maximum value given by geometrical optics.
b) using a dielectric material with larger $\varepsilon$. There have been attempts, e.g., to use particles of $TiO_2$ with $\varepsilon(\Omega) = 7.3$ and appreciable decrease in $\ell$ is claimed.[12]
c) The other possibility we propose is to use metal particles.[7,17] For metals with negative $\varepsilon(\Omega)$, $\ell$ can decrease by an order of magnitude due to the factor $(1 - \varepsilon(\Omega))/(1 + \varepsilon(\Omega))$ appearing in Eq. (5.1). It is well known from Mie theory[33,34] that for a metal sphere (i.e., with a negative dielectric constant) the em field near its surface is quite large, especially when incident frequency is close to Mie resonance (Note $c_{kl}^{E}$ has a pole for each $l$ value which corresponds to the Mie resonance; for $ka < 1$, Mie resonance is given by $\varepsilon(\Omega) = -(l+1)/l$ (see also Eq. (5.1)). This gives a large scattering cross-section from one sphere and hence $\ell$ can become sufficiently small even for small metal concentration. We have calculated $\ell$ using Eqs. (3.38) and (3.39) for $\gamma$ for a suspension of Ag particles, with radii $a = 300, 400, 500, 600, 700$ Å and $f = 0.2$. These values are given in Table 1. The dielectric function $\varepsilon(\Omega)$ of Ag is taken

Table 1. Scattering length $\ell$ (in units of $\ell_c = \lambda/2\pi$) calculated from Eqs. (3.35) and (3.38) for the optical wave through a random suspension of silver particles.

| | | $\ell/\ell_c$ | | | | |
|---|---|---|---|---|---|---|
| $\Omega\,(eV)$ | $\epsilon(\Omega)$ | $a = 300$ Å | $a = 400$ Å | $a = 500$ Å | $a = 600$ Å | $a = 700$ Å |
| 2.7 | $-7.5$ | 10.74 | 3.57 | 1.55 | 1.06 | 1.15 |
| 2.8 | $-6.5$ | 7.71 | 2.48 | 1.12 | 0.92 | 1.16 |
| 2.9 | $-6.0$ | 5.98 | 1.87 | 0.91 | 0.89 | 1.24 |
| 3.0 | $-5.0$ | 3.63 | 1.09 | 0.66 | 0.86 | 1.33 |
| 3.1 | $-4.5$ | 2.41 | 0.73 | 0.59 | 0.92 | 1.47 |
| 3.2 | $-3.5$ | 0.78 | 0.34 | 0.60 | 1.10 | 1.71 |
| 3.3 | $-3.0$ | 0.25 | 0.35 | 0.77 | 1.30 | 1.81 |

from Ref. (36). These parameters are experimentally accessible and we find that in a reasonable frequency range it is possible to satisfy the condition $(\ell \lesssim \ell_c)$ for strong localization. Similarly this condition can be satisfied in other metals like Au, Cu, etc.

One however, should note that an optical wave has to make a large number of multiple scatterings to build up strong localization. Therefore, the losses due to inelastic scattering in the dielectric should be minimum. Metals usually have large imaginary part of the dielectric function, however, for noble metals, in particular, for Ag these losses are very small in the optical frequency range. We should also mention that values for $\ell$ given in Table 1 are calculated only taking first term in Fig. 1b for $\Sigma_{ij}$ which neglects any correlation between different particles. Correlation can be important for large $n$ or $f$ and therefore other terms in Fig. 1b should also be taken into account.

Instead of using a steady state wave for the studies of Anderson localization which has been discussed in this paper, similar studies in the time domain can also be of interest.[10,37] For example, by using an optical picosecond pulse, one should be able to observe directly the building up of localization due to multiple scattering both in reflection and transmission experiments. Similar experiments should also be possible with nanosecond pulse in the microwave region.

In addition to the localization studies in 3-dimensions which we have discussed in this article, similar studies in 1- and 2-dimensions are also important. Since, there is no transition one should try indirectly to see localization effects, e.g., on the various optical properties.[7] The important field is surface enhanced optical phenomena. We know now that multiple scattering of 2-dimensional em wave (surface plasmon polariton) on a metal surface in the presence of random roughness plays crucial role in the enhancement of various optical properties, e.g., surface enhanced Raman scattering.[29,33] Experiments should also be tried to observe coherence effects in 2-dimensions similar to those in 3-dimensions.[8,9] In fact, coherence effects in 2-dimensions are expected to be much larger. Studies to observe reflection peak in the backward direction from a rough metal surface are in progress. Similar studies in 1-dimension, e.g., effects of multiple scattering due to roughness in optical fibers are also worthwhile.

In this article, we have discussed the effects of (electric) field-field correlation on em intensity in a random medium. Recently there has been interest also to study intensity-intensity correlation in a finite random medium.[18−23] This requires the calculations of average 4-photon Green's function. These studies have earlier been done in case of electron. Because of large wavelength compared to that of electron, em wave however, has many more advantages in these studies.


## Note Added in the Proof:

After we submitted this manuscript almost one year back, many research papers have appeared in this rapidly growing field. We have included some of them at the end in the reference list.[38−42] However, we would like to mention that to our knowledge there is yet no report of any successful experiment in which strong localization of the em wave in the transmission has been observed.

## ACKNOWLEDGEMENTS

This research was supported in part by FRAP CUNY-PSC and NASC.

## APPENDIX

Here we discuss the calculations of scattering $t$-matrix. For this we first write scattered field in terms of $t$-matrix using Eqs. (2.4) and (3.6). We then calculate this scattered field by solving the Maxwell equations in the presence of a dielectric sphere. Comparison of these two will give $t$-matrix.

Using Eq. (3.5) with $T_{ij} = t_{ij}$ for a single sphere in Eq. (2.4), we obtain

$$\mathbf{E}(\mathbf{r},\Omega) = \mathbf{E}^0(\mathbf{r},\Omega) + \int d^3r' d^3r'' \overleftrightarrow{d^0}(\mathbf{r}-\mathbf{r}',\Omega) \cdot \overleftrightarrow{t}(\mathbf{r}',\mathbf{r}'',\Omega) \cdot \mathbf{E}^0(\mathbf{r}'',\Omega) , \tag{A.1}$$

where

$$\mathbf{E}^0(\mathbf{r},\Omega) = \int d^3r' \overleftrightarrow{d^0}(\mathbf{r}-\mathbf{r}',\Omega) \cdot \mathbf{j}^0(\mathbf{r}',\Omega) , \tag{A.2}$$

corresponds to incident field in vacuum. The second term in Eq. (A.1) gives the scattered field $\mathbf{E}_{sc}(\mathbf{r},\Omega)$. We now calculate $\mathbf{E}(\mathbf{r},\Omega)$ in the limit $r \to \infty$, i.e., far away from the sphere. For vacuum Green's function $d^0_{ij}(\mathbf{r}-\mathbf{r}',\Omega)$, we have[35]

$$d^0_{ij}(\mathbf{r}-\mathbf{r}',\Omega) = \left(\mathbf{I} - \frac{1}{k^2}\overrightarrow{\nabla}\overrightarrow{\nabla}\right)_{ij} \frac{e^{ik|\mathbf{r}-\mathbf{r}'|}}{|\mathbf{r}-\mathbf{r}'|} , \tag{A.3}$$

which for $r > r'$ and in the limit $r \to \infty$ can be written as

$$d^0_{ij}(\mathbf{r}-\mathbf{r}',\Omega) = (\mathbf{I} - \widehat{\mathbf{k}}\widehat{\mathbf{k}})_{ij} \frac{e^{ikr}}{r} e^{-i\mathbf{k}\cdot\mathbf{r}'} , \tag{A.4}$$

where $\mathbf{k}$ is along $\mathbf{r}$.

Substituting Eq. (A.4) in (A.1) and using Eq. (3.23) for the Fourier transform of $t_{ij}$, we have

$$\mathbf{E}(\mathbf{r},\Omega) = \widehat{e}(\lambda,\mathbf{Q})e^{i\mathbf{Q}\cdot\mathbf{r}} + \frac{e^{ikr}}{r}(\mathbf{I} - \widehat{\mathbf{k}}\widehat{\mathbf{k}}) \cdot \overleftrightarrow{t}(\mathbf{k},\mathbf{q},\Omega) \cdot \widehat{e}(\lambda,\mathbf{Q}) , \tag{A.5}$$

where for incident field we have used

$$\mathbf{E}(\mathbf{r}, \Omega) = \widehat{e}(\lambda, \mathbf{Q}) e^{i\mathbf{Q}\cdot\mathbf{r}} \ . \tag{A.6}$$

We now calculate $\mathbf{E}(\mathbf{r}, \Omega)$ from the solutions of Maxwell equation in the presence of dielectric sphere at origin and with dielectric constant $\varepsilon(\Omega)$. Since $\mathbf{E}(\mathbf{r}, \Omega)$ is a solution of Maxwell equation one can expand it in terms of eigenfunctions of the Maxwell equation. For example,

$$\mathbf{E}(\mathbf{r}, \Omega) = \frac{2}{\pi} \int dk\, k^2 \sum_{lm\,\sigma} \alpha^{\sigma}_{klm} A^{\sigma}_{klm}(\mathbf{r}) \ , \tag{A.7}$$

where $\mathbf{A}^{\sigma}_{klm}(\mathbf{r})$ are the eigen solutions of Maxwell equation

$$-\vec{\nabla} \times \vec{\nabla} \times \mathbf{A}(\mathbf{r}) + \frac{\Omega^2}{c^2}\varepsilon(\Omega)\mathbf{A}(\Omega) = 0 \ , \tag{A.8}$$

$\sigma = E, M$ correspond to the electrical and magnetic modes. These solutions are known. For example,[33,35]

$$\mathbf{A}^{M}_{klm}(\mathbf{r}) = f^{M}_{l}(kr)\mathbf{X}_{lm}(\Omega) \ , \tag{A.9}$$

$$\mathbf{A}^{E}_{klm}(\mathbf{r}) = \frac{i}{k}\vec{\nabla} \times \left[ f^{E}_{l}(kr)\mathbf{X}_{lm}(\Omega) \right] \ , \tag{A.10}$$

where $\mathbf{X}_{lm}(\Omega)$ are the vector spherical harmonics. $f^{\sigma}_{l}(x)$ is a linear combination of spherical Bessel functions:

$$f^{\sigma}_{l}(kr) = j_{l}(kr) + \frac{-ic^{\sigma}_{kl}}{1 + ic^{\sigma}_{kl}} h^{1}_{l}(kr) ; \qquad r \geq a \ , \tag{A.11}$$

$$f^{\sigma}_{l}(kr) = \frac{b^{\sigma}_{kl}}{1 + ic^{\sigma}_{kl}} j_{l}(kr) ; \qquad r \leq a \ , \tag{A.12}$$

where $h^{1}_{l} = j_{l} + i\eta_{l}$ is Henkel function. The coefficients $c^{\sigma}_{kl}$ and $b^{\sigma}_{kl}$ are determined by matching the Maxwell boundary conditions at the surface of the sphere. These are[33]

$$c_{kl}^E = \frac{-\varepsilon(\Omega)j_l(k_i a)[kaj_l(ka)]' + j_l(ka)[k_i aj_l(k_i a)]'}{\varepsilon(\Omega)j_l(k_i a)[ka\eta_l(ka)]' - \eta_l(ka)[k_i aj_l(k_i a)]'} \, , \tag{A.13}$$

$$c_{kl}^M = \frac{j_l(k_i a)[kaj_l(ka)]' + j_l(ka)[k_i aj_l(k_i a)]'}{-j_l(k_i a)[ka\eta_l(ka)]' - \eta_l(ka)[k_i aj_l(k_i a)]'} \, , \tag{A.14}$$

$$b_{kl}^E = \frac{-\sqrt{\varepsilon(\Omega)}\eta_l(kR)[kRj_l(kR)]' + \sqrt{\varepsilon(\Omega)}j_l(kR)[kR\eta_l(kR)]'}{\varepsilon(\omega)j_l(k_i R)[kR\eta_l(kR)]' - \eta_l(kR[k_i Rj_l(k_i R)]'} \, , \tag{A.15}$$

$$b_{kl}^M = \frac{\eta_l(kR)[kRj_l(kR)]' - j_l(kR)[kR\eta_l(kR)]'}{-j_l(k_i R)[kR\eta_l(kR)]' + \eta_l(kR[k_i Rj_l(k_i R)]'} \tag{A.16}$$

The linear combination in Eq. (A.11) is taken so that first term corresponds to incident wave (i.e., well behaved at $r = 0$) and second term corresponds to scattered wave. Note that in the absence of dielectric sphere, $c_{kl}^\sigma = 0$. Also $h_1^1(kr) \xrightarrow{r \to \infty} \exp(ikr)/kr$ satisfies the outgoing boundary condition for the scattered wave. Substituting Eqs. (A.9), (A.10) and (A.11) in (A.7) and separating the incident and scattered part we have $(r > a)$,

$$\mathbf{E}(\mathbf{r},\Omega) = \mathbf{E}^0(\mathbf{r},\Omega) + \mathbf{E}^{sc}(\mathbf{r},\Omega) \, , \tag{A.17}$$

where

$$\mathbf{E}^0(\mathbf{r},\Omega) = \hat{e}(\lambda,\mathbf{Q})e^{i\mathbf{Q}\cdot\mathbf{r}} = \frac{2}{\pi}\int dk\, k^2 \sum_{lm\sigma} \alpha_{klm}^\sigma \mathbf{A}_{klm}^{\sigma(0)}(\mathbf{r}) \, , \tag{A.18}$$

$$\mathbf{E}^{sc}(\mathbf{r},\Omega) = \frac{2}{\pi}\int dk\, k^2 \sum_{lm\sigma} \frac{-ic_{kl}^\sigma}{1 + ic_{kl}^\sigma} \mathbf{A}_{klm}^{\sigma(sc)}(\mathbf{r})\alpha_{klm}^\sigma \, . \tag{A.19}$$

$\mathbf{A}_{klm}^{\sigma(0)}(\mathbf{r})$ and $\mathbf{A}_{klm}^{\sigma(sc)}(\mathbf{r})$ are given by Eqs. (A.9) and (A.10) with $f_l^\sigma(kr) = j_l(kr)$ and $f_l^\sigma(kr) = h_l^1(kr)$, respectively. Note that $\mathbf{A}_{klm}^{\sigma(0)}(\mathbf{r})$

makes a complete set being eigen solutions of Maxwell equation in vacuum, i.e.,

$$\int d^3 r\, \mathbf{A}^{\sigma(0)}_{klm}(\mathbf{r}) \cdot \mathbf{A}^{\sigma'(0)*}_{k'l'm'}(\mathbf{r}) = \frac{\pi}{2k^2} \delta(k-k')\delta_{\sigma\sigma'}\delta_{ll'}\delta_{mm'} \; . \tag{A.20}$$

We can thus calculate $\alpha^{\sigma}_{klm}$ from Eq. (A.18):

$$\alpha^{M}_{klm}(\lambda,\mathbf{Q}) = 4\pi i^l \mathbf{X}_{lm}(\Omega_k) \cdot \widehat{e}(\lambda,\widehat{Q})\frac{\pi}{2Q^2}\delta(k-Q) \; , \tag{A.21}$$

$$\alpha^{E}_{klm}(\lambda,\mathbf{Q}) = 4\pi i^l \left(\widehat{Q} \times \mathbf{X}_{lm}(\Omega_k)\right) \cdot \widehat{e}(\lambda,\widehat{Q})\frac{\pi}{2Q^2}\delta(k-Q) \; . \tag{A.22}$$

Substituting Eqs. (A.21) and (A.22) in Eq. (A.19) and using $h^1_\ell(kr) \overset{r\to\infty}{\sim} (-i)^{l+1}\exp(ikr)/kr$, one can write for the scattered field $\mathbf{E}^{sc}(\mathbf{r})$ far away from the sphere. We then compare Eqs. (A.17) and (A.5) and find

$$\overleftrightarrow{t}(\mathbf{k},\mathbf{k}',\Omega) = \frac{4\pi}{k} \sum_{l,m} \left[ \widehat{\mathbf{k}} \times \mathbf{X}_{lm}(\Omega_k)\frac{c^E_{kl}}{1+ic^E_{kl}}\widehat{\mathbf{k}}' \times \mathbf{X}^*_{lm}(\Omega_k) \right.$$
$$\left. + \mathbf{X}_{lm}(\Omega_k)\frac{c^M_{kl}}{1+ic^M_{kl}}\mathbf{X}^*_{lm}(\Omega_k) \right] \; . \tag{A.23}$$

# REFERENCES


1. P. W. Anderson, *Phys. Rev.* **109** (1958) 1492.
2. For example, see G. Bergmann, *Phys. Rep.* **1** (1984) 107.
3. S. John, *Phys. Rev. Lett.* **53** (1984) 2169 and *Phys. Rev.* **B31** (1985) 304.
4. K. Arya, Z. B. Su and J. L. Birman, *Phys. Rev. Lett.* **54** (1985) 1559.
5. A. A. Mc. Gurn, A. A. Maradudin and V. Calli, *Phys. Rev.* **B31** (1985) 4866.
6. P. W. Anderson, *Phil. Mag.* **B52** (1985) 505.
7. K. Arya, Z. B. Su and J. L. Birman, *Phys. Rev. Lett.* **57** (1986) 272.
8. M. P. Van Albada and Ad Lagendijk, *Phys. Rev. Lett.* **55** (1985) 2692.
9. P. E. Wolf and G. Mart, *Phys. Rev. Lett.* **55** (1985) 2696.
10. G. H. Watson, Jr., P. A. Fleury and S. L. McCall, *Phys. Rev. Lett.* **58** (1987) 945.

11. A. Z. Genack, *Phys. Rev. Lett.* **58** (1987) 2043.

12. E. Akkermans, P. E. Wolf and R. Maynard, *Phys. Rev. Lett.* **56** (1986) 1471.

13. E. Akkermans and R. Maynard, *J. Phys. Lett.* **46** (1986) L1045.

14. E. Akkermans, P. E. Wolf, R. Maynard and G. Maret, *J. Phys.* **49** (1988) 77.

15. M. J. Stephen and G. Cwilich, *Phys. Rev.* **B34** (1986) 7564.

16. M. Kaveh, M. Rosenbluh, I. Edrei and I. Freund, *Phys. Rev. Lett.* **57** (1986) 2049.

17. K. Arya, Z. B. Su and J. L. Birman, in *Proceedings of III USA-USSR Symposium on Laser Optics of Condensed Matter*, (Leningrad, 1987).

18. B. Shapiro, *Phys. Rev. Lett.* **57** (1986) 2168.

19. S. Etemad, R. Thompson and M. J. Anddeiijco, *Phys. Rev. Lett.* **57** (1986) 575.

20. Ping Sheng, Benjamin White, Zhao-Qing Zhang and G. Papanicolau, *Phys. Rev.* **B34** (1986) 4757.

21. Ping Sheng and Zhao-Qing Zhang, *Phys. Rev. Lett.* **57** (1986) 1879.

22. Ping Sheng, Zhao-Qing Zhang, Benjamin White and G. Papanicolau, *Phys. Rev. Lett.* **57** (1986) 1000.

23. M. J. Stephen and G. Cwilich, preprint.

24. S. John and M. J. Stephen, *Phys. Rev.* **B28** (1983) 6358.

25. S. John, H. Sompolinsky and M. J. Stephen, *Phys. Rev.* **B27** (1983) 5592.

26. P. A. Lee and T. V. Ramakrishnan, *Rev. of Mod. Phys.* **57** (1985) 287.

27. D. Vollhardt and P. Wolfle, *Phys. Rev.* **B22** (1980) 4666.

28. D. Vollhardt and P. Wolfle, *Phys. Rev. Lett.* **48** (1981) 699.

29. K. Arya and R. Zeyher, *Phys. Rev.* **B28** (1983) 4090.

30. B. Laks and D. L. Mills, *Phys. Rev.* **B20** (1979) 4962.

31. V. A. Davis and L. Schwartz, *Phys. Rev.* **B31** (1985) 5155.

32. M. Lax, *Rev. Mod. Phys.* **23** (1951) 287.

33. K. Arya and R. Zeyher, in *Light Scattering in Solids IV*, eds. M. Cardona and G. Guntherodt (Springer-Verlag, Heidelberg, 1984).

34. For example, see M. Born and E. Wolf, *Principles of Optics*, (Pergamon, New York, 1970).

35. J. D. Jackson, *Classical Electrodynamics* (Wiley, New York, 1975) Chap. 16.

36. P. B. Johnson and R. W. Chersty, *Phys. Rev.* **B6** (1972) 4370.

37. K. M. Yoo, K. Arya, G. C. Tang, J. L. Birman and R. R. Alfano, *Phys. Rev.* **A39** (1989) 3728.

38. S. Feng, C. Kane, P. A. Lee and A. D. Stone, *Phys. Rev. Lett.* **61** (1988) 834.

39. I. Freund, M. Rosenbluh, R. Berkovits and M. Kaveh, *Phys. Rev. Lett.*
    **61** (1988) 1214.
40. R. W. Ziolkowski, D. K. Lewis and B. D. Cook, *Phys. Rev. Lett.* **62**
    (1989) 147.
41. I. Edrei, M. Kaveh and B. Shapiro, *Phys. Rev. Lett.* **62** (1989) 2120.
42. P. D. Vries, H. D. Raedt and Ad Lagendijk, *Phys. Rev. Lett.* **62** (1989)
    2515.

# OPTICAL LOCALIZATION: CALCULATIONAL TECHNIQUES AND RESULTS


E. N. ECONOMOU

*Research Center of Crete and Department of Physics**
*P. O. Box 1527, 711 10 Heraklio, Crete, Greece*

*Ames Laboratory and Department of Physics*
*Iowa State University, Ames, Iowa 50011, USA*

and

C. M. SOUKOULIS

*Ames Laboratory and Department of Physics*
*Iowa State University, Ames, Iowa 50011, USA*



Various calculational methods for studying the problem of optical (or more generally classical wave) localization are briefly presented. Emphasis is given to a simple Coherent Potential Approximation developed by the authors. Explicit results based on this method are presented, discussed, and compared with reliable data from numerical techniques as well as with recent experimental observations.


---


*Permanent Address

# Contents


1. Introduction      406

2. Methods of Calculation      409

3. Results      415

4. Prospects for Future Research      420

Acknowledgements      421

References      421

## 1. INTRODUCTION

Anderson[1] in his classic 1958 paper considered an electron initially located at a given site and surrounded by an infinite medium of randomly placed strong scatterers. He proposed that, when the strength of the scatterers exceeds a certain critical value, the electron cannot anymore diffuse away; it will rather be confined by the repeated scattering events in the neighborhood of the initial site. This seminal idea opened up a whole new field of electronic localization. John[2] raised the question of localization of light waves (or more generally electromagnetic waves) by the same mechanism of successive strong scattering events. As we shall point out below, it is much more difficult to localize classical waves rather than electronic waves. Anyway, if light can be localized, it will show in a number of different experiments.[3,4] That is, the transmission coefficient through a non-absorbing slab of thickness $L$ would decrease exponentially with $L$, instead of the $1/L$ dependence of the ordinary diffusive regime. Several attempts have been made to fabricate composite materials capable of localizing light. Enhanced coherent backscattering (a precursor of Anderson localization) of light has been reported.[5-8] The first efforts to reach the critical region near localization have been unsuccessful.[9,10] In a recent preprint, Drake and Genack[11] reported results clearly indicating that the critical region very near localization was reached. We conclude this introductory section by pointing out the relation between electronic and classical wave propagation. Electronic propagation is governed by Schrödinger's equation which takes the form

$$\nabla^2 \psi + \frac{2m}{\hbar^2}(E - V)\psi = 0 , \tag{1}$$

where $E$ is the energy, $m$ is the electronic mass and $V$ is the (random) potential experienced by the electron. The scalar classical wave $(SCW)$ equation for a fixed frequency $\omega$ is of the form

$$\nabla^2 u + \frac{\omega^2}{c^2}\varepsilon u = 0 , \tag{2}$$

where $c$ is the wave velocity in "vacuum" and $\varepsilon$ is the dielectric function which is considered to be a random variable. The electromagnetic $(EM)$

equation for the electric field $\mathbf{E}$ is

$$\nabla^2 \mathbf{E} - \nabla(\nabla \cdot \mathbf{E}) + \frac{\omega^2}{c^2} \varepsilon \mathbf{E} = 0 . \tag{3}$$

Because of the absence of external charges, $\nabla \cdot \mathbf{D} = \nabla \cdot (\varepsilon \mathbf{E}) = 0$. However, the random dielectric function $\varepsilon$ is position-dependent (either continuously or discontinuously at the interfaces of different materials). Thus, the $EM$ equation is not of the same form as the electronic or as the classical wave equation. As a result, rigorous correspondence can be established only between Eqs. (1) and (2). In what follows, we consider, for concreteness, the particular case of a composite system consisting of spheres (average radius $\bar{a}$, total volume fraction $x$) embedded randomly in a host material. For the classical wave problem, the dielectric function of each sphere is $\varepsilon_2$, while that of the host material is $\varepsilon_1 (0 < \varepsilon_1 \leq \varepsilon_2)$. For the electronic problem, the sphere potential is $E_B$ and that of the host potential $E_A (E_A \geq E_B)$. Comparing Eqs. (1) and (2), we have the following correspondences

$$\frac{\omega^2 \varepsilon_1}{c^2} \leftrightarrow \frac{2m(E - E_A)}{\hbar^2} \tag{4}$$

$$\frac{\omega^2 \varepsilon_1 (\mu - 1)}{c^2} \leftrightarrow \frac{2m\delta}{\hbar^2} , \tag{5}$$

where $\mu = \varepsilon_2/\varepsilon_1$ and $\delta = E_A - E_B$. From Eqs. (4) and (5) it follows that every point in the $\omega, \mu$ plane is mapped onto a single point in the $E - E_A, \delta$ plane, and vice versa. In particular by dividing Eq. (4) by Eq. (5), we find that lines of constant $\mu$ (and variable $\omega^2$) are mapped onto the lines

$$\delta = (\mu - 1)(E - E_A) \tag{6}$$

and the lines of constant $\omega^2$ (and variable $\mu$) are mapped onto the straight lines $E - E_A = \text{const}$. This is shown in Fig. 1, where for any point $C$ in the $(E - E_A, \delta)$ plane the slope of $OC$ gives $\mu - 1$ and its projection to the $E - E_A$ axis, $(OC_0)$, gives $\hbar^2 \omega^2 \varepsilon_1/2mc^2$. Since $1 \leq \mu \leq \infty$, the classical wave problem is mapped onto the $\delta \geq 0$, $E - E_A \geq 0$

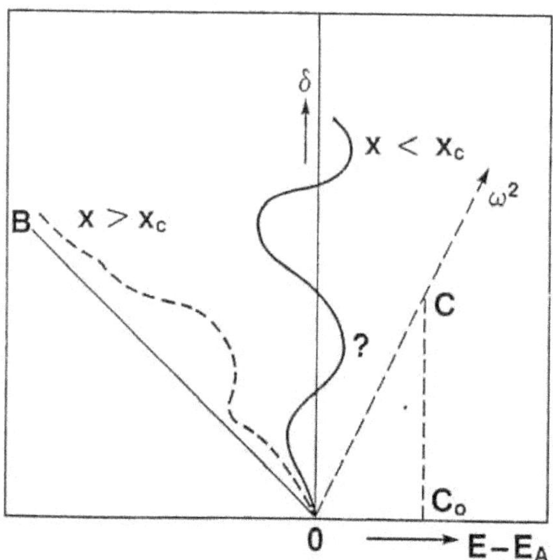



Fig. 1. Schematic diagram for a composite binary random electronic system $A_{1-x}B_x$; $E$ is the energy and $\delta = E_A - E_B \geq 0$, where $E_A(E_B)$ is the bottom of the spectrum of pure $A(B)$ material. The line $OB$ corresponds to $E = E_B$. The expected mobility edge trajectory $(MET)$ separating localized states to its left from extended states to its right, is also shown either for $x < x_c$ (heavy solid line) or for $x > x_c$ (heavy dashed line); $x_c$ is the critical percolation concentration for the $B$ material. The slope of the line $OC$ equals $\mu - 1$ and $OC_0$ determines $\varepsilon_1\omega^2/c^2$, where $\mu = \varepsilon_2/\varepsilon_1$ (see text).

electronic quarter plane. In Fig. 1, we give a schematic plot of the so-called mobility edge trajectory $(MET)$ consisting of the critical points (mobility edges, $E_c$), where the character of eigenstates changes from extended (on its right) to localized (on its left). For low $\delta$, the $MET$ begins[12] as $E_c - \bar{V} \sim -\delta^2$, where $\bar{V} = xE_B + (1 - x)E_A$. As a result for low $\delta$, the $MET$ is to the left of the $\delta$ axis as shown in Fig. 1, which implies immediately that for low $\omega^2$, classical waves are always extended (as long as $\mu$ is positive). In order to have classical wave localization, it is necessary for the MET to cross the $\delta$-axis and pass to the right of the $\delta$-axis. For very high frequencies (corresponding to very high $\delta$) the wave motion approaches that of classical particles and the question of localization reduces to that of percolation.[13] Therefore, if $x_c$ is the critical volume fraction for an infinite $B$ channel to appear, the *MET* will approach the $\delta$-axis (for $x < x_c$) or the *OB* line (for $x_c < x$) as $\delta \to \infty$. In either case, there will be no localized states reachable by finite $\mu$ and very large $\omega^2$. Thus, any chance for classical wave localization is restricted to intermediate values of $\omega^2$ and only if at some intermediate values of $\delta$ the MET crosses to the right of the $\delta$-axis.

## 2. METHODS OF CALCULATION

The problem of wave propagation in a disordered medium can be studied in two stages. The first considers the effect of the disorder on the phase of the wave, $e^{i\phi(r)}$, which changes from its ideal value, $e^{ik_0 r}$, to $\langle e^{i\phi(r)} \rangle = e^{ikr}e^{-r/2l}$, where the symbol $\langle \rangle$ denotes average value, $k$ is the renormalized propagation constant and $l$ is the phase coherence length or mean free path. One can define also the renormalized phase velocity by $v = \omega/k$ and show (assuming that the amplitude of the wave is not affected by the disorder) that the diffusion coefficient $D_0$ is given by

$$D_0 = \frac{1}{3}vl \ . \tag{7}$$

The next stage is to consider what disorder does to the amplitude of the wave. For a weak disorder and for 3-dimensional systems, it turns out that the effect is negligible, thus justifying the traditional approach which ignores any amplitude fluctuations. However, as the disorder becomes stronger, amplitude fluctuations begin appearing, the extent and the size of which are characterized by a length $\xi$. The length $\xi$ increases with increasing disorder and at the critical point for the onset of localization, $\xi$ diverges. As we enter the localized region, the wave still has large amplitude fluctuations but in addition, its envelope decays for large distances $r$ as $\exp(-r/L_c)$, thus defining the localization length $L_c$.

The appearance of amplitude fluctuations further reduces the diffusion coefficient $D$ by a factor $f^{-1}$, i.e.,

$$D = D_0 f^{-1} \ . \tag{8}$$

Many versions of an approximate theory have been developed which express the amplitude-related quantities $\xi, f^{-1}, L_c$ in terms of the phase-related quantities, $l$ and $k$. Probably the most accurate among them is the so-called Potential Well Analogy $(PWA)$[12] which determines the onset of localization, i.e., the mobility edge, by the condition $kl = 0.844$. The PWA expressions for $L_c, f$, and $\xi$ are:

$$L_c = 2l\,; \qquad\qquad\qquad 1-d \qquad\qquad (9a)$$

$$= 2.72l\exp\left[\frac{\pi}{2}kl\right]\,; \quad 2-d \qquad\qquad (9b)$$

$$= \frac{2.2 + 14.12x^2}{1-x^2}l\,; \quad 3-d \qquad\qquad (9c)$$

$$f = 1 + \frac{6}{x^2(x^2-1)}\,; \quad 3-d \qquad\qquad (10)$$

$$\xi = 2.72l\frac{f}{x^2}\,; \qquad\qquad 3-d \qquad\qquad (11)$$

where

$$x = \frac{kl}{0.844}\,. \qquad\qquad\qquad (12)$$

In order to calculate the quantities of interest according to Eqs. (7)–(12), one needs to obtain $k$ and $l$, or equivalently, the complex effective propagation constant

$$q = k + i\frac{1}{2l}\,. \qquad\qquad\qquad (13)$$

Usually perturbation theory is inadequate in the region of strong scattering. The standard way to go beyond perturbation theory is to treat each scattering center independently of all the others and then simply add the effects (complete omission of the multiscattering processes). Under this drastic assumption, the mean free path $l$ is given by

$$\frac{1}{l} = \frac{x}{V_0}\sigma_s\,, \qquad\qquad\qquad (14)$$

where $x$ is the volume fraction of the scatterers, $V_0$ is the average volume occupied by a scatterer, and $\sigma_s$ is the average total cross section of a scatterer. In order to obtain $k$, one usually finds a properly averaged phase velocity $v$ and then $k = \omega/v$. For low frequencies $v^{-2} = x/v_2^2 + (1-x)/v_1^2$. This simply corresponds in the electronic problem to average the potential, which is, of course, the first order perturbation theory result. For high frequencies, where geometric optic is valid, we have $v^{-1} = x/v_2 + (1-x)/v_1$, a result which is obtained straightforwardly by considering the average time that a light ray of a given length spends in medium 1 and medium 2, respectively. This technique for calculating $l$ and $k$ has been used by Ping Sheng and Z. Q. Zhang.[14] Similar methods have been used by other authors.[15,16]

A definite improvement over the technique just outlined is the so-called Coherent Potential Approximation $(CPA)$. The $CPA$ introduces a yet undetermined effective medium, characterized by a complex frequency-dependent propagation constant $q$. The quantity $q$ is then determined by requiring that the scattering induced as a result of the local replacement of the effective medium by the actual medium is on the average zero. In spite of its local character, the $CPA$ incorporates some of the multiscattering effects, failing only in the case where multiscattering by small specific clusters dominates the behavior of the wave. The $CPA$ has been extensively used in electronic calculations, mainly in lattice models where the differential scattering cross section is determined by as many parameters as the unknowns characterizing the effective medium. In the present case where the wave propagates in the continuum and each scatterer has finite size, the differential cross section requires an infinite number of coefficients for its complete determination, while a uniform effective medium is characterized by only two parameters, namely $k$ and $l$. As a result, when applying the $CPA$ to the present problem, a further approximation is made to determine which averaged scattering quantity should be set equal to zero. The most natural choice is to set the average forward scattering amplitude equal to zero, since this quantity is directly proportional to the total scattering cross section (when $q$ is real). The choice leads to a complicated equation for the unknown $q$. In attempting to solve this equation numerically near the critical region, one is faced with unphysical multisolutions and other misbehaviors, especially for large $x(1-x)$ (i.e., $0.2 \leq x \leq 0.8$).

Recently, Soukoulis *et al.*[17] made the choice to set the average total cross section $(TCS)$ equal to zero. The $TCS$ was defined as the total normalized flux of the outgoing spherical wave just outside the sphere, plus the normalized absorption (if any) within the sphere, minus the normalized "absorption" of the incident wave of propagation constant $q$ within a sphere of equal size. The $TCS$ for the electromagnetic case is

$$TCS = 4\pi \frac{\mathrm{Im}\, C}{\mathrm{Re}\, q} \,, \tag{15}$$

where

$$C = \frac{qa^2}{2} \sum_{l=1}^{\infty} (2l+1)[j_l^* h_l'(r_l + \bar{r}_l) + j_l' h_l^* (r_l^* + \bar{r}_l^*)] \,, \tag{16}$$

$a$ is the radius of the sphere, $j_l$ is the spherical Bessel function, $h_l$ is the spherical Hankel function of the first kind, asterisk indicates complex conjugate, prime denotes differentiation with respect to the argument which is $qa$, and $r_l$ and $\bar{r}_l$ are given by

$$r_l = \frac{[q j_l(k_i a) j_l'(qa) - k_i j_l(qa) j_l'(k_i a)]}{[q j_l(k_i a) h_l'(qa) - k_i h_l(qa) j_l'(k_i a)]} \,, \tag{17a}$$

$$\bar{r}_l = \frac{[k_i j_l(k_i a) j_l'(qa) - q j_l(qa) j_l'(k_i a)]}{[k_i j_l(k_i a) h_l'(qa) - q h_l(qa) j_l'(k_i a)]} \,. \tag{17b}$$

The propagation constant $k_i$ is either $k_1 = \omega\sqrt{\varepsilon_1}/c$ or $k_2 = \omega\sqrt{\varepsilon_2}/c$. The scalar case is recovered by putting in Eq. (16) $\bar{r}_l = 0$, multiplying its RHS by 2, and starting the $l$ summation from $l = 0$, instead of $l = 1$. For real $q, C = f(0)$ and the $TCS$ becomes $(4\pi/q)\mathrm{Im}\, f(0)$, where $f(0)$ is the forward scattering amplitude. The $CPA$ equation,

$$\langle C \rangle = 0 \,, \tag{18}$$

where the average is over the two values of $k_i$ and over the distribution of sphere radii (if any), was brought to the form $q_{n+1} = q_n + A\langle C \rangle$, where $n$ is the order of iteration and $A$ is chosen using the weak scattering limit and demanding as good a convergence as possible. We have used $A = 3/2q$. Of course, when convergence has been reached, $\langle C \rangle = 0$. We found that the equation $q_{n+1} = q_n + A\langle C \rangle$ converges to a unique solution in almost all cases and this unique solution is practically the same as the solution of the equation $\langle f(0) \rangle = 0$, when the latter does not misbehave.

It must be pointed out that the calculational techniques outlined above involve several uncontrolled approximations. Hence, it is very desirable to have a reliable technique, even of limited applicability which could serve as a test of the $CPA$-$PWA$ method. Such a technique, known as the wire or the strip method,[18,19] has already been developed for the electronic motion in simple disordered lattice models and is considered the most reliable in obtaining quantitative results. In the wire method, one considers cylinders of square cross section $M^2$ made from our random lattice model. For each cylinder of length $N$, one determines numerically the largest localization length $\lambda_M$ as $N \to \infty$. At the mobility edge,[18,19] $\lambda_M/M = 0.6$, while for extended (localized) states $\lambda_M/M$ versus $M$ increases (decreases), respectively, with increasing $M$. The electronic motion in our random lattice model is governed by a one-orbital per site tight-binding Hamiltonian of the form

$$H = \sum_n |n\rangle \varepsilon_n \langle n| + \sum_{n,m} t_{nm} |n\rangle \langle m| \, ,$$

where usually $t_{nm}$ is $t$ for nearest neighbors and zero otherwise; the disorder is introduced by making each $\varepsilon_n$ a random variable. In order for this lattice model to approximate our material consisting of $B$ spheres embedded in the $A$ host material, one should consider clusters of lattice points shaped as close to spherical as possible and take $\varepsilon_n = \varepsilon_B$ for all $n$'s belonging to the clusters and $\varepsilon_n = \varepsilon_A$ for all $n$'s not belonging to the clusters. ($E_{A(B)}$ corresponding to $\varepsilon_{A(B)} - 6|t|$). The larger the size of the clusters, the better our lattice model approximates the continuum. Actually, the approximation is very satisfactory only when $a_0 \ll \lambda$, where $a_0$ is the lattice spacing, and $\lambda$ is the wavelength. On the other hand, as we shall show below, the best chance for attaining localization is when $\lambda$ is comparable to the linear dimension $L$ of each cluster. Hence, our lattice model must have $L \gg a_0$. In actual calculations, $L/a_0$ is at most 4 or 5. Thus, the lattice model is not going to behave exactly as our actual physical system. One may argue that the main differences will be (a) the lack of sharp interfaces between the $A$ and $B$ component, and (b) the non spherical shape of the clusters that facilitates the merging of two or more clusters together, thus creating new shapes and eventually (as the concentration of $B$ lattice points increases) leading to the opening up of an infinite $B$ channel. Both of these differences make the lattice model easier for the wave propagation. Hence, if one can prove that localization sets in within our lattice model, it is fair to conclude that localization will be present to a larger extent in the actual physical system. Furthermore, the optimum concentration $x_{opt}$ for localization is expected to be lower than the critical concentration $x_c$ at which the infinite $B$ channel is formed. In a lattice model, $x_c$ can be made very low (down to about 0.25) while in a system of spheres the infinite channel does not open even near the close packing concentration ($x \simeq 0.64$). Thus, one should not be surprised if $x_{opt}$ in a lattice model is considerably lower than in our actual physical system of spheres. One last remark about the wire method: this method has been applied only to the scalar (electronic) case. Thus, its conclusions are not necessarily valid for the electromagnetic case.

We conclude this section by mentioning another system where accurate results can be obtained and is indirectly related to our composite random material. This system consists of $B$ spheres placed periodically at lattice points within the $A$ host material.[20] Of course for such periodic systems there are no localized states, only Bloch states and gaps. However, the creation of the gaps is due to strong destructive interference of multiple scattered waves, i.e., to a similar mechanism as the creation of localized states in the disordered system. Furthermore, one may consider random systems arising by a weak disordering procedure from a periodic system. This is more likely to arise for very high $x$. In this case, the *MET* will be very close to the band edge trajectories (*BET*) of the periodic system. Thus, by finding the *BET* of the system of periodically placed spheres (this can be done very accurately using standard band structure techniques), one obtains an approximate determination of the *MET* for a slightly disordered system. A systematic study along these lines was recently undertaken by Economou and Zdetsis[21] in order to check the *CPA* results.

## 3. RESULTS

In Fig. 2 we present results for low $x(x = 0.10)$. The solid line is the *MET* as determined numerically by the wire method. The dashed lines are the *MET* for the scalar and the electromagnetic case, respectively, as determined by the *CPA-PWA*. For the wire method, the $B$ clusters were taken as cubes of linear dimension $L(L/a_0 = 2, 3, 4)$ and $M$ was taken up to 12. Because of the cubic shape of the clusters, we expect localization to set in with difficulty with an $x_{opt} < x_c \simeq 0.31$. As can be seen from Fig. 2, our lattice model exhibits classical wave localization (the *MET* crosses to the right of the $\delta$ axis at $\delta \approx 5$ and $\delta \approx 10$. According to our previous arguments, we expect that the sphere system will exhibit even more localization in agreement with the prediction of the *(CPA-PWA)* technique. We consider the existence of classical wave localization in our model lattice system as the stronger evidence up to now in favor of classical wave localization in real systems. The crossings of the *MET* to the right of the $\delta$-axis (i.e., the appearance of classical wave localization) occur at values of $\delta$ where the so-called Mie resonances in the scattering cross section by a sphere (or more generally by a cluster) appear. These resonances are, in turn, associated with the values of $\delta$ which are correct for the appearance of a new bound states in a potential well of depth $\delta$ and radius $a$. For each angular momentum $l$, there are infinitely many critical values of $\delta$ given by the equation[22]

$$j_{l-1}(ka) = 0; \qquad l \geq 1, \tag{19a}$$

where $\delta = \hbar^2 k^2 / 2m$. For $l = 0$ the critical values are given by[22]

$$ka = (2n + 1)\frac{\pi}{2}; \quad l = 0 . \tag{19b}$$

Note that the $l = 0$ component appears only for the scalar case. As $x$ increases, the values of $\delta$ for which resonances occur shift, in general, towards higher values (since the effective potential outside each

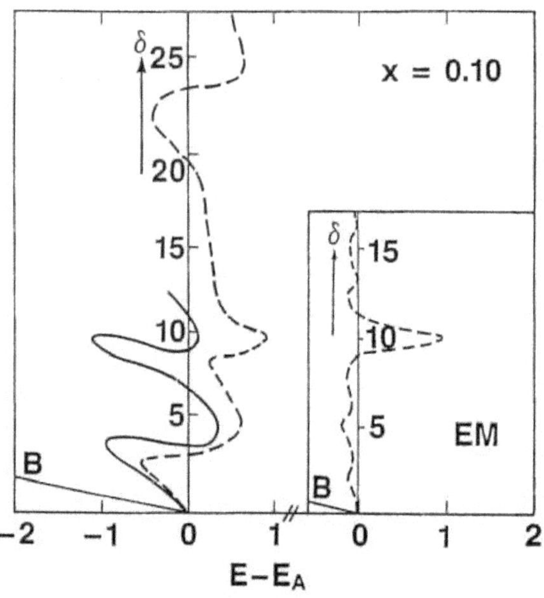


Fig. 2. Mobility edge trajectory obtained from numerical results (solid line) and *CPA* results (dashed line) for the scalar case with $x = 0.10$. In the insert the *CPA* mobility edge trajectory for the electromagnetic wave case for $x = 0.1$ is shown. The units of $\delta$ and $E - E_A$ are equal to $\hbar^2/2ma^2$. The accuracy of the correspondence of our tight-binding model with the continuum model breaks down for $\delta > 10$ (solid line).

sphere is no longer $E_A$, but lower due to the presence of the other $B$ spheres). In Table 1, we give the lowest and most important values of $\delta$ at which resonances occur (for $x \to 0^+$). As $x$ increases, the values of the resonances not only shift, but they become broader and may even mix together. A specific example of this feature is shown in Fig. 3 ($x = 0.60$, electromagnetic case, *CPA-PWA* results) where the $p1$ and the $d1$ resonances have merged together to form the broad region of localized states. The $p2$ and $d2$ resonances are also shown in Fig. 3; those resonances produce localized states to the right of the $\delta$-axis only when $x$ exceeds a particular value. The $x$ value at which a particular resonance gives classical wave localization is an increasing function of the corresponding $\delta$, i.e., the higher the resonance the larger the value of $x$.

Table 1. Values of $\delta$ for which some important Mie resonances occur. Each resonance is characterized by a letter denoting its $l$ character ($s,p,d,\ldots$ for $l = 0,1,2,\ldots$, respectively) followed by a number indicating its order. Note the almost degeneracy of the $s$, $n$, and $d$, $n-1$ resonances.

| order | $s$ ($l = 0$) | $p$ ($l = 1$) | $d$ ($l = 2$) |
|-------|---------------|---------------|---------------|
| 1     | 2.47          | 9.87          | 20.19         |
| 2     | 22.21         | 39.48         | 59.68         |
| 3     | 61.69         | 88.83         |               |

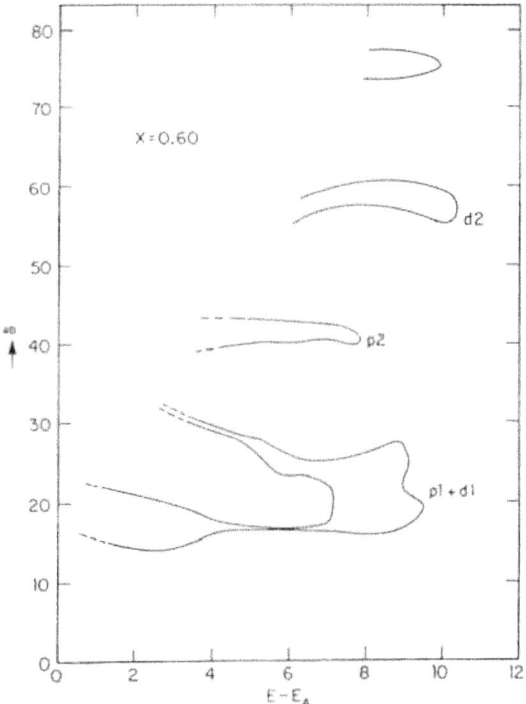


Fig. 3. Mobility edge trajectory obtained from *CPA* results for the electromagnetic wave case with $x = 0.60$. The units of $\delta$ and $E-E_A$ are equal to $\hbar^2/2ma^2$. d2, p2, d1, and p1 represent the positions of the second and first resonances of the $l = 2$ and $l = 1$ components, respectively.

As $x$ increases further, the corresponding resonance produces more localization at higher values of $E - E_A$, reaches an optimum $x_{opt}$, (which seems to be very broad) and then a further increase of $x$ beyond $x_{opt}$ leads to a recession of the corresponding localized region towards the $\delta$-axis. The basic conclusion is that at relatively low $x$, the lowest resonance dominates localization, while as $x$ increases, the importance of higher resonances is increasing even to the point where they may dominate. Our $CPA$-$PWA$ gives $x_{opt} \simeq 0.20 - 0.35$ for the lowest resonance and for the scalar case, while the corresponding value for the electromagnetic case seems to be in the range 0.40–0.65 corresponding to minimum value of $\mu \simeq 2.8$. Preliminary results[21] based on the periodic placement of the spheres for the scalar case show that for $x \simeq 0.65$, the lowest five resonances do not produce any classical wave gaps, while strong gaps remain associated with $d2, s3$, and $p3$ resonances. Thus, it is not clear whether the persistence of $p1$ and $d1$ localization for such high values of $x$ in the electromagnetic case is a real effect or an overestimation of $x_{opt}$ due to inadequacies of our $CPA$. Let us mention in this connection recent experimental results by Drake and Genack[11] showing for the first time that the critical region close to optical localization has been reached as a range of frequencies corresponding to the $p2$ and $d2$ resonances for $x$ associated with close packing of polydispersed spheres ($\sigma_a/\bar{a} \simeq 0.3$, where $\sigma_a$ is the standard deviation of the radii distribution and $\bar{a}$ its average value; one expects that $x$ must be larger than 0.637 in this case). In Fig. 4, we plot our results for the diffusion coefficient $D$ (together with the experimental points from Ref. 11) and $kl$ for values of the parameters as in Ref. 11 ($x$ was taken as 0.73, while experimental conditions suggest that $0.637 < x$). The agreement between theory and experiment is impressive, given the crudeness of our $CPA$ and the absence of any adjustable parameters (with the possible exception of $x$). We also see that $kl$ approached closely the critical value 0.844 for $\lambda_1/\bar{a} \simeq 1.71$ showing that optical localization is almost reached. As a way to obtain optical localization, we proposed[17] a less polydispersed sample ($\sigma_a/\bar{a} \simeq 0.05$ to $0.1$) and/or a lower concentration of the spheres which would shift the expected localization region from the $d2, p2$ resonances to the $d1, p1$ resonances as shown in Fig. 5, where

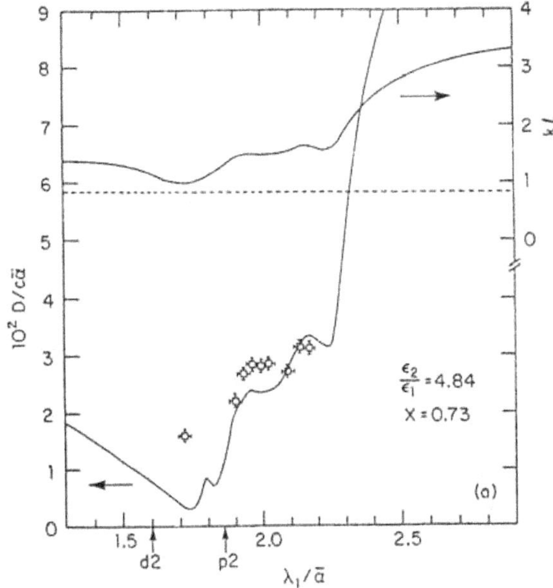



Fig. 4. (a) Diffusion constant ($10^2 D/c\bar{a}$, where $c$ is the velocity of light) and the localization parameter $kl$ as a function of the wavelength $\lambda_1/\bar{a}$, for a sample of tinania spheres ($\varepsilon_2 = 4.84$) in air ($\varepsilon_1 = 1$). The concentration of spheres is $x = 0.73$, while the standard deviation of the sphere radii distribution is $\sigma_a/\bar{a} = 0.29$, where $\bar{a}$ is the average radius of the spheres. The points ($\square$) represent the measured experimental data of Drake and Genack for the same set of parameters. The dashed line represents the value $kl \simeq 0.844$, which is the critical value of localization. The arrows $d_2$ and $p_2$ in $\lambda_1/\bar{a}$-axis represent the positions of the second resonances of the $l = 2$ and $l = 1$ components, respectively.

the quantities $D, kl, l$, calculated by the *CPA-PWA* method for $x = 0.60, \sigma_a/\bar{a} = 0.05$ and $\mu = 4.84$ are plotted vs $\lambda_1/\bar{a}$. The position of $p1, d1, p2, d2$ resonances (no $s$ resonance exists since the $l = 0$ component is absent in the electromagnetic case) are also indicated. Our calculation predicts a true localization region around $\lambda_1/\bar{a} = 3.0$ associated mainly with the $p1$ resonance. Localization (although weaker) appears also around $\lambda_1/\bar{a} = 2.1$ associated with $p2$ resonance. In this connection, let us remind the reader of our previous discussion about the possibility of our *CPA* overestimating $x_{opt}$. If this is actually the

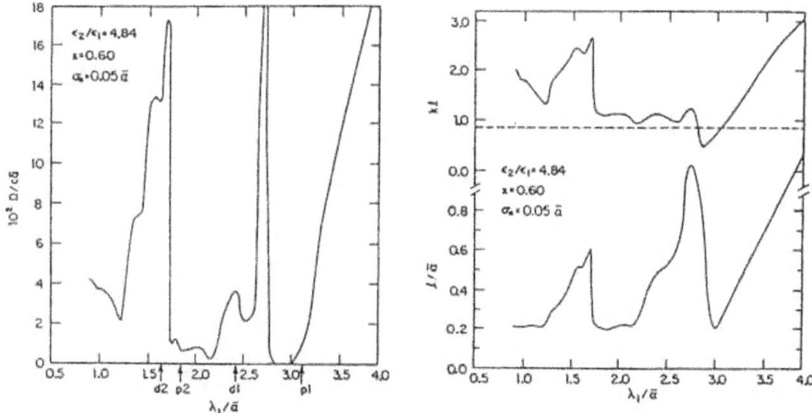


Fig. 5. Diffusion constant $(10^2 D/c\ a)$ the localization parameter $kl$ and the mean free path $l/a$ as a function of the wavelength $\lambda_1/a$ for a sample of tinania spheres $(\varepsilon_2 = 4.84)$ in air $(\varepsilon_1 = 1)$. The concentration of spheres is $x = 0.60$, while the standard deviation of the sphere radii distribution is $\sigma_a/a = 0.05$, where $a$ is the average radius of the spheres. The arrows $d2$, $p2$, $d1$, and $p1$ in the $\lambda_1/a$-axis represent the positions of the second and first resonances of the $l = 2$ and $l = 1$ components, respectively.

case, we expect the behavior shown in Fig. 4 would occur for $x < 0.73$ and in Fig. 5 for $x < 0.60$.

## 4. PROSPECTS FOR FUTURE RESEARCH

Obviously, the most important task immediately ahead is to demonstrate optical localization experimentally. Drake and Genack almost achieved this. The success of our *CPA-PWA* approach in accounting for their experimental data allows some confidence in pointing out the needed modifications in the sample preparation in order to reach the localization region: less polydispersed spheres $(\sigma_a/\bar{a} \simeq 0.1$ or even $0.05)$, lower volume fraction (which can be achieved by mixing the titania $B$ spheres (with $\varepsilon_2 = 4.84$) with spheres of about equal size made from a material with $\varepsilon \simeq 1$) and shifting the quantity $\lambda_1/\bar{a}$ to higher values $(\lambda_1/\bar{a} \simeq 3)$ which can be achieved with the same laser lines by changing $\bar{a}$ from 3000 Å to 2000 Å.

From a theoretical point of view, more work is needed in extending the wire method and the periodic placement of spheres approach in order to use them as checks of the $CPA$-$PWA$ technique, especially for a high concentration $x$. If our $CPA$ proves inadequate in this very important region, one may try to develop more sophisticated $CPA$s along the lines of the muffin-tin $CPA$ developed in the past for the electronic problem.[23]

## ACKNOWLEDGMENTS

We thank S. John, M. H. Cohen, A. Genack, and G. S. Grest for several helpful discussions. We also thank the Exxon Research and Engineering Company where part of this work was done for their hospitality. This work was partially supported by a North Atlantic Treaty Organization Grant No. RG769/87. Ames Laboratory is operated for the U.S. Department of Energy by Iowa State University under contract no. W-7405-ENG-82. This investigation was supported by the Director for Energy Research, Office of Basic Energy Sciences.

## REFERENCES

1. P. W. Anderson, *Phys. Rev.* **109** (1958) 1492.
2. S. John, *Phys. Rev. Lett.* **53** (1983) 2169.
3. P. W. Anderson, *Phil Mag.* **B52** (1985) 505.
4. S. John, Comments in *Conden. Matt. Phys.* **14** (1988) 193.
5. Y. Kuga and A. Ishimaru, *J. Opt. Soc. Am.* **A1** (1984) 831.
6. M. P. Van Albada and A. Lagendijk, *Phys. Rev. Lett.* **55** (1985) 2692; M. P. Van Albada *et al.*, *Phys. Rev. Lett.* **58** (1987) 361.
7. P. E. Wolf and G. Maret, *Phys. Rev. Lett.* **55** (1985) 2696.
8. S. Etemad, R. Thompson and M. J. Andrejco, *Phys. Rev. Lett.* **57** (1986) 574; S. Etemad *et al.*, *Phys. Rev. Lett.* **59** (1987) 1420.
9. A. Z. Genack, *Phys. Rev. Lett.* **58** (1987) 2043.
10. M. Kaveh, M. Rosenbluh, I. Edrei and I. Freund, *Phys. Rev. Lett.* **57** (1986) 2049; M. Rosenbluh *et al.*, *Phys. Rev.* **A35** (1987) 4458.
11. J. M. Drake and A. Z. Genack, *Phys. Rev. Lett.* **63** (1989) 259.
12. E. N. Economou, C. M. Soukoulis, M. H. Cohen and A. D. Zdetsis, *Phys. Rev.* **B31** (1985) 6172; E. N. Economou, C. M. Soukoulis and A. D. Zdetsis, *Phys. Rev.* **B30** (1984) 1686.
13. C. M. Soukoulis, E. N. Economou and G. S. Grest, *Phys. Rev.* **B36** (1987) 8649 and references therein.
14. Ping Sheng and Z. Q. Zhang, *Phys. Rev. Lett.* **57** (1986) 1879.
15. K. Arya, Z. B. Su and J. L. Birman, *Phys. Rev. Lett.* **57** (1986) 2725.

16. C. A. Condat and T. R. Kirkpatrick, *Phys. Rev. Lett.* **58** (1987) 226.

17. C. M. Soukoulis, E. N. Economou, G. S. Grest and M. H. Cohen, *Phys. Rev. Lett.* **62** (1989) 575; E. N. Economou and C. M. Soukoulis, *Phys. Rev. B* (1989).

18. J. L. Pichard and G. Sarma, *J. Phys.* **C14** (1981) L127; A. Mackinnon and B. Kramer, *Phys. Rev. Lett.* **47** (1981) 1546; *Z. Phys.* **B53** (1983) 1.

19. C. M. Soukoulis, E. N. Economou and G. S. Grest, *Phys. Rev.* **B36** (1987) 8649; A. D. Zdetsis *et al.*, *ibid.*, **32** (1985) 7811; C. M. Soukoulis *et al.*, *ibid.* **34** (1986) 2253.

20. S. John, *Phys. Rev. Lett.* **58** (1987) 2489; S. John and R. Rangavajan, *Phys. Rev. B* **38**, (1988) 10101.

21. E. N. Economou and A. D. Zdetsis, *Phys. Rev. B* **40** (1989) 1334.

22. L. I. Schiff, *Quantum Mechanics*, (McGraw-Hill, New York, 1968).

23. B. L. Cyorffy and G. M. Stocks in *Electrons in Disordered Metals and at Metalic Surfaces*, eds. P. Phariseau, B. L. Gyorffy and L. Scheire, (Plenum, New York, 1979); J. S. Faulkner, *Progr. Mat. Sci.* **27** (1982) 1.

# LOCALIZATION OF ACOUSTIC WAVES


C. A. CONDAT

*Institute for Physical Science and Technology*
*University of Maryland, College Park*
*Maryland 20742, USA*


T. R. KIRKPATRICK

*Department of Physics and Astronomy and*
*Institute for Physical Science and Technology*
*University of Maryland, College Park*
*Maryland 20742, USA*



This review summarizes some aspects of the theory of the Anderson localization of acoustic waves. Particular emphasis is placed on the so-called self-consistent diagrammatic approach to localization although some other theoretical methods are also briefly discussed. We also emphasize the advantages of using classical waves to study pure Anderson localization as compared to conventional electronic systems. In classical wave systems, effects analogous to electron-electron interactions in electronic systems can be carefully controlled experimentally. Another advantage of classical wave systems is that the disorder causing localization can be varied to a much greater extent and as a consequence,

there are a number of interesting resonance phenomena in classical wave
localization which are not accessible in electronic systems. The possibility of using acoustic systems with flow to study the crossover between
universality classes in Anderson localization is also reviewed. Here
we also compare theoretical predictions with available experimental results. Detailed theoretical predictions for proposed experiments are also
discussed.

# Contents


1. Introduction — 427

2. Self-consistent Diagrammatic Approach to Localization — 432
   2.1. Formulation of the problem — 433
   2.2. Transformation to multiple-scattering formalism — 438
   2.3. Diagram rules — 440
   2.4. Low density or Boltzmann approximation — 444
   2.5. Self-consistent theory of localization — 446

3. Application of the SC Method — 453
   3.1. Introduction: The models — 453
   3.2. The total scattering cross section — 455
       3.2.1. The sphere — 455
       3.2.2. The cylinder — 459
   3.3. The effective speed of sound — 460
       3.3.1. Spheres — 461
       3.3.2. Cylinders — 463
   3.4. The mean free paths — 464
       3.4.1. Spheres — 465
       3.4.2. Cylinders — 466
   3.5. Anderson localization, spherical scatterers — 468
       3.5.1. The phase boundary — 470
       3.5.2. The localization length — 474
       3.5.3. The diffusion coefficient — 475
       3.5.4. The coherence length — 477
       3.5.5. An $n^*$-independent cutoff — 479
       3.5.6. A cutoff proportional to $l_{sc}^{-1}$ — 480
       3.5.7. The case of large $n^*$ — 480
   3.6. Anderson localization, cylindrical scatterers — 482
   3.7. The effects of dissipation — 484

4. The Pseudosphere Approximation — 486
   4.1. Nature of the approximation — 486
   4.2. The optical problem: Conducting spheres — 489

| | |
|---|---|
| 4.3. Localization in a bubbly liquid | 490 |
| 4.4. Intensity correlation function | 494 |
| 4.5. The experiment of Ferrari *et al.* | 496 |
| 5. The Localization of Third Sound | 500 |
| 5.1. Third sound on superfluid helium films | 500 |
| 5.2. Striped configurations | 502 |
| 5.2.1. Van der Waals scatterers | 502 |
| 5.2.2. Index of refraction scatterers | 503 |
| 5.2.3. Averaged Green's function results | 504 |
| 5.2.4. Anderson localization | 507 |
| 5.3. Other random substrates | 513 |
| 5.3.1. One dimension: Grooved substrate | 513 |
| 5.3.2. Two-dimensions: Dusted substrate | 514 |
| 5.4. The effects of intrinsic dissipation | 516 |
| 5.5. The experiment of Smith *et al.* | 518 |
| 5.5.1. The periodic substrate | 518 |
| 5.5.2. The random substrate | 521 |
| 6. Acoustic Localization with a Flow | 522 |
| 6.1. Formulation of the problem and perturbative results | 523 |
| 6.2. Exact results for the one-dimensional case | 525 |
| 6.3. Field theory methods and acoustic localization with flow | 526 |
| 7. Conclusions | 532 |
| Acknowledgements | 535 |
| References | 536 |

# 1. INTRODUCTION

Shortly after Anderson's original work[1] on electron localization in disordered systems,[2] the problem of phonon localization in random systems was considered.[3] The closely related topic of localization of acoustic (sound) waves in macroscopic systems that are artificially constructed to be disordered has attracted attention only more recently.[4-7] In this review we discuss some of the recent work on localization of acoustic waves in disordered macroscopic systems.

In general pure Anderson localization is due to wave interference effects[2] and it is not unique to low temperature disordered materials where quantum mechanics is important. Anderson localization is a many scatterer wave coherence effect whereby a wave excitation gets trapped in a region in space typically involving many scatterers or inhomogeneities. It in general does *not* mean some kind of trapping or local binding to a particular point in space. In principle one also expects Anderson localization in any type of classical wave system if the disorder, which causes localization, is strong enough. Concrete examples where localization physics might play an important role range from geophysical systems, where an important problem is the description of elastic waves which are used for oil detection, to astrophysical problems, where electromagnetic radiation from stellar sources in general propagates in a disordered universe. Due to the ubiquitous nature of localization it is desirable to understand Anderson localization in ideal systems which can be artificially constructed experimentally. Once this is realized one can hope to attack nonideal problems involving localization with more confidence. We argue below that certain types of acoustic systems are excellent paradigms for studying pure Anderson localization.

One of the main advantages of acoustic wave localization compared to electronic localization is that the effects of the nonlinearities in the underlying hydrodynamic equations describing wave propagation in the disordered medium can be carefully controlled experimentally. In the electronic problem the analogous effect is electron-electron interaction which is much more difficult to control experimentally. Furthermore, the problem of electron localization in an interacting electron system is not yet an understood subject.[2] Consequently, it is difficult to distinguish experimentally between effects due to pure Anderson localization and "localization" effects induced by the interactions.

Another advantage of using acoustic waves to probe localization physics is that due to its macroscopic nature, the type of disorder can be experimentally controlled. Further, the possible types of disorder is much greater in acoustic wave localization than in electronic localization. For example, the simplest possibility is stationary hard sphere-like scatterers placed at random in a fluid.[6] Another, more interesting possibility might involve scatterers placed in a fluid at random with the scatterers chosen such that at a certain discrete set of frequencies they are very efficient scatterers of sound waves.[8−10] This resonance type of behavior is not accessible in the electron localization problem. In this review we place emphasis on the interplay between isolated scattering processes and many body localization physics. Since Anderson localization is in part due to the strength of individual scattering processes, the resonance phenomena discussed above have a very pronounced effect.

There are also other phenomena in localization physics that can be probed in acoustic systems but which are inaccessible in electronic systems due to competing phenomena such as electron-electron interactions, the Kondo effect, and the quantum Hall effect. Here we have in mind the crossover between the different universality classes of Anderson localization.[2,11] If a system has time reversal symmetry then its localization properties are in general different than if time reversal symmetry is somehow broken. In Sec. 6 of this review we describe a particular acoustic wave experimental arrangement that should allow the study of Anderson localization as a field which breaks time reversal symmetry is slowly turned on.[12,13]

There is a disadvantage to using acoustic systems to study localization physics. In general dissipation, which is analogous to inelastic scattering in electron systems, is greater than in low temperature electron systems. As a consequence, theoreticians must properly estimate the effects of intrinsic dissipation. Further, experimentalists must properly choose systems with small dissipation. We address this point in several places in this review.

There are several theoretical methods that have been used to study Anderson localization in general. In the bulk of this article we use the self-consistent diagrammatic approach (SCDA) to localization. Following earlier work by Götze,[14] Vollhardt and Wölfle[15] were the first to develop this approach to describe electronic localization. One of us[6] was apparently the first to use this method to describe acoustic localization. The advantages of the SCDA are that it is simple, it can be used for any spatial dimension, $d$, and it can be used for any type of scatterer or disorder. The SCDA also gives explicit theoretical predictions in the whole phase diagram and not only near a mobility edge where an Anderson transition takes place. The SCDA does have the disadvantage that it is a somewhat uncontrolled theory. In general (cf. Sec. 6) one must be careful in using this method. The SCDA has been studied in enough detail and compared with other more systematic approaches that in general one can guess apriori whether or not it will yield reasonable results.

Another, more systematic, approach to study localization is the field theory or sigma model method. This method was pioneered by Wegner,[16,17] and later by others.[18,19] The first to apply this method to phonon localization were John, Sompolinsky and Stephen.[20] Later John used this approach to study the localization of electromagnetic radiation in a disordered medium.[5,21] The advantage of the field theory approach is that it establishes a general scaling behavior near the Anderson transition and, near $d = 2$, conventional renormalization group methods can be used to characterize universal properties of the Anderson transition. The disadvantage of the field theory method is that it can only be used to calculate universal quantities like critical exponents. In practice the results are reliable only near two-dimensions.

Some estimates of the conditions needed for the observation of classical localization were made by Anderson.[22] Sornette and Souillard[23] investigated localization in media which contain scatterers with strong internal resonances. The special case of localization in one-dimensional acoustic systems has also been the subject of extensive theoretical work.[4,24-30] Finally, we mention two recent theoretical reviews of interest. Vollhardt[31] discusses the Anderson localization of electrons, light, and sound. Sornette[7] describes in detail the various regimes in which acoustic waves can occur in disordered media.

There have also been some computer experiments investigating localization in classical systems. William and Maris[32] developed a numerical method that was used to calculate the phonon density of states and the phonon localization length for a two-dimensional square lattice with mass disorder. Flesia, Johnston and Kunz[33] carried out a numerical study of the propagation of waves through a network of waveguides with a randomly varying index of refraction.

In this article we will also briefly discuss the problem of electromagnetic localization because of its close analogy with acoustic localization. There has been a considerable experimental effort concerning the weak localization (i.e., the enhanced coherent backscattering) of electromagnetic waves.[34-42] These experimental results have been understood in terms of theories built upon perturbative approaches that contain as a crucial element the set of maximally crossed diagrams.[43-49] It is important to note that the success of these experiments and of their interpretation provides a solid point of departure for both the experimental investigation of classical Anderson localization and the self-consistent theoretical formulations. Parenthetically we note that, although much less experimental work has been done on the backscattering of sound waves,[50] no substantial differences with the electromagnetic case are to be expected — except, of course, for the absence of polarization phenomena.

Considerable progress in the search for the mobility edge has been achieved lately in three-dimensional electromagnetic media. Genack used wedged random media containing titania microstructures to determine the scale dependence of transmission.[51] Measuring the total transmission and the intensity fluctuations in the transmitted light he was able to obtain the optical diffusion coefficient, the transport mean free path and the imaginary part of the dielectric function. Watson, Fleury, and McCall introduced an ingenious cylindrical configuration, with the source on the axis of the sample.[52] Although not finding evidence for Anderson localization, they were able to study the diffusive photon transport by measuring the propagation time from the source to the boundary. L. A. Ferrari, and coworkers working in the microwave range, found evidence for the proximity of the mobility edge.[53] They used copper-plated iron spheres as their scatterers. The microwave regime seems particularly promising because: (a) Dissipative effects are weaker than in the optical case. (b) It is easier to achieve "true" randomness, eliminating the clustering of scatterers that can originate secondary waves. (c) Macroscopic scatterers can be made essentially identical to each other, thus avoiding the problem of fluctuations in the scatterer sizes and shapes; these fluctuations, if present, do not allow one to take full advantage of resonant scattering.

In acoustical systems, the experimental progress has been considerable, although mostly confined to one-dimensional configurations. Possibly the first experiment specifically devoted to the study of acoustic localization was that of Hodges and Woodhouse,[54] who obtained satisfactory evidence for the presence of localized excitations using a string with masses attached in irregular positions. Depollier, Kergomard, and Laloe[55] have found evidence for localization studying acoustic waves in a long pipe. The disorder was introduced by a series of randomly ordered, resonant, closed branches that emerged perpendicularly to the main pipe. He and Maynard[56] studied the transverse excitations of a steel wire along which a set of small, identical masses were attached in random positions. They not only found strong evidence for localization but were also able to simulate the electron-photon interaction by modulating the longitudinal strain in the wire.

Some time ago, we proposed an experiment to localize third-sound waves on a helium film adsorbed on a disordered substrate.[57,58] Experiments performed by the Amherst group in the low frequency range have confirmed the localization picture.[59] The band structure predicted for the case of a periodic substrate[60] was investigated simultaneously.[59] Belzons *et al.*[61] applied a suggestion of Guazzelli, Guyon, and Souillard[62] and performed a very interesting experiment using a water tank with a rough bottom. They measured the localization length and discussed the evidence for nonlinear effects. A comparison was also made with results obtained using a periodic bottom.

The plan of this review is as follows. In Sec. 2 we review the self-consistent diagrammatic approach to localization. In Sec. 3 we apply the results of Sec. 2 to a variety of two and three dimensional disordered acoustic systems. In Sec. 4 we describe a useful approximation, which we call the pseudosphere approximation, for complicated localization problems. We then use this approximation to describe the localization of light in a disordered medium, to describe the localization of sound waves in a fluid with bubbles, and to calculate intensity correlation functions. In Sec. 5 we discuss the localization of third sound waves in helium on a disordered substrate. This is an ideal experimental configuration for observing acoustic localization because the intrinsic dissipation is very small. In Sec. 6 we review work on the problem of acoustic localization with flow. Such a system has broken time reversal symmetry and its localization properties are therefore in general very different from those of systems without flow. In Sec. 7 we conclude by mentioning a number of open problem whose solution would lead to a more detailed understanding of localization physics.

## 2. SELF-CONSISTENT DIAGRAMMATIC APPROACH TO LOCALIZATION

In this section we will review what has become known as the self-consistent ($SC$) diagrammatic approach to localization. The method is capable of describing the localization of virtually any type of wave excitation including electrons, phonons, sound waves, or photons. Its virtue is its generality and predictive power but as mentioned in Sec. 1, it is not a systematic theory. In what follows we will point out in more detail both its strengths and weaknesses.

## 2.1. Formulation of the Problem

For obvious reasons most attention has traditionally been focused on the problem of electron localization in a disordered solid. The recent advances in describing Anderson localization of nonelectronic excitations have all used concepts and methods developed for describing electron localization.[2] As first emphasized by Götze[4] the central quantity in electronic localization is the density propagator. The first to use this idea to develop a $SC$ diagrammatic theory of electron localization were Vollhardt and Wölfe ($VW$).[15] Below we use the methods of $VW$ to describe the localization of sound waves which propagate in a fluid with a random distribution of elastic stationary scattering centers.[6]

As mentioned above, the central quantity in electron localization is the density propagator or density correlation function ($dcf$). There are basically two reasons why the $dcf$ is important. First, the $dcf$ involves a disorder average of a product of an advanced and a retarded Green's function ($GF$). As a consequence, the $dcf$ has the crucial inteference effects responsible for Anderson localization. This is to be constrasted with the disorder average of a single $GF$. Wegner[63] has proven for wide class of localization problems that the averaged $GF$ is not critical at the Anderson transition or mobility edge. Physically, the effects of averaging destroys the important phase information in the averaged $GF$. It is interesting to note that the fundamental distinction between a product of averaged $GF$ and the average of a product of $GF$'s suggest that the localization problem is not self-averaging. This is a topic of considerable interest.[64] The second reason why the $dcf$ is of importance is that its long time and large distance behavior is determined by a diffusion coefficient which characterizes how the electron diffuses in the disordered solid. It is clearly physically appealing to describe the localization properties of a system in terms of transport coefficients such as the diffusion coefficient. For localized behavior the diffusion coefficient is zero (in fact, the mean square displacement is finite for long times if the electron is localized) while for nonlocalized electrons the diffusion coefficient is finite (the mean square displacement grows linearly with time).

In the sound wave localization problem the quantity analogous to the $dcf$ is the average energy density of the excited sound waves. We shall see that it also is proportional to the disorder average of a product of an advanced and retarded $GF$. The diffusive nature of the energy density is due to energy conservation. In what follows we describe the localization of sound waves in terms of a diffusion coefficient that characterizes the diffusion of energy in a liquid with stationary elastic scattering centers (cf. Fig. 1)

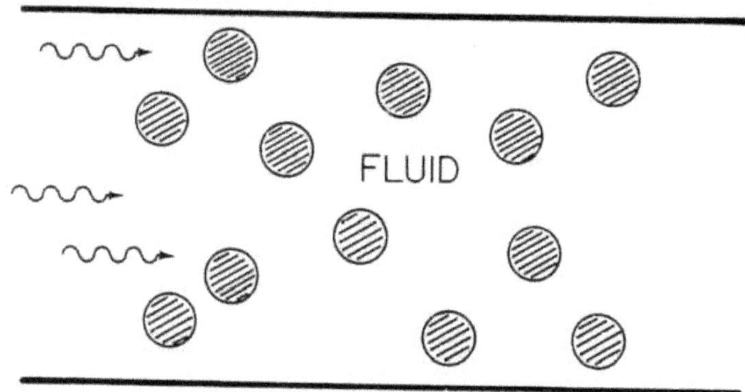


Fig. 1. Stationary scatterers (shaded circles) fixed at random positions in a fluid. The wiggly lines denotes an incident sound wave.

To formalize the problem we start with the scalar wave equation for sound propagation,

$$\partial_t^2 \phi(\mathbf{x},t) - c^2 \nabla^2 \phi(\mathbf{x},t) = 0 \,, \tag{1a}$$

in a $d$-dimensional space. Here $\phi$ is the wave amplitude, $\nabla^2$ is a $d$-dimensional Laplacian, and $c$ is the speed of sound in the fluid in the absence of any scatterers. For typical hydrodynamic applications $\phi$ is the velocity potential and $\mathbf{u} = \boldsymbol{\nabla}\phi$ is the fluid velocity or $\phi$ can also represent pressure fluctuations.[65] The Eq. (1a) is valid in the fluid regions where there are no scatterers. To specify the wave propagation completely one needs boundary conditions ($bc$) for the wave amplitude at the surface of each scatterer. In general $bc$ on both $\phi$ and on the normal derivative of $\phi$ will be needed. Here we simply denote these $bc$ by

$$\phi|_{\text{surface}} = (bc)_1 \; ; \quad \partial_n \phi|_{\text{surface}} = (bc)_2 \; . \tag{1b}$$

Particular boundary conditions will be considered in Secs. 3, 4 and 5.

Equation (1a) is a linear wave equation. We next define Green's functions for the Eqs. (1). If we imagine that a disturbance is created at time $t = 0$ at $\mathbf{x}$ then a reasonable initial condition is,

$$\phi(\mathbf{x}, t \le 0) = 0$$

and

$$\partial_t \phi(\mathbf{x}, t = 0) = f(\mathbf{x}) \; , \tag{2a}$$

where $f(\mathbf{x})$ describes the disturbances. In terms of $GF$, Eqs. (2a) are given by,

$$\phi(\mathbf{x}, t) = \int d\mathbf{x}' G(\mathbf{x}, t| \mathbf{x}', 0) f(\mathbf{x}') \tag{2b}$$

with

$$[\partial_t^2 - c^2 \nabla^2] G(\mathbf{x}, t| \mathbf{x}', 0) = 0 \tag{2c}$$

and

$$G(\mathbf{x}, t \le 0| \mathbf{x}', 0) = 0$$
$$\partial_t G(\mathbf{x}, t = 0| \mathbf{x}', 0) = \delta(\mathbf{x} - \mathbf{x}')$$
$$G(\mathbf{x}, t| \mathbf{x}', 0)|_{\text{surface}} = (bc)_1$$
$$\partial_n G(\mathbf{x}, t| \mathbf{x}', 0)|_{\text{surface}} = (bc)_2 \; . \tag{2d}$$

It is straightforward to show that the energy density, $\varepsilon$, of a linear sound wave is proportional to the kinetic energy of the wave,[65]

$$\varepsilon(\mathbf{x}, t) = \rho_0 u^2(\mathbf{x}, t) = \rho_0 [\boldsymbol{\nabla}\phi(\mathbf{x}, t)]^2 \; , \tag{3a}$$

with $\rho_0$ the equilibrium mass density of the fluid. From Eqs. (3a) and (2) and the discussion above the localization properties of a sound wave can be determined from the squared $GF$. In terms of $GF$ we define an energy density propagator by,

$$P(\mathbf{x},t|\mathbf{x}',0) = G^2(\mathbf{x},t|\mathbf{x}',0) \ . \tag{3b}$$

The energy density can be obtained from Eqs. (2) and (3) if we assume $f(\mathbf{x}')$ in Eq. (2b) is sharply peaked around $\mathbf{x}' = 0$. We further imagine that $P$ is experimentally measured at a given external frequency, $\omega$. Accordingly, we define the Laplace transform of $P$ by,

$$\begin{aligned}
P_\omega(\mathbf{x}|\mathbf{x}') &= \int_0^\infty dt\, e^{i(\omega+i0)t} G^2(\mathbf{x},t|\mathbf{x}',0) \\
&= \int_{-\infty}^{+\infty} \frac{dE}{2\pi} G_{E_+}^+(\mathbf{x}|\mathbf{x}') G_{-E_-}^+(\mathbf{x}|\mathbf{x}') \\
&= \int_{-\infty}^{+\infty} \frac{dE}{2\pi} G_{E_+}^+(\mathbf{x}|\mathbf{x}') G_{E_-}^-(\mathbf{x}'|\mathbf{x}) \ , \tag{4a}
\end{aligned}$$

where,

$$G_E^\pm(\mathbf{x}|\mathbf{x}') = \int_0^\infty dt\, e^{i(E\pm i0)t} G(\mathbf{x},t|\mathbf{x}',0) \ , \tag{4b}$$

and

$$E_\pm \equiv E \pm \frac{\omega}{2} \ . \tag{4c}$$

Here we have used the convolution property of Laplace transforms and we have defined an internal frequency, $E$, which physically signifies the $E$ component of a pulse created at $\mathbf{x}'$. The last equality in Eq. (4a) follows from time reversal symmetry and spatial isotropy. When these symmetries are broken (cf. Sec. 6) the final equality should be avoided.

Note that Eq. (4a) is the product of an advanced and retarded $GF$. As we discussed above, we are interested in the disorder average of this quantity. For our case, the disorder is introduced with $N$ elastic scattering centers located at positions $\{\mathbf{R}_i; i = 1 - N\}$ in a fluid of volume $V$ with an average number density $n = N/V$. The disorder average is then given by,

$$[f\{\mathbf{R}_i\}]_{\mathrm{av}} \equiv \lim_{\substack{N,V \to \infty \\ N/V = n}} \frac{1}{V^N} \int d\mathbf{R}^N \, F(\{\mathbf{R}_i\}) \,, \qquad (5)$$

where $f$ is an arbitrary function which depends on the locations of the $N$-impurities and we have assumed a bulk limit. In defining this average we have for simplicity neglected all correlations between scatterers so that overlapping configurations of scatterers are allowed. In general this is unphysical. We can estimate the error to be of $0(n_i^* = nv)$ where $v = (2a)^d \Omega_d/d$ is the $d$-dimensional volume excluded to a scatterer by the presence of another scatterer. Where $\Omega_d$ is the area of a $d$-dimensional unit sphere and $a$ is the geometric radius of a scatterer. In general we will find (cf. below) that our procedure is most reliable for scatterers which scatter sound waves much more efficiently than simple geometric hard sphere-like scattering. In addition, we take the point of view that the physics of localization (cf. below) is distinct from this overlap problem which we neglect. We believe that our qualitative results for the localization problem are not modified by the fact that the scatterers are correlated.

In our calculations it is convenient to work in Fourier space. We define an external wave vector $\mathbf{k}$ by,

$$[P_{\omega,k}(E)]_{\mathrm{av}} = \int d(\mathbf{x} - \mathbf{x}') \, \exp\left[-i\,\mathbf{k}\cdot(\mathbf{x} - \mathbf{x}')\right][P_{E,\omega}(\mathbf{x}|\mathbf{x}')]_{\mathrm{av}} \qquad (6a)$$

where we have used that space is homogeneous on that average so that $[P(\mathbf{x}|\mathbf{x}')]_{\mathrm{av}}$ is a function of $|\mathbf{x} - \mathbf{x}'|$ only. As we discussed above and as we show below, $[P_{\omega,k}(E)]_{\mathrm{av}}$ is proportional to a diffusive hydrodynamic pole for $k,\omega \to 0$,

$$[P_{\omega,k}(E)]_{\mathrm{av}} \sim \frac{1}{-i\omega + D(E,\omega)k^2} \,. \qquad (6b)$$

Here $D(E,\omega)$ is an internal frequency, $E$, and external frequency, $\omega$, dependent diffusion coefficient that represents, physically, the diffusion of energy. Whether or not a wave at frequency $E$ is localized is determined by the behavior of $D(E, \omega \to 0)$. For example, for $d \leq 2$ we find that for all $E$ and $n$,

$$D(E, \omega \to 0) = -i\omega \xi^2(E) + 0(\omega^2) \ . \tag{6c}$$

This implies that for $d \leq 2$.

$$[P_E(\mathbf{x}, t)]_{\text{av}} \sim \exp\left[-|\mathbf{x}|/\xi(E)\right] \ . \tag{6d}$$

does not decay in time and that it is localized in a spatial region of extent $\xi(E)$: In $d \leq 2$ dimensional random systems, waves are localized no matter how weak the disorder is and $\xi(E)$ is therefore dubbed the localization length. For $d > 2$ dimensional systems the behavior given by Eq. (6c) occurs only for a range of frequencies, $E$, and only if the system is sufficiently disordered.

Finally, it is interesting to point out that in general, localization theory seems to predict either normal diffusion or complete localization where the mean square displacement does not grow with time. The intermediate case of abnormal diffusion (which still implies zero diffusion coefficient) where the mean square displacement grows slower than linearly with time does not seem to occur.

## 2.2. Transformation to Multiple-Scattering Formalism

So far we have formulated the problem as a complicated boundary value problem. In order to use standard many-body techniques it is convenient to transform the boundary value problem into a many-body problem by using a multiple-scattering formalism.[66,67] The basic idea is that we can view the sound wave as propagating in the fluid and scattering from the randomly positioned scatterers. The scattering process can be characterized by a transition or $T$-matrix. To formalize all of the possible scattering processes we first introduce an abstract quantum-mechanical-like notation which is representation free. We define bra and ket vectors and abstract operators, denoted by circumflexes, by, for example,

$$G_E^+(\mathbf{x}|\mathbf{x}') \equiv \langle \mathbf{x}|\widehat{G}_E^+|\mathbf{x}'\rangle \ . \tag{7a}$$

In a wavenumber representation Eq. (7a) reads,

$$G_E^+(\mathbf{p}|\mathbf{p}') = \int \frac{d\mathbf{x}\,d\mathbf{x}'}{(2\pi)^d} \, \exp[-i\,\mathbf{p}\cdot\mathbf{x} + i\,\mathbf{p}'\cdot\mathbf{x}']G_E^+(\mathbf{x}|\mathbf{x}')$$

$$= \langle\mathbf{p}|\widehat{G}_E^+|\mathbf{p}'\rangle \; . \tag{7b}$$

With this notation Eq. (6a) is given by,

$$[P_{\omega,k}(E)]_{\mathrm{av}} = \frac{(2\pi)^d}{V} \int \frac{d\mathbf{p}\,d\mathbf{p}_1}{(2\pi)^d}[\langle\mathbf{p}_+|\widehat{G}_{E+}^+|\mathbf{p}_{1+}\rangle\langle\mathbf{p}_{1-}|\widehat{G}_{E-}^-|\mathbf{p}_-\rangle]_{\mathrm{av}}$$

$$\equiv \int \frac{d\mathbf{p}\,d\mathbf{p}_1}{(2\pi)^d}\phi_{\mathbf{p}\mathbf{p}_1}(\mathbf{k},\omega|E) \; , \tag{7c}$$

where

$$\mathbf{p}_\pm \equiv \mathbf{p} \pm \frac{\mathbf{k}}{2} \; . \tag{7d}$$

In giving the second equality in Eq. (7c) we have defined the function $\Phi$ and we have used that from the averaging there will be an additional factor of the volume since on the average, space is homogeneous.

To go to the multiple scattering formalism we next define a $\widehat{T}_1$ operator whose matrix elements are the transition matrix elements for a scatterer at $\mathbf{R}_1 \equiv 1$, (Ref. 66)

$$\widehat{G}_E^\pm(1) \equiv \widehat{G}_{E,0}^\pm + \widehat{G}_{E,0}^\pm \widehat{T}_1^\pm(E)\widehat{G}_{E,0}^\pm \; . \tag{8a}$$

Here $\widehat{G}(1)$ is the Green's operator when there is only one scatterer present at $\mathbf{R}_1$ and $\widehat{G}_0$ is the Green's operator in the absence of scatterers,

$$\langle\mathbf{x}|\widehat{G}_{E,0}^\pm|\mathbf{x}_1\rangle = -[c^2\nabla^2 + (E + i0)^2]^{-1}\delta(\mathbf{x} - \mathbf{x}_1) \; . \tag{8b}$$

An explicit form for the matrix elements of $\widehat{T}_1$ can be determined by solving the sound wave plus one scattering problem, with Eqs. (2d), and using Eq. (8a) (Ref. 68). In this way the boundary conditions are incorporated into the transition matrix. With this, we can in a standard way express the $N$ scatterer ($N \to \infty$) Green's operator as,[66]

$$\widehat{G}_E^\pm = \widehat{G}_{E,0}^\pm + \sum_{i=1} \widehat{G}_{E,0}^\pm \widehat{T}_i^\pm(E)\widehat{G}_{E,0}^\pm$$

$$+ \sum_{i \neq j} \widehat{G}_{E,0}^\pm \widehat{T}_i^\pm(E)\widehat{G}_{E,0}^\pm(E)\widehat{T}_j^\pm(E)\widehat{G}_{E,0}^\pm + \cdots \tag{8c}$$

i.e., an infinite series of binary interactions where no two consecutive $\widehat{T}$ operators can have the same scatterer label. Physically, it is clear that all possible scattering processes are conatined in Eq. (8c). The advantage of this technique is that it incorporates the boundary conditions into the equations of motion. With Eq. (8c) we can use the techniques developed to describe the electron localization to describe sound wave localization.

As mentioned above, the matrix elements of the $\widehat{T}$ operator can be determined from Eqs. (8a) and (2d). The explicit form depends on the boundary conditions and examples will be given in Secs. 3 and 4. Here we note that in general,

$$\langle \mathbf{p}|\widehat{T}_i^{\pm}(E)|\mathbf{p}_1\rangle = e^{-i\,\mathbf{R}_i\cdot(\mathbf{p}-\mathbf{p}_1)}\langle \mathbf{p}|\widehat{T}^{\pm}(E)|\mathbf{p}_1\rangle \,, \tag{9}$$

i.e., the dependence on the scatterer position is only through the exponential factor in Eq. (9).

### 2.3. Diagram Rules

Now that we transformed the problem into a standard many body problem we can use diagrammatic technique for calculating $[G]_{\mathrm{av}}$ and $\phi$ (cf. Eq. (7c)). In general we consider density- or disorder-expansions of $[G]_{\mathrm{av}}$ and $\phi$. To proceed we insert Eq. (8c) in Eqs. (7b) and (7c). We first consider the averaged $GF$. Since space is homogeneous, on the average, $[G]_{\mathrm{av}}$ is diagonal in the wave-vector representation and can be written,

$$\begin{aligned}
[\langle \mathbf{p}|\widehat{G}_E^{\pm}|\mathbf{p}_1\rangle]_{\mathrm{av}} &= \delta(\mathbf{p}-\mathbf{p}_1)G(p, E\pm i0) \\
&= \delta(\mathbf{p}-\mathbf{p}_1)[c^2 p^2 - (E\pm i0)^2 - \Sigma_p^{\pm}(E)]^{-1} \,,
\end{aligned}$$
$$\tag{10a}$$

where we have defined a self-energy,

$$\Sigma_p^{\pm}(E) = \Gamma_p(E) \pm i\gamma_p(E) \,,$$

which consist of a real part, $\Gamma_p$, and an imaginary part, $\gamma_p(E)$. Physically, $\Gamma_p$ will lead to renormalization to the "bare" speed of sound and in general the speed of sound will become wavevector dependent. $\gamma_p(E)$ represents dampening of the sound waves due to scattering from the randomly positioned scatterers. Diagrammatically, $\Sigma$ is given by the sum of all irreducible diagrams and the first few of them are given in Fig. 2. In this context an irreducible diagram is one which cannot be broken into two parts by severing a horizontal line. The diagram rules for the calculation of $\Sigma_p^{\pm}(E)$ are as follows. (1) With each directed line segment, $\mathbf{p}_1$, we associated a free-particle Green's function,

$$G_0(p, E \pm i0) = [c^2 p^2 - (E \pm i0)^2]^{-1} \ .$$

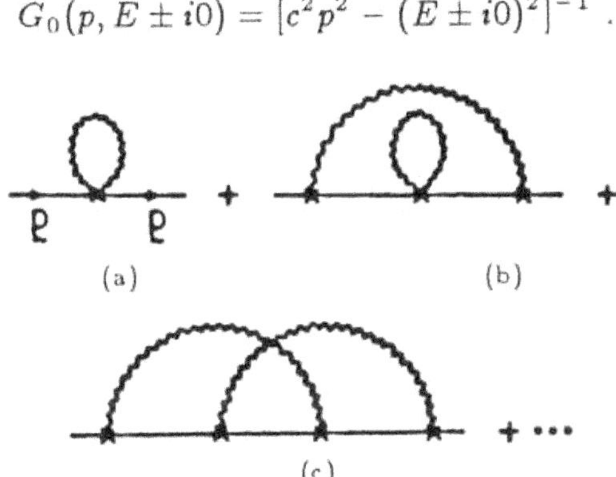


Fig. 2. First few diagrams that contribute to the self-energy $\Sigma_p$.

(2) An $x$ denotes a $\widehat{T}$ operator and

denotes the matrix element $\langle \mathbf{p} | \widehat{T}_i^{\pm}(E) | \mathbf{p}_1 \rangle$. (3) When we average over all configurations of the scatterers, the $\widehat{T}$ operators are connected by the exponential factors in Eq. (9). We represent the connection by a wavy line. (4) With each wavy line we associate a factor of $n(2\pi)^d$ times a delta-function of the sum of the momenta (or wavenumbers) entering and leaving the wavy line. The delta-function appears as a result of the $\mathbf{R}_i$ integration of the exponential factors in the $T$-matrices. (5) Diagrams with the element are not allowed due to the restrictions on the sums in Eq. (8c).

Having formulated the diagram rules for the calculation of $[G]_{\text{av}}$ as an expansion in powers of $n$, we now turn our attention to deriving an equation for $\phi$. We can represent the function $\phi_{\mathbf{p}\mathbf{p}_1}(\mathbf{k},\omega|E)$ in terms of the "bubble" diagrams given in Fig. 3. Here the directed line segments denote exact averaged Green's functions. Analytically one has,

$$\phi_{\mathbf{p}\mathbf{p}_1}(\mathbf{k},\omega|E) = G(p_+, E_+ + i0)G(p_-, E_- - i0)[\delta(\mathbf{p} - \mathbf{p}_1)$$
$$+ \Gamma_{\mathbf{p}\mathbf{p}_1}(\mathbf{k},\omega|E)G(p_{1+}, E_+ + i0)G(p_{1-}, E_- - i0)] \; . \tag{11a}$$

The complete four-part vertex function, $\Gamma_{\mathbf{p}\mathbf{p}_1}$, in Eq. (11a) can be expressed as the sum of all reducible and irreducible diagrams connecting the upper and lower parts of the diagram. The reducible diagrams can be split into two distinct parts by drawing a vertical line segment through the diagram. We can define an irreducible four-point vertex function by $U_{\mathbf{p}\mathbf{p}_1}$ that contains only irreducible diagrams by,

$$\Gamma_{\mathbf{p}\mathbf{p}_1}(\mathbf{k},\omega|E) = U_{\mathbf{p}\mathbf{p}_1}(\mathbf{k},\omega|E) + \int d\mathbf{p}' U_{\mathbf{p}\mathbf{p}'}(\mathbf{k},\omega|E)$$
$$\cdot G(p', E_+ + i0)G(p', E_- - i0)\Gamma_{\mathbf{p}'\mathbf{p}_1}(\mathbf{k},\omega|E) \; . \tag{11b}$$

With this, Eq. (11a) integrated over $\mathbf{p}_1$ can be written,

$$\phi_{\mathbf{p}}(\mathbf{k},\omega|E) \equiv \int d\mathbf{p}_1\phi_{\mathbf{p}\mathbf{p}_1}(\mathbf{k},\omega|E) = G(p_+, E_+ + i0)G(p_-, E_- - i0)$$
$$\cdot [1 + \int d\mathbf{p}_2 U_{\mathbf{p}\mathbf{p}_2}(\mathbf{k},\omega|E)\phi_{\mathbf{p}_2}(\mathbf{k},\omega|E)] \; . \tag{11c}$$

Equation (7c) is now given by,

$$[P_{\omega,k}(E)] = \int \frac{d\mathbf{p}}{(2\pi)^d} \phi_{\mathbf{p}}(\mathbf{k},\omega|E) \; . \tag{11d}$$

The first few diagrammatic contributions to $U_{\mathbf{p}\mathbf{p}_2}$ are given in Fig. 4.

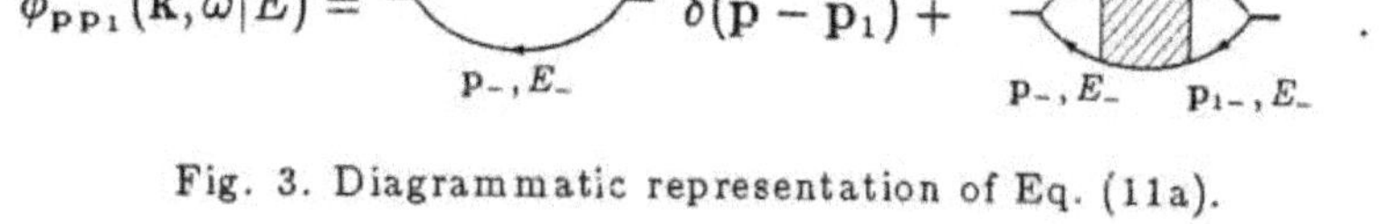

$$\phi_{\mathbf{p}\mathbf{p}_1}(\mathbf{k},\omega|E) = \quad \delta(\mathbf{p} - \mathbf{p}_1) + \quad .$$

Fig. 3. Diagrammatic representation of Eq. (11a).

Finally, Eq. (11c) can be written in the form of a generalized Boltzmann equation as, from Eq. (10a),

$$[2E\omega - 2c^2 \mathbf{k} \cdot \mathbf{p} + \Sigma^+_{p_+}(E_+) - \Sigma^-_{p_-}(E_-)]\phi_{\mathbf{p}}(\mathbf{k},\omega|E)$$

$$= \Delta G(p_+,p_-,E_+,E_-) + \Delta G(p_+,p_-,E_+,E_-)\int d\mathbf{p}_2$$

$$\cdot U_{\mathbf{p}\mathbf{p}_2}(\mathbf{k},\omega|E)\phi_{\mathbf{p}_2}(\mathbf{k},\omega|E) , \tag{12a}$$

with,

$$\Delta G(p_+,p_-,E_+,E_-) = G(p_+,E_+ + i0) - G(p_-,E_- - i0) . \tag{12b}$$

In our explicit calculations we take $E > 0$. The results for $E < 0$ can be obtained from these results.

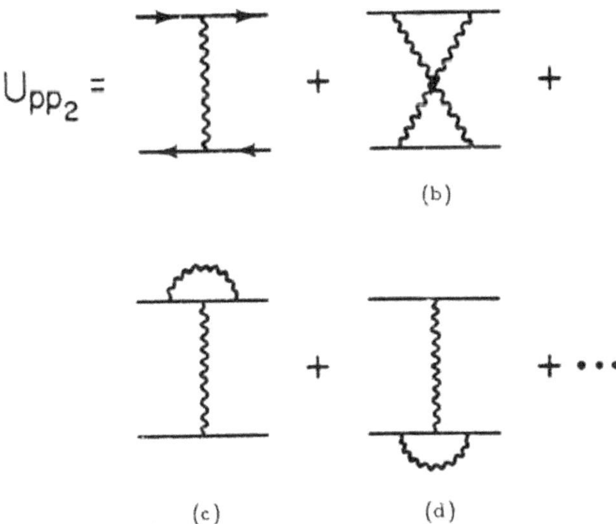

Fig. 4. First few diagrams that contribute to the irreducible four-point vertex function $U_{\mathbf{p}\mathbf{p}_2}$.

## 2.4. Low Density or Boltzmann Approximation

In order to develop the $SC$ theory of localization we first need to construct a low density theory about which we can expand. That is the aim of this subsection. In general we are interested in $\phi$, or $[P]_{\mathrm{av}}$, for small $k$ and $\omega$, or at large distances and long times. We first calculate the averaged $GF$ to lowest order in the density of scatterers. The dimensionless parameter that is assumed to be small is $n_2^* = \lambda/l_T \simeq c/El_T$. Here $\lambda(= 2\pi c/E)$ is the wavelength of the sound waves and $l_T$ is the (low density) transport mean free path of the sound wave between interactions with the randomly positioned scatterers. We have denoted this small parameter by $n_2^*$ to distinguish it from the excluded volume small parameter, $n_1^*$, introduced in Sec. 2.1. $n_2^*$ is the usual expansion parameter in localization physics. In general we will find that localization effects are important when $n_2^* \sim 0(1)$. In the electron localization literature this condition is called the Ioffe-Regel criterion.[2] Below we precisely define $l_T$ in terms of the low density diffusion coefficient.

The leading approximation to the self-energy for $n_2^* < 1$ is given in Fig. 2a, and its analytic value is,

$$\Sigma_p^{\pm}(E) = n(2\pi)^d \langle \mathbf{p}|\widehat{T}^{\pm}(E)|\mathbf{p}\rangle \equiv \Gamma_p(E) \pm i\gamma_p(E) \ . \tag{13}$$

For future reference we note that for small $k$ and $\omega$, and small $n$, the function $\Delta G$ in Eq. (12b) is given approximately by,

$$\Delta G(p_+, p_-, E_+, E_-) = 2i\pi\delta(E^2 - c^2 p^2) \ . \tag{14a}$$

We will also need,

$$\Sigma_p^{\pm}(E)|_{p=E/c} \equiv \Sigma^{\pm}(E) \equiv \Gamma(E) \pm i\gamma(E) \ . \tag{14b}$$

We next solve Eq. (12a) to lowest order in the density of scatterers. The resulting equation has the structure of a Boltzmann equation. We then derive Eq. (6b) with $D$ given by its Boltzmann equation value which we denote by $D_B$.

To lowest order in the density the irreducible four-point vertex function, $U_{\mathbf{p}\mathbf{p}_2}$, is approximated by its Boltzmann value $U^B_{\mathbf{p}\mathbf{p}_2}$, given by the diagram in Fig. 4a,

$$U^B_{\mathbf{p}\mathbf{p}_2}(\mathbf{k},\omega|E) = n(2\pi)^d \langle \mathbf{p}_+ | \widehat{T}^+(E_+) | \mathbf{p}_{2+} \rangle$$
$$\cdot \langle \mathbf{p}_{2-} | \widehat{T}^-(E_-) | \mathbf{p}_- \rangle \ . \tag{15a}$$

Also, to lowest order in the density we can neglect the $k$ and $\omega$ dependence in the $\Sigma$'s and $U$ functions in Eqs. (12), and we can use Eq. (14a). With these approximations the Boltzmann approximation for $\phi(\equiv \phi^B)$ is given by the solution of the equation,

$$[-2iE\omega + 2ic^2\mathbf{k}\cdot\mathbf{p} + 2\gamma_p(E)]\phi^B_{\mathbf{p}}(\mathbf{k},\omega|E) = 2\pi\delta(E^2 - c^2p^2)$$
$$\cdot [1 + n(2\pi)^d \int d\mathbf{p}_2 |\langle \mathbf{p}|\widehat{T}(E)|\mathbf{p}_2\rangle|^2 \phi^B_{\mathbf{p}_2}(\mathbf{k},\omega|E)] \ . \tag{15b}$$

The solution to Eq. (15b) for small $\mathbf{k}$ and $\omega$ can be readily constructed by noting that it is an integral equation only in the unit vector $\widehat{\mathbf{p}} \equiv \mathbf{p}/|\mathbf{p}|$. Defining the self-adjoint collision operator,

$$n\Lambda(\widehat{\mathbf{p}}) = n\frac{(2\pi)^{d+1}}{2c^2}\left(\frac{E}{c}\right)^{d-2}\int d\widehat{\mathbf{p}}_2\sigma(\widehat{\mathbf{p}}\cdot\widehat{\mathbf{p}}_2)[P(\widehat{\mathbf{p}}\widehat{\mathbf{p}}_2) - 1] \tag{16a}$$

where,

$$\sigma(\widehat{\mathbf{p}}\cdot\widehat{\mathbf{p}}_2) = |\langle \mathbf{p}|\widehat{T}(E)|\mathbf{p}_2\rangle|^2 \Big|_{p=p_2=E/c} \tag{16b}$$

and $P(\widehat{\mathbf{p}}\widehat{\mathbf{p}}_2)$ is a permutation operator that changes $\widehat{\mathbf{p}}$ to $\widehat{\mathbf{p}}_2$ when it acts on an arbitrary function $f(\widehat{\mathbf{p}})$. With this and defining,

$$\phi^B_{\mathbf{p}}(\mathbf{k},\omega|E) \equiv 2\pi\delta(E^2 - c^2p^2)\phi_{\widehat{\mathbf{p}}}(\mathbf{k},\omega|E) \ , \tag{16c}$$

Eq. (15b) gives,

$$[-2iE\omega + 2ic^2\mathbf{k}\cdot\mathbf{p} - n\Lambda(\widehat{\mathbf{p}})]\phi_{\widehat{\mathbf{p}}}(\mathbf{k},\omega|E) = 1 \ . \tag{16d}$$

In giving Eq. (16d) we have used an optical theorem[66] relating the imaginary part of the self-energy, $\gamma$, and the angular integral of Eq. (16b). It is this property that leads to a collision operator that has an eigenfunction, a constant, with zero eigenvalue which in turn leads to a conservation equation for $\phi^B$ and the diffusive pole result given by Eq. (6b). The generalization of this property for the exact equation given by Eq. (12a) is the Ward identity,[15]

$$\int d\mathbf{p}\, \Delta G(p_+, p_-, E_+, E_-) U_{\mathbf{p p_2}}(\mathbf{k}, \omega | E) = \Sigma^+_{p_{2+}}(E_+) - \Sigma^-_{p_{2-}}(E_-) \ .$$

$$(17)$$

Physically, Eq. (17) implies in general that there is energy conservation and that $[P_{\omega,\mathbf{k}}(E)]_{\text{av}}$ is diffusive for $k, \omega \to 0$.

For small $k$ and $\omega$, Eq. (16d) can be readily solved and one obtains,

$$\phi_{\widehat{\mathbf{p}}}(\mathbf{k}, \omega | E) \simeq \frac{1}{2E[-i\omega + D_B(E)k^2]} \tag{18a}$$

with $D_B(E)$ the Boltzmann diffusion coefficient. In $d$-dimensions,

$$D_B(E) = -\frac{2Ec^2}{d^2}[\langle \widehat{p}_x | n\Lambda(\widehat{\mathbf{p}}) | \widehat{p}_x \rangle]^{-1} \ . \tag{18b}$$

In general $D_B(E) > 0$ because the spectrum of the operator $\Lambda$ is negative semi-definite. Finally, Eqs. (16c), (18a) and (11d) yield,

$$[P^B_{\omega,k}(E)]_{\text{av}} = \frac{\pi \Omega_d}{2(2\pi)^d c^3}\left(\frac{E}{c}\right)^{d-3}\frac{1}{[-i\omega + D_B(E)k^2]} \ , \tag{19a}$$

where $\Omega_d$ is the area of a $d$-dimensional unit sphere. This result is valid only when $kl_T(E) < 1$, where $l_T(E)$ is the transport, $E$-dependent, mean free path given approximately by

$$D_B(E) = \frac{cl_T(E)}{d} \ . \tag{19b}$$

At length scales shorter than $l_T(E)$, $[P^B_{\omega,k}(E)]_{\text{av}}$ is not diffusive.

## 2.5. Self-Consistent Theory of Localization

The $SC$ theory of localization is motivated by examining density corrections to the Boltzmann diffusion coefficient given by Eq. (18b). In particular, we take into account the set of maximally crossed diagrams (and their *SC* generalizations) important for electron localization. Below we briefly discuss why these diagrams are important for localization in general. It is first convenient to approximately solve Eq. (12a), obtain Eq. (6b) in general, and also obtain a general equation for $D(E,\omega)$ directly in terms of the irreducible four-point vertex function.

The solution procedure is simple and is already in the literature for the electron localization problem,[15,69] The basic idea is to expand $\phi_{\mathbf{p}}(\mathbf{k},\omega|E)$ in terms of spherical tensors in p-space. It is easy to show that only the scalar and vector contributions are important for small $\mathbf{k}$ and $\omega$. Carrying out this procedure one finds, for $\mathbf{k}$ and $\omega$ small,

$$[P_{\omega,k}(E)]_{\mathrm{av}} = \frac{\pi\Omega_d}{2(2\pi)^d c^3}\left(\frac{E}{c}\right)^{d-3}\frac{1}{\left[-i\omega + D(E,\omega)k^2\right]} , \qquad (20a)$$

with

$$\frac{1}{D(E,\omega)} = \frac{1}{D_B(E)} + \frac{d^2}{2c\Omega_d\pi}\left(\frac{c}{E}\right)^{d+1}$$
$$\cdot \int d\,\mathbf{p}\,d\,\mathbf{p}_2 p_x p_{2x} \Delta G(p,E)\Delta G(p_2,E)\delta U_{\mathbf{p}\mathbf{p}_2}(0,\omega|E) \qquad (20b)$$

where,

$$\Delta G(p,E) \equiv \Delta G(p,p,E,E) , \qquad (20d)$$

and

$$\delta U_{\mathbf{p}\mathbf{p}_2}(0,\omega|E) = U_{\mathbf{p}\mathbf{p}_2}(0,\omega|E) - U^B_{\mathbf{p}\mathbf{p}_2}(0,\omega|E) . \qquad (20e)$$

In giving Eq. (20b) we have neglected all the $\omega$ and $k$ dependencies that are nonsingular for $k,\omega \to 0$. We have also used that the maximally crossed diagrams are not singular functions of $k$ to replace $U_{\mathbf{p}\mathbf{p}_2}(\mathbf{k},\omega|E)$ by $U_{\mathbf{p}\mathbf{p}_2}(0,\omega|E)$. This last point is actually a problem in the *SC* theory of localization. In more systematic theories of localization, $|\mathbf{k}|$ is a scaling variable and this should be reflected by singularities in $|\mathbf{k}|$ in the perturbation theory. We come back to this point in Sec. 7.

We next examine disorder or density corrections to $D$ when expanded around $D_B$ using Eq. (20b). If one assumes that a power series expansion is valid then the individual terms are found to be divergent, for $\omega \to 0$, after some order in the density that depends on the dimension of the system.[69-71] Amongst the most divergent terms are the set of maximally crossed diagrams $(MCD)$ given in Fig. 5, which are divergent for $|\mathbf{p}+\mathbf{p}_2| \to 0$. It is this class of diagrams which are assumed to cause localization in the $SC$ theory of localization. Physically, the $MCD$'s can be thought of as representing the constructive inteference of conjugate waves travelling in opposite directions along the same closed path. The probability of return of a wave to the neighborhood of the source is enhanced by these coherence effects.[2,31] These scattering processes are therefore especially relevant for the description of localization in wave systems in general.

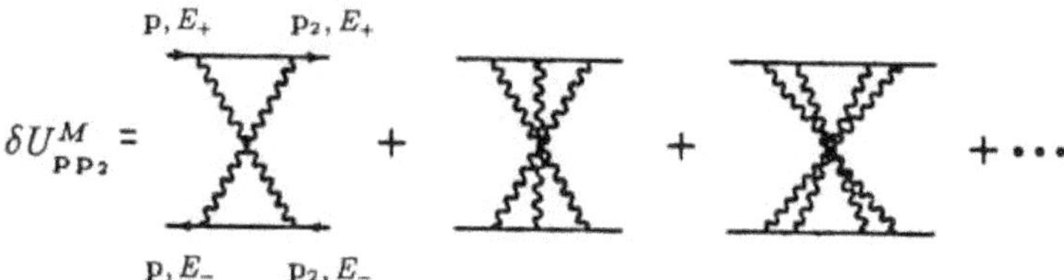



Fig. 5. The maximally crossed diagrams.

The whole class of $MCD$ can be summed by using time reversal invariance, which in this context is usually called a reciprocal relation.[72] Analytically this symmetry implies, for example,

$$[\langle \mathbf{p}|\widehat{G}^+_{E+}|\mathbf{p}_2\rangle\langle \mathbf{p}_2|\widehat{G}^-_{E-}|\mathbf{p}\rangle]_{\mathrm{av}}$$
$$= [\langle \mathbf{p}|\widehat{G}^+_{E+}|\mathbf{p}_2\rangle\langle -\mathbf{p}|\widehat{G}^-_{E-}|-\mathbf{p}_2\rangle]_{\mathrm{av}} .$$

Using this symmetry we can rotate the bottom line in Fig. 5 by $180°$ and relate the resulting infinite series to the Boltzmann diffusive pole discussed in Sec. 2.4. For small $\omega$ and $q = |\mathbf{p}+\mathbf{p}_2|$, one then finds that

$\delta U^M_{\mathbf{p}\mathbf{p}_2}$ ($M$ denotes $MCD$) is proportional to a hydrodynamic diffusive pole,

$$\delta U^M_{\mathbf{p}\mathbf{p}_2}(0,\omega|E) = \frac{2\gamma^2(E)c}{\Omega_d\pi}\left(\frac{c}{E}\right)^{d-1}$$
$$\cdot\frac{1}{-i\omega + D_B(E)(\mathbf{p}+\mathbf{p}_2)^2}[1 + 0(|\mathbf{p}+\mathbf{p}_2|^2,\omega)] \tag{21}$$

where $\gamma(E) = \gamma_{p=E/c}(E)$.

Inserting Eq. (21) into Eq. (20b) and evaluating the prefactor of the hydrodynamic pole to lowest order in $n$ yields,

$$\frac{1}{D(E,\omega)} = \frac{1}{D_B(E)} + \frac{d\gamma(E)}{\pi\Omega_d c^2}\left(\frac{c}{E}\right)^d\int_{q<Q} d\mathbf{q}\frac{1}{-i\omega + D_B(E)q^2} \ . \tag{22}$$

Here $Q$ is a hydrodynamic cutoff which we physically expect to be of the order of $l_T^{-1}(E)$. The $Q$ is needed because in giving Eq. (21) we have assumed that $|\mathbf{p}+\mathbf{p}_2|/Q < 1$. The appearance of an imposed ultraviolet cutoff is actually another problem with the $SC$ approach to localization. In a more complete theory it would arise naturally from the short distance behavior of the $GF$ and $T$-matrices. In Secs. 3 and 4 we discuss the advantages and disadvantages of various alternative cutoffs. In general critical parameters, like exponents, near the Anderson transition are independent of $Q$ but amplitudes etc. are not. We will also find that for $d \leq 2$ the localization effects are only weakly cutoff dependent and that for $d > 2$ our results are strongly cutoff dependent.

It is important to note that the second term on the right-hand side of Eq. (22) is still divergent (even after the resummation given in Fig. 5) for $d \leq 2$. The $SC$ theory of localization, for all $d$, consists of replacing the low-density hydrodynamic pole in Eqs. (21) by an exact hydrodynamic pole, with $D(E,\omega)$ replacing $D_B(E)$. Equation (22) is then replaced by,

$$\frac{1}{D(E,\omega)} = \frac{1}{D_B(E)} + \frac{d\gamma(E)}{\pi\Omega_d\tilde{c}^2}\left(\frac{\tilde{c}}{E}\right)^d\int_{q<Q} d\mathbf{q}\frac{1}{-i\omega + D(E,\omega)q^2} \ . \tag{23}$$

The speed of sound has been renormalized everywhere for self-consistency. Equation (23) is the most important equation in the $SC$ theory of localization. This equation is a closed algebraic equation for $D(E, \omega)$ in terms of which we can describe the localization properties of sound waves in disordered system. The step from Eq. (22) to Eq. (23) technically involves a further resummation of diagrams. It is at this point that the $SC$ theory of localization becomes an uncontrolled theory: We cannot justify the neglect of other diagrammatic contributions. We take the point of view that the $SC$ procedure is reasonable and we next work out the consequences of Eq. (23).

For $d = 1, 2, 3$ Eq. (23) is,

$$\frac{D(E,\omega)}{D_B(E)} = 1 - \frac{\gamma}{\pi \tilde{c} E} \left( \frac{D(E,\omega)}{-i\omega} \right)^{\frac{1}{2}} \arctan \left[ \left( \frac{D(E,\omega)}{-i\omega} \right)^{\frac{1}{2}} Q(E) \right]$$

$$(d = 1) \qquad\qquad (24a)$$

$$= 1 - \frac{\gamma(E)}{\tilde{c}^2 \pi} \left( \frac{\tilde{c}}{E} \right)^2 \log \left( \frac{Q^2(E) D(E,\omega) - i\omega}{-i\omega} \right)$$

$$(d = 2) \qquad\qquad (24b)$$

$$= 1 - \frac{3\gamma(E)}{\tilde{c}^2 \pi} \left( \frac{\tilde{c}}{E} \right)^3 Q(E)$$

$$(d = 3) \qquad\qquad (24c)$$

$$+ \frac{3\gamma(E)}{\tilde{c}^2 \pi} \left( \frac{\tilde{c}}{E} \right)^3 \left( \frac{-i\omega}{D(E,\omega)} \right)^{\frac{1}{2}} \arctan \left[ \left( \frac{D(E,\omega)}{-i\omega} \right)^{\frac{1}{2}} Q(E) \right] .$$

We next examine these equations.

For $d = 1, 2$ it is easy to show that the only physical solution to Eqs. (24a) and (24b) for $\omega \to 0$ is of the form,

$$D(E, \omega \to 0) = -i\omega \xi^2(E) + 0(\omega^2) . \qquad\qquad (25)$$

Here $\xi(E)$ is the $E$-dependent localization length that was discussed in Sec. 2.1. An equation for $\xi(E)$ can be obtained by inserting Eq. (25)

into Eqs. (24a) and (24b),

$$1 = \frac{\gamma(E)}{\pi \tilde{c} E} \xi(E) \ \arctan[\xi(E)Q(E)] \qquad (d=1) \qquad (26a)$$

$$\xi(E) = Q^{-1}(E)\{\exp[\pi E^2/\gamma(E)] - 1\}^{\frac{1}{2}} \qquad (d=2) \ . \qquad (26b)$$

From Eq. (25) we conclude that for $d \leq 2$, the sound waves are always localized no matter how weak the disorder is as long as it is finite. Also note that in the limit of zero disorder, $Q(E), \gamma(E) \to 0$ and $\xi(E) \to \infty$. In general for weak disorder the localization length can be larger than the size of an experimental system. For this case the sound waves will appear to be diffusive although in the bulk limit the theory predicts localized behavior. Since $\gamma(E)/cE$ is typically of $0[Q(E) \sim l_T^{-1}(E)]$ we note that in two-dimensions the localization length is exponentially large for weak disorder. Because $\gamma/cE \sim l_T^{-1}$ the argument of the exponential in Eq. (26b) is of $0(l_T/\lambda)$ which indicates that localization effects become important in $d = 2$ only when $n_2^* \sim 0(1)$ in agreement with the Ioffe-Regel criterion. Finally, the real part of $D(E,\omega)$ can be easily calculated and as indicated in Eq. (25) it is of $0(\omega^2)$. This is in conflict with exact $1 - d$ calculations and systematic higher dimensional results which yield Re $D(E,\omega \to 0) \sim 0[\omega^2(\log \omega)^{d+1}]$. Systematic theories predict that all states are localized for $d \leq 2$ in agreement with the $SC$ method. Exact calculations of $\xi(E)$ in one-dimensional systems give localization lengths that typically differ by a factor of two from results obtained with the $SC$ method.

For $d = 3$ the sound waves behave diffusively for weak enough disorder. For strong enough disorder there is an Anderson transition to localized behavior. To determine (within the $SC$ method) where the mobility edge is and the properties of the phase transition we first assume diffusive behavior and see where this assumption breaks down. For diffusive behavior $D(E,\omega \to 0)$ is a finite constant equal to the usual diffusion coefficient. Equation (24c) gives

$$\frac{D(E,0)}{D_B(E)} = 1 - \frac{3\gamma(E)}{\tilde{c}^2 \pi} \left( \frac{\tilde{c}}{E} \right)^3 Q(E) \ . \qquad (27a)$$

The assumption on diffusive behavior breaks down when $D(E,0) = 0$ (negative values of $D(E,0)$ are not physical). An implicit equation for the location of the mobility edge is,

$$1 = \frac{3\gamma^*(E_c)}{\tilde{c}^2\pi} \left(\frac{\tilde{c}}{E_c}\right)^3 Q^*(E_c) , \qquad (27b)$$

where $\gamma^*$ and $Q^*$ are the values of $\gamma$ and $Q$ at the mobility edge. Also note that there are two experimentally controllable parameters: The density of scatterers, equal to $n_c$ at the mobility edge, and the internal frequency, equal to $E_c$ at the mobility edge. In general Eq. (27b) has many solutions and examples will be given in Sec. 3. Near a mobility edge Eq. (27a) predicts,

$$D(E,0) \sim |E - E_c| \sim |n - n_c| , \qquad (27c)$$

depending on whether $E$ or $n$ is varied. Note that the exponent defined by Eq. (27c) is unity in agreement with more systematic theories (in $d = 3$) (Ref. 2). Finally, the second term in Eq. (27a) is of $0([\lambda/l_T]^2) = 0(n_2^{*2})$ which indicates localization effects are important when $n_2^* \sim 0(1)$ in agreement with the Ioffe-Regel criterion. We also note that in an exact perturbation expansion the leading correction to $D_B(E)$ in Eq. (27a) is of $0(n_2^*)$ (Ref. 71).

We next examine the localization length in the localized regime as the mobility edge is approached. Using Eq. (25) in Eq. (24c) yields,

$$\xi(E) = [\arctan(\xi(E)Q(E))] \frac{\tilde{c}^2\pi}{3\gamma(E)} \left(\frac{E}{\tilde{c}}\right)^3$$
$$\times \left[\frac{3\gamma(E)}{c^2\pi} \left(\frac{\tilde{c}}{E}\right)^3 Q(E) - 1\right]^{-1} . \qquad (28a)$$

Near a mobility edge Eq. (28a) predicts,

$$\xi(E) \sim |E - E_c|^{-1} \sim |n - n_c|^{-1} . \qquad (28b)$$

The exponent of unity in $d = 3$ is in agreement with more systematic theories.[2]

We conclude this section by noting that the $SC$ method predicts universal parameters like exponents as well as nonuniversal amplitudes. This is to be contrasted with the more controlled field theoretic approach which only gives exponents which are reliable only near two dimensions.

## 3. APPLICATIONS OF THE SC METHOD

### 3.1. Introduction: The Models

In Sec. 2 we presented the $SC$ method, which yields explicit results for various quantities associated with the diffusion of energy in a disordered system. Although the accuracy of the method is hard to ascertain *a priori*, the good agreement with the results obtained from other, more elaborate, procedures — in the cases for which these are available — suggests the reliability of the $SC$ predictions.

In this section we apply the $SC$ formalism to obtain concrete results for some reasonable models of acoustic disorder, making full use of the ability of the method to provide predictions in the whole frequency range. We consider a uniform lossless fluid, of mass density $\rho_0$ and compressibility $\kappa$, which contains a random array of identical spheres of radius $a$. The number density of the scatterers is $n$; hence they occupy a volume fraction $n^* = (4/3)\pi a^3 n$. (As mentioned in Sec. 2 we neglect corrections that arise due to scatterer overlap).

We will also consider the special case of propagation normal to the axes of infinite, parallel, cylindrical, scatterers. The problem is then effectively two-dimensional and the appropriate volume fraction is $n^* = \pi a^2 n$. One should also note the relationship, $n_1^* = 2^d n^*$, between the excluded volume and the volume fraction dimensionless densities.

The pressure waves can be described using a velocity potential $\phi(\mathbf{x}, t)$ which in the absence of scatterers satisfies the wave equation (1a). The speed of sound in Eq. (1a) can be expressed in terms of $\rho_0$ and $\kappa$ by $c = (1/\kappa\rho_0)^{1/2}$.

We will consider three types of scatterers[73]:

(1) Hard (Neumann) scatterers: The component of the fluid velocity normal to the surface of the scatterers must vanish, and we have the condition

$$\frac{\partial \phi}{\partial r} = 0 \text{ at } r = a .\tag{29a}$$

(2) Soft (Dirichlet) scatterers: The pressure vanishes at the surface of the scatterers. This leads to

$$\phi = 0 \text{ at } r = a .\tag{29b}$$

Note that this is also the condition satisfied by the electronic wave function at the surface of an impenetrable sphere.

(3) Permeable scatterers: They are characterized by a mass density $\rho_s$ and a compressibility $\kappa_s$. (The subscript $s$ will always indicate a magnitude associated to the scatterers.) We will also find it convenient to describe the scatterers using the density ratio, $\Delta = \rho_0/\rho_s$, and the index of refraction ratio, $M = c/c_s = (\rho_s\kappa_s/\rho_0\kappa)^{1/2}$. The boundary conditions at their surfaces are determined by the requirements of continuity in the pressure and normal component of the fluid velocity, i.e.,

$$\begin{aligned}\phi_s &= \phi\Delta\\[4pt]\frac{\partial \phi_s}{\partial r} &= \frac{\partial \phi}{\partial r}\end{aligned} \qquad \text{at } r = a .\tag{29c}$$

It should be mentioned that for solid scatterers, a description in terms of $\kappa_s$ and $\rho_s$ is not complete. In this case two different wave numbers, corresponding to compression and shear, enter the description and the formulas become considerably more involved. Our description is accurate for fluid scatterers and for solids (for example, silicone rubber), for which there is negligible shear conversion at the scatterer surface.[74]

Anderson localization arises from the interaction of the wave-like excitations in the fluid with a many-scatterer system. However, the nature of the energy diffusion (or lack thereof) in a given system is determined by the nature of the individual scattering events. Let us note, for example, that those $T$-matrix elements whose evaluation is required in order to apply the results of Sec. 2 to specific systems can be given in terms of the scattering amplitudes $\phi^+(\theta)$ for the individual scatterers. If we consider a spherical scatterer and choose the polar axis to be parallel to the direction of propagation of the incident wave, we have[75]

$$\langle \mathbf{p}|\widehat{T}^+(E)|\mathbf{p}\rangle|_{p=q=E/c} = -\frac{c^2}{2\pi^2}\phi^+(\theta) \ . \tag{30}$$

Using Eqs. (13) and (30) and applying the optical theorem,

$$S(E) = \frac{4\pi c}{E}\,\operatorname{Im}\,\phi^+(\theta) \ , \tag{31}$$

we can express the imaginary part of the self-energy in terms of the total cross section $S(E)$:

$$\gamma(E) = cnES(E) \ . \tag{32}$$

Since $\gamma(E)$ plays a crucial role in the localization equations, it is necessary to carefully compute $S(E)$ for the cases of interest. For a given value of $n^*$, this quality also provides a measure of the effective strength of the disorder, and a description of its properties leads to a better physical intuition of some of the processes involved.

In the next three subsections we will evaluate $S(E)$ and two other physically meaningful quantities that appear in the expression obtained in Sec. 2: the renormalized speed of sound $\tilde{c}(E)$ and the transport mean free path $l_T(E)$.

## 3.2. The Total Scattering Cross Section

We consider separately the problem of the sphere and the infinitely long cylinder.

3.2.1. *The Sphere.* Let us first define some auxiliary functions:

$$\delta_l(\eta_s) = M\Delta j_l'(\eta_s)[j_l(\eta_s)]^{-1} \ , \tag{33a}$$

$$\alpha_l = j_l'(\eta) - \delta_l(\eta_s)j_l(\eta) \ , \tag{33b}$$

and

$$f_l = y_l'(\eta) - \delta_l(\eta_s) y_l(\eta) \ . \tag{33c}$$

Here $\eta = Ea/c, \eta_s = M\eta, j_l$ and $y_l$ are spherical Bessel functions and the primes denote derivatives with respect to the argument.

The angle distribution factor (or scattering amplitude) $\phi^+(\theta)$ for a plane wave scattered by a single permeable scatterer can be obtained by solving a standard boundary-value problem. It is not difficult to show that,[76,77]

$$\phi^+(\theta) = \frac{i}{k} \sum_{l=0}^{\infty} \frac{(2l+1)\alpha_l}{\alpha_l + if_l} P_l(\cos\theta) \ . \tag{34}$$

where $P_l$'s are Legendre polynomials. The total scattering cross section can be obtained immediately by the use of Eq. (34) and the optical theorem, Eq. (31). We get,

$$S(\eta) = \frac{4\pi a^2}{\eta^2} \sum_{l=0}^{\infty} \frac{(2l+1)\alpha_l^2}{\alpha_l^2 + f_l^2} \equiv \frac{4\pi a^2}{\eta^2} S_l(\eta) \ . \tag{35}$$

The hard and soft sphere cases can be formally obtained by taking special values of the parameter $M\Delta = (\rho_0 \kappa_s / \rho_s \kappa)^{1/2}$:

  a) Neumann $(N)$ spheres: $M\Delta = 0$. This corresponds to infinitely dense or totally incompressible spheres.

  b) Dirichlet $(D)$ spheres: $M\Delta \to \infty$. This corresponds to empty or infinitely compressible spheres.

Except for the $D$ case, for which $S_D(\eta \to 0) = 4\pi a^2$, the cross section vanishes in the long-wavelength (Rayleigh) limit:

$$S(\eta) = \frac{4\pi a^2 \eta^4}{9} \left[ (M^2\Delta - 1)^2 + \frac{3(1-\Delta)^2}{(2+\Delta)^2} \right] \ . \tag{36}$$

In the $N$ problem, $S_N \simeq (7\pi/9)\pi a^2 \eta^4$. Note that in the $N$ and $D$ cases the $\eta \to 0$ limit must be taken after setting $M\Delta = 0$ or $M\Delta = \infty$.

The total scattering cross section is plotted in Fig. 6 for several types of scatterers. The smooth behavior of the Dirichlet and Neumann cross sections (repesented by the curves labelled $D$ and $N$, respectively) constrasts with the peaks that appear in the curves corresponding to compressible scatterers. The numbers above the peaks in the $M = 2$ curve denote that value of $l$ for the term that originates the peak. The peaks occur when the denominator of one of the terms in the series in Eq. (35) becomes suddenly small. The second term in these denominators is usually much larger than the first, and when it vanishes the magnitude of the scattering of the corresponding partial wave may grow enormously. Thus, the condition for the existence of a peak or resonance is $f_l(\eta) = 0$. The first peak in the curve for $\Delta = 1, M = 2$ in Fig. 6 is an exception: it is generated by the $l = 0$ partial wave even if $f_0(\eta)$ does not vanish. The reason for this maximum is the simultaneous rapid increase in $|\alpha_0|$ and decrease in $|f_0|$. The rounded peaks at $\eta = 6.34$ and $\eta = 6.92$ correspond to the second resonances of the 6th and 7th partial waves, respectively. The multiple resonances that may be present for a given $l$ are related to the successive radial nodes in each partial wave. There is only one peak corresponding to each of $l = 1, \ldots, 5$, but the second zeroes of $f_1, \ldots, f_5$ generate changes in the slope at the right of some of main peaks.

It can also be seen[78] that an increase in $M$ will cause the first peak in $S(\eta)$ not only to become taller, but also to move towards lower frequencies. Since localization is strengthened at longer wavelengths, this characteristic will be seen to be crucial in establishing that high $M$ scatterers are to be preferred in the search for localization effects.

The strongest effect is obtained for low speed ($M > 1$) scatterers: The higher the $M$, the more abundant, taller and thinner the resonances. The effect of high speed ($M < 1$) scatterers is much weaker and probably quite uninteresting from the viewpoint of localization-related phenomena. It is also generally true that fluctuations in the compressibility cause stronger scattering than fluctuations in the density.

The connection between the scatterer resonances, the scatterer eigenmodes, and the creeping waves on the scatterer surfaces have

                 *C. A. Condat & T. R. Kirkpatrick*

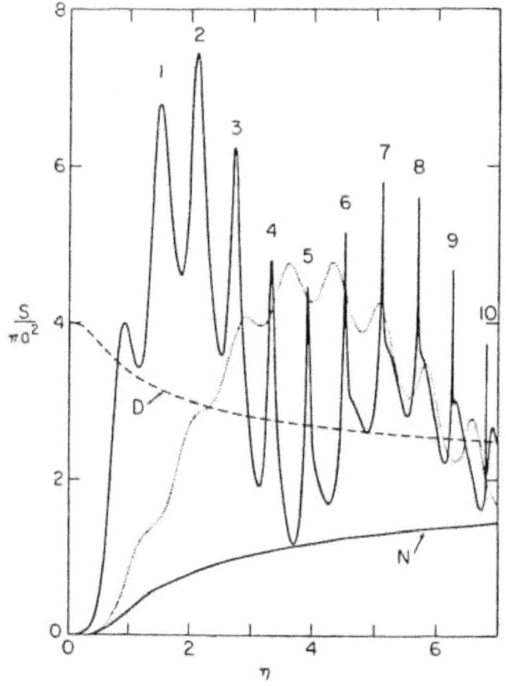

Fig. 6. Total cross section for a sphere as a function of frequency. The solid and dotted oscillating lines correspond to $M = 2$ and $M = 1.5$, respectively; $\Delta = 1$ in both cases.

been investigated in a series of papers by Überall, Flax, and co-workers.[74,79-83] They find that the resonant behavior can be described in terms of the constructive superposition of circumferential waves. Equation (31) suggests that the peaks in $S(\eta)$ are due to the constructive interference between the surface wave and the forward diffracted wave.

The problem of resonant electromagnetic scattering has a longer history.[84-86] Detailed analyses of the structure and origin of the resonances can be found in the papers by Chýlek *et al.*[87] and Probert-Jones.[88] The cases of the $D, N$, and electromagnetic scattering by cylinders, sphere, and other shapes are considered in the book edited by Bowman, Senior, and Uslenghi.[89]

3.2.2. *The Cylinder.* In this case it is convenient to define,

$$\Delta_l(\eta_s) = M\Delta J_l'(\eta_s)[J_l(\eta_s)]^{-1} \,, \tag{37a}$$

$$\beta_l = J_l'(\eta - \Delta_l(\eta_s)J_l(\eta) \,, \tag{37b}$$

and

$$g_l = Y_l'(\eta) - \Delta_l(\eta_s)Y_l(\eta) \,. \tag{37c}$$

Here $J_l$ and $Y_l$ are cylindrical Bessel functions and the primes denote derivatives with respect to the argument. The formulas for the $N$ cases are obtained by writing $\beta_l = J_l'$ and $g_l = Y_l'$; those for the $D$ case by letting $\beta_l = J_l$ and $g_l = Y_l$.

The total cross section per unit length for a plane wave normally incident on an infinitely long cylinder is[77]

$$S(\eta) = \frac{4a}{\eta} \sum_{l=0}^{\infty} (2 - \delta_{l,0}) \frac{\beta_l^2}{\beta_l^2 + g_l^2} \equiv \frac{4a}{\eta} T_1(\eta) \,. \tag{38}$$

Its long wavelength limit is

$$S(\eta) \simeq \frac{a\pi^2\eta^3}{4} \left[ (M^2\Delta - 1)^2 + \frac{2(1 - \Delta)^2}{(1 + \Delta)^2} \right] \,. \tag{39}$$

In particular, in the $N$ case, $S_N = (3\pi^2/4)a\eta^3$. For large values of the product $M^2\Delta$ the $\eta^3$ behavior may be apparent only at very long wavelengths, since to obtain Eq. (39) we have used the condition $\eta^2 \ln \eta \ll 2(M^2\Delta)^{-1}$. In the Dirichlet case one obtains the limit form $S_D = \pi^2 a[\eta \ln^2(\eta)]^{-1}$. The reason for the divergence in $S_D$ is discussed by Morse and Feshbach.[72]

Equation (38) closely resembles its three-dimensional counterpart, Eq. (35). The analogy becomes evident when looking at Fig. 7, where we have represented the same cases as in Fig. 6. The number above each peak in the $(\Delta = 1, M = 2)$ curve denotes the value of $l$ for the term that originates the peak. The peaks for the purely compressible $(\Delta = 1)$ scatterers are due to zeroes in the functions $g_l(\eta)$. Again, the first peak in the $M = 2$ case is an exception, being due to the same origin as its $3 - d$ counterpart. In Fig. 7 we have also represented the curve corresponding to $(\Delta = 0.5, M = 2^{1/2})$ (i.e., $\rho_s = 2\rho_0, \kappa_s = \kappa$), which shows a broad maximum around $\eta \simeq 4$. The small value of $S$ for $\eta \sim 1$ will certainly imply very long localization lengths for random arrays of these "pure density" scatterers.

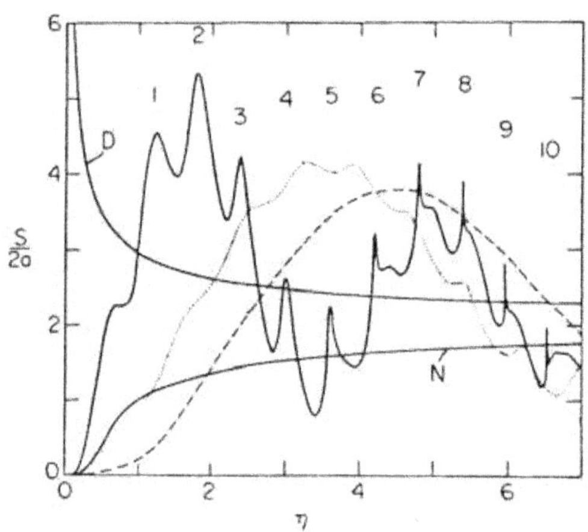

Fig. 7. Total cross section per unit length for a cylinder of radius $a$ and normal incidence. The solid, dotted and dashed lines represent the results for $(\Delta = 1, M=2)$, $(\Delta = 1, M = 1.5,)$, and $(\Delta = 0.5, M = 2^{1/2})$, respectively.

Note that, as it is the case for the scattering by a sphere, the center of gravity of the curves for $S(\eta)$ always goes to twice the geometrical cross section in the high $\eta$ limit.

## 3.3. The Effective Speed of Sound

In Sec. 2 we have mentioned [see the discussion following Eq. (10a)] that the real part of the self-energy renormalizes the speed of sound. The resulting effective speed of sound $\tilde{c}$ can be evaluated at low $n^*$ considering only the first diagram in Fig. 2. We obtain,

$$\frac{1}{\tilde{c}^2} = \frac{1}{c^2} \left[ 1 - \frac{2d\pi^{d-1}n^*}{c^2\eta^2 a^{d-2}} \operatorname{Re}\langle \mathbf{p}|\widehat{T}^+(E)|\mathbf{p}\rangle|_{p=E/c} \right] . \tag{40}$$

Let us now specialize to the cases of spheres and cylinders.

3.3.1. *Spheres.* The explicit form for Eq. (40) is obtained by writing the diagonal elements of the scattering matrix in terms of the forward scattering amplitude. We do this with the help of Eq. (30). This yields,

$$\frac{1}{\tilde{c}^2} = \frac{1}{c^2}\left[1 + \frac{3n^*}{\eta^3}\sum_{l=0}^{\infty}\frac{(2l+1)\alpha_l f_l}{\alpha_l^2 + f_l^2}\right].\tag{41}$$

Since the denominators coincide with those of Eq. (35), we expect the structure of the function $\tilde{c}(\eta)$ to closely resemble that obtained for the total cross section. However, there is an important difference: The functions $f_l(\eta)$ appear now also as factors in the numerator. Thus, instead of a peak, we find a minimum immediately followed by a maximum, the contribution of the "critical $l$" term vanishing in between. These features become rapidly narrower as we increase $l$ at fixed $m$. Their width follows the pattern of the widths of the resonances in $S(\eta)$.

We have plotted $\tilde{c}(\eta)$ in Fig. 8 for propagation in a medium containing a random distribution of the spherical scatterers considered in Fig. 6. It is clear that $\tilde{c} \to c$ at high values of $\eta$. In the limit $\eta \to 0$, we find

$$\tilde{c}(0) = c\left\{1 + n^*\left[M^2\Delta - 1 + \frac{3(1-\Delta)}{2+\Delta}\right]\right\}^{-1/2}.\tag{42}$$

In particular, in the $N$ case, $\tilde{c}_N(\eta = 0) = c(1 + n^*/2)^{-1/2}$. The $D$ case must be computed separately, yielding for small $\eta$

$$\tilde{c}_D \sim c(1 - 3n^*/\eta^2)^{-1/2}.\tag{43}$$

We see that for $\eta < (3n^*)^{1/2}, \tilde{c}_D$ is not real. This would imply the existence of a gap in the effective propagation spectrum: excitations with wavelengths larger than $\lambda_c = 2\pi a(3n^*)^{-1/2}$ cannot propagate. However, we have to point out that: (a) The strength of the low frequency $D$ scattering makes our low $n^*$ approximation inaccurate, even for very small values of $n^*$. (b) It is not clear that the interpretation of $\tilde{c}_D$ as a speed of sound is physically correct: for example, using Eq. (43) into

(10a) we can see that a wave equation is not recovered in the $\eta \to 0$ limit. In spite of these difficulties, we are still calling $\tilde{c}_D$ a "renormalized speed of sound" in order to compare and show explicitly the crucial importance of the choice of the boundary conditions. It is also possible to argue that the nodes in the wave function at the surface of the Dirichlet spheres will make long-wavelength transport impossible.

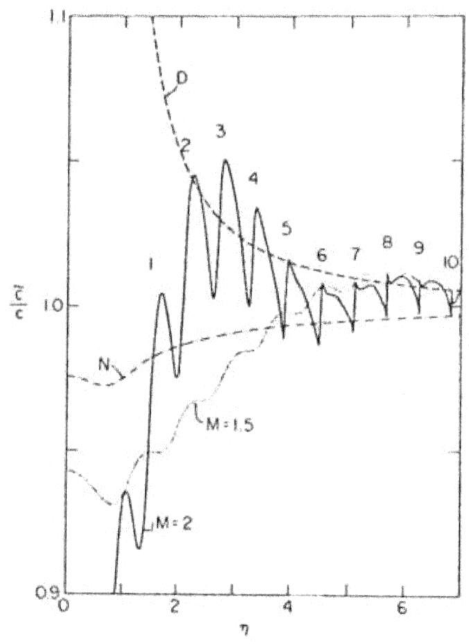

Fig. 8. Renormalized speed of sound as a function of frequency in a medium containing a random distribution of spheres for the cases considered in Fig. 6. The scatterer volume fraction is $n^* = 0.1$.

Let us apply Eq. (42) to the case of bubbly water, for which $\kappa_s \gg \kappa$ and $\rho_s \ll \rho$. We assume that the bubbles are not too small, so surface tension and non-adiabaticity effects can be ignored. If $n^*$ is small enough Eq. (42) can be further approximated by

$$\tilde{c}(0) \simeq c \left[1 - \frac{n^* \kappa_s}{2\kappa}\right] = c \left[1 - \frac{3n^* c^2}{2a^2 E_R^2}\right] , \qquad (44)$$

where $E_R$ is the resonant frequency of the bubbles. Equation (44) is the known result for the effective speed of sound in bubbly water at low frequencies, $E \ll E_R$ (Ref. 90).

An important point should be stressed here. In the present approximation for the self-energy, the dependence on the volume fraction is linear and thus a variation in $n^*$ will modify only the magnitude of the features in the curves for $\tilde{c}(\eta)$. However, as $n^*$ is increased, we should also expect a continuous smoothing out of these features; the reason for this is that, with increasing $n^*$, the scattering mean free path decreases and more and more scattering events occur in the "near zone" of previous events. Our linear approximation cannot adequately account for these processes.

3.3.2. *Cylinders.* In this case, Eq. (40) yields

$$\frac{1}{\tilde{c}^2} = \frac{1}{c^2} \left[ 1 + \frac{4n^*}{\pi \eta^2} \sum_{l=0}^{\infty} \frac{(2 - \delta_{l,0})\beta_l g_l}{\beta_l^2 + g_l^2} \right] . \tag{45}$$

We have plotted $\tilde{c}(\eta)$ in Fig. 9 for the scatterers studied in Fig. 7. We have eliminated for clarity the uninteresting ($\Delta = 0.5, M = 2^{1/2}$) curve. Note that, as in the case of the spheres, the $N$ curve has a rounded minimum below $\eta = 1$. Again, $\tilde{c}(\eta) \to c$ at high $\eta$ in all cases. The $\eta = 0$ limit is obtained easily:

$$\tilde{c}(0) = c \left\{ 1 + n^* \left[ M^2 \Delta - 1 + \frac{2(1 - \Delta)}{1 + \Delta} \right] \right\}^{-1/2} , \tag{46}$$

with $\tilde{c}_N(0) = (1 + n^*)^{-1/2}$. In the Dirichlet case we get, for small $\eta$,

$$\tilde{c}_D \simeq \left( 1 + \frac{2n^*}{\eta^2 \ln \eta} \right)^{-1/2} , \tag{47}$$

reflecting again the strong long-wavelength scattering. The total scattering cross section is indeed essentially proportional to the excitation wavelength. For scatterers that are less compressible than the background, ($\Delta = 1, M < 1$), the ratio $\tilde{c}/c$ has a small maximum at $\eta = 0$ and then decreases gently towards unity.

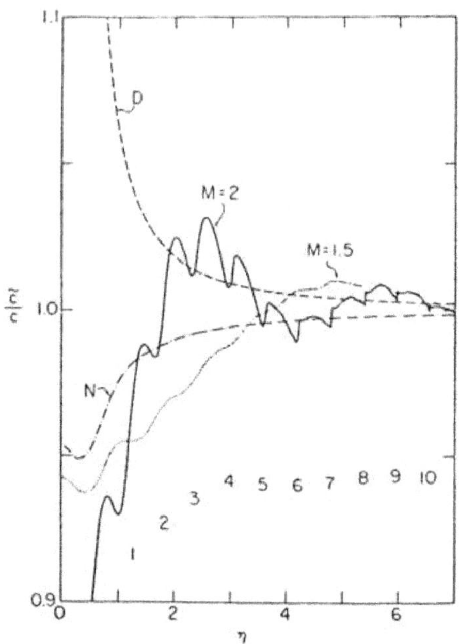

Fig. 9. Renormalized speed of sound for a medium containing a volume fraction $n^* = 0.1$ of the cylinders whose total cross sections per unit length are shown in Fig. 7.

## 3.4. The Mean Free Paths

An estimate of the possibilities of observing Anderson localization of sound in fluids containing disordered arrays of the various types of scatterers can be obtained by a simple application of the Ioffe-Regel criterion.[91] This criterion states that the excitations will be localized if $El_T/c \lesssim 1$, or

$$\left[\frac{n^* l_T(\eta)}{a}\right]\left(\frac{\eta}{n^*}\right) \lesssim 1 \ . \tag{48}$$

Here $l_T(\eta) = dD_B(\eta)/c$ is the transport mean free path (TMFP). We have already seen (Sec. 2) that the TMFP can be used to define the cutoff of the momentum integral generated by the SCT.

The TMFP is not the only MFP of interest. We can also define a scattering mean free path (SMFP) $l_{sc}(\eta)$ in the usual form,

$$l_{sc}(\eta) = [nS(\eta)]^{-1} . \tag{49}$$

In their work on electromagnetic localization, Arya, Su, and Birman[92] have chosen the SMFP to define the SCT cutoff. In a later section, we will see that $l_{sc}$ is also relevant for the determination of typical sizes in speckle patterns.

Next we make a comparative study of the TMFP and the SMFP.

3.4.1. *Spheres.* Using Eqs. (18b), (19b), and (40) we can obtain an explicit form for the three-dimensional TMFP:

$$l_T = \frac{2a^3}{3n^*} \left[ \int_0^\pi d\theta \sin\theta (1 - \cos\theta)|\phi^+(\theta)|^2 \right]^{-1} . \tag{50}$$

This is the expression for the mean free path of particles used in conventional transport theory.[93] Inserting Eq. (34) and (50), we can compute the integral explicitly. We obtain

$$l_T(\eta) = \frac{a\eta^2}{3n^*} \left\{ \sum_{l=0}^\infty \left[ \frac{(2l+1)\alpha_l^2}{\alpha_l^2 + f_l^2} - \frac{2(l+1)\alpha_l\alpha_{l+1}(\alpha_l\alpha_{l+1} + f_l f_{l+1})}{(\alpha_l^2 + f_l^2)(\alpha_{l+1}^2 + f_{l+1}^2)} \right] \right\}^{-1}$$

$$\equiv \frac{a\eta^2}{3n^* S_2(\eta)} . \tag{51}$$

The low frequency behavior of this equation is given by

$$l_T(\eta) \simeq \frac{3a}{n^*\eta^4} \left[ (M^2\Delta - 1)^2 + \frac{3(1-\Delta)^2}{(2+\Delta)^2} - \frac{2(M^2\Delta - 1)(1-\Delta)}{(2+\Delta)} \right]^{-1} . \tag{52}$$

In particular, for $N$ scatterers, $l_T(\eta+0) = (33a/4n^*)\eta^{-4}$. This divergent behavior was to be expected on the basis of the $\eta^4$ vanishing of the total cross section. It is an indication of the extended nature of the long wavelength excitations.[6,9] The exception is the $D$ case, where $l_T(0) = a/3n^*$, a result compatible with wave localization.

Combining Eqs. (35) and (49) we see that the SMFP can be obtained from Eq. (51) if only the first term inside of the square brackets is taken into account. Consequently, the long wavelength form of $l_{sc}(\eta)$ is given by Eq. (52) if we omit the last term inside of the square brackets. Therefore, when the scatterers have the same density ($\Delta = 1$) or compressibility ($M^2\Delta = 1$) as the background, $l_{sc}$ and $l_T$ have the same long wavelength limit.

We have plotted the functions $n^* l_{sc}(\eta)/a$ and $n^* l_T(\eta)/a$ in Figs. 10 and 11. From Fig. 10 we see that in the $D$ case $l_T(0) = l_{sc}(0)$, but $l_T > l_{sc}$ for all $\eta \neq 0$, with $l_T \simeq 2l_{sc}$ for $\eta \gtrsim 3$. In the $N$ case $l_{sc}$ decreases smoothly with increasing $\eta$, while $l_T$ has some structure and becomes larger than $l_{sc}$ only if $\eta \gtrsim 1.9$. The relatively large size of $l_T$ for $\eta \sim 1$ is responsible for the predicted impossibility of observing the localization of acoustical excitations by a random array of hard sphere scatterers.[9] On the other hand, in the $D$ case, condition (48) is well satisfied for low values of $\eta$.

The case ($\Delta = 1, M = 2$) is depicted in Fig. 11. The minima of $l_{sc}$, located at the frequencies of the maxima of $S(\eta)$, are labelled by the values of $l$ corresponding to the partial waves responsible for them. Since $\rho_s = \rho_0, l_{sc}(0) = l_T(0)$, and both curves are essentially indentical for $\eta < 1$. Although, at higher values of $\eta, l_T$ can be several times as large as $l_{sc}$, both functions show similar structures of maxima and minima. Naturally, the maxima in the mean free paths (larger diffusion) correspond to the minima in the scattering cross sections and vice versa. From Fig. 11 and Eq. (48) we would expect localization at $\eta \simeq 0.8$ and $n^* \sim 0.3$ when $\Delta = 1$ and $M = 2$. In the next subsection we will see that this is indeed the case.

3.4.2. *Cylinders.* In this case Eq. (51) must be replaced by

$$l_T(\eta) = \frac{\eta a\pi}{4n^*}\left\{\sum_{l=0}^{\infty}\left[\frac{(2-\delta_{l,0})\beta_l^2}{\beta_l^2+g_l^2} - \frac{2\beta_l\beta_{l+1}(\beta_l\beta_{l+1}+g_lg_{l+1})}{(\beta_l^2 g_l^2)(\beta_{l+1}^2+g_{l+1}^2)}\right]\right\}^{-1}$$

$$\equiv \frac{\eta a\pi}{4n^* T_2(\eta)}\,. \tag{53}$$

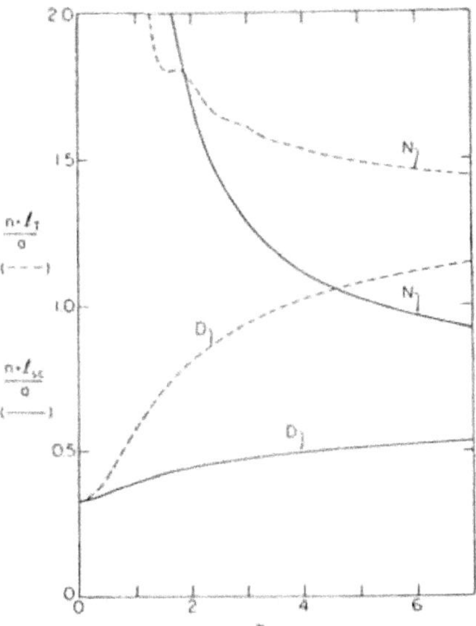

Fig. 10. Scattering (solid line) and transport (dashed line) mean free paths for the Dirichlet (D) and Neumann (N) spheres.

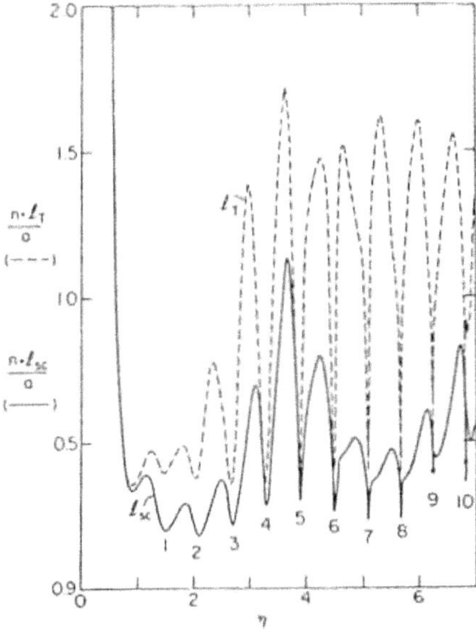

Fig. 11. Scattering (solid line) and transport (dashed line) mean free paths for the ($\Delta = 1$, $M = 2$) spheres. The big dots give the values of the two minima in the curve for $l_T$.

For low frequencies we have the approximate behavior

$$l_T(\eta) \simeq \frac{4a}{\pi n^* \eta^3} \left[ (M^2\Delta - 1)^2 + \frac{2(1-\Delta)^2}{(1+\Delta)^2} - \frac{2(1-\Delta)(M^2\Delta - 1)}{1+\Delta} \right]^{-1} . \tag{54}$$

For $N$ scatterers, $l_T(\eta \to 0) = (4a/5\pi n^*)\eta^{-3}$. As in the case of the spheres, the low $\eta$ behavior of $l_{sc}(\eta)$ coincides with that of $l_T(\eta)$, except that the last term inside the brackets must be omitted. Consistently with the diverging scattering cross section, both the transport and scattering MFPs for Dirichlet scatterers vanish at low frequencies; $l_{sc} \simeq l_T \simeq (a/\pi n^*)\eta ln^2(\eta)$.

The MFPs for the Dirichlet, Neumann, and $(\Delta = 1, M = 2)$ problems have been plotted in Ref. 78. The resulting curves are qualitatively very similar to their three-dimensional counterparts, although, of course, they are different in the $\eta \to 0$ limit.

## 3.5. Anderson Localization, Spherical Scatterers

In the preceding subsection we mentioned how the Ioffe-Regel criterion can be employed to obtain a rough estimate of the conditions for the onset of acoustical wave localization. We will now use the SCT results to obtain more quantative information.[10] In particular, we will get explicit forms for:

a) The phase boundary separating regions corresponding to localized and extended states in the $(\eta, n^*)$ plane.

b) The localization length $\xi(\eta)$ in the localized phase. This function determines the volume in which the excess energy resulting from an initial disturbance is confined.

c) The diffusion coefficient $D(\eta, \omega \to 0)$ for the extended states. This function describes the diffusion of the excess energy in the disordered medium a long time after the creation of the disturbance.

In addition, we will plot $D(\eta, \omega \to 0)$ against the Ioffe-Regel parameter $El_T/c$. Not only will this give us a better understanding of the onset of localization, but it will also lend justification to the use of the Ioffe-Regel criterion for "quick estimates" in acoustical problems. In the frequency domains where the excitations are localized, $D(\eta, \omega \to 0)$ vanishes linearly with $\omega$. Otherwise the excitations will be delocalized. In the delocalized regime, the ratio $D(\eta, \omega \to 0)/D_B(\eta)$ yields a convenient measure of the contribution of coherent effects. When this ratio is close to unity, the diffusion is mostly incoherent and the Boltzmann description is adequate. For values of $\eta$ such that $D(\eta, \omega \to 0)/D_B(\eta) \ll 1$, coherent effects are strong and a mobility edge cannot be far away.

The diffusion coefficient was shown in Sec. 2 to satisfy an algebraic equation, Eq. (23). To solve it as a function of frequency, we need to evaluate the functions $\gamma(E)$ and $D_B(E)$. If $d = 3$, we have, from Eqs. (32) and (35),

$$\gamma(\tilde{\eta}) = -4\pi \tilde{c}^2 n \ \mathrm{Im}\ \phi^+(0)$$
$$= -\frac{3n^*}{\tilde{\eta}} \left(\frac{\tilde{c}}{a}\right)^2 S_1(\tilde{\eta}) , \tag{55}$$

and, from Eq. (51),

$$D_B(\tilde{\eta}) = \frac{\tilde{c}a\tilde{\eta}^2}{9n^* S_2(\tilde{\eta})} . \tag{56}$$

Here $\tilde{\eta} = Ea/\tilde{c}$. These formulas are valid for permeable scatterers. The Neumann and Dirichlet cases are obtained by letting $\alpha_l = j_l', f_l = y_l'$ and $\alpha_l = j_l, f_l = y_l$, respectively.

Although the theory does not provide the value of the ultraviolet cutoff $Q$, it is reasonable to assume that $Q \sim l_T^{-1}$, since this is the wavenumber scale where the $q^2$ approximation for the diffusion pole breaks down. However, even if we accept that $Q \sim l_T^{-1}$, the proportionality constant is not known. We will write $Q = rdl_T^{-1}$ and analyze how our results depend on the value of $r$. We will briefly discuss other possible cutoffs later.

We have seen in Sec. 2 that there is always a region in the $(\eta, n^*)$ plane where Eq. (24c) yields a finite value for $D(\tilde{\eta}, 0) \equiv D(\tilde{\eta}, \omega \to 0)$.

This value is given by Eq. (27a), which can be rewritten as,

$$D(\widetilde{\eta},0) = D_B(\widetilde{\eta})\left[1 - \frac{81rn^{*2}S_1(\widetilde{\eta})S_2(\widetilde{\eta})}{\pi\widetilde{\eta}^6}\right] . \tag{57}$$

There is also often a region where $D(\widetilde{\eta},\omega \to 0) \sim \omega$. In this case the function $\xi(\widetilde{\eta})$, defined by Eq. (25), is finite and satisfies Eq. (28a) or, more compactly,

$$\left[1 - \frac{\pi\widetilde{\eta}^4}{9n^*QaS_1}\right]Q\xi = \ \mathrm{arctg}(Q\xi) . \tag{58}$$

Substituting Eq. (25) into the equation for the energy density propagator, Eq. (20a), we see that $\xi(\widetilde{\eta})$ is the characteristic size of the region where the energy is confined; $\xi(\widetilde{\eta})$ is therefore the localization length.

3.5.1. *The Phase Boundary.* The right-hand side of Eq. (57) vanishes upon a curve $n_c^* = n_c^*(\widetilde{\eta})$; this curve defines the phase boundary separating localized and extended states in the $(\eta, n^*)$ plane and is given by the equation,

$$81n_c^{*2}rS_1(\widetilde{\eta})S_2(\widetilde{\eta})/\pi\widetilde{\eta}^6 = 1 . \tag{59}$$

Since in the figures we have preferred to use as a variable the dimensionless source frequency $\eta$, Eq. (59) must be solved by writting

$$\widetilde{\eta} = \widetilde{\eta}(n_c^*,\eta) .$$

We start the analysis of Eq. (59) by noting that, if the renormalization of the speed of sound is neglected, $n_c^* \sim r^{-1/2}$: an increase in the cutoff decreases the minimum scatterer volume fraction compatible with localization. In Fig. 12 we show the phase boundary for three values of $r$ when the scatterers are of higher compressibility than the surrounding medium ($\Delta = 1, M = 2$). The numbers below the curves indicate the position of the features corresponding to the peaks in $S(\eta)$ (see Fig. 6). The minima corresponding to the first maxima in $S(\eta)$

are indicated by $M$. Although the curves for the various values of $r$ have the same general shape, they are not obtained from each other by simply multiplying by a constant. The reason for this is that $\tilde{c}$ contains very rapid variations (see Fig. 8) and so does the "rescaling" from $\tilde{\eta}$ to $\eta$. Even if this effect complicates the picture, it does not obscure the main features. We also note that $r = 2\pi/3$ corresponds to the choice of Sheng and Zhang.[8]

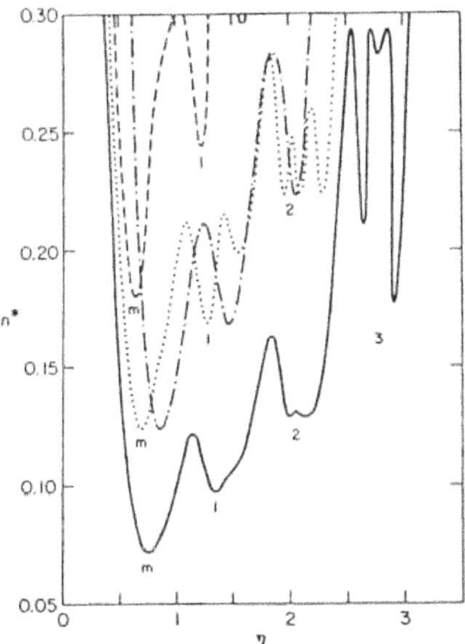

Fig. 12. Phase boundary for the case of $(\Delta = 1, M = 2)$ scatterers. Three values were chosen for the cutoff; the solid line corresponds to $r = 2\pi$, the dotted line to $r = 2\pi/3$, and the dashed line to $r = 1$. The results obtained for $r = 2\pi/3$ using the unrenormalized speed of sound are represented by the dot-dashed curve.

Equation (59) also implies that,

$$n_c^* \sim \tilde{\eta}^3 (D_B/S)^{1/2} \ . \tag{60}$$

The factor $\tilde{\eta}^3$ is responsible for the delocalization of the high frequency excitations. The minima of $D_B$ (see Fig. 11) occur at frequencies close to those of the maxima in $S$; this confirms the intuitively obvious conclusion that localization is favored by large cross sections. The renormalization of the $c$ causes the structure of $n_c^*$ not to always follow that of $S$. For the sake of comparison we include in Fig. 12 a curve (for $r = 2\pi/3$) obtained using the unrenormalized speed of sound.

At low frequencies, Eq. (59) formally (the physical value of $n^*$ cannot go above that for random close-packing) to the limit

$$n_c^* \sim \widetilde{\eta}^{-3} \left\{ \frac{r}{\pi} \left[ (M^2\Delta - 1)^2 + \frac{3(1-\Delta)^2}{(2+\Delta)^2} \right] \right. $$
$$\left. \cdot \left[ (M^2\Delta - 1)^2 + \frac{3(1-\Delta)}{(2+\Delta)^2} - \frac{2(M^2\Delta - 1)(1-\Delta)}{2+\Delta} \right] \right\}^{-1/2} \tag{61}$$

For Neumann scatterers, $n_c^* \sim (16\pi/77r)^{1/2}\widetilde{\eta}^{-3}$; for Dirichlet scatterers, $n_c^* \sim (\pi/81r)^{1/2}\widetilde{\eta}$. These results indicate the localization of low frequency excitations in the Dirichlet problem and the delocalization that occurs in all the other cases.

The phase boundaries for $\Delta = 1$ and three values of $M$, which label the corresponding curves, are shown in Fig. 13. The "most favorable" $(r = 2\pi)$ cutoff has been selected. Estimates for other values of $r$ can be obtained by multiplying the curves by the factor $(2\pi/r)^{1/2}$. It is evident that an increase in the index of refraction is followed by a fast increase in the size of the localization region. In Figs. 12 and 13 we have considered scatterers having the same density but higher compressibilities than the background medium. Scatterers having lower compressibilities than the background are much less efficient and generally do not lead to localization. This property, as well as the increase in the size of the localization region with increasing $M$, were first discussed by Sheng and Zhang in Ref. 8. As we found in Ref. 9, Neumann scatterers are also inefficient and do not lead to localization. If we select scatterers that have the compressibility of the background but different density, the localization effects are weak. In Fig. 14 we show the phase boundary for the $\Delta = 0.25, M = 2$ case, which corresponds to $\rho_s = 4\rho_0$.

The localized states are present only at high scatterer densities, even for the most favorable ($r = 2\pi$) choice for the cutoff. In the same figure we present the Dirichlet phase boundary, which encompasses the low frequency localized region.

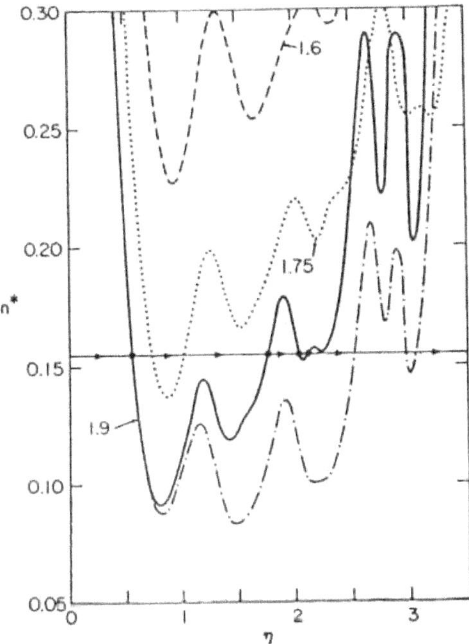


Fig. 13. Phase boundaries for scatterers characterized by $\Delta = 1$ and three values of $M$. We have chosen $r = 2\pi$. The horizontal line at $n^* = 0.155$ represents the evolution of the parameter $\eta$ as it generates the trajectory depicted in Fig. 17b. The dot-dashed curve is the pseudosphere approximation for $M = 1.9$.

It must not be understood that the perspectives of observing localization will increase monotonously with the value of the index of refraction. Although for large values of $M$ the resonances become taller, they also become, as mentioned before, much thinner (for example, see Figs. 2 and 4 in Ref. 78). Hence, it seems that due to the smoothing out caused by proximity effects and the variations in scatterer shape and size, not much will be gained by going beyond $M \sim 3$. We note, however, that there is one feature that makes high $M$ appealing: since at lower frequencies waves tend to localize more easily, the fact that the first resonance moves to lower frequencies with increasing $M$ should be always borne in mind.

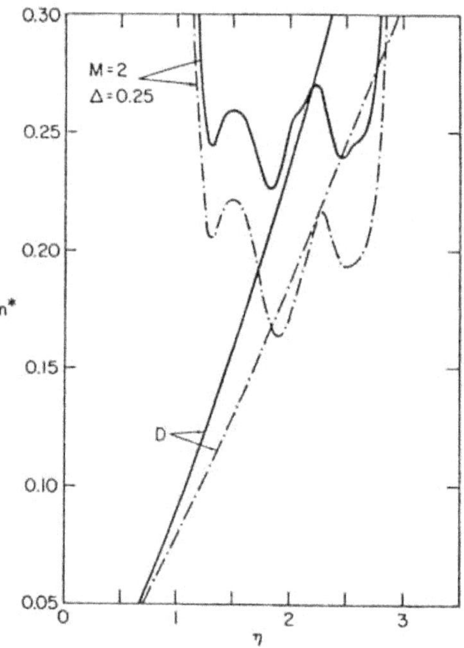


Fig. 14. Phase boundaries for the Dirichlet (D) problem and for high density $(\rho_s = 4\rho)$ scatterers. The localized states lie above the respective curves. We have taken $r = 2\pi$. The dot-dashed lines are the pseudosphere approximation results.


### 3.5.2. *The Localization Length.*

We next investigate the behavior of the localization length in the region where it takes finite values. Since $\xi(\eta)$ is determined by a superposition of rational combinations of Bessel functions, a series expansion around the mobility edges is possible — in principle — and leads to $\xi \sim |\eta - \eta_c|^{-1}$, if we approach the mobility edge $\eta_c$ keeping $n^*$ fixed. This is the behavior obtained in Refs. 6, 8–10; a numerical evaluation shows that it can be observed over s sizable (typically $\Delta\eta \sim 0.01$) neighborhood of $\eta_c$. The diffusion coefficient also vanishes with an exponent of unity: $D(0) \sim |\eta - \eta_c|$. From Eqs. (57)

and (58) we find that the same exponents occur if we approach the mobility edge "vertically" (i.e. varying $n^*$ while keeping $\eta$ fixed).

The localization length is plotted in Fig. 15 for $(\Delta = 1, M = 1.75)$ scatterers. The value of $\xi$ decreases extremely fast as we move away from the phase boundary and into the localization region. The strong effect of varying the scatterer volume fraction $n^*$ is also evident from the figure: a relatively small increase in $n^*$ leads to a considerable shortening of the localization length. In the higher density cases, the value of $\xi/a$ decreases below unity. This unphysical result is due to the large size of the cutoff for those densities and frequencies. Since $\xi$ cannot be smaller than the interscatterer distance, it appears that a MFP-related cutoff is inappropriate for states well inside the localization region. At this point it is not clear to us how this difficulty can be overcome without making ad-hoc assumptions. The whole problem of the choice of the cutoff is a delicate one and needs further theoretical work.

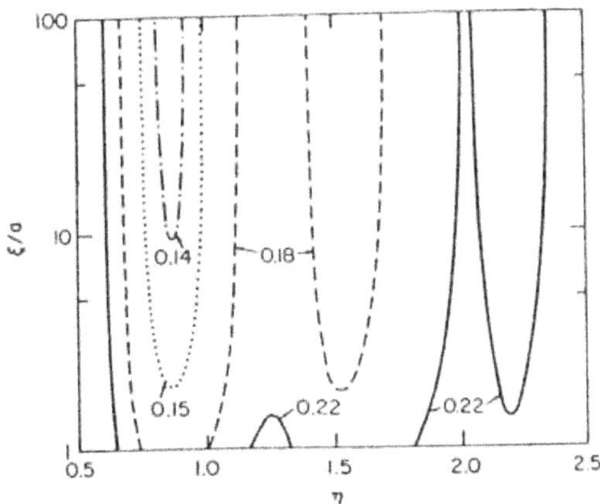


Fig. 15. Localization lengths for the volume fractions specified next to the corresponding curves. We have taken $\Delta = 1$, $M = 1.75$, and $r = 2\pi$.

3.5.3. *The Diffusion Coefficient.* The diffusion of the extended excitations at long times is described in Fig. 16, where we have plotted $D(\tilde{\eta}, 0)$ in units of the Boltzmann diffusion coefficient. The various curves have the same features, which become intensified as the density is increased. This result was to be expected, since the location of these features is determined by the properties of the individual scatterers. Two mobility edges are present for $n^* = 0.14$ and four for $n^* = 0.18$.

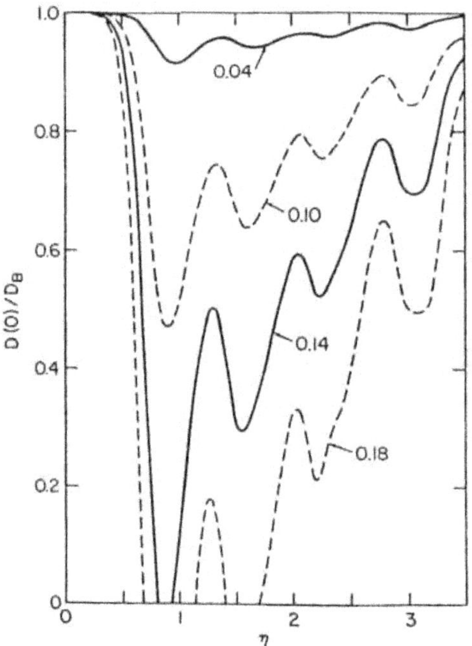

Fig. 16. Diffusion coefficient, scaled by $D_B(\eta)$, for the case of Fig. 15. The numbers next to the curves are the volume fractions $n^*$.

It is also interesting to plot the diffusion coefficient as a function of the Ioffe-Regel parameter $El_T/c$. We do this using $\eta$ as a parameter and observing the evolution of the curve representing $D(\tilde{\eta}, 0)/D_B$ as $\eta$ is increased. We show the results for the Dirichlet and Neumann cases in Fig. 17(a). In the Dirichlet problem there is one mobility edge above which the diffusion coefficient increases monotonically, in agreement with our previous discussions. The onset of localization occurs for $El_T/c \sim 1.8$, which shows that the Ioffe-Regel criterion is not too far off the mark. In the Neumann case, the $D(\tilde{\eta}, 0)/D_B$ ratio is always close to unity, an indication of the relative weakness of the coherent processes. The multi-valuedness of $D(\tilde{\eta},0)$ is typical of the situations when these processes are strongest in an intermediate frequency range.

More interesting is Fig. 17(b), where we have plotted the results for compressible scatterers; for a full comprehension of this figure, one should look at it in conjunction with Fig. 13 where the horizontal line at $n^* = .155$ indicates the evolution of $\eta$ that causes the trajectory in Fig. 17(b). First, we note the presence of four mobility edges, located at values of $El_T/c$ somewhat higher than the Ioffe-Regel estimate. These edges are depicted by big dots in Fig. 13. The loops in the branch corresponding to $\eta > 2.13$ are related to variations in the vertical distance from the $n^* = .155$ line to the phase boundary. The segments composed of three thin lines describe regions where the trajectory comes back approximately on itself and then proceeds along its original direction of evolution. For example, the first "triple" region, located in the neighborhood of the fourth mobility edge, is due to the "kink" in the phase boundary at $\eta = 2.25$. We have depicted only the portion of the curve that corresponds to $\eta < 3.77$. Although for higher values of $\eta$ the trajectory reenters the region $El_T/c < 7$, we have chosen not to present the portion of the curves corresponding to $\eta > 3.77$ in order not to obscure the figure.

3.5.4. *The Coherence Length.* It is important to stress that the diffusion coefficient $D(\tilde{\eta},0)$ will be observed only at distances from the source larger than the coherence length[22,8]

$$\varsigma(\eta) = l_T(\eta)D_B(\eta)/D(\eta,0) . \tag{62}$$

The corresponding "coherence volume" around the source point grows as we move closer to the phase boundary. Correspondingly, in a transmission experiment the effective diffusion will be larger than predicted by $D(\eta,0)$ if the observation is made at distance $X \equiv |\mathbf{x}-\mathbf{x}'| < \varsigma$ from the source. As $X$ decreases the measured diffusion coefficient $D_X(\eta,0)$ should increase, reaching a value of about $D_B(\eta)$ when $X \sim l_T$. This expected behavior is roughly sketched in Fig. 18, where we

     *C. A. Condat & T. R. Kirkpatrick*

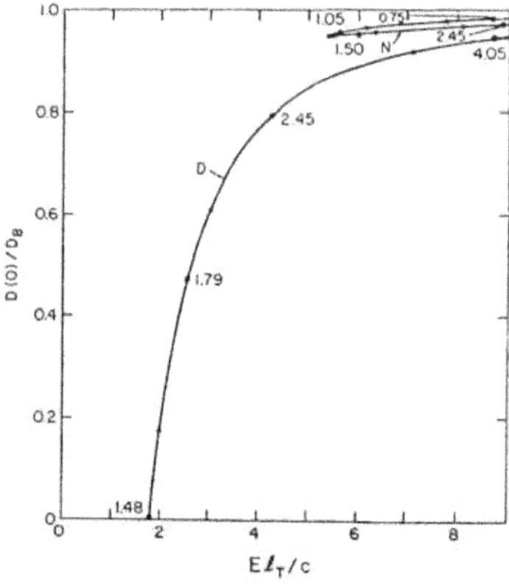

Fig. 17(a)

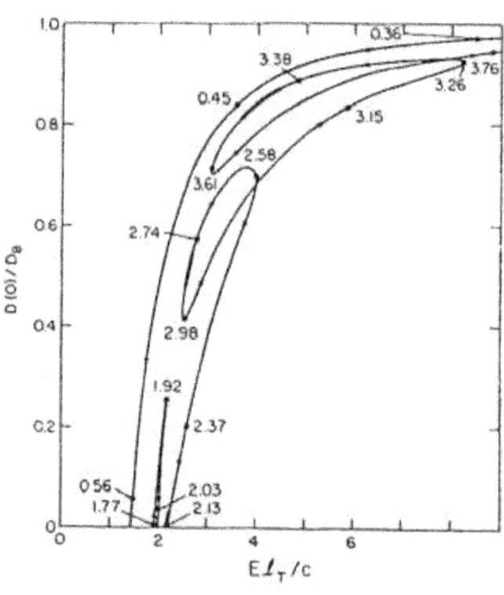

Fig. 17(b)

Figs. 17(a) and (b). Evolution of the diffusion coefficient as a function of the Ioffe-Regel product $El_T/c$ for (a) the $N$ and $D$ problems and (b) the ($\Delta=1$, $M = 1.9$) case. We have chosen $n^* = .155$ and $r = 2\pi$. The arrows denote the direction of growth of the parameter $\eta$. Some values of $\eta$ have been specified to the curves. The meaning of the three-line segments is discussed in the text.

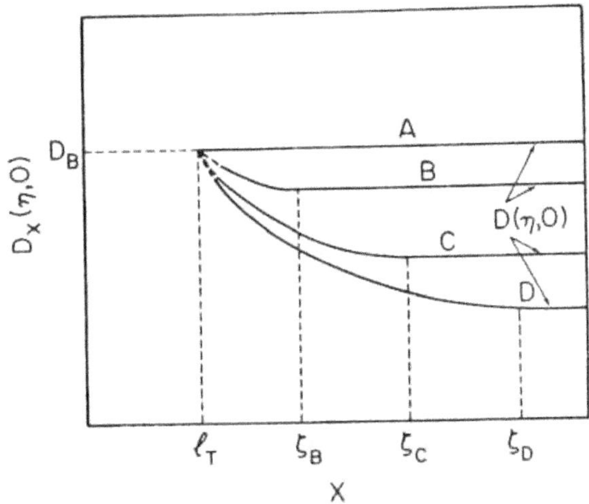


Fig. 18. Sketch of the expected behavior of $D(\eta,0)$ as the field point is moved away from the source. Curve $A$ corresponds to a point in the phase diagram far removed from the mobility edge. Curves $B$, $C$, and $D$ correspond to points successively closer to the edge. The corresponding coherence lengths are $\zeta_B$, $\zeta_C$, and $\zeta_D$.

have neglected the differences between the values of $D_B(\eta)$ for the various cases.

3.5.5. *An $n^*$-Independent Cutoff.* Even if it is appealing to use the transport MFP to define the momentum cutoff, this choice leads to results that are at variance with the leading volume-fraction corrections obtained using perturbation theory.[71] The selection of a cutoff independent of the volume fraction would remove this disagreement. This suggests the possibility of choosing a cutoff proportional to the inverse of the scatterer radius, the only $n^*$-independent length that can be defined in the problem. We will now write down the formulas resulting from the option $Q = r/a$. First we observe that, since $D_B$ enters the localization equations principally through the cutoff, it will not appear here (except as the obvious prefactor in the diffusion coefficient). The phase boundary is given by

$$n_c^* = \pi\tilde{\eta}^4/9rS_1 \ . \tag{63}$$

Choosing $Q \sim l_T^{-1}$ yielded [see Eq. (61)] $n_c^* \sim \widetilde{\eta}^3 (rS_1 S_2)^{-1/2}$, a result less sensitive to the value selected for $r$. The localization length solves Eq. (58), with $Q = r/a$. The correction to the diffusion coefficient is now linear in the volume fraction,

$$D(\widetilde{\eta},0) = D_B \left[ 1 - \frac{9rn^* S_1}{\pi \widetilde{\eta}^4} \right] . \tag{64}$$

Aside from the modifications just discussed, the qualitative features of the graphs resulting from the choices $Q = r/a$ and $Q = rcD_B^{-1}$ are generally similar.

3.5.6. *A Cutoff Proportional to $l_{sc}^{-1}$.* As we mentioned before, the cutoff was taken in Ref. 92 to be proportional to $l_{sc}^{-1}$. In the pseudo-sphere approximation (PA), to be considered in a later section, we also effectively replace $l_T$ by $l_{sc}$ in the cutoff. Since $l_{sc}^{-1}$ is usually longer than $l_T^{-1}$ (see Figs. 10 and 11), the replacement of $l_T$ by $l_{sc}$ in the cutoff implies that more wavenumbers contribute to the integral in Eq. (23). As a result, stronger localization effects are predicted. Two examples are shown in Fig. 14. Although qualitatively similar, the phase boundaries occur at appreciably smaller volume fractions; this effect is more noticeable in the $\eta > 1$ range. If we use the subscript $(sc)$ to indicate the volume fraction obtained choosing $l_T (l_{sc})$ for the cutoff, it is easy to verify that the corresponding phase boundaries are related by

$$n_{c,sc}^*(\eta) = [l_{sc}(\eta)/l_T(\eta)]^{1/2} n_{c,T}^*(\eta) . \tag{65}$$

3.5.7. *The Case of Large $n^*$.* Let us consider a composite material where the value of $n^*$ for the minority component is near or beyond the precolation threshold, and there is a strong admixture of two waves, each propagating mainly in one of the components. In this case, the structure in the functions describing localization will be smoothed out, even if one of the components is present only under the form of identical spherical particles: the wave that propagates mostly in these spherical inclusions will be scattered by the interstices, which do not have spherical shape. Therefore, the best conditions for the observation of the structures would occur when the scatterers are identical and $n^*$ is reasonably high, but not so much that the wave characteristic of the minority component makes a significant contribution. In this "best" case, we believe our approach gives an essentially correct description, although a marked smoothing out of sharp features should be expected in view of the occurrence of many scattering events in the near zone or preceding scatterers.

To investigate the case when both components contribute, Sheng and Zhang formulated an effective medium approximation:[8] they treated both components on an equal footing, i.e. each is simultaneously the "medium" and the "scatterer" for the wave propagating in the second "medium". They further assumed that the two types of scatterers were spherical. As a consequence, they obtain an equation for $l_T$ that is identical to our Eq. (50), except that in the denominator the following replacement must be made:

$$n^* |\phi^+(\theta)|^2 \to n^* |\phi_1^+(\theta)|^2 + (1 - n^*)|\phi_2^+(\theta)|^2 , \tag{66}$$

where $\phi_{1(2)}^+(\theta)$ is the scattering amplitude for a wave incident on a sphere of component 1(2) and $n^*$ is the volume fraction occupied by component 2. The effective speed of sound is obtained using an interpolation formula that gives the correct frequency and compositional limits:

$$\frac{1}{\tilde{c}} = \frac{1 - n^*}{\tilde{c}_1} + \frac{n^*}{\tilde{c}_2} . \tag{67}$$

Here $\tilde{c}_1(\tilde{c}_2)$ is the renormalized speed of sound, considering propagation in medium 1(2) and scattering by medium 2(1).

Using the SCT, Sheng and Zhang found that localization islands appear in the $(n^*, \eta)$ diagram when the ratio $M$ between the indices of refraction of the two media is high enough ($M > 2.11$ if $r = 2\pi/3$).

We believe, however, that at high values of $n^*$ only two mobility edges would be observed. One reason is the already mentioned non-spherical shape assumed by one of the components. Another is that essentially all scattering events occur in the near zone of preceding scatterers.

## 3.6. Anderson Localization, Cylindrical Scatterers

The diffusion coefficient $D(E, \omega \to 0)$ is always zero in two dimensions, and all states in an effectively two-dimensional system are localized. However, it is possible that the localization occurs only on a very large spatial scale; if this scale is bigger than the size of the system or the inelastic attenuation length, the localization effects will not be very relevent. The single-scatterer resonances lead to a considerable shortening of the localization length and suggest the frequency ranges when localization will be strongest. The explicit form for $\xi(\eta)$ is obtained from Eq. (26b):

$$\xi(\eta) = \frac{\pi a \tilde{\eta}}{8 r n^* T_2(\tilde{\eta})} \left\{ \exp\left[ \frac{\pi^2 \tilde{\eta}^2}{4 n^* T_1(\tilde{\eta})} \right] - 1 \right\}^{1/2}. \tag{68}$$

Although the right-hand side is evaluated using $\tilde{\eta} = Ea/\tilde{c}$, in the graphs we will again plot $\xi$ as a function of the source frequency $\eta$.

The factor preceding the square root in Eq. (68) is $l_T/r$. This direct proportionality between the localization length and the transport mean free path is a consequence of our choice for the cutoff. Had $l_{sc}$ been chosen instead of $l_T$, as in Ref. 92, or the pseudosphere approximation made as in Ref. 9 the only resulting modification would be the replacement of $l_T$ by $l_{sc}$ in the pre-factor in Eq. (68). We also see that $r$ appears only as a multiplicative factor. Therefore, the choice of cutoff is less important than in the truly 3-$d$ problems.

On the other hand, the appearance of $n^*$ in the exponent generates enormous changes in $\xi$ when the density is modified. This can be seen from Fig. 19, where we have plotted the localization length for the case $(\Delta = 1, M = 2)$ (i.e., $\rho_s = \rho_0, \kappa_s = 4\kappa$) and several values of $n^*$. All the curves are seen to have the same general shape, which is determined by single-scattering processes. The minima (maxima) in Fig. 19 are obviously related to the maxima (minima) in Fig. 7. The

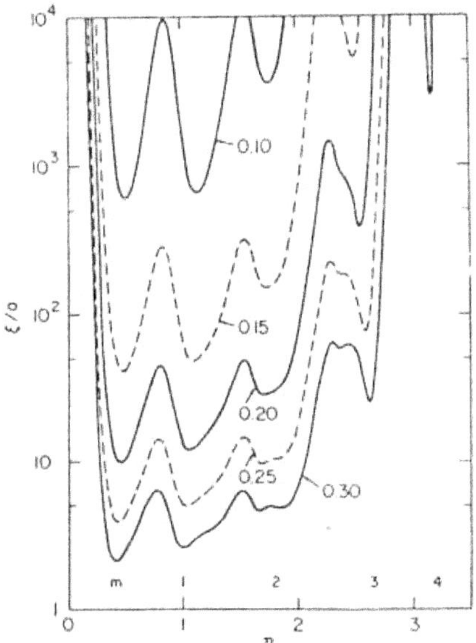

Fig. 19. Localization length for $(\Delta = 1, M = 2)$ scatterers. The numbers next to the curves are the corresponding volume fractions $n^*$. The numbers below indicate which of the resonances in Fig. 7 generates the corresponding minimum. We have chosen $r = 2\pi$.

small distortions, which grow with $n^*$, are due to the renormalization of the speed of sound.

Other cases are plotted in Fig. 20. The localization length for the Dirichlet problem shows the expected behavior: the large value of the low-frequency cross-section $S(\eta)$ generates strong localization in that range. The localization length for the $(\Delta = 1, M = 1.5)$ case turns out to be very long, showing that even a large maximum in the total cross section (cf. Fig. 7) is not conducive to observable localization effects if it occurs at relatively high frequencies. Similarly, the curve for the $(\Delta = 0.5, M = 2^{1/2})$ (i.e., $\rho_s = 2\rho_0$) problem does not appear in the graph. In spite of the large size of the cross section near its maximum at $\eta = 4.5$, there will be no observable localization. For these short wavelengths the excitation propagates from scatterer to scatterer; the strong interactions with the individual scatterers will simply lead to "normal" diffusion, described by $D_B$, over scales larger than $l_T$. If we choose very dense $(\rho_2 = 4\rho_0)$ scatterers, then the maximum in $S$ is shifted to lower frequencies and we get smaller values for the localization length. For the sake of comparison we have also plotted in Fig. 19 the results obtained using the PA. As in the problem of the spheres, the PA overestimates the localization effects but generates the appropriate structure of maxima and minima. The value of $\xi$ for $N$ scatterers is too long to appear in the graph.

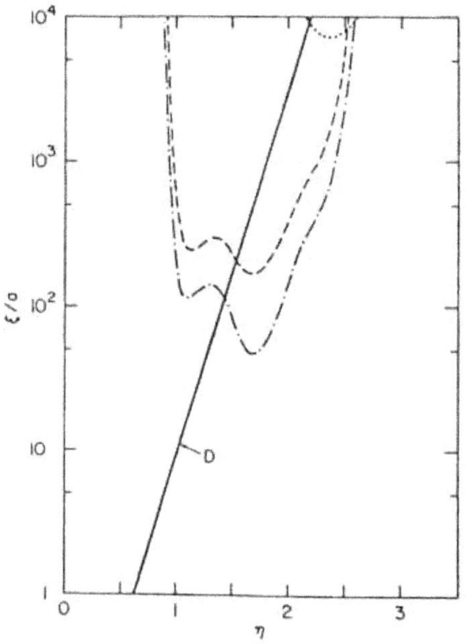

Fig. 20. Localization length for $D$ and permeable scatterers. The dotted line corresponds to $(\Delta = 1, M = 1.5)$, the dashed line to $(\Delta = 0.25, M = 2)$, and the dot-dashed line to $(\Delta = 0.25, M = 2)$ in the PA approximation. We have taken $n^* = 0.2$ and $r = 2\pi$.

## 3.7. The Effects of Dissipation

In any experimental sample there is an inelastic MFP $l_i$ due to the dissipative processes occurring in the background medium and (for penetrable scatterers) in the scatterers themselves. We can make an estimate of the importance of the total inelastic effects by considering separately a background inelastic MFP $l_1$ and a scatterer inelastic MFP $l_2$. Let us write

$$L^{-1} = l_1^{-1} + l_2^{-1} \; . \tag{69}$$

Defining $N = L/l_T$ and considering the diffusive nature of the excitations, the inelastic MFP is approximately given by $l_i = N^{1/2} l_T$. ($l_i$ is measured radially and not along the actual path of the excitation.)

In the cases of interest we will have generally $l_2 < l_1$, but when $n^* \ll 1$, the contribution of the background may be relatively large and should be included. Let us consider it first.

We assume that the thermal conductivity $\kappa$ and the bulk and shear vicosities, $\varsigma_B$ and $\varsigma_s$, are small but finite. (This corresponds to short but finite thermal and viscous relaxation times.) Under these conditions the sound wave becomes attenuated over a distance[65]

$$L_{at} = \left\{ \frac{E^2}{2\rho c^3} \left[ \frac{4\varsigma_s}{3} + \varsigma_B + \left( \frac{1}{C_v} - \frac{1}{C_p} \right) \kappa \right] \right\}^{-1} \; . \tag{70}$$

Here $C_v$ and $C_p$ are the specific heats at constant volume and pressure, respectively. In our problem we have to introduce a factor of $(1 - n^*)$ to account for the proportion of space filled by the background medium. Then

$$l_1^{-1} = (1 - n^*) L_{at}^{-1} \; . \tag{71}$$

We have effectively assumed that we have plane waves propagating on a fraction $(1 - n^*)$ of the system. This is a reasonable assumption, at least for low $n^*$.

The dissipation in the scatterers can be described by introducing a complex acoustic admittance $i\delta_l$ (Ref. 77) ($i\Delta_l$ if $d = 2$) instead of the purely imaginary form we used before. Nevertheless, it is simpler to define $l_2$ directly in terms of the experimental absorption cross section $S_a = \pi a^2 Q_a$ for a single scatterer. In the three-dimensional case,

$$l_2 \simeq 4a/3n^* Q_a \; . \tag{72}$$

The effects of dissipation have not been taken into account in the description of localization presented here. They have been considered, from different viewpoints, in Refs. 5, 21 and 22. John[5,21] used field theory and a Gaussian noise description to find that near the mobility edge the strong wave interference should lead to an anomalously large value of the absorption coefficient. On the other hand, Anderson[22] argues that finite absorption will make localizing systems more reflecting. He finds that the reflection coefficient will change from a $1 - (l_T/l_i)^{1/2}$ behavior far from the mobility edge to a $1 - (l_T/l_i)^{2/3}$ form the close to the edge.

Sornette[7] has estimated the value of the transmitted intensity, $T$, through a slab of a disordered medium of thickness $L$. Predictions are made for the form of $T(L)$ in the various three-dimensional regimes: Boltzmann-diffusive (or "classical"), near a mobility edge (or "critical"), and localized.

## 4. THE PSEUDOSPHERE APPROXIMATION

### 4.1. Nature of the Approximation

For some kinds of scatterers the calculation of the Boltzmann diffusion coefficient is very complicated, and an approximation that eliminates the need for its explicit computation would be desirable. In this section we develop such an approximation, which is based on the Gaussian noise (GN) model; it is suggested by the fact that neither $n^*$ nor $\eta$ is integrated (or differentiated) upon in the development of the formalism presented in Sec. 2.

The GN model has been often used to describe random media. It is a standard procedure that simplifies the derivations, although it gives a faithful representation of the scattering only in the long-wavelength limit. In this model the scatterers are represented by Gaussian random functions $\psi(\mathbf{x})$. Instead of the boundary-value problem of Sec. 2, the scalar potential satisfies the perturbed wave equation,

$$\partial_t^2 \phi(\mathbf{x},t) - c^2 \nabla^2 \phi(\mathbf{x},t) - \psi(\mathbf{x}) \nabla^2 \phi(\mathbf{x},t) = 0 . \tag{73}$$

The correlator for the Gaussian random functions is

$$[\psi(\mathbf{x})\psi(\mathbf{x}')]_{\mathrm{av}} = \mu\delta(\mathbf{x} - \mathbf{x}') . \tag{74}$$

The localization problem is analyzed in exactly the same form as in Sec. 2, computing the energy-density propagator and starting from identical initial conditions, i.e. Eqs. (2a). Our basic SCT results, Eqs. (20) and (23) are still valid, except that $\gamma(E)$ and $D_B(E)$ must be written in terms of the correlation strength $\mu$: (Ref. 9).

$$\gamma(E) = \frac{\mu\pi^{1-d}\Omega_d E^{d+2}}{2^{d+1}c^{4+d}} , \tag{75a}$$

and

$$D_B(E) = \frac{2^{d+1}\pi^{d-1}\tilde{c}^{6+d}}{\mu d\Omega_d E^{d+1}} . \tag{75b}$$

(Note that the values of $\mu$ and $\Omega_d$ here are respectively, $(2\pi)^{-d}$ times and twice those used in Ref. 9).

If $d = 2$ all states are localized and the localization length is formally given by Eq. (26b).[9] In three dimensions there is a single mobility edge, located at

$$E_c = \pi^{1/2}\tilde{c}^{7/3}(4/3\mu)^{1/3}r^{-1/6} \tag{76}$$

(choosing the cutoff $Q = rl_T^{-1}(E)$). For $E < E_c$ the excitations are delocalized, the diffusion coefficient being $D(E,\omega \to 0) = D_B(E)[1 - (E/E_c)^6]$. If $E > E_c$ all states are localized, $\xi(E \to E_c)$ diverging with a critical exponent equal to unity, as in the more realistic models.

We will now show how the GN model can be used to make reasonable predictions for realistic (i.e. finite size) scatterers. The procedure, which we call the pseudosphere approximation (PA) has its underpinnings in the observation made at the end of the first paragraph of this section. It consists of writing an effective $\mu = \mu(n^*,\eta)$ by equating the imaginary parts of the self-energies corresponding to the GN and the finite-size problems. The imaginary part of the self-energy is related to the total cross section of the spherical (or disk-like) scatterers through Eq. (32), i.e.,

$$\gamma(\eta) = \left( \frac{dc^2 n^* \eta}{\Omega_d a^{d+1}} \right) S(\eta) \;. \tag{77}$$

The PA allows us to equate the right hand sides of Eqs. (75a) and (77). Hence an effective noise correlation strength $\mu(n^*, \eta)$ can be defined in terms of the total scattering cross section of the spherical scatterers:

$$\mu(\eta, n^*) = \frac{d2^{d+1}\pi^{d-1}ac^4 n^* S(\eta)}{\eta^{d+1}\Omega_d^2} \;. \tag{78}$$

Substitution of $\mu(n^*, \eta)$ for $\mu$ in the localization formulas permits us to obtain in a straightforward fashion explicit results for complicated spherical scatterers whenever the total cross section $S(\eta)$ is known. For example, if $d = 3$, the equation for the phase boundary can be found combining Eqs. (76) and (78):

$$n_c^*(\widetilde{\eta}) = \frac{4\pi a^2 (\pi/r)^{1/2}\widetilde{\eta}}{9S(\widetilde{\eta})} \;. \tag{79}$$

Here the speed of sound has been again renormalized everywhere in accordance with the self-consistence ideas.

In Ref. 9 this approximation was used to show that a random array of hard spheres is incapable of localizing acoustical excitations. Here we have plotted some results obtained for the phase boundary and the localization length in Figs. 13 and 20, respectively. The cases of Dirichlet and high density ($\rho_s = 4\rho$) permeable scatterers have been considered and compared to the "exact" (in the SCT) results. Observing the figures, we conclude that, although the magnitude of the localization effects is overestimated, the structure generated by the PA is essentially the same as that obtained from the "exact" calculation of Sec. 3. This confirms the usefulness of the PA: it yields a reasonable estimate for problems involving finite-size scattering centers while taking advantage of the simplicity afforded by a Gaussian noise model of the inhomogeneities. We next use the PA to make specific predictions for some relevant problems.

## 4.2. The Optical Problem: Conducting Spheres

If polarization is neglected, the problem of electromagnetic (EM) localization in an optically disordered system can be treated in the framework of a scalar wave equation.[5,21] Conducting spheres are specially efficient scatterers for EM waves. At high frequencies the total cross section for a single conducting sphere of radius $a$ tends to $2\pi a^2$, the same value as in the acoustical problem. At low $\eta$, it vanishes following the Rayleigh law $S(\eta) \to (10\pi/3)a^2\eta^4$; we note that the coefficient of $\eta^4$ is 7.5 times larger than in the case of hard acoustic spheres. It is then conceivable that in the crucial region $\eta \sim 1$ the total cross section will be large enough as to make possible the observation of Anderson localization.

Using the complete expression for the total cross section for a conducting sphere,[72]

$$S(\eta) = \frac{2\pi a^2}{\eta^2} \sum_{l=1}^{\infty} (2l+1) \left[ \frac{j_l'^2}{j_l'^2 + y_l'^2} + \frac{(j_l + \eta j_l')^2}{(j_l + \eta j_l')^2 + (y_l + \eta y_l')^2} \right] . \quad (80)$$

We can now easily obtain explicit expressions for the effective correlation strength $\mu(\eta, n^*)$ and the phase boundary by the use of Eqs. (78) and (79), respectively. In similar manner, we can also obtain expressions for the localization length, coherence length, etc. We also note that $n_c^*$ and $r$ enter the equation for the phase boundary, Eq. (79), only through the product $n_c^* r^{1/2}$. Consequently, the same curve yields the phase boundary for arbitrary values of $r$ provided that we rescale the volume fraction appropriately.

The phase boundary is shown in Fig. 21a. The localization length $\xi(\eta)$ is plotted in Fig. 21b for several values of the scatterer density. The speed of light has not been renormalized in these curves. We see that there is a frequency window in which EM localization phenomena should be observable. When we increase the scatterer density, $\xi$ decreases abruptly and the sample must become rapidly opaque. The localization length is also quickly reduced as we move away from the boundary and into the shaded region of Fig. 12a. Note that if we used the Rayleigh limit for $S(\eta)$, we would obtain for the wavelength at the phase boundary $\lambda^* = a(15n^*/2)^{1/3}(r/\pi)^{1/6}$, a result consistent with that of John.[21]

In Fig. 22 we have plotted the phase diagram for a cutoff $r = \pi$. We have also plotted the coherence length $\varsigma(\eta)$ (see paragraph 3.4.5) for various values of the scatterer volume fraction. We observe that a noticeable bump in $\varsigma(\eta)$ occurs only for the volume fractions quite close to the mobility edge. [Far from the edge, $\varsigma(\eta) \approx l_T(\eta)$]. Hence, the presence of a marked maximum in $\varsigma(\eta)$ should tell the experimentalist he is on the right path towards Anderson localization.

On account of the foregoing comparison between "exact" and PA results, it is reasonable to assume that the actual phase boundary will be located at values of the volume fraction somewhat higher than those indicated in Fig. 21a. On the other hand, it seems certain that the optimum frequency range to search for localization effects is the neighborhood of $\eta = 1$. An experiment that lends itself well to the PA description will be presented in subsection 4.5.

## 4.3. Localization in a Bubbly Liquid

We have seen in Sec. 3 how the presence of scattering resonances enhances the possibilities of observing Anderson localization in some frequency ranges. Some systems, such as gas bubbles in a fluid, show a more spectacular resonant behavior; therefore, it is natural to consider them as promising candidates for the study of Anderson localization. Sornette and Souillard[23] have analyzed the localization of sound waves in a bubbly fluid and also in a gas containing a disordered collection of identical Helmholtz oscillators.[94] In both cases, there is a resonant frequency at which the scattering cross section can become much larger than the geometric cross section. They carried out their evaluations for the position of the mobility edge using the Ioffe-Regel criterion, whose applicability to acoustic systems is buttressed by the calculation presented in subsection 3.5. The PA is well-suited to explore these

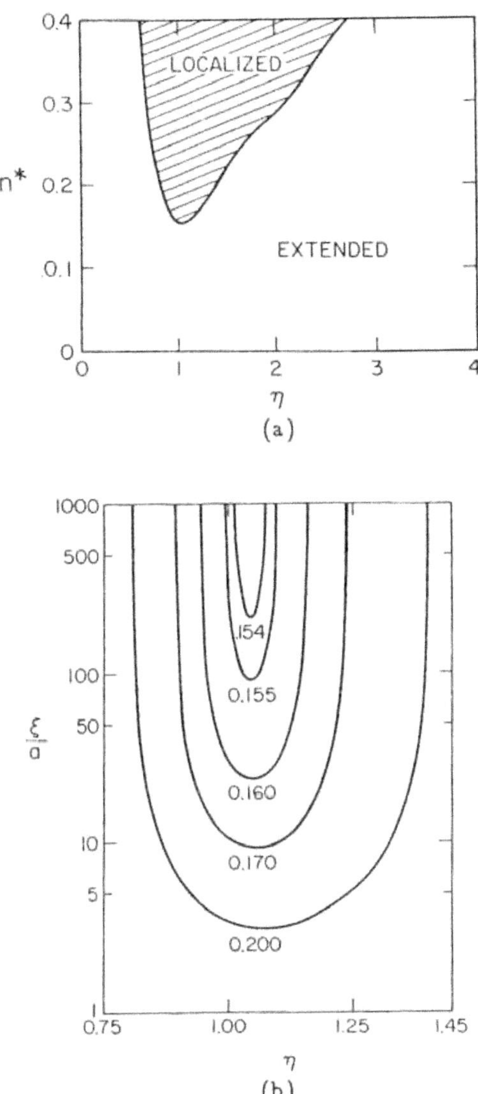



Fig. 21(a) Phase diagram for electromagnetic localization. The scatterers are conducting spheres and $Q = 2\pi l_T^{-1}$. If $r \neq 2\,\pi$, the vertical scale should be modified according to the discussion in the text. (b) Localization length for states in the shaded region of (a). The volume fractions occupied by the scatterers are indicated next to the corresponding curves. The speed of light has not been renormalized here.

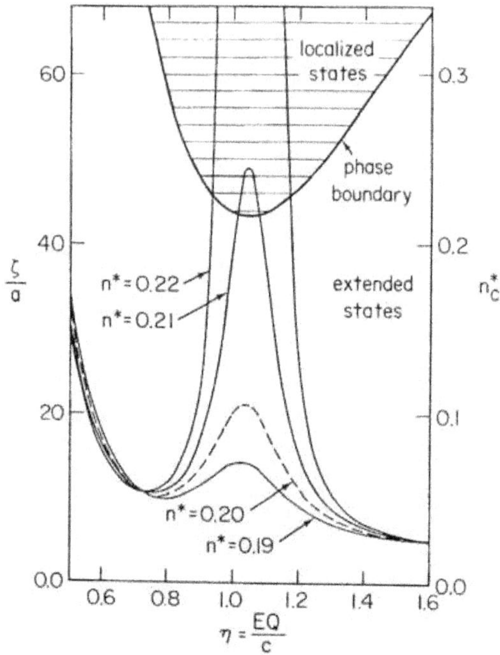



Fig. 22. Phase diagram and coherence lengths for EM localization with $Q = \pi l_T^{-1}$. The volume fractions are indicated next to the corresponding $\zeta(\eta)$ curves.

problems. Here we apply it to the case of sound in a bubbly liquid.

The theory of the scattering of sound waves by bubbles is described for example, in the textbook by Clay and Medwin.[95] Assuming that the wavelength $\lambda$ of the incident radiation is much larger than the radius $a$ of the bubble, and neglecting nonlinearities, the scattering cross section is given by

$$S(E) = 4\pi a^2 \left\{ \left[ \left( \frac{E_R}{E} \right)^2 - 1 \right]^2 + \delta^2(E) \right\}^{-1}, \qquad (81)$$

with $E_R = (3\gamma P/a^2 \rho_0)^{1/2}$ being the natural pulsation frequency of the bubble and $\delta$ a damping function. Here $\gamma$ is the specific heat ratio and $\rho_0$ and $P$ are the ambient fluid density and hydrodynamic pressure, respectively.

The form given for $E_R$ is valid provided that the bubble, although much smaller than the wavelength, is itself not too small. To derive it, it was assumed that the gas vibrates adiabatically and that the only restoring force is the volume stiffness. For very small values of $a$, thermal equilibration effects and the surface tension must be taken into account. Fortunately, since both processes counterbalance each other to some extent[95] the resulting condition on $a$ is not very restrictive. For example, in the case of air bubbles at sea level, they can reasonably be neglected when $a$ is larger than a few microns.

Let us turn now to the damping function $\delta(E)$. It is formed by three contributions:[95]

$$\delta = Ea/c + \delta_t(E) + \delta_v(E) , \tag{82}$$

where $Ea/c$ is the contribution due to reradiation and $\delta_t$ and $\delta_v$ are the damping functions originating in the thermal conductivity and in the viscosity, respectively. Inserting Eq. (81) into Eq. (79), and writing the result in terms of the frequency relative to resonance, $y = E/E_R$, we obtain the following form for the mobility edge equation:

$$n_c^*(y) = (\pi/r)^{1/2}(\tilde{\eta}_R/9y^3)\{[1 - y^2]^2 + y^4(y\tilde{\eta}_R + \delta_D)^2\} , \tag{83}$$

with $\tilde{\eta}_R = E_R a/\tilde{c}$ and $\delta_D \equiv \delta_t(E) + \delta_v(E)$. It is easy to verify that, in the Rayleigh limit, $y \ll 1$, Eqs. (81) and (83) reduce to the limits

$$S \simeq 4\pi a^2 \eta^4 (M^2\Delta)^2/9 , \tag{84a}$$

and

$$n_c^* \simeq (\pi/r)^{1/2}\tilde{\eta}^{-3}(M^2\Delta)^{-2} , \tag{84b}$$

respectively. These are exactly the limiting expressions found in the case of compressible spheres [See Eqs. (36) and (61)], when we take $M^2\Delta \gg 1$.

We have plotted $n_c^*(y)$ in Fig. 23. We have chosen $\gamma = 1.3, c = 1.5 \times 10^5$ cm/sec, $P = 10^6$ g/cmsec$^2$, and $\rho_0 = 1$g/cm$^3$. This yields

$E_R \approx (2000/a)$ cm/sec and, choosing $\tilde{c} \approx c, \tilde{\eta}_R \approx 0.013$. To obtain estimates of the effects of damping we have replaced $\delta_D(y)$ by a constant (say, its magnitude at resonance) and then evaluated Eq. (83) for several arbitrary values of this constant. We observe that, although for $\delta_D = 0$ the tip of the phase boundary occurs at bubble volume fractions as low as $n^* = 2.5 \times 10^{-7}$, the addition of a nonzero damping increases considerably the minimum required volume fraction. A typical "favorable" value can be obtained from Fig. A6.1.2 in Ref. 95. For $E_R \simeq 6.3 \times 10^4$ sec$^{-1}$, $\delta_D \simeq 0.05$. Therefore, it is probably realistic to expect the onset of localization at values of $n_c^*$ slightly below $10^{-5}$. The observation of localization would then be relatively simple, if only the bubbles could be made all of the same size!

Bubble size ploydispersity is, however, a big handicap. From the data of Gazanhes *et al.*,[90] Sornette and Souillard[23] estimate that the fraction of resonant bubbles is of the order of $10^{-2}$. Unless polydispersity can be reduced, a successful localization experiment will require a total bubble fraction of the order of $10^{-3}$. (This may not be too difficult; see Ref. 97).

We further note that the specific value adopted for the cutoff is not crucial here, and will, at most, change the predictions by a factor of 2. In Fig. 23 we chose $r = \pi$. More important corrections can arise from the use of a correctly renormalized speed of sound. As discussed in Sec. 3, these corrections are, at low values of $n^*$, proportional to $n^*$. On the other hand, the dispersion increases steeply near the resonance[98,99] and can become very important when the total bubble fraction is that required to observe the mobility edge. Since it could appreciably shift the position of the edge, a more refined calculation would be useful. Another area that should be explored is that of the influence of bubble motion upon localization.

## 4.4. Intensity Correlation Function

Shapiro[100] and Stephen and Cwilich[101] have recently analyzed the intensity fluctuations occuring in a random medium when a wave-

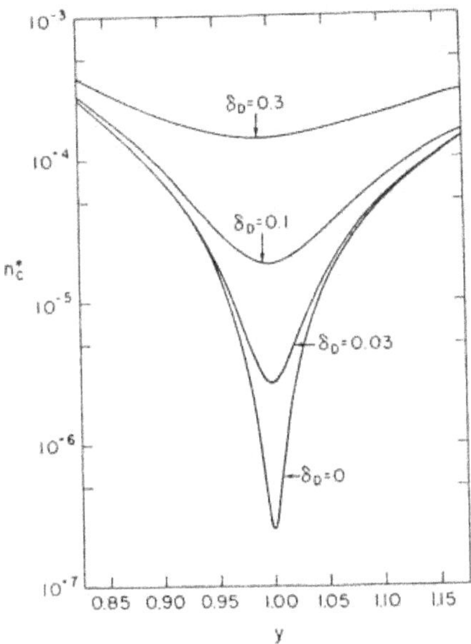

Fig. 23. Phase boundary as a function of the relative frequency $y = E/E_R$ for a bubbly liquid. The various curves correspond to the indicated values of the damping $\delta_D$. We have chosen $r = \pi$ and approximated $\tilde{c} = c$; the values of the physical parameters are indicated in the text. In each case the localized states lie above the corresponding curve.

like excitation is present. These studies, made in the framework of the GN model, involved the calculation of the intensity-intensity correlation function $c(\mathbf{x},\mathbf{x}')$. Performing a diagrammatic expansion, Shapiro[100] showed that, for a source of fixed frequency, $c(\mathbf{x},\mathbf{x}') \sim \exp[-|\mathbf{x} - \mathbf{x}'|/l_{sc}]$. Later, Stephen and Cwilich[101] included more diagrams in the calculation. They found that, for $|\mathbf{x} - \mathbf{x}'| > l_{sc}$, the decay of the correlation function crosses over to a power law:

$$c(\mathbf{x},\mathbf{x}') \sim |\mathbf{x} - \mathbf{x};|^{-3} \ .$$

The SMFP there plays an important role in the analysis of the speckle pattern: it gives the size of the exponential region, which is also, if the long algebraic tail is neglected, the typical size of the fluctuations. Accepting the PA ideas, it is possible to determine the evolution of this size as a function of the frequency in the case of discrete scatterers.[102] From Fig. 10 we see that the fluctuation size should decrease (increase) monotonously with frequency in the case of Neumann (Dirichlet) scatterers. According to Fig. 11, a sweep in frequencies would show an oscillatory behavior in the fluctuation size in the case of compressible scatterers.

### 4.5. The Experiment of Ferrari *et al.*

In this subsection we describe a recent experiment on microwave transmission in a medium containing a random distribution of metallic spheres.[53] Although this is not an acoustical experiment, it is included here because we consider that: (i) It is an important step in the quest for the mobility edge in disordered classical systems. (ii) The elements used in the analysis are similar to those that would be required to describe an acoustic localization experiment.

The transmission measurement were performed using several microwave frequencies in the range [18.15 GHz, 24.15 GHz]. The sample investigated consists of 29 stacked slots of paraffin, each of which containing 1024–1026 randomly-located copper-plated spheres, with $a = 0.2184 \pm 0.0006$ cm. Each slab is 12.7 cm $\times$ 12.7 cm in cross section and approximately 1.5cm thick. The average packing fraction of the spheres is $n^* = -0.186$. The sample is surrounded on four sides with reflecting copper walls. The effective sample length $L$ was varied by substracting some slabs.

In order to obatin an appropriate configurational average for each value of $L$ the sample was shuffled repeatedly; the wire probe was also placed at several (from ten to thirty) points on the exit face of the sample. At each point the wave intensity was recorded as the frequency was varied by about 300 MHz in steps of width $\sim 5.6$ MHz. All the data thus obtained were then averaged and the average entered as a point in a log-log graph of the recorded transmitted intensity $I$ vs. $L$. The procedure was repeated for several values of $L$ in each of the chosen frequency bands. See Figs. 24 and 25, where the error bars are the

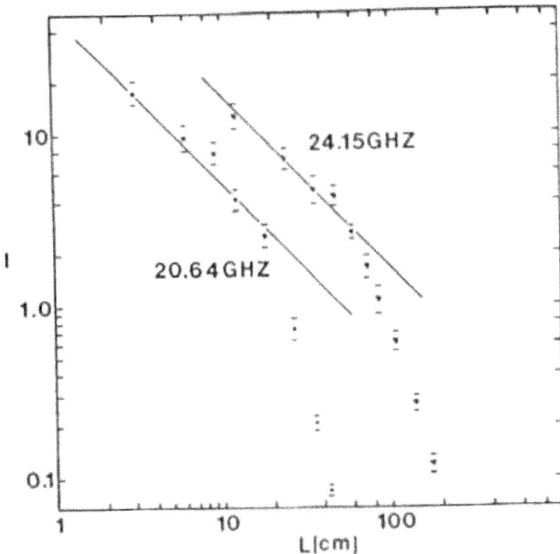



Fig. 24. Log-log plot of the average microwave intensity $I$ versus sample thickness $L$. The straight lines drawn through the data have slope $s = -1$. The $L$ and $I$ scales for the 24.15 GHz data have been adjusted in order to display the data points clearly [$L$(plot) = $4L$ (actual)]. (After L. A. Ferrari et al.[53]).

standard deviations of the mean.

The theory of Secs. 2 and 3 was developed for an infinite medium. In order to account for the finite size ($L$) effects, Eq. (62) has to be generalized. This was done by Ferrari et al.[53] by invoking scaling ideas. For the diffusion coefficient measured in a transmission experiment performed with a sample of length $L$, they write

$$D(\eta,0;L) = D_B(\eta)l_T(\eta)\left[\frac{1}{L} + \frac{1}{\varsigma(\eta)}\right] . \qquad (85)$$

Compare with the discussion for an infinite medium in paragraph 3.5.4. Many of the wave trajectories that contribute to $D_X(\eta,0)$ are terminated at $L$ in the case of a finite sample. In the experiment, $l_T$ was typically of the order of two centimeters. We note that, for extended states, $\varsigma \sim l_T$ and $D(\eta,0;L) \simeq D_B(\eta)$, but near a mobility edge $\varsigma$ grows rapidly and $D(\eta,0;L) \sim D_B(\eta)[l_T(\eta)/L]$.

Although the nature of the absorption processes is not yet well understood, it is clear that they are important. The measured effective absorption length $L_a$ turns out to be in all cases of the order of 8 cm. Following Anderson,[22] Ferrari *et al.*[53] work out a steady state solution of the one-dimensional diffusion equation. They predict that in a plot of the transmitted intensity vs. $L$, an $L^{-2}$ falloff should be observed when $L < \varsigma$. For $L > \varsigma$ one should observe an $L^{-1}$ falloff $(L \ll L_a)$ or an exponential decay $(L \gg L_a)$.

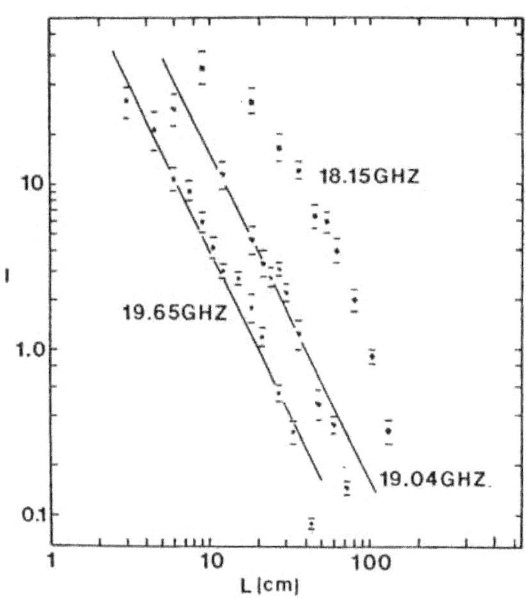


Fig. 25. Log-log plot of the average microwave intensity $I$ versus sample thickness $L$. The straight lines through the 19.04 GHz and 19.65 GHz data have slope $s = -2$. For the 18.15 GHz data $L$ (plot) $= 3L$ (actual), and for the 19.04 GHz data $L$ (plot) $= 2L$ (actual). (After L. A. Ferrari *et al.*[53]).

The data for five values of the frequency are plotted in Figs. 24 and 25. Lowering the frequency is equivalent to moving to the left on a horizontal line $(n^* = 0.186)$ in Fig. 21. At the highest frequencies, reported in Fig. 24, the measured intensity is proportional to $L^{-1}$ at small values of $L$ and then decays exponentially for larger values of $L$. In these cases

$\varsigma \sim l_T$ and the energy diffusion is "classical" (i.e. Boltzmann-like). For the two intermediate frequencies (19.04 GHz and 19.65 GHz) the transmission at large $L$ is still exponential, but at smaller values of $L$ it varies as $1/L^2$, which clearly suggests the onset of the critical regime. The transition between the $1/L^2$ and the exponential dependences occurs at a scale $L = 10$–15 cm. For the lowest (18.15 GHz) frequency studied the behavior of the transmitted intensity is found to be an admixture of those for the highest and the intermediate frequencies.

Transforming to dimensionless variables, the data suggests that the point of closest approach between the experimental path (the $n^* = 0.186$ line) and the phase boundary occurs at $\eta \approx 1.3$. This value is somewhat higher than that predicted by Figs. 21 and 22 ($\eta \approx 1$). It is argued in Ref. 53, using an analogy with wavelength transmission, that this difference might be reduced if one takes into account the renormalization of the wavenumber in the presence of the conducting spheres.

The data for the intermediate frequencies suggest $\varsigma \sim 10$–15 cm. Comparing with Fig. 22, it is apparent that the experimental path has gone below (although not very far from!) the lowest tip $n_c^*$ (min) of the phase boundary. We can reasonably estimate that: (1) $0.186 < n_c^*(\text{min}) \lesssim 0.20$. (2) If the form $Q = r l_T^{-1}$ is chosen, the "best $r$" is probably somewhat larger than $\pi$. We remind the reader that the PA may overestimate the localization effects and so, even if the "true" phase boundary is located at values of $n^*$ higher than those predicted in Fig. 21, the "most favorable" cutoff ($r = 2\pi$) may also be in general the best choice.

We can reasonably expect that a measurement carried out at a slightly higher value of $n^*$ will show a marked increase in the value of $\varsigma$. (Compare also with Fig. 6 in Ref. 10). Hopefully, the mobility edge might be observed before dissipation washes away the coherence. Sornette's detailed predictions[7] could help in the identification of the regime (classical, critical, localized) in which such an experiment is performed.

# 5. THE LOCALIZATION OF THIRD SOUND

## 5.1. Third Sound on Superfluid Helium Films

An appropriate system for the investigation of wave localization is a film of superfluid helium adsorbed on a specially prepared substrate. The excitations to be localized in such a systen are third-sound waves. (See Fig. 26.) Since the wavelength of these waves is large in comparison with the film thickness, the theory of third sound is similar to the theory of shallow water waves. The restoring force, however, is not provided by gravity but by the van der Waals interaction between the helium film and the atoms in the substrate.[103]

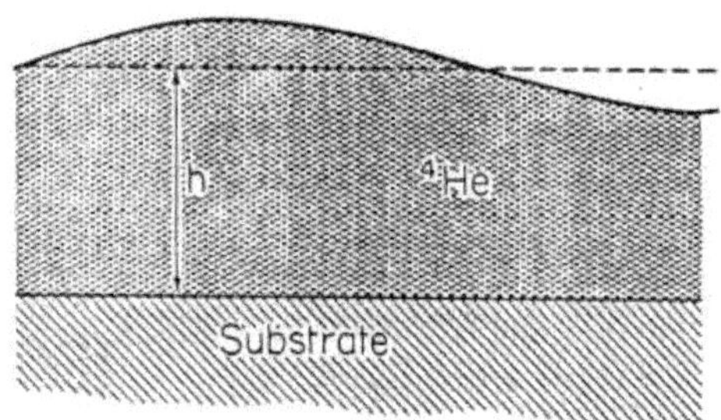


Fig. 26. A third-sound wave on a thin helium film whose equilibrium thickness is $h$.

The force on the helium atoms near the surface of a thin film has the form

$$f = \frac{3\alpha}{h^4} \, , \tag{86}$$

where $\alpha$ is a substrate-dependent constant and $h$ is the equilibrium film thickness. The speed of sound is given by $c = (fh)^{1/2}$. Given the relatively long wavelength of the third-sound waves we neglect surface tension.

Considering a quasi-one-dimensional configuration, the velocity potential for waves propagating on a uniform substrate satisfies the differential equation

$$\partial_t^2 \phi(x,t) - c^2 \partial_x^2 \phi(x,t) + \Gamma \partial_t \phi(x,t) = 0 \, . \tag{87}$$

Here $\Gamma$ is a phenomenological damping constant.[104,105] which accounts for the residual ($T \to 0$) damping of third-sound excitations and which has been interpreted by Rutledge and Mochel as a surface friction.[104] Experiments performed using a uniform vitreous quartz substrate yield $\Gamma h \approx 1.0$ atomic layers/sec.[104] Since this is a very small value we leave the discussion of the effects of intrinsic dissipation for subsection 5.4 and set $\Gamma = 0$ in the bulk of our analysis.

There are several reasons why this is a convenient system:

(A) At sufficiently low temperatures ($T \lesssim 1\text{K}$) the intrinsic attenuation length for excitations propagating on uniform substrates can be made extremely large (indeed, much larger than any possible experimental substrate). The absence of dissipation is equivalent to the disappearance of the electron-phonon interaction in the problem of an electron in a disordered lattice. Consequently, the conditions for a third-sound equivalent of the electronic "resonant tunnelling"[106,107] are well-satisfied in the one-dimensional configurations.

(B) The interaction between excitations (expressed through the nonlinear terms in the hydrodynamic equations) can be made arbitrarily small be controlling their amplitude.

(C) The scatterers are macroscopic and can be built essentially identical to each other. As a consequence there is a closer correspondence between theory and experimental reality.

(D) Since a stable uniform flow field can be created in a superfluid film,[108] it is possible to investigate how the flow and the concomitant destruction of the time-reversal invariance modify the localization phenomena. This aspect of the problem will be discussed at length in Sec. 6.

The one-dimensional electronic problem has been the subject of a great amount of analytical effort, as attested in several reviews.[109-111] The prediction by Mott and Twose[112] that all states would be localized for any amount of disorder has been rigorously proved.[113-115] At most, it is possible to have a set of measure zero of delocalized states. We will see, however, that it is possible to construct a strongly disordered system whose spectrum will effectively have wide transmission bands located in the vicinity of the elements of a discrete set of delocalized states. Therefore, the set of measure zero can have physical significance.

## 5.2. Striped Configurations

5.2.1. *van der Waals Scatterers.* The properties of the third-sound excitations are substantially modified if we place on the original substrate an array of parallel identical strips of a second substrate. (See Fig. 27.) Since the equilibrium film thickness, which is determined by the van der Waals interaction, is modified when going from one substrate to the other[116] the portion of the film located on a strip acts effectively as a scatterer for the third-sound waves. To a good approximation, the equilibrium thickness on a strip, $h_2$, is related to that on the original substrate, $h_1$, by the equation $h_2/h_1 = (\alpha_2/\alpha_1)^{1/3}$, where $\alpha_2$ and $\alpha_1$ are the respective van der Waals force constants. The velocity of third sound is the same on both regions. We will call these strips van der Waals (VDW) scatterers.

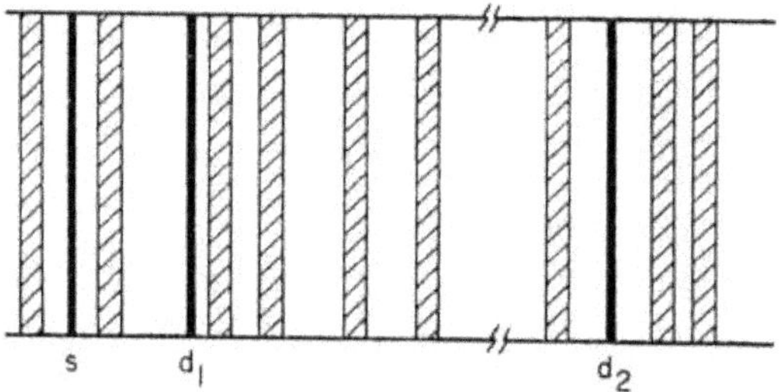


Fig. 27. Sketch of the random arrangement of parallel strips of a modified substrate. The positions of a detector $d_1$ close to the source $s$ and of the second detector $d_2$ far from it are indicated.

The boundary conditions can be derived by noting that the region where the film thickness changes is very narrow for thin films.[116] If $d$ is the distance over which the thickness changes, $\lambda$ the third-sound wavelength, and $a$ the strip half-width, we will have, in general, $d \ll \lambda$

and $d \ll a$. Consequently, the waves effectively react to a sudden variation in the film thickness at the edges of the strips. If $X_j$ is the position of the center of the $j$th strip, the boundary conditions are

$$\phi_{j-1}(x,t) = \phi_j^s(x,t) \text{ at } x = X_j - a , \tag{88a}$$

$$h_1 \frac{\partial \phi_{j-1}}{\partial x}(x,t) = h_2 \frac{\partial \phi_j^s}{\partial x}(x,t) \text{ at } x = X_j - a , \tag{88b}$$

$$\phi_j^s(x,t) = \phi_j(x,t) \text{ at } x = X_j + a , \tag{88c}$$

$$h_2 \frac{\partial \phi_j^s}{\partial x}(x,t) = h_1 \frac{\partial \phi_j}{\partial x}(x,t) \text{ at } x = X_j + a . \tag{88d}$$

Here, $\phi_j^s$ is the velocity potential for the film on the $j$th strip, and $\phi_{j-1}$ and $\phi_j$ are the potentials corresponding to the regions at the left and the right of that strip, respectively. The conditions on the derivatives are obtained by using mass conservation.

If the strips form an ordered array $X_j = jb$, with $b$ being the distance between strip centers, the behavior of the third-sound excitations can be described by elementary methods. A simple band structure with alternating allowed and forbidden regions for the wave number was obtained in Ref. 60. In that reference, a comparison was also made between the boundary conditions (88) and those occurring in a square-well electronic Kronig-Penney potential.

The initial conditions are taken again to have the form specified by Eqs. (2a). The strip density is $n$ (volume fraction $n^* = 2an$) and our treatment allows overlapping strip configurations.

5.2.2. *Index of Refraction Scatterers.* An alternative disordered substrate can be created by "roughening" some pieces of an otherwise smooth substrate. The effective speed of sound $c_s$ in the pieces of the film on the roughened parts can be substantially smaller than that on the smooth substrate. These pieces of film therefore act as very efficient scatterers (indices of refraction $N = c/c_s$ as high as 5.8 for roughened Si have been reported).[117]

The rough regions may be distributed as a random array of strips similar to that in Fig. 27. We call these strips index-of-refraction (IR) scatterers. It is not difficult to include both possibilities for the scatterers (i.e., changes in the film thickness and the speed of sound) simultaneously in the formalism. The velocity potential satisfies a wave equation with speed of sound $c$ on the unperturbed substrate and $c_s$ on the strips, and the boundary conditions (88) must be satisfied.

5.2.3. *Averaged Green's Function Results.* The formalism described in Sec. 2 can be used to compute the averaged Green's function and the energy density propagator. The calculations have been presented in detail in Ref. 58, so we will be very brief here. We will assume that the two substrates differ both in composition and roughness. The VDW and IR cases can then be obtained by taking the $R \equiv h_2/h_1 = 1$ and $N = 1$ limits, respectively.

The complete expression for the $T$-matrix is quite complicated.[58] Its elements satisfy the following identities:

$$\langle p|\widehat{T}^\pm(E)|p_1\rangle = \langle p|\widehat{T}^\mp(E)|p_1\rangle^* \,, \tag{89a}$$

$$\langle p|\widehat{T}^\pm(-E)|p_1\rangle = \langle p|\widehat{T}^\pm(E)|p_1\rangle^* \,, \tag{89b}$$

$$\langle p|\widehat{T}^\pm(E)|p\rangle = \langle -p|\widehat{T}^\pm(E)|-p\rangle \,, \tag{89c}$$

$$\langle p|\widehat{T}^\pm(E)|-p\rangle = \langle -p|\widehat{T}^\pm(E)|p\rangle \,, \tag{89d}$$

and the optical relation

$$\mathrm{Im}\langle p|\widehat{T}^\pm(E)|p\rangle = \pm(\pi/2Ec)(|\langle p|\widehat{T}^\pm|-p\rangle|^2 + |\langle p|\widehat{T}^\pm|-p\rangle|^2) \,, \tag{90}$$

for $p = E/c$. Defining $\eta = 2Ea/c$ and

$$W(x) = x[4x^2 + (x^2 - 1)^2 \sin^2(\eta N)]^{-1} \,, \tag{91}$$

the evaluation of the self-energy yields for the renormalized speed of sound $\tilde{c}$,

$$\tilde{c}^2(E) = c^2\{1 - 4na\eta^{-1}W(RN)[(RN + 1)^2 \sin(\eta(N - 1)) + (RN - 1)^2$$
$$\times \sin(\eta(N + 1))]\} \,, \tag{92a}$$

with its long wavelength ($\eta \ll 1$) limit

$$\tilde{c} \to c[1 - (na/R)(R^2 N^2 - 2R + 1)] \ . \tag{92b}$$

For the imaginary part of the self-energy we obtain

$$\gamma(E) = (nc^2 \eta/a)\{1 - W(RN)[(RN + 1)^2 \cos(\eta(N - 1)) - (RN - 1)^2$$
$$\times \cos(\eta(N + 1))]\} \ , \tag{93a}$$

with its long wavelength limit

$$\gamma \to (nc^2/4aR^2)\eta^3(R^4 N^4 - 2R^3 N^2 + 2R^2 - 2R + 1) \ . \tag{93b}$$

The averaged Green's function decays exponentially in space, with an attenuation length $L_a = \eta c^2/a\gamma$.

We note that $R$ and $N$ always appear in the combination $RN$, except in the arguments of the circular functions. These arguments are modified because of the phase lag caused by an index of refraction different from unity, an effect that is not present in the VDW scatterer case. The variable $E$ always appears forming the product $\eta = 2Ea/c$, which is $2\pi$ times the ratio between the strip width and the wavelength $\lambda = 2\pi c/E$.

The renormalized speed of sound and the disorder-induced attenuation parameter $\tilde{\Gamma}(E) = E\gamma(E)$ are plotted in Fig. 28 for VDW scatterers in the case $R = 1.26$, corresponding to a ratio of about two between the van der Walls constants. We have chosen a relatively high strip density ($n = 40$ strips/cm). [The magnitudes we actually graph are $(\tilde{c} - c)/2\pi ac$ and $\tilde{\Gamma}/2nc$ since, except for negligible corrections, they are functions of $\eta$ and $R$ alone]. The renormalized speed of sound for IR scatterers is graphed in Fig. 29 for several values of the index. The values for $c$ have been taken to be approximately those corresponding to clean Si substrates.[117] The curves for $N = 3$ correspond to $c = 2040$ cm/sec (film thickness: about 7 layers), that for $N = 2$ corresponds to

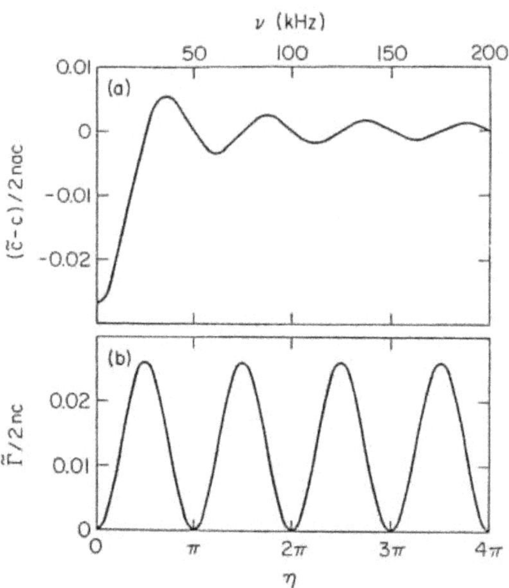

Fig. 28. (a) Renormalized speed of sound and (b) effective attenuation for van der Waals scatterers with $R = 1.26$. In the lower scale we only specify $\eta = 2Ea/c$, while in the upper scale we give the values of $\nu = E/2\pi$ corresponding to $a = 0.005$ cm and $c = 10^3$ cm/sec.

$c = 900$ cm/sec (thickness: about 13 layers), and that for $N = 1.5$ corresponds to $c = 750$ cm/sec (thickness: about 16 layers). In all figures we take the strip width to be $2a = 10^{-2}$ cm.

We note that at low frequencies the waves are slowed down by the scatterers. However, as we increase the frequency, $\tilde{c}$ oscillates between values higher and lower than $c$, becoming more and more insensitive to the presence of the strips. For relatively weak scatterers $\tilde{c}$ is not very different from the speed of sound on the original substrate, but for stronger scatterers the derivations are quite large at low frequencies. This can be seen to occur in the $N = 3$ curves in Fig. 29 at frequencies $\nu \lesssim 5 \text{kHz}$ if $n = 10 \text{ cm}^1$. Indeed, under those conditions the approximation of using only the lowest order diagram to evaluate the self-energy does not yield accurate results, and higher order diagrams must be included. We believe that the contribution of these diagrams will be to decrease the difference between $c$ and $\tilde{c}$ at low frequencies. At least in the VDW case it is not difficult to verify, taking the limit $E \to 0$, that the contributions of the first and third diagrams in Fig. 2 have opposite signs, while the second diagram is irrelevant.

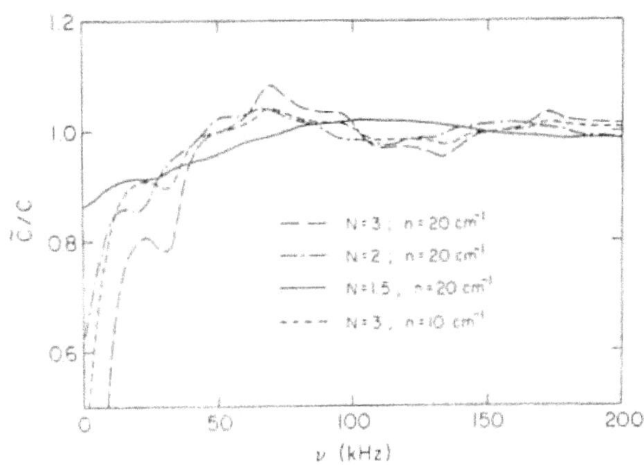


Fig. 29. Renormalized speed of sound for index of refraction scatterers. The unrenormalized speeds of sound $c$ correspond to experimental values on smooth substrates (see text). The frequency $\nu$ is $\nu = E/2\pi$.

5.2.4. *Anderson Localization.* The localization length for wave-like excitations in a random one-dimensional medium is given by Eq. (26a). In this equation we must insert the value of $\gamma(E)$ given by Eq. (93a). We also select the cut-off $Q = r\tilde{c}D_B^{-1}$, the Boltzmann diffusion coefficient being given by

$$D_B(E) = \left[\frac{\tilde{c}}{2n}\right] \left[\frac{4R^2 N^2 + (R^2 N^2 - 1)^2 \sin^2(\tilde{\eta}N)}{(R^2 N^2 - 1)^2 \sin^2(\tilde{\eta}N)}\right] , \qquad (94)$$

with $\tilde{n} = 2Ea/\tilde{c}$. The speed of sound has again been renormalized everywhere for self-consistency. It is easy to see from Eq. (26a) that the value of the localization length will be only weakly dependent on the specific choice of the cutoff. In the figures we have always chosen $r = 1$.

Once the abovementioned substitutions are made, Eq. (26a) is solved numerically. We first consider VDW scatterers. The lengths $\xi$ and $L_a$ are shown in Fig. 30 for $R = 1.26$ and $n = 40$ strips/cm. They are periodic functions of the frequency and $\xi = 1.15 L_a$ everywhere. The most interesting features in Fig. 30 are the wide transmission resonances, which occur at values of the third sound wavelength $\lambda = c/\nu$ that satisfy $j\lambda/2 = 2a$, for $j = 1, 2, 3 \ldots$ These values correspond to the eigenfrequencies of an isolated piece of film located on a strip of width $2a$. For these frequencies the transmission coefficient for an isolated scatterer is unity; since the scatterers are identical it follows that the random substrate is transparent to the third-sound excitation.

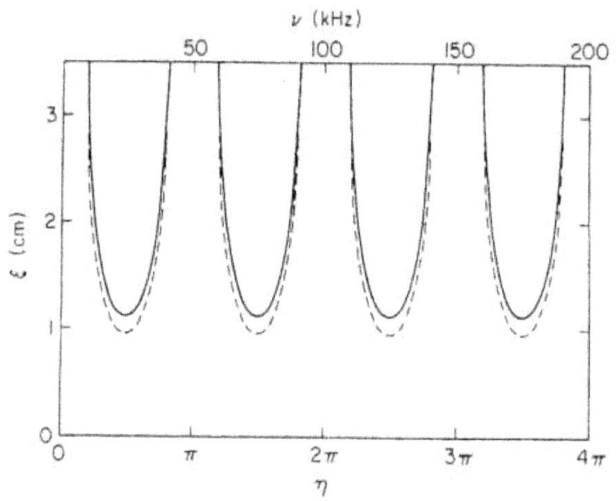

Fig. 30. Localization (soild line) and attenuation (dashed line) lengths for VDW scatterers with $R = 1.26$. The horizontal scales are identical to those in Fig. 28.

The resonances are obtained only if we perform an exact evaluation of the $T$ matrix. A long-wavelength approximation to the $T$ matrix, as in Ref. 57, or a white-noise representation of the scattering, as in Ref. 20, do not give any information about resonant behavior.

It is important to remark that reasonably short localization lengths ($\xi \sim 1$ or 2cm) can be obtained at frequencies as low as 20kHz, well within the usual experimental range.[117,118] In the long-wavelength limit we obtain

$$\xi(E) = \frac{1}{n} \left( \frac{\tilde{c}}{Ea} \right)^2 \frac{R^2}{(R^2+1)(R-1)^2} \ . \tag{95}$$

The $E^{-2}$ behavior agrees, as it should, with the GN results.[13,20]

Near the resonances, the localization length can be easily shown to diverge as $|\nu - \nu_j|^{-2}$, where $\nu_j$ is the frequency corresponding to the $j$th resonance. Defining the resonance width $\delta\nu$ as the length of the neighborhood of $\nu_j$ such that $\xi(\nu) > \xi_0$ if $|\nu - \nu_j| < \delta\nu$, we obtain

$$\delta\nu \sim \tilde{c}R[2\pi a|R-1|]^{-1}[(R^2+1)n\xi_0]^{-1/2} \ . \tag{96}$$

Here $\xi_0$ is a length that may be arbitrarily chosen. The width of the resonances is such that they can be considered truly "pass" bands. The resonances become sharper if $R$ is increased.

Next we discuss the more complicated case of IR scatterers. The different effective speed of sound on the strips generates a phase change that is the source for a richer structure in the localization length. We plot $\xi(E)$ and $L_a(E)$ in Fig. 31 for several values of the index of refraction and $n = 20$ strips/cm. The unrenormalized speeds of sound and the indices of refraction are again approximately those obtained by Smith and Hallock using Si substrates.[117] We observe the following:

(1) All the resonances occur at values of the frequency such that the wavelength $\lambda_s = c_s/\nu$ for the third-sound on the strip satisfies

$$j\lambda_s/2 = 2a \ , \tag{97}$$

for $j = 1, 2, 3, \ldots$

(2) There are two types of resonances. While at the thinner ones (type I) the localization diverges as $\xi(\nu) \sim |\nu - \nu_j|^{-1}$, at the wider ones (type II), it diverges as $\xi(\nu) \sim |\nu - \nu_j|^{-2}$. The typical width of a type-II resonance is

$$\delta\nu_{\mathrm{II}} \sim 0.2\tilde{c}(a|N^2-1|)^{-1}(n\xi_0)^{-1/2} \ , \tag{98}$$

while type-I resonances have a width $\delta\nu_{\text{I}} \sim A_j (n\xi_0)^{-1/2}\delta\nu_{\text{II}}$, where $A_j$ is a coefficient of the order of one, whose value depends on the particular resonance.

(3) For a type-II resonance to occur, $N$ must be a rational number; if $N = P/Q$, the following condition must be satisfied, in addition to Eq. (97):

$$k\lambda_s = 4a/m \, , \tag{99}$$

where $m$ is the integer appearing in the denominator that results from the reduction of the fraction $(P - Q)/2P$ to its lowest terms, and $k = 1, 2, 3, \ldots$ It is easy to see that $m = 6$ in the case of Fig. 31(a), $m = 4$ in the case of Fig. 31(b), and $m = 3$ in the case of Fig. 31(c).

(4) Experimentally, it is impossible to fix $N$ to be a rational number but we expect that a $|\nu - \nu_j|^{-2}$ behavior will be observed whenever Eq. (99) is approximately satisfied. For irrational indices of refraction, however, the $|\nu - \nu_j|^{-2}$ behavior will cross over to a $|\nu - \nu_j|^{-1}$ divergence sufficiently close to the resonance.

The attenuation length is shorter than the localization length (except near the type-II resonances) and does not show type-I resonances. This last result will be shown below to be an artifact of the averaging procedure.

It is interesting to analyze in more detail the reasons for the different behavior of $\xi$ and $L_a$, for this will explicitly show why the calculation of the averaged Green's function can lead to wrong predictions. Using the results for the $T$ matrix given in Ref. 58, it is easy to verify that the matrix elements $\langle p|\widehat{T}^+|p\rangle$ and $\langle p|\widehat{T}^+|-p\rangle$ evaluated at $p = E/c$ vanish at type-II resonances (as well as for the VDW scatterers), while for type-I resonances $\langle p|\widehat{T}^+|-p\rangle = 0$ but $\langle p|\widehat{T}^+|p\rangle = (Eci/\pi)[1 - (-1)^j \exp(-2iEa/c)]$. Although the ratio between the transmitted and incident amplitudes is unity for type-II resonances, it is

$$F/A = (-1)^j \exp(-2iEa/c) \, , \tag{100}$$

for type-I resonances. We conclude that the excitation does not interact

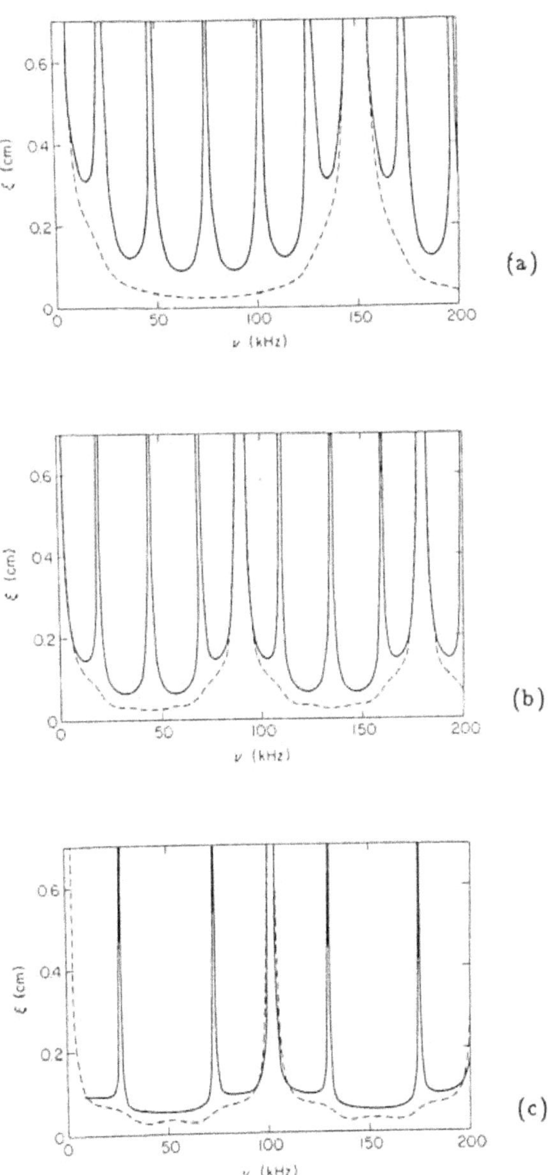

Fig. 31. Localization (solid line) and attenuation (dashed line) lengths for IR scatterers with (a) $N = 1.5$ and $c = 750$ cm/sec, (b) $N = 2$ and $c = 900$ cm/sec, and (c) $N = 3$ and $c = 2040$ cm/sec.

at all with the scatterers when the frequency is at a type-II resonance: neither the amplitude nor the phase are modified. At type-I resonance there is a phase change even if the transmission coefficient is unity.

To understand the reason for the behavior we have found for the attenuation length, we consider the nature of the configurational average. Since the number of strips between the source and the detector depends on the particular element of the ensemble of substrates we consider, the total phase lag varies from element to element. Thus, the spatial average involves a sum over random phases, and we conclude that although there is resonant transmission for all members of the esemble, the propagation on the "average" substrate is strongly attenuated. This unphysical behavior is due to the interactions between different systems in our esemble or configurational average. For the same reason, we conclude that our value for $L_a$ underestimates the real attenuation length for a given experimental system, even far from the resonances. This effect becomes more pronounced when there is none or very little destructive forward-wave interference on each element of the ensemble, a condition that occurs in the neighborhood of type-I resonances.

The localization of the excitations is due to coherent backscattering. The calculation of $\xi$, involving the average of the squared Green's functions, is free of the problem of artificial phase cancellations. A discussion of the relation between $\xi$ and $L_a$ for the case of an electron in a weakly disordered lattice was given by Thouless.[119] Lu, Nelkin and Arita dealt with the same problem for a glasslike disordered elastic chain.[120]

In conclusion, it has been shown that transmission resonances should occur at a discrete set of frequencies; they correspond to a set of measure zero of extended states in the disordered system. Far from the resonances, the localization length can be very short, especially for IR scatterers. We believe that, although the excitations are mostly confined to a small region containing only a few scatterers, their tails should be able to sense the disorder at distances beyond $\xi$. As we move toward the resonances, the localization weakens continuously, until the excitations can be considered to be essentially extended sufficiently close to the resonances. For typical values of the experimental parameters, the strongly frequency-dependent localization length should be readily observable with present-day techniques. Due to the simplicity of the system, we expect that the analysis of the experimental data can be made following closely the proposed theoretical model.

The apparition of a discrete, infinite spectrum of delocalized eigenstates in a random one-dimensional system was noticed a few years ago by Denbigh and Rivier, who consider the problem of an electron in a potential consisting of identical, randomly-spaced rectangular barriers.[121]

In the case of third-sound excitations on striped substrates, the presence of overlapping scatterers — which will make the effective width of some scatterers to be larger than 2a — as well as microscopic differences between the strips will probably cause a lowering of the resonance peaks. However, we still expect the bands of enhanced transmission to be clearly present.

## 5.3. Other Random Substrates

The van der Waals interaction makes the helium atoms prefer substrate regions having negative curvature. Consequently the film will be thicker in regions where the substrate is concave and thicker where it is convex. (To accurately obtain the film profile the surface tension must also be taken into consideration). We will now discuss two possible configurations where the substrate geometry seems appropriate for localization studies.

### 5.3.1. *One Dimension: Grooved Substrate.*

Generazio and co-workers[118,122] have studied experimentally the effects that an isolated "groove" or "edge" has on the propagation and attenuation of third-sound waves on an otherwise smooth substrate. A sketch of the film profile on a groove is shown in Fig. 32. This suggests the possibilities of considering a collection of parallel, identical grooves distributed at random distances from each other on a given substrate. Since the isolated scatterer effects can be computed by the methods used in Ref. 118, it should be possible to analyze quantitatively the modifications introduced by the disorder in the distance between irregularities.

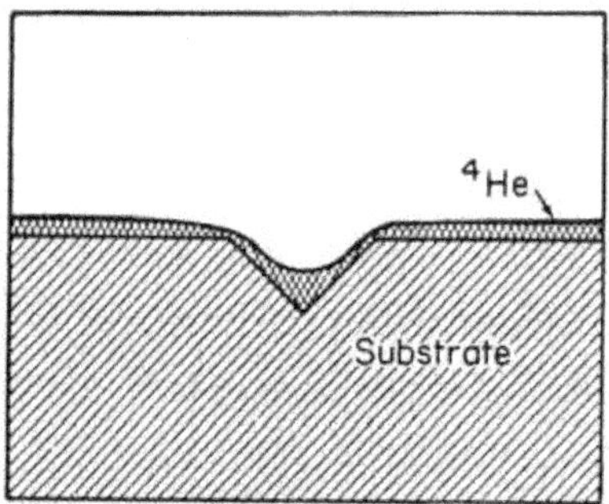


Fig. 32. Modification of film thickness due to a substrate groove.

As for the result of a third-sound experiment carried out on a substrate with a random distribution of grooves, we expect the $(nE^2)^{-1}$ dependence of the localization length to be valid for frequencies such that the wavelength is larger than the groove size. At higher frequencies a more complicated behavior, resulting from interaction with the eigenmodes of the puddles at the bottom of the grooves is to be expected. In any case, the more acute the angles at the edges and bottom of the groove, the more intense is the expected scattering and the shorter the localization length.

5.3.2. *Two Dimensions: Dusted Substrate.* It is possible to create substrates containing random two-dimensional distributions of VDW or IR scatterers. These can be made forming circles, squares, etc., which would allow to study the influence of scatterer shape and size on the third-sound excitations. A different type of random two-dimensional system can be obtained by dusting an otherwise smooth substrate. This idea, put forward by Cohen and Machta,[123] was the first proposal that included third-sound waves as the excitations to be localized. As shown in Fig. 33, a large amount of helium is collected near the base of each dust particle, which then makes a very efficient scatterer. The effective area $\alpha$ of each scatterer (i.e. the area of the region where the amount of fluid has been enhanced) will be typically much larger than the geometric cross section of a dust particle.

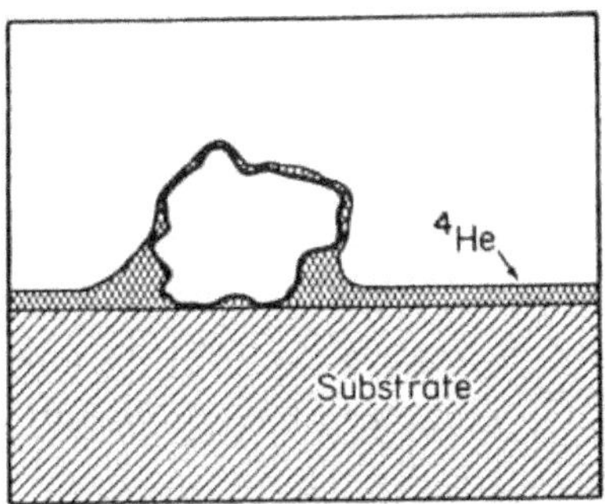



Fig. 33. Modification of film thickness due to a particle of dust.

Using a Gaussian noise representation for the scatterers, which are assumed to be distributed with an average area density $n$, Cohen and Machta found that the renormalized speed of sound has the form[123]

$$\tilde{c} = c(1 + n\alpha)^{-1/2} \, , \tag{101}$$

where $c$ is the speed of third sound on a clean substrate. Equation (101) allows one to define an index of refraction $N = c/\tilde{c}$ of the dusted substrate. All states were found to be localized, with a localization length

$$\xi(E) = l_T(E) \, \exp[(E_0/E)^2] \, , \tag{102}$$

where the frequency $E_0$ is given by

$$E_0 = (2\pi n)^{1/2} Nc/(N^2 - 1) \, . \tag{103}$$

[Cohen and Machta actually took $Q = L_a^{-1}(E)$ as the momentum cutoff, obtaining the "attenuation length" $L_a(E)$ instead of $l_T(E)$ as the prefactor in Eq. (102). As pointed out before, the cutoff appears only as multiplicative factor in the expression for the localization length.]

Equation (102) is only valid for frequencies $E < E_0$, i.e., roughly for wavelengths longer than the perimeter of the effective scatterer area. At higher frequencies, we expect $\xi(E)$ to level off and finally, for $E \gg E_0$, in the "geometrical optics" limit, it should increase again. Since the scatterer shapes will not be uniform, we do not expect any resonant behavior in the localization length.

## 5.4. The Effects of Intrinsic Dissipation

Next we will consider some of the consequences of introducing a finite dissipation constant $\Gamma$, as in Eq. (87). Although there is not much experimental data on $\Gamma$, it is reasonable to assume it is a function of the substrate.[104] Indeed, it is quite possible that it is considerably higher for a film on a rough substrate. A complete theory should introduce two $\Gamma$'s, one accounting for dissipation on the original substrate, and the other accounting for dissipation on the strips. However, since intrinsic dissipation effects at low temperatures are expected to be small, it seems sensible to take a single effective damping constant valid everywhere.

It is not difficult to calculate the modifications suffered by the averaged Green's function in the presence of a finite value of $\Gamma$. For simplicity, we will restrict ourselves to the case of VDW-strip scatterers. First, we note that although the symmetry relations (89b) and (89c) are still valid, $\langle p | \widehat{T}^+ | p_1 \rangle$ and $\langle p | \widehat{T}^- | p_1 \rangle$ are no longer the complex conjugates of each other. The optical theorem is also modified because of the dissipation inside of the scatterer. Equation (90) should be replaced by

$$
\begin{aligned}
\mathrm{Im}\langle p | \widehat{T}^\pm | p \rangle = {}&\pm (\pi/4Ec)(1 + R^2)^{-1}[(1 + R)^2 \\
&\times \exp(\pm \Gamma a/c) + (1 - R)^2 \exp(\mp \Gamma a/c)] \\
&\times (\langle p | \widehat{T}^+ | p \rangle \langle p | \widehat{T}^- | p \rangle + \langle p | \widehat{T}^+ | -p \rangle \langle -p | \widehat{T}^- | p \rangle) \,,
\end{aligned}
\tag{104}
$$

for $p = E/c$. If $\Gamma = 0$, there is energy conservation and Eq. (90) is recovered. An explicit calculation[58] of the self-energy yields, for the renormalized speed of sound $\tilde{c}$ and renormalized dissipation constant

$$\tilde{\Gamma}[= \Gamma + E^{-1}\gamma(E)],$$

$$\tilde{c} = c\left[1 - \frac{na(R-1)^2 H_- \sin(2\eta)}{2\eta[4R^2 + (R^2-1)^2 \sin\eta]}\right] , \qquad (105a)$$

and

$$\tilde{\Gamma} = \Gamma + \frac{4cnH_+(R-1)^2 \sin^2\eta}{(H_-)^2 + 4(R^2-1)^2 \sin^2\eta} \,. \qquad (105b)$$

Here $\eta = 2Ea/c$ and we have introduced the auxiliary functions

$$H_\pm = (1+R)^2 \exp(\Gamma a/c) \pm (1-R)^2 \exp(-\Gamma a/c) , \qquad (106)$$

which for small values of $\Gamma$ can be approximated by $H_+ \to 2(1+R^2)$ and $H_- \to 4R$.

The energy density propagator in the case of finite $\Gamma$ was computed in Ref. 13 starting from a Gaussian noise model and using the Berezinskii technique.[124] If $t > \Gamma^{-1}$, one finds

$$\langle P(x,t|x',0)\rangle \sim e^{-\Gamma t} e^{-|x-x'|/\xi(E)} . \qquad (107)$$

We note that the temporal decay of the localized excitations has the same form as the attenuation of the propagating modes on a "smooth" substrate. This is not surprising since in the model we used the damping processes are associated with the film itself or with its interaction with the substrate; the scatterers do not play any role as specific sources of dissipation.

We believe it should not be too difficult for an experimentalist to decide whether he is observing an exponential decay due to localization or an exponential decay due to dissipation. On one hand, the frequency dependence of both processes will be certainly different. On the other hand, a signature of the localized excitations is its permanence: if the source is stopped at $t = 0$, a detector located at a distance $d_1 \lesssim \xi$ from the source (see Fig. 27) should record variations in film thickness for times as large as $\Gamma^{-1}$. These times will be typically of the order of seconds, i.e., several orders of magnitude longer than the time $\tau = d_1/\tilde{c}$

it will take a propagating wave to reach — and leave behind — the detector. We also note that if the localization length is not much shorter than the length of the random system, the energy leaks at the ends of the random region will introduce another time-decay constant.

## 5.5. The Experiment of Smith *et al.*

In this subsection we describe an experiment very recently carried out at the University of Massachusetts.[59] The propagation of third sound waves on helium films adsorbed on silicon substrates was investigated. For disordered substrates, the results are in substantial agreement with the predictions of subsection 5.2.

A configuration similar to that shown in Fig. 27 was created on a circular silicon platelet. Two IR substrates were used; in both of them 100 abraded strips having width $2a = 16 \pm 0.6 \, \mu$m and and length 1.5 cm were arranged, parallel to each other, over a length of 5 cm. In the first substrate the arrangement is periodic and the distance between strips is $l = 500 \pm 4.3 \, \mu$m, while in the second the arrangement is random and the strip density is $n = 20 \; \text{cm}^{-1}$. Two superconducting detectors $d_1$ and $d_2$ were positioned, respectively, at distances of 0.25 cm and 1.0 cm from the generator $s$. The sensitivity at high frequencies is limited by the finite width of the detectors. As a consequence, only the range $\nu < 60$ kHz was investigated.

Since the thickness of the film can be controlled, the speed of sound and the index of refraction can be modified. Experimental runs were made for thickness varying between 6 and 20 atomic layers. The amplitudes recorded for both types of substrates are shown in Fig. 34 for a film thickness of 8.3 atomic layers. Let us discuss separately the results obtained for both configurations.

5.5.1. *The Periodic Substrate.* As mentioned before, the properties of third-sound propagation on a periodic VDW substrate were studied in Ref. 60. A similar analysis for the IR substrate is straightfoward. Defining

$$\chi_1 = \cos \eta \cos(N\eta) + \beta \sin \eta \sin(N\eta)$$

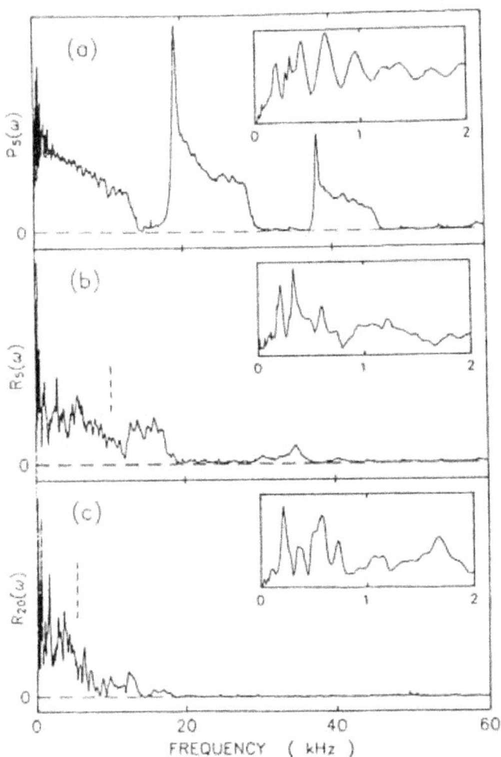

Fig. 34. Amplitude (arbitrary units) vs. frequency for a film thickness of 8.3 atomic layers ($c = 1945$ cm/sec). $P_s$ is the amplitude measured at $d_1$ in the case of the periodic substrate (the distance between $S$ and $d_1$ is of $5l$). $R_5$ and $R_{20}$ are respectively the amplitudes recorded by $d_1$ and $d_2$ in the case of the random substrate. The vertical dashed lines indicate the values of scaling frequencies generated by the data analysis (see Ref. 59). Expanded views of the amplitude for $\nu < 2$ kHz are shown in the insets. (After D. T. Smith *et al.*[59]).

$$\text{and} \tag{108}$$

$$\chi_2 = sin\eta \cos(N\eta) - \beta \cos\eta \sin(N\eta) \, ,$$

with $\beta = (N^2 + 1)/2N$, the eigenvalues of the transmission matrix have the form

$$P_{\pm} = H(E) \pm [H^2(E) - 1]^{1/2} \, , \tag{109}$$

where

$$H(E) = \chi_1 \cos(El/c) + \chi_2 \sin(El/c) \ . \tag{110}$$

The form of Eq. (109) implies the existence of a band structure: there are transmission gaps for the values of $E$ such that $|H(E)| > 1$. The band edges $E_j$ are given by the solutions of $|H(E_j)| = 1$. It is not difficult to verify that, for $a \ll l$, the lower band edges of the second, third,... bands occur at $E_j l/c = \pi, 2\pi, \ldots$ (under certain conditions some gaps may be missing).

In the interior of the bands we can define the Bloch wavevector $K(E)$ through

$$\cos K(E) = H(E) \ . \tag{111}$$

The number of states per unit length and unit range of frequency, $\rho(E)$, can thus be obtained as

$$\rho(E) = \frac{1}{\pi} |\frac{dE}{dK}|^{-1} \ . \tag{112}$$

A complicated expression — but one that contains no adjustable parameters — is then obtained for $\rho(E)$. We will only mention that near the band edges a singularity of the form $\rho(E) \sim |E - E_j|^{-1/2}$ is predicted.

Using $l = 500$ $\mu$m and $c = 1945$ cm/sec, we can see that the values $E_j = \pi c/l$ and $2\pi c/l$ give quite well the positions of the lower edges of the second and third bands in Fig. 34(a). While the divergence is clearly visible at the lower band edges, it has completely disappeared from the upper band edges. A simple qualitative explanation for this phenomenon can be obtained from the obvious analogy between the IR system and a diatomic linear lattice, the sections of film adsorbed on the abraded regions corresponding to the heavy masses. Since the strips are not totally identical to each other, there is some site disorder. Some of the strips will be less (more) efficient than the average; in our analogy one of these corresponds to a slightly lighter (heavier) impurity replacing one of the heavy masses. In the case of a replacement by a heavier impurity nothing dramatic occurs, but the replacement by a lighter impurity leads to localized states splitting off the top of the acoustical and optical bands[125,126] and to the removal of the singularity at the upper edges. The singularity at the bottom of the optical band is preserved. This is related to the fact that at the upper edges of the bands the heavy masses are in motion.[127] On the contrary, at the bottom of the optical band, the heavy masses are at rest, and the replacement of one of them by an impurity causes no important effects.[128]

5.5.2. *The Random Substrate.* The amplitudes measured by the detectors $d_1$ and $d_2$ for a particular value of the film thickness are shown in Figs. 34 (b) and (c), respectively. We first note that, according to Eq. (97), the first transmission resonance is predicted to occur at a frequency of about $1.8 \times 10^5$ Hz. (Assuming $N \simeq 3.4$, as found by Smith *et al.*). The localization length is thus a monotonically decreasing function of the frequency in the range investigated, and the data analysis was performed using the long-wavelength approximation for the localization length,

$$\xi_1(E) \simeq \frac{5.2\theta_1\theta_2}{n} \left(\frac{c}{2Ea}\right)^2 \frac{1}{(N^2 - 1)^2} . \tag{113}$$

The magnitude of the factors $\theta_1$ and $\theta_2$ is close to, but probably slightly below, unity. The value of $\theta_1$ depends on the choice of the cutoff $r$, varying continuously from $\theta_1 = 1$ for $r = 1$ to $\theta_1 \simeq 0.81$ for $r = 2\pi$. The factor $\theta_2$ appears because it is reasonable to assume that there will be a partial renormalization of the speed of sound. Therefore, $\theta_2 = (c_{\text{eff}}/c)^2$. Since $na \simeq 0.016$, we expect $c_{\text{eff}}$ not to be very different from $c$. (In Ref. 59 the choice $5.2\theta_1\theta_2 = 4$ was made).

In an ideal experiment, one should build an ensemble of substrates and then average the results. However, in this case this can only be a theoretician's dream! The analysis was performed instead using simultaneously the data obtained for several film thicknesses adsorbed on the same subtrate. The data was shown to be consistent with Eq. (113) verifying both the proportionality of $\xi_1(E)$ to $E^{-2}$ and to $(N^2 - 1)^2$. The site disorder mentioned in 5.5.1 is also present here. As discussed by Smith *et al.*, it is responsible for a reduction of the value of the localization length below what is predicted by Eq. (113): Since both types of disorder are independent, one can write $\xi^{-1} = \xi_1^{-1} + \xi_2^{-1}$, where $\xi_2$ is the localization length associated with site disorder. As a consequence, if the frequency were to be increased up to the value of the first resonance predicted in Eq. (97), one would not observe $\xi \to \infty$ but $\xi \to \xi_2$.

In conclusion, the results of this experiment lend, in the frequency range investigated, qualitative and quantitative support for the predictions of the SCT of localization presented in subsection 5.2.

## 6. ACOUSTIC LOCALIZATION WITH A FLOW

In general the qualitative features of the Anderson transition are modified by an external field which breaks time reversal invariance.[2] Such a perturbation destroys the singular nature of the dominant coherent backscattering mechanism (MCD) responsible for localization in systems with time reversal invariance. There are however, other scattering processes that lead to a new mobility edge and a new set of critical exponents even in the presence of this external field.

In this section we discuss the problem of acoustic localization in the presence of a uniform flow field which acts as an external field that breaks time reversal symmetry.[12,13] As mentioned in Sec. 5, the particular experimental realization we have in mind is a flow field in helium on a disordered substrate and the excitations exhibiting Anderson localization are third-sound waves. The formalism, however, is more general.

The techniques involved in solving this problem are outside the main thrust of this review article and as a consequence of this, the discussion here will be brief and somewhat sketchy. For the details we refer elsewhere. We first review how the structure of the perturbations theory changes when there is an external field which breaks time reversal symmetry (TRS). Following this we report exact results for one-dimensional systems with broken TRS. Finally, we indicate how field theory methods can be used to describe the crossover between universality classes near $d = 2$ when a TRS breaking field is applied.

## 6.1. Formulation of the Problem and Perturbative Results

To simplify our discussion we use a Gaussian noise similar to the one introduced in Sec. 4. In the presence of a uniform flow the equation describing the propagation of third sound on a substrate with random scatterers is,[12]

$$\left\{ \frac{D}{Dt}[1 + \psi(\mathbf{x})]\frac{D}{Dt} - c^2 \nabla^2 \right\} \phi(\mathbf{x},t) = 0 \ . \tag{114}$$

Here $\phi$ is the velocity potential, $\psi(\mathbf{x})$ is a Gaussian random function with correlation given by Eq. (74), $c$ is the effective speed of sound on the substrate and

$$\frac{D}{Dt} = \frac{\partial}{\partial t} + \mathbf{u} \cdot \nabla \tag{115}$$

is the convective derivative with $\mathbf{u}$ the flow field velocity.

Using Eq. (114) as the basic equation the steps outlined in Sec. 2, where $\mathbf{u} = 0$, can be repeated. After some lengthy but straightfoward algebra the following results emerge:[13]

(1) The energy density propagator, $[P_{\omega,k}(E)]_{\mathrm{av}}$, is still diffusive. For example, in the one-dimensional case the Boltzmann result is,

$$[P^B_{\omega,k}(E)]_{\mathrm{av}} = \frac{\tilde{c}^2 - u^2}{2E^2 \tilde{c}^3} \frac{1}{\left[-i\omega + D_B(E,u)k^2\right]} \ . \tag{116a}$$

Here $D_B(E,u)$ is the $u$-dependent Boltzmann diffusion coefficient,

$$D_B(E,u) = 2(\tilde{c}^2 - u^2)^3/\mu E^2 \ . \tag{116b}$$

Physically this implies the expected result that energy is conserved even in the presence of a flow field.

(2) The infrared singularity in the MCD is cutoff due to the flow field. For example, in $d = 1$, the sum of MCD is (compare Eq. (21)),

$$\delta U^M_{pp_2}(0,\omega|E) = \frac{\mu^2 E^2 p(p - l)p_2(p_2 - l)C_+^4 (4\pi\tilde{c}^3 C_-^3 C_u)^{-1}}{-i\omega + \tilde{D}(E,u)l'^2 + 2\mu E^2 u^2 C_+^2 (\tilde{c}C_-^3 C_u)^{-1}} \tag{117}$$

where, $l = |p + p_2|$,

$$l' = l + 2uE(\tilde{c} - u^2)^{-1} \tag{118a}$$

and

$$\tilde{D}(E, u) = 2\tilde{c}C_-^5 (\mu E^2 C_u)^{-1} \tag{118b}$$

$$C_\pm = \tilde{c}^2 \pm u^2 \tag{118c}$$

$$C_u = \tilde{c}^4 + u^4 + 6\tilde{c}^2 u^2 \ . \tag{118d}$$

Examining Eqs. (117) and (118) we see that the presence of a uniform flow field brings about the following modifications to $\delta U^M$.

(a) The maximum contribution occurs for $l = E[(\tilde{c} + u)^{-1} - (\tilde{c} - u)^{-1}]$. This means that the constructive interference essential for localization occurs between excitations whose wavenumbers are Doppler shifted.

(b) In the absence of flow, the coefficient of $l^2$ coincides with $D_B(E,0)$. If $u \neq 0$ the coefficient of $l'^2$ is smaller than $D_B(E, u)$ by a factor $C_-^2 C_u^{-1}$.

(c) As already mentioned, the most important consequence of flow is that the hydrodynamic pole has been cutoff by a term of the order of $\mu u^2$. If the MCD were the only set of diagrams responsible for localization, we would conclude that the flow destroys localization. The same conclusion would be reached if we accepted the premise that the $\omega \to 0$ behavior of the MCD is typical of that of the other sets of diagrams contributing to localization. However, such a conclusion is not warranted. As Eq. (116) indicates, the singularity in the ladder diagrams is *not* cutoff by the flow field. At higher order, beyond the MCD, the ladder diagrams contribute singular contributions to $\delta U$. Therefore we can only anticipate that in general flow modifies but does not necessarily destroy localization.

Finally, we mention that technically the differences between zero flow and nonzero flow occurs because the reciprocal relation above Eq. (21) is not satisfied for $u \neq 0$. This in turn reflects broken TRS.

## 6.2. Exact Results for the One-Dimensional Case

Berezinskii[124] has introduced an exact low-density analysis of localization in the one-dimensional electron localization problem. In this subsection we briefly comment on his ideas and how they can be modified to study one dimensional acoustic localization with flow.

The crucial step in Berezinskii's solution technique is the introduction of diagrams that are explicitly ordered in coordinate space. For the one-dimensional problem this is extremely useful because the coordinate dependence in the free Green's function $G_{E,0}(x_i.x_j)$ uniquely factorizes for $x_i$ $x_j$. With this factorization, all factors in diagarmmatic development can be moved to the vertices. Berezinskii[124] has determined all of the relevant topologically distinct vertices and has shown how all of the relevant diagram elements can be resumed.

Berezinskii's ideas can be adopted to the acoustic localization with flow problem with only minor modifications.[13] The most important one being that the flow introduces an asymmetry so that the directions in which the GF propagate must be taken into account. The easiest quantity to calculate within the Berezinskii formalism is the long time limit of the energy density propagator. After some lengthy asymptotic analysis (following Ref. 129) we obtain, in the region $\mu E^2 |x - x'| \gg c^2(c^2 - u^2)^2$,

$$[P_E(x, x'|t \to \infty)]_{\text{av}} \simeq \frac{\mu \pi^{5/2}}{2048 c^5 (c^2 - u^2)} \frac{l_u^{3/2}}{|x - x'|^{3/2}} \exp[-|x - x'|/l_u] .$$

$$(119a)$$

The new length we have introduced,

$$l_u = 16 c^2 (c^2 - u^2)^2 / \mu E^2 \qquad (119b)$$

is clearly the localization length of the excitations.

Examining Eqs. (119) the following comments should be made:

(1) Equation (119) indicates that even in the presence of flow the acoustic excitations are localized. In the absence of flow the asymptotic form of the energy propagator is given by Eq. (119) with $u = 0$.

(2) This indicates that the perturbative approach in one-dimensional systems can be very misleading. Localization is not destroyed by the flow and, in fact, Eq. (119b) indicates that for $u \neq 0$ the localization length is smaller by a factor $(1 - u^2/c^2)^2$. In the next subsection we argue that this conclusion is very special to one-dimensional systems. For higher dimensional systems, the presence of flow drastically modifies the localization properties. Technically the crucial difference is that in one-dimensional systems there are diagrams other than the MCD which contribute to localization to the same order in the scatterer density. This is not the case in higher dimensions where well defined density expansions are possible.

(3) In the absence of flow the Eqs. (119), with $u = 0$, indicate that the perturbative approach used in the bulk of this review is a reasonable approach. The localization length given by Eq. (119b) is approximately equal to twice that obtained using the self-consistent diagram approach. The general conclusion is that when TRS is valid the self-consistent formalism is qualitatively correct in one-dimensional systems. This is important because, due to the lack of a well-defined density expansion, the application of the self-consistent formalism is more suspect in one-dimensional than for higher dimensionality systems.

## 6.3. Field Theory Methods and Acoustic Localization with Flow

The field theory approach to the localization problem[16-19] starts with the assumptions that the Anderson transition is continuous and that all of the important physics is associated with the hydrodynamic poles from ladder diagrams and MCD. Given this, the first step in the field theoretic approach is to construct a field theory for these long wavelength interacting hydrodynamic poles.

Technically, one begins the derivation of the field theory by using the replica trick to write energy density propagator in terms of functional integrals.[16-19] For the acoustic problem with flow one finds,[12]

$$[P_\omega(\mathbf{x}, \mathbf{x}'|E)]_{\mathrm{av}}$$
$$= \lim_{n \to 0} \int [d\phi]\phi_{1+}^*(\mathbf{x})\phi_{1+}(\mathbf{x}')\phi_{1-}^*(\mathbf{x}')\phi_{1-}(\mathbf{x})\exp\{-H[\phi]\} , \tag{120a}$$

with,

$$H[\phi] = \int d\mathbf{x} \sum_\alpha \phi_\alpha^*(\mathbf{x})s_\alpha[D_\alpha^2 + c^2\nabla^2]\phi_\alpha(\mathbf{x})$$
$$- \frac{\mu}{2}\int d\mathbf{x} \sum_{\alpha\beta} s_\alpha s_\beta |D_\alpha\phi_\alpha(\mathbf{x})|^2 |D_\beta\phi_\beta(\mathbf{x})|^2 , \tag{120b}$$

where $D_\alpha = E + is_\alpha\omega/2 + i\mathbf{u}\cdot\nabla$. The index $\alpha$ has two components: $\alpha = (a, p)$ with $a = 1, 2, \ldots, n$ the replica index and $p = +, -$ indicating advanced or retarded Green's functions. $[d\phi]$ indicates functional integration over all fields and $s_\alpha \equiv ip$ ensures the convergence of the functional integrals.

Except for the $s_\alpha\omega$ term the Hamiltonian is invariant under pseudounitary, $U(n, n)$, transformations of the fields. Note that as $u \to 0$, the real and imaginary parts of $\phi_\alpha$ can be treated as independent components and the Hamitonian has a larger pseudoorthogonal, $O(2n, 2n)$, symmetry.

Here we outline how to construct a field theory which incorporates the orthogonal and unitary symmetries and the crossover between them. To this end we choose real fields having three indices, $\alpha = (a, p, \sigma)$, with $a$ and $p$ as before and $\sigma = \uparrow, \downarrow$. The complex fields are replaced by two real fields with $\phi_\uparrow$ replacing $\mathrm{Re}\phi$ and $\phi_\downarrow$ replacing $\mathrm{Im}\phi$. Complex operators in Eq. (120b) are written as $2 \times 2$ matrices acting on the spin components of the fields according to the prescriptions,[12]

$$0 \to \mathbf{0} = \tau_0 \times \mathbf{1}(\mathrm{Re}\,0) - i\tau_2 \times \mathbf{1}(\mathrm{Im}\,0) , \tag{121}$$

where $\tau_j, j = 1, 2, 3$ are the Pauli matrices, $\tau_0$ is the $2 \times 2$ identity matrix and $\mathbf{1}$ is the $2n \times 2n$ identity matrix. The multiplication signs in

Eq. (121) indicate Kronecker products. In this way $H$ is transformed into a $4n \times 4n$ real representation. For future use we note that general $4n \times 4n$ real matrices can be rewritten as $2n \times 2n$ quanternion matrices and quanternion having only $\tau_0$ and $\tau_2$ components represent complex quantities. This is the key idea in describing the crossover from a real (orthogonal) field theory to a complex (unitary) field theory.

The next step in the construction of the appropriate field theory is to decouple the quartic term in Eq. (120b) by using a Gaussian transformation,[16-19]

$$\exp\left\{\frac{\mu}{2}\int d\mathbf{x}[(\mathbf{D}\boldsymbol{\phi})^T\mathbf{S}\,\mathbf{D}\,\boldsymbol{\phi}]^2\right\}$$
$$= \int [dQ]\,\exp\left\{-\int d\mathbf{x}[\frac{1}{2}\,\mathrm{Tr}\,\mathbf{Q}^2 - \mu^{1/2}\boldsymbol{\phi}^T\mathbf{S}^{1/2}\mathbf{D}\,\mathbf{Q}\,\mathbf{D}\,\mathbf{S}^{1/2}\boldsymbol{\phi}]\right\}, \tag{122}$$

where the functional integral is over the space of real $4n \times 4n$ symmetric matrices. $\mathbf{S}$ is a diagonal matrix having $s_\alpha$ along the diagonal. At this stage a gauge transformation can be used to eliminate the u-dependence from the quadratic part of Eq. (120b).

With Eq. (122) the $\phi$ integration can be done yielding a field theory of interacting $\mathbf{Q}$-fields with Hamiltonian,

$$H[Q] = \frac{1}{2}\{\mathrm{tr}\,\mathbf{Q}^2 + \mathrm{trlog}\,\mathbf{C}[\mathbf{Q}]\}\,, \tag{123a}$$

where,

$$\mathbf{C}[\mathbf{Q}] = E^2\mathbf{1} + c^2\nabla^2\mathbf{1} + i\omega E\mathbf{S} - \mu^{1/2}\mathbf{D}\,\mathbf{Q}\,\mathbf{D}\,, \tag{123b}$$

and "tr" indicates a trace over discrete and continuous labels. All quantities of interest can be related to the average of powers of $Q$ weighted by $\exp\{-H\}$.

To obtain a tractable field theory from Eq. (123) the first step is to obtain a saddle point solution to Eq. (123a) and then expand about the saddle point to second order in the fluctuations, $\delta Q = Q - Q_{sp}$. Examining the structure of the theory for $\mathbf{u} = 0$, one finds that there are both

Goldstone modes, reflecting the broken symmetry as $\omega \to 0$ between advanced and retarded Green's function, and massive modes.[130] Assuming the Anderson transition is controlled by the Goldstone modes, the massive fluctuations can be integrated out in general. For long wavelengths and small frequencies the final field theory for the Goldstone modes (which are just the ladder and MCD hydrodynamic poles) is described by the Hamiltonian,

$$H[\mathbf{Q}] = \frac{1}{2t} \int d\mathbf{x}\ \mathrm{Tr}[\boldsymbol{\nabla}\mathbf{Q}\cdot\boldsymbol{\nabla}\mathbf{Q} + h\,\mathbf{S}\,\mathbf{Q} + g(\mathbf{Q}^2 - \mathbf{Q}\tau_2\mathbf{Q}\tau_2)] \ . \quad (124)$$

The constants, $t, h,$ and $g$ can be explicitly calculated when $u/c$ and $\mu(E/c)^d$ are small and one finds,[12]

$$t = dE^2\mu/c^2 \qquad\qquad\qquad (125a)$$

$$h = t\omega \qquad\qquad\qquad (125b)$$

$$g/2t = 2u^2 q_0^2/dc^2 \qquad\qquad\qquad (125c)$$

where $q_0 = \mu^{1/2}\pi\Omega_d E^d/4(2\pi c)^d$. In the field theory, $t \sim \mu$ is the coupling constant, $h \sim \omega$ is analogous to a magnetic field in usual phase transition problems, and $g \sim u^2$ is like a perturbing anisotropy. The fields in Eq. (124) have been rescaled, $\mathbf{Q} \equiv \delta\,\mathbf{Q}/q_0$, and integrating out the massive fluctuations leads to the constraints

$$\begin{aligned} \mathbf{Q}^2 &= -1 \\ \mathrm{Tr}\,\mathbf{Q} &= 0 \ . \end{aligned} \qquad\qquad (126)$$

In actual calculations it is convenient to use a quanternion representation,[12]

$$\mathbf{Q} = \frac{1}{\sqrt{2}} \sum_j \tau_j Q^j \ . \qquad\qquad (127a)$$

Note that

$$\frac{1}{2}\,\mathrm{Tr}(\mathbf{Q}^2 - \mathbf{Q}\,\tau_2\,\mathbf{Q}\,\tau_2) = \mathrm{Tr}[(Q^1)^2 + (Q^3)^2] \ . \qquad (127b)$$

This indicates that the anisotropy term couples only to the $\tau_1$ and $\tau_3$ quanternion components of $\mathbf{Q}$. The coefficient in Eq. (124) of the fields $Q^0$ and $Q^2$ is the inverse of the sum of the ladder diagrams in a perturbative calculation of the product of two Green's functions. The coefficient of the $Q^1$ and $Q^3$ fields is the inverse of the sum of the MCD. Thus the anisotropy corresponds to the cutoff, due to broken TRS, in the $k \to 0$ divergence of the MCD.

The field theory defined by Eqs. (124)–(127) is in a form where standard renormalization group (RG) techniques can used to describe the Anderson transition. For small $t$ the RG flows are,[12,13]

$$\kappa\frac{\partial t}{\partial \kappa} = \varepsilon t - \frac{t^2}{8\pi(1+g)} + 0(t^3) \quad (\omega = 0\,, g \text{ finite})$$

$$(128a)$$

$$= \varepsilon t - \frac{t^3}{2(8\pi)^2} + 0(t^4) \quad (\omega = 0\,, g = \infty) \tag{128b}$$

$$\kappa\frac{\partial \omega}{\partial \kappa} = -d\,\omega + 0(t^2) \tag{128c}$$

$$\kappa\frac{\partial g}{\partial \kappa} = -2g + 0(t^2)\,, \tag{128d}$$

when $\kappa$ is an arbitrary momentum scale, $\varepsilon = d - 2$, and we denote the RG parameter space by $\mu = (t,\omega,g)$. The fixed points of Eqs. (128) and their stability properties characterize the Anderson transition in acoustic systems with flow.

For $d < 2$ there is a single nontrivial fixed point (FP) at $t = 0$ and the RG flow of $t$ near the FP are unaffected by the velocity field $(\sim g)$. This result is consistent with the discussion in Sec. 6.2 and with the fact that all states are localized in $d = 1$.

For $d > 2$ there are two nontrivial fixed points,

$$\mu_1^* = (8\pi\varepsilon, 0, 0) \tag{129a}$$

and

$$\mu_2^* = (8\pi\sqrt{2\varepsilon}, 0, \infty)\,. \tag{129b}$$

These results are valid only for small $\varepsilon$. The $g = 0$ FP is the orthogonal FP which is the one relevant for systems with TRS. The $g = \infty$ FP is appropriate for systems with broken TRS and it is called the unitary FP. Note that in $d = 2 + \varepsilon$ the disorder at the unitary FP is much greater than at the orthogonal. Physically this implies that a larger scatterer density $(\sim t)$ is needed to localize the acoustic waves when $u \neq 0$.

For finite $g$ the RG flows describe the crossover from orthogonal to unitary localization. The crossover exponent near the orthogonal FP is obtained from the eigenvalues of the RG equations linearized about the orthogonal FP. We obtain,[12]

$$\phi = \nu[2 + 0(\varepsilon^2)]$$
$$\nu = \frac{1}{\varepsilon} + 0(1) , \tag{130}$$

where $\nu$ is the localization length exponent. The exponent $\phi$ describes the crossover behavior of, for example, $\xi$,

$$\xi(t,g) = (\Delta t)^{-\nu} F(g/\Delta t^\phi) . \tag{131}$$

Here $\Delta t = |t - t_c|$ is the distance from the Anderson transition, $F$ is a scaling function, and all lengths are measured in units of the mean free path $l_0 \sim \left(\frac{c}{E}\right)^{d+1} \mu^{-1}$.

For $d = 2$ the Eqs. (128) indicate that all states are localized with and without flow. This follows because the renormalized coupling $t$ in Eqs. (128) always increases for $\varepsilon = 0$ with increasing length scale. The behavior of $\xi$ for small $t$ can be obtained by integrating the RG flows. The localization length is defined as the length scale $L$ at which the coupling constant is of $0(1)$ (Ref. 20). Near the orthogonal FP$(u/c \ll 1)$ the result is,[12]

$$\xi(t,g) = \xi(t,0)F[\xi^2(t,0)g] , \tag{132a}$$

where $\xi(t,0) = \exp[8\pi/t]$ is the orthogonal localization and the crossover function is given by

$$F(x) = 1 + \frac{x}{2} + 0(x^2) . \tag{132b}$$

For $\xi^2(t,0)g \gg 1$, the localization length is controlled by the unitary FP. From Eq. (128b) we obtain,[12]

$$\xi(t, g \gg 1) = \exp[(8\pi/t)^2] . \tag{132c}$$

For third sound propagating on a disordered substrate both the weak $[\xi(t,0) \gg 1]$ and strong $[\xi(t,0) \approx 1]$ localization regimes are predicted to be experimentally accessible in the absence of flow. In the presence of flow the theory outlined here predicts a dramatic increase in the localization length from $\xi \sim \exp(E_0/E)^2$ to $\xi \sim \exp(E_0/E)^4$ as the flow field is turned on. Here $E_0$ is a fixed frequency scale. If such experiments can be carried out then they will provide an important test of localization theory is difficult to observe in electronic systems because of competing phenomena such as electron-electron interactions, the quantum Hall effect, and the Kondo effect.[2]

Finally, we note that the Eqs. (128) in the absence of flow $(g = 0)$ predict that the diffusion coefficient in $d = 2 + \varepsilon$ dimensions satisfies the scaling equation,

$$D(\Delta t, \omega) = b^\varepsilon D(b^{1/\varepsilon}\Delta t, b^d \omega) , \tag{133a}$$

or

$$D(\Delta t, 0) \sim \Delta t \tag{133b}$$

$$D(0, \omega) \sim \omega^{\varepsilon/d} . \tag{133c}$$

These results are consistent with the self-consistent approach to localization described in the bulk of this review.

## 7. CONCLUSIONS

We conclude this review by mentioning a number of problems whose solution would lead to a more detailed understanding of localization physics.

(1) Virtually all method used to study Anderson localization neglect any scatterer or disorder correlation. This is a physically in-

correct approximation when the volume fraction of scatterers becomes too large. It would be an interesting problem to study the interplay between localization physics and the short range order caused by scatterer correlation.[132]

(2) In the self-consistent approach to localization the appearance of an artificial cutoff is unsatisfactory. In principle a natural cutoff is inherent in the theory if all wavenumber dependences in the Green's functions and $T$-matrices in the four-point vertex $U$ are retained. In practice it is difficult although not impossible to retain this wavenumber dependence. A more fundamental problem is that the proper self-consistent behavior at large wavenumber, or small length scales, is not clear. If $k$ is the wavenumber, $a$ is the scatterer diameter, and $l_T$ is a mean free path than these question are relevant both when $kl_T \sim 1$ and, probably more importantly, when $ka \sim 1$.

(3) As seen from subsections 4.3 and 5.5 in many experiments fluctuations in the scatterer shapes and compositions are unavoidable. This is particularly important in the case of resonant scattering, due to the small width of the resonances. The effects of size polydispersity should also be included in the calculations. In the description we have presented, the scatterers have been considered to be identical — or nearly so — in composition, shape and size. Variations in scatterer radius generate a rescaling of the dimensionless frequency $\eta$ and consequently the position of the peaks in $S(\eta)$ changes from scatterer to scatterer. As the distribution $P(a)$ of the radii is widened, the structures of $n_c^*(\eta), \xi(\eta)$, etc., become more and more smeared out, and localization weakens. Fluctuations in scatterer shapes also lead to a smearing out of localization effects; in this case, the $S(\eta)$ for different scatterers cannot be obtained from each other by means of a simple frequency rescaling. Although a similar prediction can be made if one has a distribution of values of $\kappa_s$, the results shown in Fig. 13 indicate that some structure will remain even for a relatively wide distribution.

(4) It would be interesting to explore the possibility of observing localization related effects in elastic waves. This could be done by

extending the calculations presented here to the case in which the scatterers are "real" solids, such that energy conversion between pressure and shear waves is allowed.[73]

(5) The self-consistent approach predicts that near a mobility edge that the diffusion coefficient, $D(\Delta t, k, \omega)$, is a scaling function of $\Delta t$ and $\omega$ but not of the wavenumber, $k$. Here $\Delta t = |t - t_c|$, with $t$ a measure of the disorder ($\sim n$) and $t_c$ the critical amount of disorder for an Anderson transition in $d > 2$ dimensions. The field theory method, on the other hand, predicts that $k$ is a scaling variable with, for example, $D(0, k, 0) \sim k^{d-2}$ as $k \to 0$. This field theory result is probably correct and it is an open problem on how to build this scaling behavior into a self-consistent approach. Beyond the MCD it is easy to identify diagrammatic contributions to $U$ that are singular functions of $k$, but it is not clear how to incorporate them into a self-consistent solution.

(6) It is clear that the lower critical dimension for an Anderson transition is $d = 2$. However, the proper upper critical dimension (UCD) is not clear. For some recent work on this problem see Ref. 133. If an UCD can be established then it would be useful to perform $\varepsilon$-expansions around that dimension and compare the results with expansions around $d = 2$. In this context it is interesting to note that very high order corrections to Eqs. (128) and (133) have been calculated and to $0[(d - 2)^4]$ they are found to vanish.[27] This suggests that if there are corrections to Eq. (133) then they are nonperturbative around $d = 2$. The self-consistent approach suggests an UCD of $d = 4$, although it is possible that an UCD does not exist in the conventional sense.[133]

(7) We have stressed that some acoustic systems are ideal for paradigms for studying pure localization effects. One motivation for this is that effects analogous to electron-electron interaction in electronic systems can be carefully controlled experimentally by adjusting the amplitude or strength of the incident sound waves. An interesting problem would be to study the localization properties of large amplitude sound waves in a disordered medium. Interactions between the sound

waves would then be important although the physics of this interaction problem would be distinct from the interaction problem in electronic systems. For example, a high frequency sound wave can decay into two (or several) lower frequency sound waves. It is unclear whether this phenomena simply leads to a decay of localized high frequency sound waves to extended low frequency sound waves or whether the physics is more interesting.

(8) An open subject is the contribution (or lack thereof) of the localized or nearly localized acoustic modes to the low-temperature thermodynamic properties of glasses and aggregates.[134-139]

In particular, assuming that the long-wavelength acoustic phonons are the dominant heat carriers, the thermal conductivity, $K(T)$, can be estimated by[136]

$$K(T) = \frac{c}{3} \int_0^{\widehat{E}} C(E,T) l(E) dE \ ,$$

where $T$ is the temperature, $c$ the speed of sound, $C$ the heat capacity of the thermal carriers and $\widehat{E}$ is cutoff frequency. If the phonon mean free path is computed using the diffusion coefficient provided by the SCT theory, it is clear that a band of localized states will generate a decrease in the slope of $K(T)$ as a function of $T$, in the thermal region where the more energetic modes are localized. More calculations are needed to properly estimate the magnitude of these effects.

## ACKNOWLEDGEMENTS

It is with great pleasure that we thank Scott Cohen and Jon Machta who collaborated in part of the research reviewed here. We would also like to thank Smith *et al.* and Ferrari *et al.* for allowing us to use their experimental data prior to publication. CAC acknowledges useful discussions with R. Hallock and L. Ferrari. This work was supported by National Science Foundation grants Nos. DMR-86-07605, PHY-83-51473, and CHE-86-09722 and by the Presidential Young Investigators Program.

# REFERENCES


1. P. W. Anderson, *Phys. Rev.* **109** (1958) 1492.
2. For a review of electron localization see, P. A. Lee and T. V. Ramakrishnan, *Rev. Mod. Phys.* **57** (1985) 287.
3. P. Dean and M. D. Bacon, *Proc. Phys. Soc.* **81** (1963) 642.
4. C. H. Hodges, *J. Sound Vib.* **82** (1982) 411.
5. S. John, *Phys. Rev. Lett.* **53** (1984) 2169.
6. T. R. Kirkpartick, *Phys. Rev.* **B31** (1985) 5746.
7. For a recent review with additional references see, D. Sornette, preprint (1987).
8. P. Sheng and Z. Q. Zhang, *Phys. Rev. Lett.* **57** (1986) 1879.
9. C. A. Condat and T. R. Kirkpatrick, *Phys. Rev. Lett.* **58** (1987) 226.
10. C. A. Condat and T. R. Kirkpatrick, *Phys. Rev.* **B36** (1987) 6782.
11. F. J. Wegner, *Nucl. Phys.* **B270** [FS16] (1986) 1.
12. S. M. Cohen, J. Machta, T. R. Kirkpatrick and C. A. Condat, *Phys. Rev. Lett.* **58** (1987) 785.
13. C. A. Condat, T. R. Kirkpatrick and S. M. Cohen, *Phys. Rev.* **B35** (1987) 4653.
14. W. Götze, *Solid State Comm.* **27** (1978) 1393.
15. D. Vollhardt and P. Wolfle, *Phys. Rev.* **B22** (1980) 4666.
16. F. J. Wegner, *Z. Phys.* **B35** (1979) 207.
17. L. Schäfer and F. J. Wegner, *Z. Phys.* **B38** (1980) 113.
18. A. J. McKane and M. Stone, *Ann. Phys.* (N.Y.) **131** (1981) 36.
19. K. B. Efetov, A. I. Larkin and D. E. Kheml'nitskii, *Zh. Eksp. Teor. Fiz.* **79** (1979) 1120, [*Sov. Phys. JEPT* **52** (1980) 568].
20. S. John, H. Sompolinsky and M. J. Stephen, *Phys. Rev.* **B27** (1983) 5592.
21. S. John, *Phys. Rev.* **B31** (1985) 304.
22. P. W. Anderson, *Phil. Mag.* **B52** (1985) 505.
23. D. Sornette and B. Souillard, preprint (1987).
24. H. Levine and J. Willemsen, *J. Acoust. Soc. Am.* **73** (1983) 32.
25. V. Baluni and J. Willemsen, *Phys. Rev.* **A31** (1985) 3358.
26. J. B. Pendry and P. D. Kirkman, *J. Phys.* **C17** (1984) 6711.
27. C. Dépollier, J. Kergomard, J. C. Lesueur and F. Laloe, *Rev. Cethedec* **79** (1984) 67.
28. J. Jäckle, *Solid State Commun.* **39** (1981) 1261.
29. A. R. Wenzel, *Wave Motion* **5** (1983) 215.
30. A. R. Wenzel, *Wave Motion* **7** (1985) 589.
31. D. Vollhardt, to appear in *Festkörperprobleme* **27**, ed. P. Grosse (Vieweg, Wiesbaden, 1987).
32. M. L. Williams and H. J. Maris, *Phys. Rev.* **B31** (1985) 4508.

33. C. Flesia, R. Johnston and H. Kunz, *Europhys. Lett.* **3** (1987) 497.

34. Y. Kuga and A. Ishimaru, *J. Opt. Soc. Am.* **A1** (1984) 831.

35. M. P. van Albada and A. Lagendijk, *Phys. Rev. Lett.* **55** (1985) 2692.

36. P. E. Wolf and G. Maret, *Phys. Rev. Lett.* **55** (1985) 2696.

37. A. Lagendijk, M. P. van Albada and M. B. van der Mark, *Physica* **140A** (1986) 183.

38. S. Etemad, R. Thompson and M. J. Andrejco, *Phys. Rev. Lett.* **57** (1986) 575.

39. M. Kaveh, M. Rosenbluh, I. Edrei and I. Freund, *Phys. Rev. Lett.* **57** (1986) 2049.

40. M. P. van Albada, M. B. van der Mark and A. Lagendijk, *Phys. Rev. Lett.* **58** (1987) 361.

41. S. Etemad, R. Thompson, M. J. Andrejco, S. John and F. C. MacKintosh, *Phys. Rev. Lett.* **59** (1987) 1420.

42. G. Maret and P. E. Wolf, *Z. Phys.* **B65** (1987) 409.

43. A. A. Golubentsev, *Zh. Eksp. Teor. Fiz.* **86** (1984) 47 [*Sov. Phys. JETP* **59** (1984) 26].

44. L. Tsang and A. Ishimaru, *J. Opt. Soc. Am.* **A2** (1985) 1331.

45. E. Akkermans, P. E. Wolf and R. Maynard, *Phys. Rev. Lett.* **56** (1986) 1471.

46. M. J. Stephen and G. Cwilich, *Phys. Rev.* **B34** (1986) 7564.

47. I. Edrei and M. Kaveh, *Phys. Rev.* **B35** (1987) 6461.

48. G. Cwilich and M. J. Stephen, *Phys. Rev.* **B35** (1987) 6517.

49. M. P. van Albada and A. Lagendijk, *Phys. Rev.* **B36** (1987) 2353.

50. Yu A. Kravtsov and A. I. Saichev, *Usp. Fiz. Nauk.* **137** (1982) 501. [*Sov. Phys. Usp.* **25** (1982) 494]. This is an excellent review of the theoretical and experimental status of the subject of enhanced back-scattering by 1982.

51. A. Z. Genack, *Phys. Rev. Lett.* **58** (1987) 2043.

52. G. H. Watson Jr., P. A. Fleury and S. L. McCall, *Phys. Rev. Lett.* **58** (1984) 945.

53. L. A. Ferrari, N. Garcia, J. Zhu and A. Z. Genack, preprint.

54. C. H. Hodges and J. Woodhouse, *J. Acoust. Soc. Am.* **74** (1983) 894.

55. C. Dépollier, J. Kergomard and F. Laloe, *Ann. Phys.* (Paris) **11** (1986) 457.

56. S. He and J. D. Maynard, *Phys. Rev. Lett.* **57** (1986) 3171.

57. C. A. Condat and T. R. Kirkpatrick, *Phys. Rev.* **B32** (1985) 495.

58. C. A. Condat and T. R. Kirkpatrick, *Phys. Rev.* **B33** (1986) 3102.

59. D. T. Smith, C. P. Lorenson, R. B. Hallock, K. R. McCall and R. A. Guyer, *Phys. Rev. Lett.* **C1**, (1988) 1286.

60. C. A. Condat and T. R. Kirkpatrick, *Phys. Rev.* **B32** (1985) 4392.

61. M. Belzons, P. Devillard, F. Dunlop, E. Guazzelli, O. Parodi and B. Souillard, *Europhys. Lett.* **4** (1987) 909.

62. E. Guazzelli, E. Guyon and B. Souillard, *J. Phys. Lett.* (Paris) **44** (1983) 837.

63. F. J. Wegner, *J. Phys.* **C13** (1980) L45.

64. See, for example, P. A. Lee, A. D. Stone and H. Fukuyama, *Phys. Rev.* **B35** (1987) 1039.

65. See, for example, L. D. Landau and E. M. Lifshitz, *Fluid Mechanics* (Pergamon Press, New York).

66. M. Goldberg and K. Watson, *Collision Theory* (Wiley, New York, 1964), p. 750.

67. A. Ishimaru, *Wave Propagation and Scattering in Random Media* (Academic, New York, 1978).

68. For an example see, T. R. Kirkpatrick, *J. Chem. Phys.* **76** (1982) 4255.

69. T. R. Kirkpatrick and J. R. Dorfman, in *Fundamental Problems in Statistical Physics VI*, ed. E. G. D. Cohen (North-Holland, Amsterdam, 1985).

70. J. S. Langer and T. Neal, *Phys. Rev. Lett.* **16** (1966) 984.

71. T. R. Kirkpatrick and D. Belitz, *Phys. Rev.* **B34** (1986) 2168.

72. P. M. Morse and H. Feshbach, *Methods of Theoretical Physics* (McGraw-Hill, New York, 1953).

73. The different possibilities for the scattering of acoustic, *EM*, and elastic waves by discrete objects are catalogued in *Acoustic, Electromagnetic and Elastic Wave Scattering — Focus on the T-Matrix Approach*, ed. V. K. Varadan and V. V. Varadan (Pergamon, New York, 1980).

74. C. M. Davis, L. R. Dragonette and L. Flax, *J. Acoust. Soc. Am.* **63** (1978) 1694.

75. U. Frisch, in *Probabilistic Methods in Applied Mathematics*, ed. A. T. Bharucha-Reid (Academic, New York, 1968) Vol. I.

76. K. Yosioka and Y. Kawasima, *Acustica* **5** (1955) 167.

77. P. M. Morse and K. U. Ingard, *Theoretical Acoustics* (McGraw-Hill, New York, 1968).

78. C. A. Condat, *J. Acoust. Soc. Am.*, to be published (1988).

79. H. Überall, L. R. Dragonette and L. Flax, *J. Acoust. Soc. Am.* **61** (1977) 711.

80. L. Flax, L. R. Dragonette and H. Überall, *J. Acoust. Soc. Am.* **63** (1978) 723.

81. G. C. Gaunard and H. Überall, *J. Acoust. Soc. Am.* **63** (1978) 1699.

82. J. D. Murphy, J. George, A. Nagl and H. Überall, *J. Acoust. Soc. Am.* **65** (1979) 368.

83. L. R. Dragonette and L. Flax, in Ref. 73, p. 401.

84. G. Mie, *Ann. Phys.* **25** (1908) 377.

85. H. C. van de Hulst, *Light Scattering by Small Particles* (Wiley, New York, 1957).

86. M. Kerker, *The Scattering of Light* (Academic, New York, 1969).

87. P. Chýlek, J. T. Kiehl, M. K. W. Ko and A. Ashkin, in *Light Scattering by Irregularly Shaped Particles*, ed. D. W. Schuerman (Plenum, New York, 1980).

88. J. R. Probert-Jones, *J. Opt. Soc. Am.* **A1** (1984) 822.

89. *Electromagnetic and Acoustic Scattering by Simple Shapes*, eds. J. J. Bowman , T. B. A. Senior and P. L. Uslenghi, (North-Holland, Amsterdam, 1969).

90. C. S. Clay and H. Medwin, *Acoustic Oceanography: Principles and Applications* (Wiley, New York, 1977) p. 471.

91. A. F. Ioffe and A. R. Regel, *Prog. Semicond.* **4** (1960) 237.

92. K. Arya, Z. B. Su and J. L. Birman, *Phys. Rev. Lett.* **57** (1986) 2725.

93. See, for example, J. M. Ziman, *Electrons and Phonons* (Oxford U. Press, London, 1960) Chaps. 7 and 8.

94. L. E. Kinsler and A. R. Frey, *Fundamentals of Acoustics* (Wiley, New York, 1962) 2nd edn., Chap. 8.

95. See Ref. 90, Chap. 6 and Appendix A6.

96. C. Gazanhes, P. Arzeliès and J. Léandre, *Acustica* **55** (1984) 113.

97. E. Silberman, *J. Acoust. Soc. Am.* **29** (1957) 925.

98. V. K. Varadan, V. V. Varadan and Y. Ma, *J. Acoust. Soc. Am.* **78** (1985) 1879.

99. U. Pollmann, *Acustica* **62** (1987) 211.

100. B. Shapiro, *Phys. Rev. Lett.* **57** (1986) 2168.

101. M. J. Stephen and G. Cwilich, *Phys. Rev. Lett.* **59** (1987) 285.

102. C. A. Condat, *Phys. Rev. Lett.* **59** (1987) 606.

103. See, for example, K. R. Atkins and I. Rudnick in *Progress in Low Temperature Physics*, ed. C. T. Gorter (North-Holland, Amsterdam, 1970) Vol. 6.

104. J. E. Rutledge and J. M. Mochel, *Phys. Rev.* **B30** (1984) 2569.

105. R. A. Guyer, *Phys. Rev.* **B31** (1985) 2713.

106. I. N. Lifschitz and V. Ya Kirpichenkov, *Zh. Eksp. Teor. Fiz.* **77** (1979) 989 [*Sov. Phys. JETP* **50** (1980) 499].

107. M. Ya Azbel, *Solid State Commun.* **37** (1981) 789 and *Phys. Rev.* **B28** (1983) 4106.

108. See, for example, D. T. Eckholm and R. B. Hallock, *Phys. Rev.* **B21** (1980) 3902 and references cited therein.

109. K. Ishii, *Suppl. Prog. Theor. Phys.* **53** (1973) 77.

110. A. A. Gogolin, *Phys. Rep.* **86** (1982) 1.

111. P. Erdös and R. C. Herndon, *Adv. Phys.* **31** (1982) 65.

112. N. F. Mott and W. D. Twose, *Adv. Phys.* **10** (1961) 107.

113. I. Ya Gol'dshtein, S. A. Molchanov and L. A. Pastur, *Funct. Anal. Appl.* **11** (1977) 1.

114. R. Carmona, *Duke Math. J.* **49** (1982) 191.

115. F. Deylon, Y. Lévy and B. Souillard, *J. Stat. Phys.* **41** (1985) 375.

116. M. W. Cole and E. Vittoratos, *J. Low Temp. Phys.* **22** (1976) 223.

117. D. T. Smith and R. B. Hallock, *Phys. Rev.* **B34** (1986) 226.

118. E. R. Generazio and R. W. Reed, *J. Low Temp. Phys.* **56** (1984) 355.

119. J. Thouless, *J. Phys.* **C6** (1973) L49.

120. M. S. Lu, M. Nelkin and M. Arita, *Phys. Rev.* **B10** (1974) 2315.

121. J. S. Denbigh and N. Rivier, *J. Phys.* **C12** (1979) L107.

122. E. R. Generazio, R. W. Reed and H. M. Frost, *Phys. Rev. Lett.* **50** (1983) 174.

123. S. M. Cohen and J. Machta, *Phys. Rev. Lett.* **54** (1985) 2242.

124. V. L. Berezinskii, *Zh. Eksp. Teor. Fiz.* **65** (1973) 1251 [*Sov. Phys. JETP* **38** (1974) 620].

125. A. A. Maradudin, E. W. Montroll, G. H. Weiss and I. P. Ipatova, *Solid State Phys.*, Suppl. 3., 2nd ed. (Academic, New York, 1971) p. 374.

126. D. Prato and C. A. Condat, *Am. J. Phys.* **51** (1983) 140.

127. P. Brüesch, *Phonons: Theory and Experiments I* (Springer, Berlin, 1982) p. 24.

128. A specific calculation for the third sound problem has been carried out by R. A. Guyer and K. R. McCall (unpublished).

129. A. A. Gogolin, V. I. Mel'nikov and E. I. Rashba, *Zh. Eksp. Teor. Fiz.* **69** (1975) 327 [*Sov. Phys. JETP* **42** (1976) 168].

130. A. M. M. Pruisken and L. Schäfer, *Nucl. Phys.* **B200** [FS4] (1982) 20.

131. S. M. Cohen, J. Machta and T. R. Kirkpatrick, to be published.

132. For some recent work in this direction see, D. E. Logan and P. G. Wolynes, *Phys. Rev.* **B31** (1985) 2437.

133. J. B. Marston and I. Affleck, *Nucl. Phys.* **B290** [FS20] (1987) 137.

134. E. Akkermans and R. Maynard, *Phys. Rev.* **B32** (1985) 7850.

135. J. E. Graebner, B. Golding, *Phys. Rev.* **B34** (1986) 5788.

136. J. E. Graebner, B. Golding and L. C. Allen, *Phys. Rev.* **B34** (1986) 5696.

137. M. Randeria and J. P. Sethna, preprint.

138. J. Freeman and A. C. Anderson, *Phys. Rev.* **B34** (1986) 5684.

139. D. G. Cahill and R. O. Pohl, *Phys. Rev.* **B35** (1987) 4067.

# LOCALIZATION OF SURFACE GRAVITY WAVES ON A RANDOM BOTTOM


MAX BELZONS*, ELISABETH GUAZZELLI* and
BERNARD SOUILLARD**‡

*Département de Physique des Systèmes Désordonnés
UA 857 du CNRS, Université de Provence
Centre de Saint-Jérôme, F-13397
Marseille Cedex 13, France


** Centre de Physique Théorique, Ecole Polytechnique
F-91128 Palaiseau Cedex, France
‡ X-RS, X-Recherche Service, 22, rue Emile Baudot
F-91120 Palaiseau, France


This paper reviews the status of the theoretical and experimental works
on localization of surface gravity waves on a random bottom. As such
it summarizes works by *P. Devillard, F. Dunlop, E. Guyon, O. Parodi*
and the present authors. Localization of surface waves was proposed
in 1983 as a possible experimental realization of Anderson localization
with classical waves. First evidences of the localization of water gravity
waves on a random bottom were later obtained through a very precise
experimental setting. These experimental results do agree well with the
predictions obtained through an extension of the theory of localization

to the full potential theory of gravity waves.  This system provided
also for the first time a direct observation of the stochastic resonant
modes due to the disorder.  It also opens the way to much theoretical
and experimental works in the field of nonlinear wave excitations in the
presence of a random environment.

# Contents


1. Introduction 544

2. Experimental Techniques 546

3. Theoretical Models 550
   3.1. Shallow water theory 550
   3.2. Full potential theory 551

4. Physical Limitations 554
   4.1. The surface tension 554
   4.2. The limits of the linear theory 554
   4.3. The effects of viscosity 555
   4.4. Finite size effects 556

5. Experimental Results 557

6. Conclusions 560

References 561

## 1. INTRODUCTION

The localization phenomenon, which has been discovered in the study of electric conductivity of disordered solids, is not specific to electrons or phonons. It is rather a genuine wave interference effect, and potentially may concern any *linear* wave, described by a Schrödinger equation, Helmoltz equation or a similar wave equation, propagating in a *static disordered* medium. The wave undergoes many partial reflections and transmissions from the scattering centers in the disordered medium and the various transmitted and scattered waves, which have random dephasing, interfere with each other. The basic result is that, if the disorder is strong enough, the wave will not propagate in the medium. More precisely, the stationary solutions or eigenmodes are exponentially localized, i.e. they decay exponentially along the medium with a rate of decay called the localization length $\xi$. In a one-dimensional medium, any disorder is strong enough to induce the exponential localization phenomenon.

The possible relevance of localization, as a genuine interference phenomenon, to any linear wave equation, was presented in Ref. 1, leading to the suggestion of studying the localization of classical waves. At about the same time, a British group was pursuing the same line of reasoning and eventually started the study of localization of acoustic waves.[2] In France, besides the possibility to study localization for acoustic or electromagnetic waves,[1] Guyon, Guazzelli and Souillard[3] suggested to study the case of gravity waves at the surface of water or another liquid. Here the medium itself is not random, but the surface waves interact with the bottom in such a way that steps on the bottom act as scatterers for the wave, providing an essential ingredient for the apparition of localization.

For review, introduction and references to localization in general, as well as to localization of classical waves (electromagnetic waves in plasmas, acoustic waves,...) we refer to Ref. 4, as well as to the papers in the present book. We will stick here as much as possible to the specific case of hydrodynamic waves on a random bottom.

As localization theory is concerned, such a realization has provided us with one of the first experimental realization of Anderson localization for classical waves, at least for one-dimensional systems, and indeed a very pedagogical one. In addition it has provided the first direct observation of stochastic resonances due to disorder. On the theoretical point of view, it has called for an extension of localization theory in a new context, the one of a potential theory. Finally it yields precise and stimulating theoretical and experimental challenges to the case of nonlinear waves in the presence of disorder.

On the other hand, the interaction of surface gravity waves with a rough bottom is a topic of importance in hydrodynamics and oceanography. Some preexisting topography may provide a mechanism for coastal protection and for possible dune growth on a fully erodible sand bed. In recent years, the behavior of surface water waves incident on a region with a periodic bottom topography has been studied, both experimentally and theoretically (see, for example, Ref. 5). Here, if the surface wavelength ($\lambda$) is twice the bed wavelength ($\Lambda$), a resonant reflection can arise in which a significant proportion of the incident wave energy may be reflected by the undulations of the bottom. This resonance phenomenon, which is due to the multiple interference of the waves from the periodic scattering centers, has been well known in solid-state physics as Bragg reflection for about sixty years. It corresponds to the first forbidden band found in the quantum theory of solids (see, for example, Ref. 6). Although the correspondence with solid state physics has been pointed out in some of these oceanographic works, none of them considers the crucial modifications which happen when passing from the case of a periodic bottom to a random one. So, our study of the localization of gravity waves throw new light on a classical problem in hydrodynamics which, possibly, can provide new practical solutions to oceanographers.

In Sec. 2, the experimental techniques, used in order to realize one-dimensional types of experiments on surface waves, are described. In Sec. 3, the theoretical models and their treatment are presented. In Sec. 4, their relevance to the experimental situation and their limitations are discussed as well as the way to overcome them. The experimental results are presented in Sec. 5 and compared with the theories. Section 6 summarizes the results and raises a certain number of questions.

Since most of this work has already been published, we emphasize
the main ideas and results in their connection to localization theory.
We refer the reader to the papers for more details:

• Ref. 3 is the original paper presenting the idea to study the
localization of surface waves and giving some feasible estimations in
this direction; it considers the shallow water case.

• Ref. 7 compares the theoretical predictions obtained from the
shallow water theory as well as those developed in the framework of
the full potential theory with the results obtained by the very precise
experimental set-up built in Marseille.

• Ref. 8 presents the complete description of the experimental
set-up and the experimental results.

• Ref 9 presents the theoretical estimations in the shallow water
theory and develops the theoretical study for the full potential theory.

## 2. EXPERIMENTAL TECHNIQUES

The experiments were carried out in a glass-walled wave tank
(length $= 4$ m and width $= 0.39$ m). A schematic diagram of it is
presented in Fig. 1.

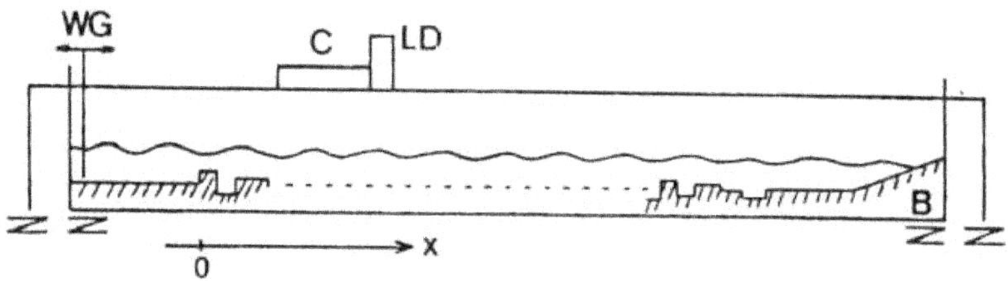


Fig. 1. Schematic diagram of the wave tank: it shows the positions of the
wave generator $WG$, the absorbing beach $B$, the carriage $C$, the linear detector
$LD$ and the random bed.

The average water depth $H_0$ can be varied between 1 and 4 cm.
At the left end of the wave tank, a piston-type wave generator cre-
ates a monochromatic sinusoidal wave with amplitude $A_i(x) \leq 1$ mm.
The frequencies $f = \omega/2\pi$ can be chosen between 1 and 5 Hz (the corresponding wavelengths $\lambda = 2\pi/k$ range from 6 to 60 cm). At the opposite
end, a wave-absorbing beach was built: its role is to prevent waves from

being back-reflected by the end of the tank onto the random bottom. The beach is indeed not perfectly absorbing and the back-scattering due to the beach introduces a small uncertainty into the experimental measurements in the low frequency range: the reflection coefficient of the beach (see the definition below) below 2 Hz are of the order of 0.1–0.2 because the absorbing beach is not long enough in this range of long wavelengths; above 2 Hz the reflection by the beach is negligible.

The bottoms were composed of periodically or randomly organized steps which were built into a false flat bottom with the mean water depth $H_0$. The heights of the bottoms were varied only in the $x$-direction along the tank so that, apart from weak edge effects, the motion of the wave was one-dimensional. Two main bottoms were used in the experiments and are displayed in Fig. 2 (other bottoms were also used but will not be described here; for more details see Ref. 8):

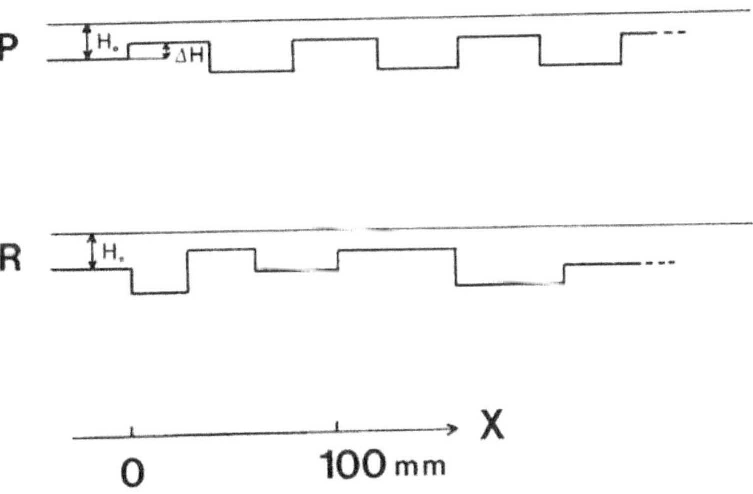


Fig. 2. Schematic diagram of the variable bottoms $(R)$ and $(P)$ used in the experiments.

• The second bottom $(R)$ is a random one consisting of 58 steps. The heights and lengths of the steps are randomly chosen with uniform distribution respectively in $H_0 \pm \Delta H$ and $L_0 \pm \Delta L$ with $\Delta H = 1.25$ cm, with $\Delta L = 2.0$ cm and $L_0 = 4.1$ cm. We have also constructed other random bottoms by different permutations of the random steps.

- The first one is the periodic bottom $(P)$ which has the same mean characteristics as the previous random one. The length of the steps is $L_0$, the heights of the steps are successively $H_0 + \sigma_H$ and $H_0 - \sigma_H$ where $\sigma_H = 0.75$ cm is the standard deviation of the previous random height and, finally, the two bottoms have the same total length: the wavelength of $(P)$ is $\Lambda = 8.2$ cm.

The interaction of the waves with the variations of depth due to the steps is responsible for the scattering of the waves and eventually for localization. It is thus interesting to note here that diminishing the mean depth of water correlatively increases the amplitude of the bottom modulations and thus of the disorder which is characterized by $\Delta H/H_0$. One can thus very easily vary the disorder by diminishing the water depth and in principle one may easily turn to arbitrarily large disorder (the percolative limit is attained when the steps emerge from water!).

The height of the waves is small (less than one millimeter): this is not due to intrinsic limitations, but is assured in order to avoid non-linear effects which happen for larger amplitudes at such wavelength. A very precise disposal for measuring the height of the wave is thus necessary. A measurement of the wave elevation $\eta(x,t)$ with absolute accuracy better than 40 $\mu$m were achieved using an optical detection technique: a laser beam is incident at right angle on the free water surface, the diffusion spot is focused on a linear detector (a 256 pixels-linear-camera) and one measures the displacement of the image spot. A carriage supports the optical device and moves along the tank.

The whole experiment was conducted by an Apple IIe microcomputer:
- determination and control of the frequency of the wave generator,
- motion of the carriage which supports the optical device along the tank,
- lecture of the linear detector,
- accumulation of data over several periods (typically 20) of the wave generator.

The motion of the wave generator and of the carriage were realized by stepping motors in order to insure a great reproducibility of the measurements.

From the records on the wave elevation $\eta(x,t)$ measured at a position $x$ along the tank, the wave amplitude $A(x)$ is deduced at each position from the relation:

$$\eta(x,t) = A(x)\cos(\omega t + \varphi(x)) ,$$

where $\varphi(x)$ is the phase of the wave (the reference phase is the phase of the wave generator). In order to check the linearity of the wave, we also examined the wave elevation more precisely by fitting it to the relation:

$$\eta(x,t) = \sum_n A_n(x)\cos(n\omega t + \varphi_n(x))$$

thus obtaining, the amplitudes $A_n(x)$ and phases $\varphi_n(x)$ of the fundamental wave and its harmonics.

For any given frequency, we obtained, the modulus of the reflection amplitude $R'$ which is defined as the quotient of the reflected wave amplitude to the incident wave amplitude. The interference of the reflected and incident waves induces a modulation of the resulting wave amplitude $A(x)$ which varies between $A_{\max} = 1 + |R'|$ and $A_{\min} = 1 - |R'|$. From the measurement of $A(x)$ on the up-wave side of the variable bottom region, we deduce the rate of stationary waves

$$A_{\max}/A_{\min} = (1 + |R'|)/(1 - |R'|)$$

and thus the reflection coefficient $|R'|^2$.

The wave amplitude is measured throughout the whole variable bottom region. This leads to the measurement of an "attenuation" of the amplitude of the type

$$A(x) \sim \exp(-x/l) .$$

The attenuation length $l$ is obtained by fitting $A(x)$ using a least-squares fit of $\ln A(x)$ to $x$; in Fig. 4 later, the correlation coefficient lies between .85 and .95.

## 3. THEORETICAL MODELS

### 3.1. Shallow Water Theory

In the simplest theory, the wave for a one-dimensional channel is governed by the linear long wavelength equation of the shallow water theory:

$$\eta_{tt}(x,t) = g[H(x) \cdot \eta_x(x,t)]_x \; ,$$

where $g$ is the acceleration due to gravity, the horizontal $x$-axis is taken at the mean free surface level $(y = 0)$, $\eta(x,t)$ is the vertical displacement of the free surface and $y = -H(x)$ defines the bed which varies only in the $x$-direction.

The drastic physical limitations of the linear shallow wave theory can be stated as follows (see for example Ref. 10):

- the wavelength is large with respect to the depth of water: $k \cdot H_0 \ll 1$,
- the amplitude of the wave is small so that the nonlinear terms remain small: $k \cdot \eta \ll 1, \eta \cdot H_0^{-1} \ll 1, k^{-2} \cdot \eta \cdot H_0^{-3} \ll 1$,
- there is a small bottom slope: $H_x \ll 1$,
- and a small bottom amplitude: $H(x)/H_0 \approx 1$, in order to ensure that the velocity is approximately constant as a function of depth for given $x$, which is the basis of shallow water approximation.

For this simple theory, it is quite easy to demonstrate a close analogy with the quantum theory of solids. The corresponding stationary equation for waves of frequency $f = \omega/2\pi$ is:

$$-\omega^2 \cdot \eta(x) = g[H(x) \cdot \eta_x(x)]_x \; .$$

In the case of a horizontal flat bottom $H(x) = H_0$, the stationary solutions are plane waves with velocity $(g \cdot H_0)^{1/2}$. In the case of a sinusoidal bottom, this equation can be transformed into a Matthieu equation which is equivalent to the Schrödinger equation with a sinusoidal one-dimensional potential; the forbidden bands corresponding to a strong reflection of the incident wave appear for $2k/K = 1, 2, 3, 4, \ldots$ where $k$ is the wave vector of the wave and $K$ the wave vector of the bottom.[11] In the case of a random bottom, if we apply the concepts of one-dimensional localization theory,[4] all stationary solutions are exponentially decaying (actually the long wavelength equation is the continuum analog of the equation describing phonons in condensed matter physics). If the bottom is built from steps with random heights and lengths, the theoretical localization length $\xi(\omega)$ can be computed from a product of $2 \times 2$ random transfer matrices, which are obtained by writing the continuity conditions for the amplitude and for the flux at each step. The resulting localization length can be studied by several analytical and numerical methods (see Ref. 4). For a given disorder, it diverges for small frequencies as $\omega^{-2}$ and goes to a constant (which can be explicitly computed) for large $\omega$.

## 3.2. Full Potential Theory

The shallow water theory is certainly false for abrupt fluctuations of the bottom which we want to consider as the strong disorder situation, and in any case it is false for small wavelengths: in this later case, it is intuitive that the interaction of the wave with the bottom for a given depth of water decreases to zero with the wavelength and we would thus expect intuitively that a localization length should vanish for large $\omega$ instead of going to a constant. Actually for other reasons connected with the dissipation phenomenon discussed later, it was also necessary to go beyond the shallow water regime.

Still restricting to the linear regime by imposing small wave amplitudes, we have thus to consider the linear potential theory of waves: it had never been studied from the localization point of view.

The linear potential theory is given by the Laplace equation $\Delta \Phi = 0$ where $\Phi$ is the potential for the velocity: the velocity of the fluid $v(x, y, t)$ at a point of abscissa $x$ and depth $y$ and at time $t$ is given by

$$v = e^{i\omega t} \cdot \mathrm{grad}\Phi .$$

In addition there are boundary conditions

$$\Phi_y + \omega^2 g^{-1} \Phi = 0 \text{ on the free surface } y = 0$$
$$\Phi_y + \phi_x \cdot H_x = 0 \text{ on the bottom } y = -H(x) .$$

The physical limitations of the linear potential theory can be stated as follows (see for example Ref. 10):

- small wave amplitude $\eta \cdot H_0^{-1} \ll 1$,
- small wave steepness $k \cdot \eta \ll 1$,
- small Stokes parameter $k^{-2} \cdot \eta \cdot H_0^{-3} \ll 1$,

in order to insure that the nonlinear terms remain small.

We consider first the case of a flat horizontal bottom, then we will treat the case of a succession of random horizontal steps. The difference with the shallow water theory readily appears: the general solution on a region with constant depth is expanded on a complete *infinite* set of explicit solutions, two of them being propagative solutions, all the other ones being non propagative, in contrast to the case of the shallow water theory where the general solution was expandable on two propagative solutions, like in the case of the Schrödinger equation. So, in the potential theory, the general stationary solution at frequency $\omega$ on a piece of a flat bottom at depth $H$ is:

$$\Phi(x, y) = (Ae^{ikx} + Be^{-ikx})\chi(y) + \sum_n C(\kappa_n)e^{\kappa_n x}\phi(\kappa_n, y)$$

$$+ \sum_n D(\kappa_n)e^{-\kappa_n x}\phi(\kappa_n, y)$$

where $ik$ and $\kappa_n$ $(n = 1, 2, 3, \ldots)$ are solutions of the dispersion relation

$$\kappa \tan(\kappa H) = -\omega^2 g^{-1}$$

and

$$\chi(y) = F(ik, y)$$
$$\phi(\kappa_n, y) = F(\kappa_n, y)$$

with

$$F(\kappa, y) = \sqrt{2}(H - \omega^{-2}g \sin^2 \kappa H)^{-1/2} \cos(\kappa(H - y)) \, .$$

If we now consider a succession of two flat steps, it is possible to write the matching conditions at their boundary, which express the continuity of the potential $\Phi$, and the continuity of the velocity $\partial\phi_x\Phi$. This yields a transfer matrix which is now infinite dimensional and where the propagative modes on the second step are now coupled not only to the propagative modes of the first step but also to the infinite set of non propagative ones.

A first approach to this problem is to truncate the sum over $\kappa_n$ at some order $N$, obtaining thus a transfer matrix of order $(2N + 2) \times (2N + 2)$. For a succession of random steps, one could then compute in principle $N+1$ positive Lyapunov exponents, the smallest of them being the inverse localization length. However such an approach, reminiscent from the techniques used in the study of quasi-onedimensional systems, is heavy if many non-propagating modes have to be taken into account.

The approach which has been chosen[7,9] takes into account specific features of the non propagating modes of the potential theory. In particular it can be checked that these non propagating modes decay uniformly with the wavelength away from the edges of the steps. So that we have restricted our attention to the case which corresponds to the situation experimentally studied, and where the lengths of the steps is much larger than the depth of water: this insures that the non-propagating modes originating at one step are negligible when they reach the next step.

We still have an infinite number of modes coupled on the edge of each step, but we can now incorporate Miles' approximate variational solution,[12] which is a good approximation at all wavelengths. We do not describe here the details of this piece of work, which can be found in Ref. 9, but we mention the result: we are now set back to study the case of product of random renormalized $2 \times 2$ transfer matrices. That is, in this treatment, the effect of the infinite number of non-propagating modes coupled at each step amounts to renormalize the $2 \times 2$ transfer matrix of the shallow water theory.

We can now study the behavior of this random renormalized transfer matrix product by usual techniques. At large wavelengths, we recover of course the shallow water transfer matrices and its behavior, but for small wavelengths, the Lyapunov exponent, that is the inverse localization length, vanishes in contrast to the shallow water theory.

Actually the precise behavior can be predicted as

$$\xi \propto \langle \exp\{-4\omega^2 g^{-1} H(x)\}\rangle^{-1} \, ,$$

where the brackets denote the averaging over the disorder.

As a consequence the localization length is evaluated and it appears that localization is unobservable for large and small wavelengths for which the localization length becomes larger than the length of the tank, but can possibly be observed in the middle range frequencies. This holds, if the conclusions of the above theory are not invalidated by the other physical phenomenon which have not been taken into account in this theory and will limit its applicability. This is the subject of the next section.

## 4. PHYSICAL LIMITATIONS

In order to compare these theoretical predicitions with the experimental results, it is necessary to investigate the physical limitations to the linear potential theory. We discuss them successively.

### 4.1. The Surface Tension

The effects of the surface tension have been neglected in the previous discussion. They are certainly relevant when $\lambda$ is small. How small? Actually the wavelength has to be compared with the capillarity length of the fluid. It is computed that the error due to the surface tension is small as long as $\lambda > 1.7$ cm which is satisfied in the experiments.

It was also checked experimentally that the surface tension was irrelevant by verifying the dispersion relation on a flat bottom. The agreement was very good.

### 4.2. The Limits of the Linear Theory

Since the experimental system allows one to resolve very small wave amplitudes, it was possible to restrict to small wave amplitude such that $A/H$ was at most 0.05, the wave steepnesses $A \cdot k$ were in the range 0.01 to 0.1 and that the values of the Stokes parameter $A \cdot k^{-2} \cdot H^{-3}$ were in the range 0.03 to 0.8, ensuring the linearity of the waves from the theoretical criteria. The remaining criterion involves the bed modulation $\Delta H/H_0$. Its largest value was not sufficient to induce nonlinearities of the waves.

The waves examined in the experiments were thus, from the theoretical criteria, monochromatic linear gravity waves in shallow water or in water of intermediate depth. The same conclusions arise from experimental studies in two respects. Again the dispersion relation is found in good agreement with the theory. Also it is verified that more than 95% of the wave elevation was at the fundamental frequency, the remaining 5% being in the first harmonic so that the waves are linear to a very good approximation.

Actually we may also ask whether the flow above the bed is perfectly potential. This is not absolutely the case: small vortices are generated at each step and shed into the flow, especially in the case of abrupt steps as those of Fig. 2. This phenomenon cannot be taken into account in the present day theory. Although occurring, it seems to remain sufficiently small to be in fact smaller than the other causes of errors described below and which are finite size effects and viscosity effects.

## 4.3. The Effects of Viscosity

It is crucial to discuss the dissipative effects due to viscosity: this is one of the main limitations to the observation of the localization phenomenon. Viscous dissipation implies an exponential fall-off of the amplitude of the wave along the tank such that the wave amplitude behaves roughly as: $A(x) \approx \exp(x/d)$ where $d$ is a dissipation length. This exponential fall-off is of a different nature than the one predicted by localization theory. If the viscous dissipation length is much smaller than the localization length, the wave will be attenuated beyond experimentally accessible values before the interference effects related to localization are felt by the wave so that localization would not be practically observable.

Estimations indicate that the dissipation length is smaller than the localization length for large and small wavelengths. We anyway restrict ourselves to the intermediate frequencies region for other reasons.

More seriously, dissipation is not yet fully understood and it is actually known to be experimentally much larger than theories predict. Dissipation is due in particular to the friction on the bottom and the side-walls of the tank but also to the generation of vortices at the edges of each step. Since the introduction of viscous effects in the computation is beyond present knowledge, we have chosen to estimate the viscous damping experimentally. We assumed that dissipation depends only on the average characteristics of the bottom and checked this assumption on different periodic bottoms (for more details, see Ref. 8). We, therefore, estimate now the dissipation length for a random bottom from the dissipation length for the associated periodic bottom which has the same average characteristics. Identical average characteristics mean:

- a step length which is the mean value of the step lengths of the random bed,
- the same mean water depth $H_0$ as that above the random bed,
- step heights alternatively $H_0 + \sigma_H$ and $H_0 - \sigma_H$, where $\sigma_H$ is the standard deviation of the step heights for the random bed, and
- the same total length.

From this analysis it follows that in the intermediate range of frequencies which are experimentally accessible, the viscous dissipation length is larger than, or of the order of, the localization length, so that the observation of localization is possible at least when the depth of water is not too small.

## 4.4. Finite Size Effects

The experimental bottom is finite: the experimental bottom $(R)$ corresponds to 58 steps. In the best range of conditions (frequencies and mean depth) for the experiments, the length of the experimental bottom is of the same order as the localization length. This leads to the appearance of fluctuations of the transmission coefficient (or the reflection coefficient) which are not present in very large systems. They are associated to the apparition of stochastic resonances at sample dependent frequencies[4,13,14]: the eigenmodes which are exponentially localized for an infinite system are expected to be turned into resonances for a finite system, and the resonant frequencies will depend on the realization of the medium. Also the transmission is enhanced for the corresponding resonant frequencies.

## 5. EXPERIMENTAL RESULTS

Now, we present the main experimental results. We first compare the experimental results for random beds and for the associated periodic beds. We present the results for the reflection coefficient in the case of a strong modulation of the bed (Fig. 3).

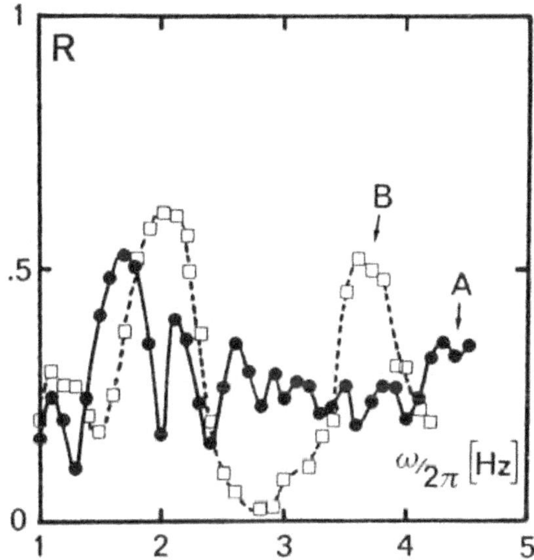


Fig. 3. Results for the reflection coefficients of the periodic bed $(P)$ (curve a) and random bed $(R)$ (curve b) with $H_0 = 1.75$ cm $(\Delta H/H_0 = 0.71$ for the random bed $(R))$.

In the periodic case (curve a), the plot of $R$ versus the wave frequency $f$ displays the first two forbidden bands corresponding to two strong maxima located about frequencies such that $\lambda = 2\Lambda$ (Bragg condition) for the first forbidden band and $\lambda = \Lambda$ for the second one. In the random case (curve b, $\Delta H/H_0 = 0.71$), the spectrum of $R(f)$ is completely modified as compared with that for the periodic case: the passing bands have disappeared and in addition huge fluctuations as a function of the frequency are present which correspond to resonant frequencies.

This behavior is associated to the appearance of resonant modes of the random bed. Our precise experimental set-up provides a direct way to observe these resonant modes by allowing measurement of the wave amplitude all along the tank as shown in Fig. 4. For a resonant frequency (Fig. 4a), the wave does not decay monotonically along the random bottom but increases in its middle region in contrast to the non resonant case (Fig. 4b). It has been checked experimentally that these resonances occur for frequencies corresponding to a decrease in the reflection coefficient: the wave can be better transmitted with the help of resonant states located near the middle of the bottom. The occurrence of these resonances is also sample dependent. Experiments on different random realizations of the bottom $(R)$ obtained by permuting the random steps in a random way, show that, for a given frequency, the resonances strongly depend on the random realization (Figs. 4a and c).

In addition to these qualitative observations, we measure the decrease of the amplitude of the wave, and deduce its rate of exponential decay $l$. We can extract from it, the localization length, by writing for the wave amplitude over a random bed, the relation: $A(x) \approx \exp(-x/l) \approx \exp(-x/\xi) \cdot \exp(-x/d)$. A rough estimate of the localization length can thus be obtained: $\xi^{-1} = l^{-1} - d^{-1}$ where $d$ is the viscous attenuation length measured for the associated periodic bed (and extrapolated when necessary in the forbidden bands). This formula gives an approximation for $\xi$ when $l < d$ (at least 10% lower), in the case of a strong disorder $(\Delta H/H_0 = 0.71)$. For smaller disorder, the disorder is not sufficient to produce a significant difference between $l$ and $d$.

In Fig. 5, the dots are the experimental values for $\langle \xi^{-1} \rangle^{-1}$ where the brackets denote the average over 5 realizations of the random bottom $(R)$, in the case of strong disorder $\Delta H/H_0 = 0.71$; the dotted and full lines are the numerical results of $\xi$ for the shallow water and full potential theories respectively for an infinite bottom. In the frequency range 1.2 to 2.5 Hz, the experimental estimations are in good agreement with the theoretical predictions of the full potential theory. Moreover, one obtains the same minimal value of $\xi$, approximately 2 m, for the same frequency 1.7 Hz. The standard deviations become very impor-

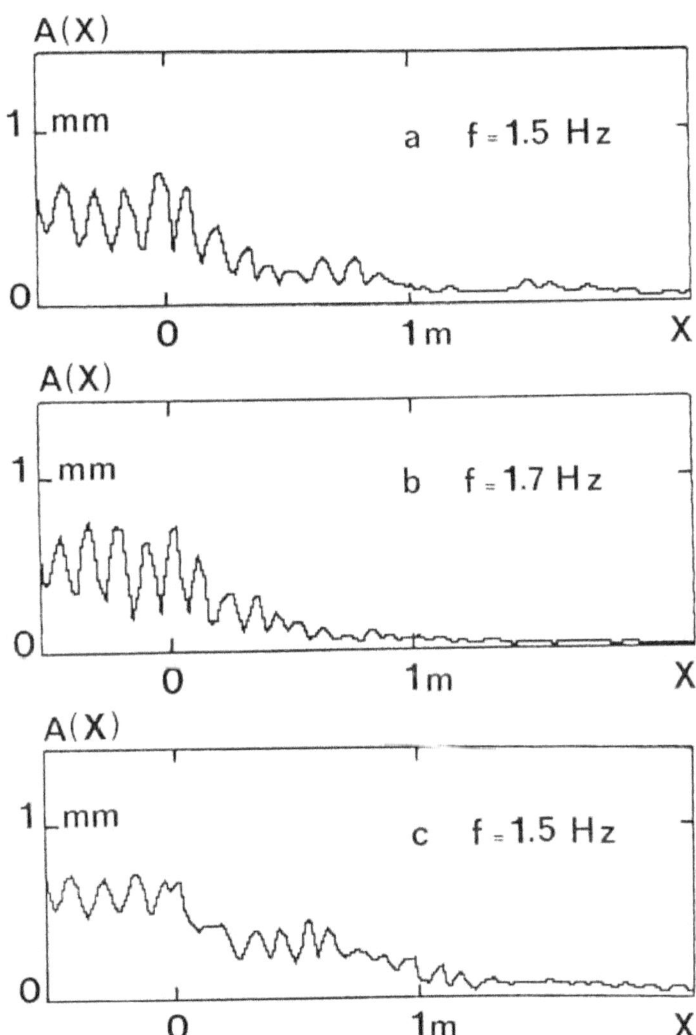



Fig. 4. Variation of the amplitude of the wave elevation $A(x)$ along the wave tank for the random bed $(R)$ with $\Delta H/H_0 = 0.71$. Abscissa $O$ corresponds to the beginning of $(R)$. Curves $a$ $(f = 1.5$ Hz$)$ and $b$ $(f = 1.7$ Hz$)$ represent respectively a resonant and a non resonant case for the same realization of the bottom. Curve $c$ $(f = 1.5$ Hz$)$ is obtained for another realization of the bottom.

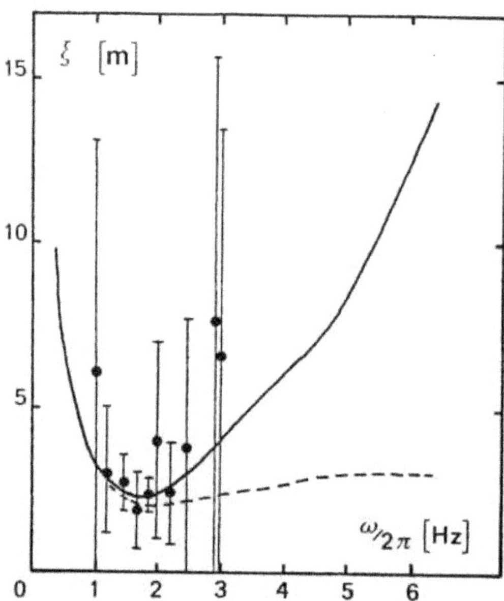

Fig. 5. Variation of the localization length as a function of the wave frequency for the random bed ($R$) with $\Delta H/H_0 = 0.71$. The dots are the experimental estimations where the average has been performed over 5 realizations of the random bed ($R$) and the full vertical lines show the standard deviations. The dotted line and full line curves are the numerical results for the shallow water and the potential theories, respectively.

tant for frequencies outside this range: this is in good agreement with the fact that the localization length as computed theoretically becomes large with respect to the length of the random bottom and thus the experimental estimation of $\xi$ does not have any meaning anymore. In addition, at high frequencies, viscous effects become dominant.

## 6. CONCLUSIONS

We have presented the one dimensional localization phenomenon in the context of hydrodynamical surface waves. On the theoretical side, we have provided the first study of localization in the context of the full linear potential theory. We estimated numerically the localization lengths within this theory and predict the possible observation of localization in some appropriate range of frequencies. On the experimental side, measurements of linear gravity waves on random bottoms in a one-dimensional channel give evidence of the localization pheonomenon. Firstly, it has been demonstrated that, in the case of strong disorder, the passing bands found in the periodic case have disappeared and thus the transmission is diminished. Secondly, we have observed directly for the first time the resonant modes predicted by the localization theory. Moreover, it has been checked experimentally that fluctuations in wave transmission are clearly related to the occurrence of these resonances. In other experiments on localization, resonances have only been seen indirectly. Finally, we have estimated the localization lengths and they present good agreement with theoretical predictions if finite size effects are taken into account.

It is interesting to note that the experimental set up allows easily to turn the linear waves that we have studied in the present paper into nonlinear waves simply by increasing the amplitude of the wavemaker and thus of the waves. One can then wonder what becomes of localization when the wave becomes nonlinear or alternatively what happens to a nonlinear wave in a random medium. But this is another, very interesting and quite open, problem ... much beyond the present paper.

# REFERENCES


1. B. Souillard, *Workshop on Macroscopic Random Media*, (Carry-le-Rouet, 1982).
2. C. H. Hodges, *J. Sound Vib.* **82** (1982) 411; C. H. Hodges and J. Woodhouse, *J. Acoust. Soc. Am.* **74** (1983) 894.
3. E. Guazzelli, E. Guyon and B. Souillard, *J. Phys. (Paris) Lett.* **44** (1983) L837.
4. B. Souillard, *Waves and Electrons in Inhomogeneous Media*, in Les Houches Summer School Proceedings *Chance and Matter*, J. Souletie, J. Vannimenus and R. Stora, eds. (North-Holland, 1987).
5. A. G. Davies and A. D. Heathershaw, *J. Fluid Mech.* **144** (1984) 419.
6. C. Kittel, *Introduction to solid state physics*, 5th edn. (John Wiley and Sons, New-York, 1976).
7. M. Belzons, P. Devillard, F. Dunlop, E. Guazzelli, O. Parodi and B. Souillard, *Europhys. Lett.* **4** (1987) 909.
8. M. Belzons, E. Guazzelli and O. Parodi, *J. Fluid Mech.* **186** (1988) 539.

9. P. Devillard, F. Dunlop and B. Souillard, *J. Fluid Mech.* **186** (1988) 521.

10. G. Witham, *Linear and Nonlinear Waves*, (John Wiley and Sons, New-York, 1974).

11. A. G. Davies, E. Guazzelli and M. Belzons, *Phys. of Fluids*, to appear.

12. J. Miles, *J. Fluid Mech.* **28** (1967) 755.

13. U. Frisch, C. Froeschle, J. Scheidecker and P. L. Sulem, *Phys. Rev.* **A8** (1973) 1416.

14. M. Azbel, *Solid State Comm.* **45** (1983) 527.

# WAVE LOCALIZATION AND MULTIPLE SCATTERING IN RANDOMLY-LAYERED MEDIA

PING SHENG, BENJAMIN WHITE, ZHAO-QING ZHANG*

*Exxon Research and Engineering Company, Route 22 East*
*Clinton Township, Annandale, NJ 08801*

and

GEORGE PAPANICOLAOU

*Courant Institute of Mathematical Sciences*
*New York University, New York, NY 10012*

We present a tutorial review of the theoretical developments in the calculation of localization length and pulse backscattering statistics for a randomly-layered medium. By using the approach of stochastic differential equations, it is shown that the localization length $l(\omega) \sim \omega^{-2}$ at low frequencies and is non-decreasing at high frequencies. Analytical expressions for $l(\omega)$ in the two limits are compared with numerical simulation results for three different types of layered models. Excellent agreement is obtained. Furthermore, for oblique-incident electromagnetic waves it is found that the localization length for one of the

---

*Present address:* Institute of Physics, Academia Sinica, Beijing, People's Republic of China.

polarizations can increase by orders of magnitude (and sometimes diverges) at a particular angle, a phenomenon that may be directly related to the Brewster effect for reflection from a plane interface. For pulse backscattering from a randomly stratified halfspace, the calculation of the power spectrum $S$ for the multiply-backscattered signal train is formulated in terms of the correlation of the reflection coefficients at two different frequencies. Analytical solution of this correlation function in the white-noise limit (correlation length of the material parameter fluctuations$\rightarrow 0$) gives the result that $S = |f(\omega)|^2 \mu(\chi)/\tau$, where $\chi =$ (distance traveled by the pulse at time $\tau$)/(the frequency-dependent localization length $l(\omega)$), $f(\omega)$ is the pulse frequency spectrum, and $\mu = \chi/(1+\chi)^2$ and $4\chi$ for the matched-impedance and total-reflection boundary conditions at the halfspace interface, respectively. Numerical simulations give excellent support to the analytical results even for model parameter values much beyond the white noise regime. We offer a plausible explanation for the robust nature of the analytical solution and discuss its physical implications for the time-domain measurement of $l(\omega)$.

# Contents


1. Introduction    566

2. Statement of Results    567

3. Model Description    569

4. Localization Length of a Randomly-Layered Medium    572
   4.1. Stochastic differential equation approach to the calculation of $l(\omega)$    572
   4.2. Low frequency solution    575
   4.3. High frequency solution    582
   4.4. Oblique incidence    586
   4.5. Numerical simulations    590

5. Localization Characteristics in the Time Domain    599
   5.1. Formulation    599
   5.2. Solution in the white noise limit    603
   5.3. Numerical simulations    611
   5.4. Physical implications    616

6. Concluding Remarks    617

Appendix    618

References    619

## 1. INTRODUCTION

The study of wave propagation and scattering in randomly-layered media has long been a subject of active interest. As a reasonable approximation to earth's stratified lithology, the layered-medium model is basic to the geophysical calculations involving forward modeling and inversion of seismic data.[1] The simplicity of the layered geometry, on the other hand, also offers a convenient setup for working out the physics of localization phenomenon.[2] In recent years, significant progress has been made in clarifying the statistical character of the multiply-scattered wave in a randomly-layered medium.[3-15] In particular, it was found that besides the usual localization behavior for the single-frequency wave in the form of exponential decay with a frequency-dependent decay (localization) length $l(\omega)$, there are additional generic time-domain characteristics associated with the scattering and localization of a pulse.[3-5] Since the concept of time-domain localization is relatively new but yet important for many applications, below we describe in more detail both the phenomenon and its potential implications.

There are many industrial and geophysical examples where the probing of a medium is accomplished not by continuous harmonic waves but rather by transient pulses. In these cases the medium response is in the form of a time trace for the backscattered or transmitted signals. When the medium is random, these time-dependent signals are generally characterized by the presence of a long noisy tail, known as coda, arising from multiple scattering of the pulse. A good example of this can be found in the coda following the arrival of each seismic pulse after it has traversed the inhomogeneous earth.[7] While the structure of this tail follows deterministically from the structure and properties of the earth, the possibility of recovering this type of detailed information from the coda is nevertheless rather remote since the information is thoroughly scrambled by multiple scattering. Two relevant questions are then: (1) What information, if any, can be extracted from this type of noise? (2) Is there a generic localization characteristic in the time-domain response to a transient pulse?

It is the purpose of the present chapter to review the recent theoretical progress in the understanding of wave and pulse localization in randomly-layered media. In order to have an overall picture of what follows, we will first summarize the key results as well as answer the two questions posed above before delving into more detailed developments.

## 2. STATEMENT OF RESULTS

Except for polarization-induced effects in vector waves,[9] it is well-known that all waves localize in a randomly-stratified medium when the propagation direction is normal to the layers. The relevant question is therefore the value of the localization length and its frequency dependence. At low frequencies, where the wavelength $\lambda$ far exceeds the correlation length of the inhomogeneities, the wave essentially "sees" a nearly homogeneous effective medium and the localization length $l(\omega)$ can be shown to diverge as $\gamma_1^{-1}\omega^{-2}$. The proportionality constant, $\gamma_1^{-1}$, is particular to the medium and is dependent on the integral of the correlation function for the random material parameters.[6] At high frequencies, it can be shown that $l(\omega)$ is non-decreasing.[6] For a model with sharp inter-layer boundaries, $l(\omega)$ saturates to a constant $c_1^{-1}$. But since at high frequencies the wave can resolve the details of the inhomogeneities, the value of $c_1$ is dependent on the nature of the interfaces. In particular, for models with continuous variation of the material parameters (or fuzzy interfaces betweeen layers) $c_1 = 0$ and $l(\omega)$ diverges[6] as $\omega \to \infty$. However, as long as there is an interface region, then for wavelength larger than the interface region $l(\omega)$ may be accurately described by the interpolation relation

$$l(\omega) = c_1^{-1} + \frac{1}{\gamma_1 \omega^2} \ . \tag{1}$$

Besides their model dependence, the parameters $c_1$ and $\gamma_1$ can also vary as a function of the angle of incidence $\theta$. A striking example may be found in the case of the electromagnetic wave when the transverse magnetic field vector is polarized parallel to the layerings and perpendicular to the plane of incidence. The localization length is found to increase by orders of magnitude at a particular angle of propagation, an effect directly traceable to the well-known Brewster effect for reflection from a plane interface.

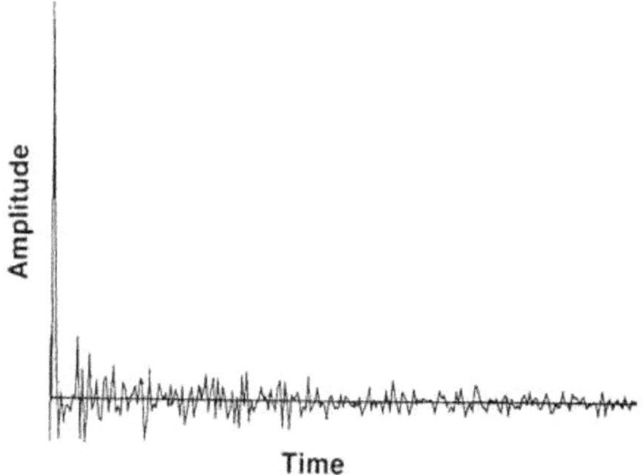


Fig. 1. Numerically simulated time-dependent response to a pressure pulse incident on a randomly-layered halfspace. The observer is at the interface between the homogeneous medium and the randomly-layered medium.

In the case of a pulse, the localization behavior has to be described in quite different terms. Figure 1 shows the numerically calculated response, at the interface between a randomly-layered halfspace and a homogeneous halfspace, resulting from an incident (plane-wave) pulse. It is seen that the initial injection of the pulse is followed by a long tail arising from multiply-backscattered wave field. The time series is obviously non-stationary in the sense that the statistics of both the magnitude and the frequency content of the signal changes with time. That means if one calculates the power spectrum $S$ of the noisy tail, then $S$ is not only a function of the frequency $\omega$, but also depends on the choice of the time window in which the frequency spectrum is calculated, i.e. $S = S(\tau, \omega)$, where $\tau$ denotes the center of the time window. Our analytical and numerical simulation[3-5] results indicate that for an incident pulse with frequency spectrum $f(\omega)$, $S(\tau, \omega) = |f(\omega)|^2 \tau^{-1} \mu(\chi)$, where $\mu$ is an universal function independent of material parameters, $\chi = \tau v_0 / l(\omega)$, $v_0$ is the effective-medium speed of the medium, and $l(\omega)$ is the frequency-dependent localization length. We have also determined the explicit forms of $\mu$ as $\chi / (1 + \chi)^2$ and $4\chi$, for the transmitting and the reflecting boundary conditions at the halfspace interface, respectively. The existence of a universal function independent of the statistics of the model as well as its material parameters clearly indicates that the power spectrum has a generic character which is a signature of the localization effect in the time domain. This fact therefore answers one of the questions posed earlier. As to the other question regarding the information extractable from back-scattered coda, we note that since $l(\omega)$ is the only quantity in $S(\tau,\omega)$ that is particular to the medium, it thus represents the maximum statistical information obtainable. The power spectrum function is therefore the composition of a generic function $\mu$ with a material-dependent function $l^{-1}(\omega)$. Prior knowledge of both functions would open the possibility for constructing statistical filters for enhancing signals, i.e. scattering from target objects embedded in a random medium, through multiple-scattering noise suppression.

In what ensues, the results described above and their implications will be developed in more detail. Description of the three different types of layered models is given in Sec. 3. This is followed by the consideration of the frequency dependence of the localization length $l(\omega)$ in Sec. 4. The pulse backscattering problem and its physical implications are addressed in Sec. 5, and Sec. 6 concludes with a discussion of further topics.

## 3. MODEL DESCRIPTION

Consider a randomly-layered halfspace in the region $0 \le z < \infty$ characterized by spatially varying density $\rho(z)$, elastic bulk modulus $K(z)$, dielectric constant $\varepsilon(z)$, and magnetic permeability $\mu(z)$. The elastic wave equation is given by

$$\rho\dot{w} = -\frac{\partial p}{\partial z}\,, \tag{2a}$$

$$\dot{p} = -K\frac{\partial w}{\partial z}\,, \tag{2b}$$

where $w$ is the displacement velocity, $p$ the pressure, and the over-dot denotes time derivative. The equation for electromagnetic wave propagation normal to the layering direction can be obtained from Eq. (2) by making the following substitutions: $\varepsilon/c \to \rho, c/\mu \to K$, transverse electric field $\to w$, transverse magnetic field $\to p$. Here $c$ denotes the speed of light in vacuum. Since the two equations are identical under the transformation, the results obtained in one case are assured to be applicable in the other case as well. In this article we will use the acoustic notations except for the discussion of oblique incidence in Sec. 4.4, where the electromagnetic case differs from the acoustic one.

The random quantities in Eq. (2) are $\rho(z)$ and $K(z)$, the materials parameters. We consider three types of random models. In model I, $\rho(z)$ and $K(z)$ are piecewise constant in layers of equal thickness $\bar{a}$:

$$\rho_i = \rho_0 \left[1 + 2\sigma_\rho \, \xi_1(z)\right] , \tag{3a}$$

$$K_j^{-1} = K_0^{-1}\left[1 + 2\sigma_K \, \xi_2(z)\right] , \tag{3b}$$

where $i$ (or $j$) denotes the $i$th (or $j$th) layer, $\xi_1, \xi_2$ are random step functions which jump simultaneously at layer interfaces and whose values are independent and uniformly distributed in the interval $[-\frac{1}{2}, \frac{1}{2}]$, and $0 \le \sigma_{\rho(K)} < 1$ specifies the amount of randomness as a fraction of the mean. In model II, $\rho(z)$ and $K(z)$ have the same piecewise-constant profiles as given by Eq. (3), but the layer thickness $a$ is now a random variable with an exponential distribution $\exp(-a/\bar{a})/\bar{a}$, where $\bar{a}$ denotes the mean layer thickness. Model III differs from the first two in that $\rho(z)$ and $K(z)$ are continuous functions of $z$. More precisely, if we let $\overline{m}_i$ denote the material parameter $\rho$ or $K^{-1}$ of model II and $m_i$ denotes that of model III, then

$$\frac{dm_i(z)}{dz} = \overline{m}_i(z) - m(z) , \tag{4a}$$

or

$$m_i(z) = \exp\left(-z\right) \otimes \overline{m}(z) = \exp\left(z_1 - z\right)m_i(z_1) + \left[1 - \exp\left(z_1 - z\right)\right]\overline{m}_i , \tag{4b}$$

where $z_1 < z$ denotes the coordinate for the lower end of the layer, and the symbol $\otimes$ denotes convolution. The three models are schematically depicted in Fig. 2.

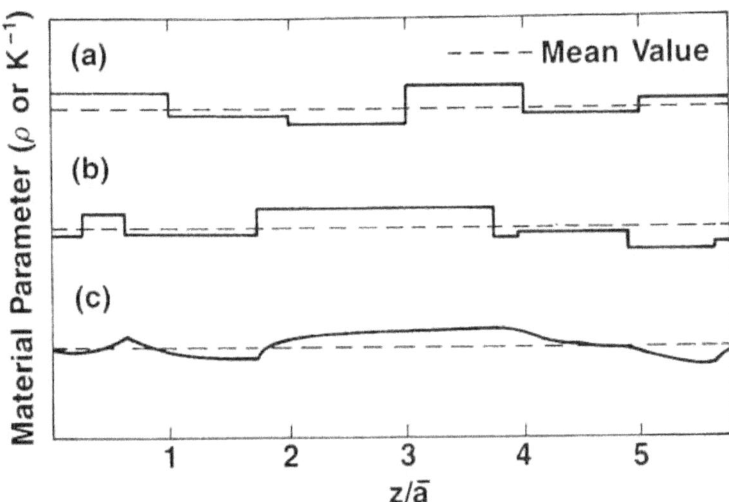



Fig. 2. Three types of randomly-layered models. (a) Model I has equal layer thickness and random material properties from one layer to the next. (b) Model II has variable layer thickness with an exponential distribution as well as random material properties from layer to layer. (c) Model III has continuous but random variation of the material properties.

While for the localization length calculation the boundary condition at $z = 0$ does not matter since $l(\omega)$ is an intrinsic property of the random medium, for the pulse reflection problem the form of the power spectrum function does depend on whether the interface at $z = 0$ is transmitting or reflecting. We will consider two choices for the region $-\infty < z \leq 0$. The first one is a homogeneous medium with $\rho = \rho_0, K = K_0$, and mean velocity $v_0 = (K_0/\rho_0)^{\frac{1}{2}}$. In this case the mean impedance of the two half-spaces are matched in the effective-medium sense so that the boundary at $z = 0$ is transmitting to the pulse. This is denoted as the matched-medium boundary condition. The second choice is for the random medium to border a vacuum so that the interface is totally reflecting to the backscattered wave. This is denoted as the reflecting boundary condition. Effects arising from the different boundary conditions will be addressed in Sec. 5.

## 4. LOCALIZATION LENGTH OF A RANDOMLY-LAYERED MEDIA

### 4.1. Stochastic Differential Equation Approach to the Calculation of $l(\omega)$

We start by rewriting Eq. (2) in the frequency domain, letting $\exp(-i\omega t)$ be the time dependence of $w$ and $p$:

$$\frac{d}{dz}\begin{pmatrix} p \\ u \end{pmatrix} = \omega \begin{bmatrix} 0 & \rho(z) \\ -K^{-1}(z) & 0 \end{bmatrix}\begin{pmatrix} p \\ u \end{pmatrix} , \qquad (5)$$

where $u = iw$ is defined so that Eq. (5) is completely real. Localization means that as $z \to \infty$, the envelope describing the magnitude of the physical solution $(p, u)$ must decay as $\exp(-\gamma z)$, where $\gamma = l^{-1}(\omega)$. Since Eq. (5) represents two coupled first-order differential equations, there must be a second independent solution. Due to the fact that the $2\times 2$ matrix in Eq. (5) has zero trace, it may be simply demonstrated (see Appendix) that the second solution must exhibit an exponentially growing behavior, $\exp(\gamma z)$, with exactly the same $\gamma$. For a set of arbitrary initial conditions (not necessarily physical) at $z = 0$, the solution to Eq. (5) is expressible as the sum of two linearly independent solutions. Provided that the coefficient of the exponentially growing solution is not identically zero, it will always dominate in the $z \to \infty$ limit. Our strategy for calculating $\gamma$ is therefore to choose a set of convenient, arbitrary initial conditions and then to calculate $\gamma$ for random realizations of $\rho(z), K^{-1}(z)$ by measuring the rate of their exponential growth. To implement this, we would like to first make a polar-coordinate transformation defined by $p = r\cos\phi$, $u = r\sin\phi$ so that the magnitude information may be inferred from $r$ alone. The new equations take the form

$$\frac{d\phi}{dz} = -\omega[\rho(z)\sin^2\phi + K^{-1}(z)\cos^2\phi] , \qquad (6a)$$

$$\frac{dr}{dz} = (r\omega/2)\sin 2\phi[\rho(z) - K^{-1}(z)] . \qquad (6b)$$

The $\phi$ equation now is noted to be decoupled from the $r$ equation, which can therefore be integrated to obtain

$$\ln\left(\frac{r}{r_0}\right) = \frac{\omega}{2}\int_0^z \sin 2\phi\{\rho[\xi_1(z)] - K^{-1}[\xi_2(z)]\}dz , \qquad (7)$$

where $r_0 = r(z = 0)$. Since it is expected that $\ln r \simeq \ln(\text{constant}) + \gamma z$ as $z \to \infty$, that means

$$\lim_{z \to \infty} \frac{\ln r}{z} = \gamma = \frac{\omega}{2} \left\{ \lim_{z \to \infty} \frac{1}{z} \int_0^z \sin 2\phi \{\rho[\xi_1(z')] - K^{-1}[\xi_2(z')]\} dz' \right\} . \tag{8}$$

We recognize from Eq. (8) that $\gamma$ is essentially $\omega/2$ times the spatial average of $\sin 2\phi[\rho - K^{-1}]$ in the limit of $z \to \infty$. If the joint stationary distribution function $P(\phi, \xi_1, \xi_2)$ is known, then the spatial averaging that appears in Eq. (8) can be rewritten as

$$\gamma = \frac{\omega}{2} \int_0^{2\pi} d\phi \iint_{-\frac{1}{2}}^{\frac{1}{2}} d\xi_1 d\xi_2 P(\phi, \xi_1, \xi_2) \sin 2\phi [\rho(\xi_1) - K^{-1}(\xi_2)] . \tag{9}$$

To determine $P$, we just note that if $(\phi, \xi_1, \xi_2)$ represents a Markov process, then $P$ must satisfy the stationary Fokker-Planck equation

$$L^* P(\phi, \xi_1, \xi_2) = 0 , \tag{10a}$$

where

$$L = Q + \frac{d\phi}{dz} \frac{d}{d\phi} , \tag{10b}$$

$L^*$ is the adjoint of $L$, and $Q$ is the infinitesimal generator associated with the random jump processes $\xi_1$ and $\xi_2$. Since the operator $Q$ underlies many of the calculations in the rest of this work, we wish to make a little digression here in order to define it more concretely in terms of the three random material models.

A infinitesimal generator is defined as the operation of taking the stochastic derivative[16]:

$$Qf(\xi_0) = \lim_{\Delta z \to 0} (\Delta z)^{-1} \{ E[f(\xi(\Delta z))|\xi(0) = \xi_0] - f(\xi_0) \} , \tag{11}$$

where the notation $E[\ ]$ denotes the operation of taking the expectation value. The knowledge of $Q$ directly generates the dynamical equation $\partial u/\partial z = Qu$ (also known as the backward Kolmogorov equation). For Markovian processes $\xi_1$ and $\xi_2$, such as Model II, we can calculate $Qf(\xi_1, \xi_2)$ as follows. Since $\xi_1$ and $\xi_2$ jump simultaneously, the probability density of no jump in an increment $\Delta z$ is given by $\exp(-\Delta z/\bar{a})$, where $\bar{a}$ is the mean separation between two consecutive jumps. The expectation value is therefore the probability density of no jump $(\exp(-\Delta z/\bar{a}))$ times $f(\xi_1^{(0)}, \xi_2^{(0)})$, plus the probability density of a jump, $1 - \exp(-\Delta z/\bar{a})$, times the integral of $f(\xi_2, \xi_2)$ over the square $[-\frac{1}{2}, \frac{1}{2}] \times [-\frac{1}{2}, \frac{1}{2}]$ since once the jump occurs, $\xi_1$ and $\xi_2$ can take any new value within the square. In short,

$$
\begin{aligned}
Qf(\xi_1^{(0)}, \xi_2^{(0)}) &= \lim_{\Delta z \to 0} (\Delta z)^{-1} \{ E[f(\xi_1(\Delta z), \xi_2(\Delta z)) | \xi_1(0) = \xi_1^{(0)}, \\
&\quad \xi_2(0) = \xi_2^{(0)}] - f(\xi_1^{(0)}, \xi_2^{(0)}) \} \\
&= \lim_{\Delta z \to 0} (\Delta z)^{-1} \{ \exp(-\Delta z/\bar{a}) f(\xi_1^{(0)}, \xi_2^{(0)}) \\
&\quad + [1 - \exp(-\Delta z/\bar{a})] \iint_{-\frac{1}{2}}^{\frac{1}{2}} d\xi_1 d\xi_2 f(\xi_1, \xi_2) \\
&\quad - f(\xi_1^{(0)}, \xi_2^{(0)}) \} \\
&= (\bar{a})^{-1} \int_{-\frac{1}{2}}^{\frac{1}{2}} d\xi_1 \int_{-\frac{1}{2}}^{\frac{1}{2}} d\xi_2 [f(\xi_1, \xi_2) - f(\xi_1^{(0)}, \xi_2^{(0)})] \, .
\end{aligned}
\tag{12a}
$$

If we let $\overline{P}(\xi_1, \xi_2) = 1$ denote the stationary distribution of $(\xi_1, \xi_2)$, it is clear that $Q^* \overline{P} = 0$, i.e. $\overline{P}$ satisfies the stationary Fokker-Planck equation for $(\xi_1, \xi_2)$. In other words, $\overline{P}$ constitutes the null space of $Q^*$.

When $(\xi_1, \xi_2)$ represents jumps at equally-spaced intervals as that in model I, the process is only discretely Markovian, and the formalism based on continuous Markov process as described above is not applicable in the strict sense. However, we will argue on physical grounds that the results obtained at high and low frequencies may still pertain, provided we make certain assumptions. Quantitative results thus obtained are verified by numerical simulations. However, in the intermediate frequency range model I does differ significantly from model II as will be seen in Sec. 4.5.

For model III, the generators of the Markov process are the $\xi_i$'s plus the $\eta_i$'s, defined by

$$
\frac{d\eta_i}{dz} = (\xi_i - \eta_i), \quad i = 1, 2
$$

as specified by Eq. (4a), where $\rho = \rho_0(1 + 2\sigma_\rho \eta_1)$ and $K^{-1} = K_0^{-1}(1 + 2\sigma_K \eta_2)$ are the material parameters of model III. The infinitesimal generator $Q$ is therefore

$$Q = Q_\xi + \frac{d\eta_1}{dz}\frac{\partial}{\partial \eta_1} + \frac{d\eta_2}{dz}\frac{\partial}{\partial \eta_2}$$
$$= Q_\xi + (\xi_1 - \eta_1)\frac{\partial}{\partial \eta_1} + (\xi_2 - \eta_2)\frac{\partial}{\partial \eta_2} , \tag{12b}$$

where $Q_\xi$ denotes that given by Eq. (12a). In what follows, formal considerations will only be given to the case of model II, where $Q = Q_\xi$. The $\eta_i$'s will be noted only in regard to their effect on the final results for model III.

Now back to Eq. (10b), we see that by using Eq. (6a) for $(d\phi/dz)$, the operator $L$ may be written as

$$L = Q - \omega(\rho \sin^2 \phi + K^{-1} \cos^2 \phi)\frac{\partial}{\partial \phi} , \tag{13}$$

and the stationary Fokker-Planck equation for $P(\phi, \xi_1, \xi_2)$ is

$$Q^* P + \omega \frac{\partial}{\partial \phi}[(\rho \sin^2 \phi + K^{-1} \cos^2 \phi)P] = 0 . \tag{14}$$

At present, no general solution of Eq. (14) is known. Below we present solutions of Eq. (14) in both the low- and the high-frequency limits and thereby get the behavior of the localization length in these regimes.

## 4.2. Low Frequency Solution

By assuming $\omega$ to be a small parameter, we expand $P$ as

$$P = P^{(0)} + \omega P^{(1)} + \omega^2 P^{(2)} + \cdots \tag{15}$$

From Eq. (9) that means $\gamma = \gamma_0 \omega + \gamma_1 \omega^2 + \dots$, where $\gamma_n$ is determined by the respective $P^{(n)}$. Denoting the quantity $(\rho \sin^2 \phi + K^{-1} \cos^2 \phi)$ as $F$, we get from Eqs. (14) and (15)

$$Q^* P^{(0)} + \omega Q^* P^{(1)} + \omega^2 Q^* P^{(2)} + \dots = -\omega \frac{\partial}{\partial \phi} F P^{(0)} - \omega^2 \frac{\partial}{\partial \phi} F P^{(1)} \dots ,$$
$$\tag{16a}$$

or, by equating terms with the same power of $\omega$,

$$Q^* P^{(0)} = 0 , \tag{16b}$$

and

$$Q^* P^{(n)} + \frac{\partial}{\partial \phi} F P^{(n-1)} = 0; \quad n \geq 1 . \tag{16c}$$

Starting with $P^{(0)}$, we see from Eq. (16b) that $P^{(0)}$ must be proportional to the null space vector of $Q^*$, i.e.

$$P^{(0)} = \overline{P}(\xi_1, \xi_2) g_0(\phi) . \tag{17}$$

In order to get the form of $g_0(\phi)$, let us consider the next order equation

$$Q^* P^{(1)} = -\frac{\partial}{\partial \phi} F P^{(0)} . \tag{18}$$

Since $Q^*$ has a non-trivial null-space, it is a singular operator and Eq. (18) will be solvable only if its right-hand side, $-\frac{\partial}{\partial \phi} F P^{(0)}$, is orthogonal to the null space of $Q^*$. This is so because otherwise $(Q^*)^{-1}$ would not be well-defined. The so-called "solvability condition" means that

$$\int d\boldsymbol{\xi} \frac{\partial}{\partial \phi} F P^{(0)} = 0 , \tag{19}$$

(here and in the rest of this work the region of integration for $\int d\boldsymbol{\xi}$ will be understood to be square $[-\frac{1}{2}, \frac{1}{2}] \times [-\frac{1}{2}, \frac{1}{2}]$) because the integration with respect to the left-hand side of Eq. (18) yields

$$\int Q^* P^{(1)} d\boldsymbol{\xi} = \int P^{(1)} Q \cdot 1 d\boldsymbol{\xi} = 0 ,$$

i.e. integrate with respect to $\boldsymbol{\xi} = (\xi_1, \xi_2)$ is equivalent to taking the inner product with the null-space vector $\overline{P} = 1$. Substitution of Eq. (17) into Eq. (19) gives

$$\frac{\partial}{\partial \phi} \overline{F} g_0(\phi) = 0 , \tag{20a}$$

where

$$\overline{F} = \rho_0 \sin^2 \phi + K_0^{-1} \cos^2 \phi \tag{20b}$$

is obtained by averaging $F$ with respect to $\boldsymbol{\xi}$ (integrating $F$ with $\overline{P}$). Equation (20a) gives the solution of

$$g_0(\phi) = \frac{(\text{const.})}{\overline{F}} \, . \tag{21a}$$

By requiring the normalization conditions that $g_0(\phi)$ must integrate to 1, we get const. $= (2\pi v_0)^{-1}$ with $v_0 = \sqrt{K_0/\rho_0}$. Therefore,

$$P^{(0)}(\phi, \xi_1, \xi_2) = \frac{1}{2\pi v_0} \frac{\overline{P}(\xi_1, \xi_1)}{\rho_0 \sin^2 \phi + K_0^{-1} \cos^2 \phi} \tag{22}$$

is the solution for $P(\phi, \xi_1, \xi_2)$ to the leading order. With the knowledge of $P^{(0)}$ we can now calculate, via Eq. (9), the inverse localization length $\gamma = \gamma_0 \omega$ to the leading order:

$$
\begin{aligned}
\gamma_0 &= \frac{1}{2} \int_0^{2\pi} d\phi \iint_{-\frac{1}{2}}^{\frac{1}{2}} d\xi_1 d\xi_2 \, P^{(0)}(\phi, \xi_1, \xi_2) \sin 2\phi [\rho(\xi_1) - K^{-1}(\xi_2)] \\
&= \frac{1}{4\pi v_0} (\rho_0 - K_0^{-1}) \int_0^{2\pi} d\phi \frac{\sin 2\phi}{\rho_0 \sin^2 \phi + K_0^{-1} \cos^2 \phi} \\
&= 0 \, . 
\end{aligned}
\tag{23}
$$

That is, $P^{(0)}$ gives zero contribution to $\gamma$. In order to get the leading-order finite contribution, we must therefore calculate $P^{(1)}(\phi, \xi_1, \xi_2)$.

Let us return to Eq. (18) but now write it in a slightly different form:

$$Q^* P^{(1)} = -\frac{\partial}{\partial \phi} F P^{(0)} = -\frac{\partial}{\partial \phi} \overline{F} P^{(0)} - \frac{\partial}{\partial \phi} \widehat{F} P^{(0)} \, , \tag{24a}$$

where we have expressed $F$ as the sum of its mean and a fluctuating part $\widehat{F}$ defined as

$$\widehat{F} = \rho \widehat{\rho}_0 \sin^2 \phi + K_0^{-1} \widehat{K}^{-1} \cos^2 \phi \, ,$$

with $\widehat{\rho} = (\rho - \rho_0)/\rho_0 = 2\sigma_\rho \xi_1$ and $\widehat{K}^{-1} = (K^{-1} - K_0^{-1})/K_0^{-1} = 2\sigma_K \xi_2$ denoting the normalized fluctuation in $\rho$ and $K^{-1}$ as given by Eq. (3). Now from Eqs. (17) and (21a), we have $\partial \overline{F} P^{(0)}/\partial \phi = 0$. Therefore,

$$Q^* P^{(1)} = -\frac{\partial}{\partial \phi} \widehat{F} P^{(0)} \, . \tag{24b}$$

Since $Q^*$ is a singular operation, we can formally write the solution $P^{(1)}$ as

$$P^{(1)} = -(Q^*)^{-1}\frac{\partial}{\partial\phi}\widehat{F}P^{(0)} + \overline{P}(\xi_1,\xi_2)g_1(\phi) \ . \tag{25}$$

Here the notation $(Q^*)^{-1}$ means the generalized inverse of $Q^*$ defined on the space of vectors that is orthogonal to the null space (the fact that $\partial\widehat{F}P^{(0)}/\partial\phi$ is indeed orthogonal to the null space is guaranteed by the solvability condition for Eq. (18)). For the part of $P^{(1)}$ that is proportional to the null vector $\overline{P}(\xi_1,\xi_2)$, we have denoted the function of proportionality as $g_1(\phi)$. To get $g_1(\phi)$, we must use the solvability condition for the next order equation

$$Q^*P^{(2)} = -\frac{\partial}{\partial\phi}FP^{(1)} \ , \tag{26}$$

given by the condition

$$\int d\pmb{\xi}\frac{\partial}{\partial\phi}FP^{(1)} = 0 \ , \tag{27a}$$

or

$$\begin{aligned}
\beta &= \int d\pmb{\xi}FP^{(1)} \\
&= \int d\pmb{\xi}\left\{(\overline{F}+\widehat{F})\left[-(Q^*)^{-1}\frac{\partial}{\partial\phi}\widehat{F}P^{(0)}\right] + F\overline{P}(\xi)g_1(\phi)\right\} \\
&= \overline{F}g_1(\phi) + \int d\pmb{\xi}\left[-\widehat{F}(Q^*)^{-1}\frac{\partial}{\partial\phi}\widehat{F}P^{(0)}\right] \ . \tag{27b}
\end{aligned}$$

Here $\beta$ is to be determined by the normalization condition for $g_1(\phi)$. In Eq. (27b) we have used the fact that the integration of $\overline{F}$ with $(Q^*)^{-1}\partial\widehat{F}P^{(0)}/\partial\phi$ gives zero since $\overline{F}$ is independent of $\pmb{\xi}$, and $(Q^*)^{-1}\partial\widehat{F}P^{(0)}/\partial\phi$ is orthogonal to $\overline{P}$. By the definition of adjoint operator, this last integral in Eq. (27b) may also be written as

$$\int d\pmb{\xi}\left(-\widehat{F}(Q^*)^{-1}\frac{\partial}{\partial\phi}\widehat{F}P^{(0)}\right) = \int d\pmb{\xi}\left(-\frac{\overline{P}}{2\pi v_0}\frac{\partial}{\partial\phi}\frac{\widehat{F}}{\overline{\overline{F}}}Q^{-1}\widehat{F}\right) \ . \tag{28}$$

Now we define $Q^{-1}$ formally as

$$Q^{-1} = -\int_0^\infty ds \, \exp\,(sQ) \,, \qquad (29)$$

where $\exp(sQ)$, representing the formal solution to the backward Kolmogorov equation $\partial_s u = Qu$, may be viewed as a stochastic shift operator. For example,

$$\exp\,(sQ)\hat{\rho}[\xi(0)] = E\{\hat{\rho}[\xi(s)] | \hat{\rho}[\xi(0)]\} \,, \qquad (30)$$

where the right-hand-side stands for the expectation value of $\hat{\rho}$ at $z = s$, *given* the value of $\hat{\rho}$ at $z = 0$. Provided that the systems we are trying to model have finite correlation length, $\exp(sQ)\widehat{F} = 0$ for $s \to \infty$. That is why the upper limit of the integration for Eq. (29) always gives zero. Now let us return to Eq. (28) and substitute Eq. (29) for $Q^{-1}$. The result is

$$\int d\boldsymbol{\xi}\left[-\frac{\overline{P}}{2\pi v_0}\left(\frac{\partial}{\partial\phi}\overline{F}^{-1}\right)\widehat{F}Q^{-1}\widehat{F}\right]$$

$$+ \int d\xi\left[-\frac{\overline{P}}{2\pi v_0}\overline{F}^{-1}\left(\frac{\partial}{\partial\phi}\widehat{F}\right)Q^{-1}\widehat{F}\right]$$

$$= (2\pi v_0)^{-1}\left\{\left(\frac{\partial}{\partial\phi}\overline{F}^{-1}\right)\int d\boldsymbol{\xi}\,\overline{P}\int_0^\infty ds\langle\widehat{F}(0)\widehat{F}(s)\rangle\right.$$

$$\left. + \overline{F}^{-1}\int d\boldsymbol{\xi}\,\overline{P}\int_0^\infty ds\langle\frac{\partial\widehat{F}(0)}{\partial\phi}\widehat{F}(s)\rangle\right\}$$

$$= (2\pi v_0^2\overline{F}^2)^{-1}\sin 2\phi[\rho_0\alpha_{\rho\rho}\sin^2\phi - K_0^{-1}\alpha_{KK}\cos^2\phi$$

$$+ (K_0^{-1}\cos^2\phi - \rho_0\sin^2\phi)\alpha_{\rho K}] \,. \qquad (31)$$

580          *Ping Sheng et al.*

Here we have defined

$$\alpha_{\rho\rho} = \int_0^\infty ds \langle \rho(0)\rho(s)\rangle \, , \tag{32a}$$

$$\alpha_{KK} = \int_0^\infty ds \langle K^{-1}(0)K^{-1}(s)\rangle \, , \tag{32b}$$

$$\alpha_{\rho K} = \int_0^\infty ds \langle \widehat{\rho}(0)\widehat{K}^{-1}(s)\rangle = \alpha_{K\rho} \, . \tag{32c}$$

where $\langle \ \rangle$ denotes taking the expectation value with respect to $\boldsymbol{\xi}$. It is recognized that $\alpha$'s are the integrals of the autocorrelation and correlation functions for $\widehat{\rho}$ and $\widehat{K}^{-1}$. Combining Eqs. (31) and (27b) yields

$$g_1(\phi)$$
$$= \frac{\beta}{\overline{F}} - \frac{\sin 2\phi[\rho_0\alpha_{\rho\rho}\sin^2\phi - K_0^{-1}\alpha_{KK}\cos^2\theta + (K_0^{-1}\cos^2\phi - \rho_0\sin^2\phi)\alpha_{\rho K}]}{2\pi v_0^2 \overline{F}^3} \tag{33}$$

The knowledge of $g_1(\theta)$, plus the definition of $Q^{-1}$ by Eq. (29), enable us to evaluate $\gamma = \gamma_1\omega^2$:

$$\gamma_1 = \frac{1}{2}\int_0^{2\pi} d\phi \int d\boldsymbol{\xi} P^{(1)}\sin 2\phi[\rho - K^{-1}]$$

$$= \frac{1}{2}\left\{ \int_0^{2\pi} d\phi \int d\boldsymbol{\xi}\sin 2\phi[\rho - K^{-1}]\int_0^\infty ds\, \exp{(sQ^*)}(2\pi v_0)^{-1}\overline{P}\frac{\partial}{\partial\phi}\frac{\widehat{F}}{\overline{\overline{F}}} \right.$$

$$\left. + \int_0^{2\pi}d\phi\int d\boldsymbol{\xi}\overline{P}(\xi_1,\xi_2)[\rho(\xi_1) - K^{-1}(\xi_2)]\sin 2\phi\, g_1(\phi) \right\} \, . \tag{34a}$$

It is noted that the term $\beta/\overline{F}$ in $g_1(\phi)$ integrates to zero for the same reason as $\gamma_0 = 0$ and therefore does not contribute. After carrying out the operations and simplifying the remaining, we get the final answer[12] as

$$\gamma = \gamma_1\omega^2 = \frac{1}{4}(\alpha_{\rho\rho} + \alpha_{KK} - 2\alpha_{\rho K})\frac{\omega^2}{v_0^2} \, . \tag{34b}$$

Equation (34b) tells us that to the leading order, the localization length diverges as $\omega^{-2}$ when $\omega \to 0$. Physically, this quadratic frequency dependence may be attributed to the $\omega^2$ variation of the Rayleigh scattering in one dimension. The proportionality constant, $4v_0^2/(\alpha_{\rho\rho} + \alpha_{KK} -$

$2\alpha_{\rho K}$), on the other hand, does depend on both the mean property of the medium, $v_0 = (K_0/\rho_0)^{\frac{1}{2}}$, as well as the correlations of the fluctuating parts. For specific models, the $\alpha$'s may be evaluated explicitly. If we let $\widehat{\rho} = 2\sigma_\rho \xi_1(z)$ and $\widehat{K}^{-1} = 2\sigma_K \xi_2(z)$, then

$$\alpha_{\rho\rho} = \int_0^\infty ds \langle \rho(s)\rho(0) \rangle = 4\sigma_\rho^2 \int_0^\infty ds \langle \xi_1(s)\xi_1(0) \rangle \ . \qquad (35a)$$

Now for model I where the layer thickness is constant $= \bar{a}$, we will make the assumption that since at low frequencies the wave cannot resolve the microstructures, the continuous Markovian model may still apply provided we allow the process $\boldsymbol{\xi}$ to start at random positions inside the first layer. The quantity $\langle \xi_1(s)\xi_1(0) \rangle$ can then be evaluated as follows. For $s > \bar{a}$, $\langle \xi_1(s)\xi_1(0) \rangle = 0$, and for $s < \bar{a}$

$$\langle \xi_1(s)\xi_1(0) \rangle = \left(1 - \frac{s}{\bar{a}}\right)\langle \xi_1^2 \rangle = \frac{1}{12}\left(1 - \frac{s}{\bar{a}}\right) , \qquad (35b)$$

where $\left(1 - \dfrac{s}{\bar{a}}\right)$ gives the probability that two points, separated by $s$, can still lie in the same layer. Substitution of Eq. (35b) into Eq. (35a) gives

$$\alpha_{\rho\rho} = \frac{1}{6}\sigma_\rho^2 \bar{a} \ . \qquad (35c)$$

Similarly, $\alpha_{KK} = \sigma_K^2 \bar{a}/6$ and $\alpha_{\rho K} = 0$ because $\xi_1$ and $\zeta_2$ are independent (although they jump simultaneously). That means

$$\gamma_1 = \frac{(\sigma_\rho^2 + \sigma_K^2)\bar{a}}{24v_0^2} \qquad (36)$$

for model I. For model II, the only difference is that the layers have an exponential thickness distribution with mean $\bar{a}$. The continuous Markovian model is strictly applicable, and

$$\langle \xi_1(s)\xi_1(0) \rangle = \frac{1}{\bar{a}}\exp\left(-\frac{s}{\bar{a}}\right)\langle \xi_1^2 \rangle = \frac{1}{12\bar{a}}\exp\left(-\frac{s}{\bar{a}}\right) \ . \qquad (37)$$

Substitution of Eq. (37) into Eq. (35a) then yields $\alpha_{\rho\rho} = \sigma_\rho^2 \bar{a}/3$. That means for model II, $\gamma_1$ is given by

$$\gamma_1 = \frac{(\sigma_\rho^2 + \sigma_K^2)\bar{a}}{12v_0^2} , \qquad (38)$$

which differs from Eq. (36) by a factor of two. This prediction is indeed verified by numerical simulations as will be seen in Sec. 4.4.

For model III, the material parameters $\rho$ and $K^{-1}$ are obtained from model II by an exponential filter $E(z) = \exp(-z)(z \geq 0, E(z) = 0$ for $z < 0)$, i.e. $m(z) = E(z) \otimes \overline{m}(z)$, where $\overline{m}$ and $m$ denote the material parameters for models II and III, respectively, and $\otimes$ means convolutions. Since the $\alpha$'s may be alternatively expressed as the power spectrum $S_m$ of $m$ at zero spatial frequency, i.e.

$$\int_0^\infty \langle m(s)m(0)\rangle ds = S_m(0) , \tag{39a}$$

it is clear that the low frequency localization length of model II and model III must be the same because

$$S_m(0) = S_{\overline{m}}(0) \cdot |J(0)|^2 = S_{\overline{m}}(0) , \tag{39b}$$

where $J(i\nu) = (1 + i\nu)^{-1}$ is the Fourier transform of $E(z)$, with $\nu$ denoting spatial frequency, and $|J(0)|^2$ is noted to be 1.

### 4.3. High Frequency Solution

In the limit of high frequencies, we expand $P(\phi, \xi_1 \xi_2)$ as

$$P = P^{(0)} + \frac{1}{\omega}P^{(1)} + \frac{1}{\omega^2}P^{(2)} + \ldots \tag{40}$$

Substitution of Eq. (40) into Eq. (9) gives

$$\gamma = c_0\omega + c_1 + \frac{c_2}{\omega} + \ldots , \tag{41a}$$

where

$$c_n = \frac{1}{2}\int_0^{2\pi} d\phi \int d\xi P^{(n)}(\phi, \xi) \sin 2\phi [\rho(\xi_1) - K^{-1}(\xi_2)] . \tag{41b}$$

The Fokker-Planck equation, Eq. (14), requires that

$$Q^* P^{(0)} + \frac{1}{\omega}Q^* P^{(1)} + \frac{1}{\omega^2}Q^* P^{(2)} + \ldots = -\omega\frac{\partial}{\partial\phi}F P^{(0)} - \frac{\partial}{\partial\phi}F P^{(1)}$$

$$-\frac{1}{\omega}\frac{\partial}{\partial\phi}F P^{(2)} + \ldots \tag{42a}$$

By equating terms with the same powers of $\omega$, we get

$$\frac{\partial}{\partial\phi}FP^{(0)} = 0 \, , \tag{42a}$$

and

$$Q^{*}P^{(n)} = -\frac{\partial}{\partial\phi}FP^{(n+1)} \, ; n \geq 0 \, . \tag{42b}$$

Here $F = \rho\sin^2\phi + K^{-1}\cos^2\phi$ as defined before. Equation (42a) tells us that

$$P^{(0)} = g_0(\boldsymbol{\xi})F^{-1} \, , \tag{43}$$

where $g_0(\boldsymbol{\xi})$ is determined by the requirement that

$$\int_0^{2\pi} d\phi P(\phi,\boldsymbol{\xi}) = \overline{P}(\boldsymbol{\xi}) \, . \tag{44}$$

Since $P^{(0)}$ is the only term in $P$ that is independent of $\omega$, the normalization condition on $P$ is equivalent to the normalization conditions on $P^{(0)}$. By substituting Eq. (43) into Eq. (44), we get $g_0(\boldsymbol{\xi}) = (2\pi)^{-1}(\rho K^{-1})^{\frac{1}{2}}\overline{P}(\boldsymbol{\xi})$, or

$$P^{(0)} = \sqrt{\frac{\rho(\xi_1)}{K(\xi_2)}}\frac{\overline{P}(\xi_1,\xi_2)}{2\pi[\rho(\xi_1)\sin^2\phi + K^{-1}(\xi_2)\cos^2\phi]} \, . \tag{45}$$

It is now immediately obvious that $c_0 = 0$ because the $\phi$ integration of the product of $\sin 2\phi$, an odd function of $\phi$, with $F^{-1}$, an even function of $\phi$, is zero. This result is important because *it constitutes a proof that the localization length is non-decreasing as $\omega \to \infty$*. Since $\gamma = c_1 + c_2/\omega + \ldots$, the localization length can either saturate to a constant value $(c_1 \neq 0)$, or increase with frequency $(c_1 = 0)$. To calculate $c_1$, we use $P^{(0)}$ as the input to the next-order equation

$$\frac{\partial}{\partial\phi}FP^{(1)} = Q^{*}P^{(0)} \, , \tag{46a}$$

or

$$P^{(1)} = -\frac{1}{2\pi F}Q^{*}\sqrt{\frac{\rho(\xi_1)}{K(\xi_2)}}\overline{P}(\xi_1,\xi_2)$$

$$\times \int_0^{\phi} \frac{d\phi'}{[\rho(\xi_1)\sin^2\phi' + K^{-1}(\xi_2)\cos^2\phi']} + g_1(\boldsymbol{\xi})F^{-1} \, . \tag{46b}$$

Here $g_1(\boldsymbol{\xi})$ can be obtained by the condition that

$$\int_0^{2\pi} d\phi\, P^{(1)}(\phi, \xi_1, \xi_2) = 0 \ . \tag{46c}$$

However, there is no need to determine $g_1(\boldsymbol{\xi})$ because for the calculation of $\gamma$ the term $g_1(\boldsymbol{\xi})/F$ will integrate to zero for the same reason that $c_0 = 0$. Now the constant $c_1$ may be expressed as[6]

$$c_1 = -\frac{1}{4\pi} \int d\boldsymbol{\xi} \int_0^{2\pi} d\phi \frac{\sin 2\phi [\rho(\xi_1) - K^{-1}(\xi_2)]}{[\rho(\xi_1)\sin^2\phi + K^{-1}(\xi_2)\cos^2\phi]}$$

$$\cdot \left\{ Q^* \overline{P}(\xi_1, \xi_2) \sqrt{\frac{\rho(\xi_1)}{K(\xi_2)}} \int_0^\phi \frac{d\phi'}{[\rho(\xi_1)\sin^2\phi' + K^{-1}(\xi_2)\cos^2\phi']} \right\}$$

$$= -\frac{1}{4\pi} \int d\boldsymbol{\xi}\, \overline{P}(\xi_1, \xi_2) \sqrt{\frac{\rho(\xi_1)}{K(\xi_2)}} \int_0^{2\pi} d\phi \sin 2\phi$$

$$\times \int_0^\phi \frac{d\phi'}{\rho\sin^2\phi' + K^{-1}\cos^2\phi'} Q\left[ \frac{\rho(\xi_1) - K^{-1}(\xi_2)}{\rho(\xi_1)\sin^2\phi + K^{-1}(\xi_2)\cos^2\phi} \right] , \tag{47}$$

where we have used the adjoint property of the operator $Q^*$. By carrying out the $\phi'$ integration explicitly, we get

$$c_1 = -\frac{1}{4\pi} \int d\boldsymbol{\xi}\, \overline{P}(\xi_1, \xi_2)$$

$$\times \int_0^{2\pi} d\phi \arctan(\sqrt{\rho(\xi_1)K(\xi_2)}\tan\phi)\sin 2\phi$$

$$\cdot Q\left[ \frac{\rho(\xi_1) - K^{-1}(\xi_2)}{\rho(\xi_1)\sin^2\phi + K^{-1}(\xi_2)\cos^2\phi} \right] , \tag{48}$$

where the arctan function is understood to be the branch which is zero when $\phi = 0$, and is monotonically increasing. In contrast to the low-frequency case, here we cannot go any further without explicitly evaluating the $Q$ operation. This is physically reasonable because at high frequencies the wave would be able to resolve the inhomogeneities of the medium, and the value of $c_1$ is therefore dependent on the local details of the model as expressed by the operator $Q$. That means there should be significant difference between the three models. However, for models I and II if one makes the assumption that at high frequencies the scatterings at interfaces occur incoherently, then the two models should be indistinguishable in that limit.

To evaluate $c_1$, we note that the integration implicit in the operator $Q$ can be performed analytically. The actual numerical calculation therefore involves only a double integral. For model III, however, the $Q$ operator is defined by Eq. (12b), with $\rho$ and $K^{-1}$ the functionals of $\eta_1$ and $\eta_2$, respectively. Therefore,

$$
c_1 = -\frac{1}{4\pi} \int d\boldsymbol{\xi} d\boldsymbol{\eta}\, \overline{P}(\boldsymbol{\xi}, \boldsymbol{\eta}) \int_0^{2\pi} d\phi \arctan\left(\sqrt{\rho(\eta_1) K(\eta_2)} \tan\phi\right) \sin 2\phi
$$

$$
\times\, Q\left[\frac{\rho K - 1}{\rho K \sin^2\phi + \cos^2\phi}\right], \tag{49a}
$$

where

$$
Q = Q_{\boldsymbol{\xi}} + (\xi_1 - \eta_1)\frac{\partial}{\partial \eta_1} + (\xi_2 - \eta_2)\frac{\partial}{\partial \eta_2}
$$

$$
= Q_{\boldsymbol{\xi}} + f\frac{\partial}{\partial \rho} + g\frac{\partial}{\partial K^{-1}}, \tag{49b}
$$

with $f = d\rho/dz = \rho_{\mathrm{II}} - \rho_{\mathrm{III}}$ and $g = dK^{-1}/dz = K_{\mathrm{II}}^{-1} - K_{\mathrm{III}}^{-1}$, and $\overline{P}$ is the joint stationary distributions for $\boldsymbol{\xi}$ and $\boldsymbol{\eta}$. From Eq. (49a) it is easy to see that the effect of $Q_{\boldsymbol{\xi}}$ is to give zero since the operand is independent of $\boldsymbol{\xi}$. The remaining derivative operations may be carried out explicitly to obtain

$$
c_1 = -\frac{1}{4\pi} E\left[\int_0^{2\pi} d\phi \arctan\left(\sqrt{\rho K} \tan\phi\right)\right.
$$

$$
\left. \times\, \frac{\sin 2\phi}{\rho K \sin^2\phi + \cos^2\phi} \rho K (f\rho^{-1} - gK)\right]. \tag{50a}
$$

Here $E[\ ]$ denotes the operation of taking the expectations value with respect to the variables $\eta$ and $\xi$. We note, however, that $f\rho^{-1} = (d\rho/dz)\rho^{-1} = d\ln\rho/dz$ and $gK = -d\ln K/dz$. Therefore $\rho K(f\rho^{-1} - gK)$ may be expressed as $\rho K\, d\ln(\rho K)/dz = d(\rho K)/dz$. Equation (50a) may be rewritten as

$$c_1 = -\frac{1}{4\pi}E\left[\frac{d}{dz}\int_0^{\rho K}\int_0^{2\pi}\arctan(\sqrt{u}\tan\phi)\frac{\sin\phi}{u\sin^2\phi + \cos^2\phi}\,d\phi\,du\right].$$

$$(50b)$$

By interchanging the order of the $d/dz$ operation and the $E[\ ]$ operation, we get $c_1 = 0$ because the stationary expectation value is independent of $z$ by definition.

The fact that $c_1 = 0$ for continuous random models means that $l(\omega)$ diverges as $\omega \to \infty$. This difference in the qualitative behaviors between models I, II and III may be understood physically as follows. For the discontinuous models, the reflection and transmission coefficients at a sharp interface are frequency independent. When the phase coherence between the different scattering events is lost at high frequencies, the localization length must also be frequency independent as shown. However, when the interface is fuzzy and continuous, the reflection coefficient at each interface is expected to decrease as the wavelength becomes smaller than the size of the interface region. This is because the wave would begin to perceive the environment as slowly varying and the wave transmission is consequently increased.

### 4.4. Oblique Incidence

At non-normal incidence, it becomes important to distinguish between the electromagnetic wave and the elastic wave, and the behavior of differing polarizations could be dramatic.[9] In this section we will consider only the electromagnetic case, since the elastic case would require significant modification of the formalism derived so far.

If we let $H$ denote the $y$-component of the magnetic field, $E$ the $x$-component of the electric field, $\theta$ the incident angle with respect to te normal ($z$-axis) and $\varepsilon_0, \mu_0$ the mean value of the dielectric constant and the magnetic permeability, respectively, then the wave equation may be written as

$$\frac{d}{dz}\begin{pmatrix} H \\ E \end{pmatrix} = \frac{\omega}{c}\begin{bmatrix} 0 & \varepsilon(z) \\ -\mu(z) + \dfrac{\mu_0\varepsilon_0 \sin^2\theta}{\varepsilon(z)} & 0 \end{bmatrix}\begin{pmatrix} H \\ E \end{pmatrix} \tag{51a}$$

for the polarization where the $xz$ plane is the plane of incidence so that the $H$ vector is perpendicular to it. We will denote this configuration the $H$-polarization. Analogously, we define the $E$-polarization as that configuration when the $yz$ plane is the plane of incidence. In that case the wave equation is given by

$$\frac{d}{dz}\begin{pmatrix} H \\ E \end{pmatrix} = \frac{\omega}{c}\begin{bmatrix} 0 & \varepsilon(z) - \dfrac{\mu_0 \sin^2\theta}{\mu(z)} \\ -\mu(z) & 0 \end{bmatrix}\begin{pmatrix} H \\ E \end{pmatrix}. \tag{51b}$$

In the above we have assumed that the wave is incident from a homogeneous halfspace with material parameters $\varepsilon_0$ and $\mu_0$. Comparison of Eq. (51) and Eq. (5) shows that if we make the identification

$$\rho(z) \rightarrow \frac{\varepsilon(z)}{c}, \tag{51c}$$

$$K^{-1}(z) \rightarrow \frac{\mu(z)}{c} - \frac{\mu_0\varepsilon_0 \sin^2\theta}{c\varepsilon(z)}, \tag{51d}$$

for the $H$-polarization and

$$\rho(z) \rightarrow \frac{\varepsilon(z)}{c} - \frac{\mu_0\varepsilon_0 \sin^2\theta}{c\mu(z)}, \tag{51e}$$

$$K^{-1}(z) \rightarrow \frac{\mu(z)}{c}, \tag{51f}$$

for the $E$-polarization, then the low- and the high-frequency localization length solutions developed in the previous sections may be carried through with minimal changes. However, the important point to note here is that the fluctuating part of $\rho(= \rho_0(1 + \hat{\rho}))$, $\hat{\rho}$, and the fluctuating part of $K^{-1}(= K_0^{-1}(1 + \hat{K}^{-1}))$, $\hat{K}^{-1}$, are no longer independent as before but are correlated. For example, in the case of $H$-polarization if we write $\varepsilon = \varepsilon_0(1 + \widehat{\varepsilon})$, $\varepsilon^{-1} = \bar{\varepsilon}_{\mathrm{inv}}(1 + \widehat{\varepsilon}_{\mathrm{inv}})$, and $\mu = \mu_0(1 + \widehat{\mu})$, then

$$\widehat{\rho} = \widehat{\varepsilon} \,, \tag{52a}$$

$$\widehat{K}^{-1} = \frac{\widehat{\mu} - \varepsilon_0 \bar{\varepsilon}_{\mathrm{inv}} \sin^2 \theta \; \widehat{\varepsilon}_{\mathrm{inv}}}{1 - \varepsilon_0 \bar{\varepsilon}_{\mathrm{inv}} \sin^2 \theta} \,. \tag{52b}$$

Since $\varepsilon_0^{-1} \neq \bar{\varepsilon}_{\mathrm{inv}}$ in general, $\widehat{\varepsilon}$ and $\widehat{\varepsilon}_{\mathrm{inv}}$ cannot be independent, and we have a case of built-in correlation between $\widehat{\rho}$ and $\widehat{K}^{-1}$. By expressing

$$\widehat{\varepsilon} = 2\sigma_\varepsilon \xi_1 \,, \tag{53a}$$

$$\widehat{\mu} = 2\sigma_\mu \xi_2 \,, \tag{53b}$$

we get

$$\bar{\varepsilon}_{\mathrm{inv}} = \int_{-\frac{1}{2}}^{\frac{1}{2}} \frac{d\xi_1}{\varepsilon_0(1 + 2\sigma_\varepsilon \xi_1)} = \frac{1}{2\sigma_\varepsilon \varepsilon_0} \ln\left(\frac{1 + \sigma_\varepsilon}{1 - \sigma_\varepsilon}\right) \,, \tag{53c}$$

and

$$\widehat{\varepsilon}_{\mathrm{inv}} = \frac{2\sigma_\varepsilon}{(1 + 2\sigma_\varepsilon \xi_1)} \left[\ln\left(\frac{1 + \sigma_\varepsilon}{1 - \sigma_\varepsilon}\right)\right]^{-1} - 1 \,. \tag{53d}$$

For the low frequency limit, the calculation of $\alpha_{\rho\rho}$, $\alpha_{KK}$, and $\alpha_{\rho K}$ for model I leads to

$$\gamma^{(H)} \simeq \gamma_1^{(H)} \omega^2 = \frac{1}{24}\left[\sigma_\varepsilon^2 + \frac{\sigma_\mu^2 + M \sin^4 \theta}{(1 - C \sin^2 \theta)^2} + \frac{N \sin^2 \theta}{1 - C \sin^2 \theta}\right]$$
$$\times (1 - C \sin^2 \theta)\frac{\varepsilon_0 \mu_0}{c^2}\omega^2 \,, \tag{54a}$$

where the constants $C, M, N$ are defined by

$$C = \frac{1}{2\sigma_\varepsilon} \ln\left(\frac{1 + \sigma_\varepsilon}{1 - \sigma_\varepsilon}\right) \,, \tag{54b}$$

$$M = 3\left[\frac{1}{1 - \sigma_\varepsilon^2} - C^2\right] \,, \tag{54c}$$

$$N = 6(1 - C) \,. \tag{54d}$$

Results for models II and III can be obtained from Eq. (54a) by multiplying $\gamma_1^{(H)}$ by a factor of 2. It is seen that at $\theta = 0$, Eq. (54a) recovers the known result, Eq. (36), and for non-magnetic material $(\sigma_\mu = 0)$ and $\sigma_\epsilon \ll 1$, Eq. (54a) simplifies as

$$\gamma^{(H)} \simeq \gamma_1^{(H)} \omega^2 = \frac{\sigma_\epsilon^2}{24}(1 - \tan^2 \theta) \cos^2 \theta \frac{\varepsilon_0 \mu_0}{c^2} \omega^2 \tag{55}$$

to order $\sigma_\epsilon^2$. We note that at $\theta = 45°$, $\gamma_1^{(H)} = 0$ so that the localization length actually diverges faster than $\omega^2$. This phenomenon has its origin in the well-known Brewster effect[9,17] for reflection from a plane interface. Namely, when $\sigma_\epsilon \ll 1$ the incident wave at $45°$ is roughly perpendicular to the reflected wave as shown in Fig. 3. That means the $E$-vector of the incident wave is parallel to the direction of the reflected wave. If one regards the reflected wave as generated by dipoles excited by the incident $E$-field, then there can be no wave amplitude in the reflection direction due to the fact that dipoles do not radiate in the direction of its oscillation. When $\sigma_\epsilon$ is not small, Eq. (54a) generally has a zero at some value of $\theta_B \neq 45°$. The actual value of the localization length usually does not diverge at such angle, since true divergence would require $\gamma_n^{(H)} = 0$ to all orders of $n$. However, the value of the localization length can be expected to be orders of magnitude larger at $\theta_B$ than that at normal incidence. This effect is denoted as the Brewster anomaly in accordance with its physical origin. An exception to the above description is that of the binary random medium, where true divergence of $l$ can be achieved at certain incident angle. This is due to the fact that if $\theta_B$ is the Brewster angle for a wave going from medium 1 to medium 2, then the refracted angle in medium 2 is also exactly the Brewster angle in going from medium 2 to medium 1. Therefore this is a case where the deterministic Brewster effect completely overcomes the stochastic effect of localization.

For the $E$-polarization, $\gamma_1^{(E)}$ has exactly the same form as that of $\gamma_1^{(H)}$, Eq. (54a), except the roles of $\sigma_\epsilon$ and $\sigma_\mu$ are now interchanged. Therefore, $\gamma_1^{(E)}$ would have the same Brewster anomaly for a random magnetic material. However, if $\sigma_\mu = 0$, then

$$\gamma \simeq \gamma_1^{(E)} \omega^2 = \frac{\sigma_\epsilon^2}{24} \sec^2 \theta \frac{\omega^2}{v_0^2} , \tag{56}$$

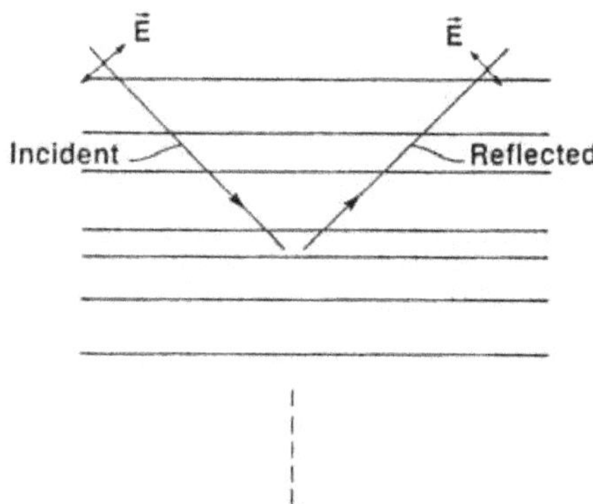


Fig. 3. Schematic diagram illustrating the $H$-polarized incident and scattered beams in a randomly-layered medium with small material fluctuations. Brewster effect occurs when the incident **E** field is parallel to the scattering direction.

for model I, and $\gamma_1^{(E)}$ is a factor of 2 larger for models II and III. It is seen that the angular dependence of $\gamma_1^{(E)}$ is perfectly regular. At high frequencies, the same formalism as before carries through with the different definitions now for $\rho$ and $K^{-1}$. Since one cannot get explicit formulas for $c_1$ anyway, the correspondence with $\rho$ and $K^{-1}$ is as far as one can go without getting into actual numerical computations.

## 4.5. Numerical Simulations

To summarize the results of the previous sections, we see that to the leading order, the frequency dependence of the localization length may be expressed as $l \sim \omega^{-2}$ for $\omega \to 0$ and $l \sim$ constant or increases as a power of $\omega$ for $\omega \to \infty$, where the proportionality constants are calculable from material parameters and model statistics. There is also strong polarization dependence at oblique incidence. To verify this form for $l(\omega)$ and to actually see the behavior at intermediate frequencies where the wavelength is comparable to the correlation length of the inhomogeneities, we use the transfer-matrix algorithm for numerically calculating the transmission coefficient $T$ through a finite segment $L$ of the random medium. Since $T \sim \exp(-\gamma L)$, the localization length is obtained as $l = -\langle \ln|T|/L \rangle^{-1}$, where $\langle \ \rangle$ now denotes configurational averaging. From the ergodic theorem, we expect this to be the same as taking the expectations value with respect to $\xi$.

The transfer-matrix algorithm is based on the assumption that the layered medium is composed of homogeneous layers within each of which the general solution to the wave equation may be expressed as

$$p_n(z) = A_n \exp\left[ik_n\left(z - z_n + \frac{a_n}{2}\right)\right] + B_n \exp\left[-ik_n\left(z - z_n + \frac{a_n}{2}\right)\right],$$
$$(57)$$

where $p_n$ denotes pressure in the $n$th layer, $a_n$ denotes the layer thickness, $z_n$ is the coordinate of the interface between the $n$th and the $(n+1)$th layers, $k_n = \omega/(K_n/\rho_n)^{\frac{1}{2}}$, and $A_n$ and $B_n$ are the coefficients for the forward-traveling and the backward-traveling waves in the $n$th layer, respectively. If the material parameters $\rho$ and $K^{-1}$ are continuously-varying, such as in model III, the medium may be approximated by fine, piecewise-constant layerings so that Eq. (57) is approximately valid in each layer. To relate the solution in the $n$th layer to that of the $(n+1)$th layer requires that the boundary conditions, $p_n(z_n) = p_{n+1}(z_n)$ and $u_n(z_n) = i(\omega\rho_n)^{-1}\partial p_n/\partial z = i(\omega\rho_{n+1})^{-1}\partial p_{n+1}/\partial z = u_{n+1}(z_n)$, be satisfied at the interface $z = z_n$. The net result is that the coefficients $A_n, B_n$ are related to coefficients $A_{n+1}, B_{n+1}$ through the following matrix equation:

$$\begin{pmatrix} A_{n+1} \\ B_{n+1} \end{pmatrix} = t_n \begin{pmatrix} A_n \\ B_n \end{pmatrix},$$
$$(58a)$$

$$t_n = \frac{1}{2}\left[\begin{array}{c} \left(1 + \frac{I_{n+1}}{I_n}\right)\exp\left[i(\nu_n + \nu_{n+1})\right] \quad \left(1 - \frac{I_{n+1}}{I_n}\right)\exp\left[-i(\nu_n - \nu_{n+1})\right] \\ \left(1 - \frac{I_{n+1}}{I_n}\right)\exp\left[i(\nu_n - \nu_{n+1})\right] \quad \left(1 + \frac{I_{n+1}}{I_n}\right)\exp\left[-i(\nu_n + \nu_{n+1})\right] \end{array}\right]$$
$$(58b)$$

where $I_n = \sqrt{\rho_n K_n}$ and $\nu_n = k_n a_n/2$. If there are a total of $N$ layers in distance $L$, then

$$\begin{pmatrix} T \\ 0 \end{pmatrix} = t \begin{pmatrix} 1 \\ R \end{pmatrix}.$$
$$(59a)$$

$$t = t_N t_{N-1} \ldots t_1 t_0.$$
$$(59b)$$

Here $R$ is the reflection coefficient and $T$ the transmission coefficient. By labeling the elements of the $t$ matrix as

$$t = \begin{pmatrix} t_{11} & t_{12} \\ t_{12}^* & t_{11}^* \end{pmatrix},$$

we get

$$T = (|t_{11}|^2 - |t_{12}|^2)/t_{11}^* \ . \tag{60}$$

By averaging $-\ln|T|/L$ over many configurations, a numerical value of the localization length is then obtained as $-\langle \ln|T|/L \rangle^{-1}$.

In Fig. 4 the computed $l(\omega)$ is plotted versus the wavelength $\lambda = 2\pi\sqrt{K_0/\rho_o}/\omega$ for model I (equal layer thickness) at two distinct values of $\sigma = \sigma_\rho = \sigma_K = 0.3, 0.8$. Each data point represents the result of averaging over 40 configurations. The dashed lines denote the low-frequency and the high-frequency limiting behaviors evaluated from the analytical formulas presented in the last sections. Good agreement is clearly seen. In the intermediate frequency regime, the numerical simulation results exhibit pronounced oscillations. They are interpretable as a remnant of a periodic system where there are bands of delocalized states. In this regard, we wish to note that if instead of a continuous distribution for $\rho$ and $K^{-1}$ we have a binary system (i.e. $\rho = \rho_1, \rho_2, K = K_1, K_2$), then the system may not be regarded as random (in the sense of having a finite localization length at all finite frequencies) unless the layer thickness of *both* components are allowed to vary randomly. This is due to the fact that if one component has a fixed thickness, then total transmission through that component can be achieved whenever the 1/4-wavelength condition is satisfied (in exact analogy with the anti-reflection coating condition). The wave would then be traveling in an equivalent one-component medium and the localization length diverges at these specific frequencies.

In Fig. 5 the computed $l(\omega)$ for model II is plotted as a function of $\lambda$ for three values of $\sigma = \sigma_\rho = \rho_K = 0.1, 0.3, 0.9$. Each data point is the result of averaging over 40 configurations. The dashed lines again represent the predicted asymptotic behaviors. They are seen to agree very well with the simulated values in the high- and low-frequency limits. In particular, we note that the prediction of a factor of 2 difference in the low frequency values of $l(\omega)$ for models I and II is indeed verified (comparing the case for $\sigma = 0.3$). Except for statistical fluctuations, the behavior of $l(\omega)$ in this case is essentially a smooth interpolation between the two limits. In fact, the form $c_1^{-1} + 6v_0^2/(\sigma^2\omega^2\bar{a})$ is an accurate representation for $l(\omega)$ over the entire frequency range.

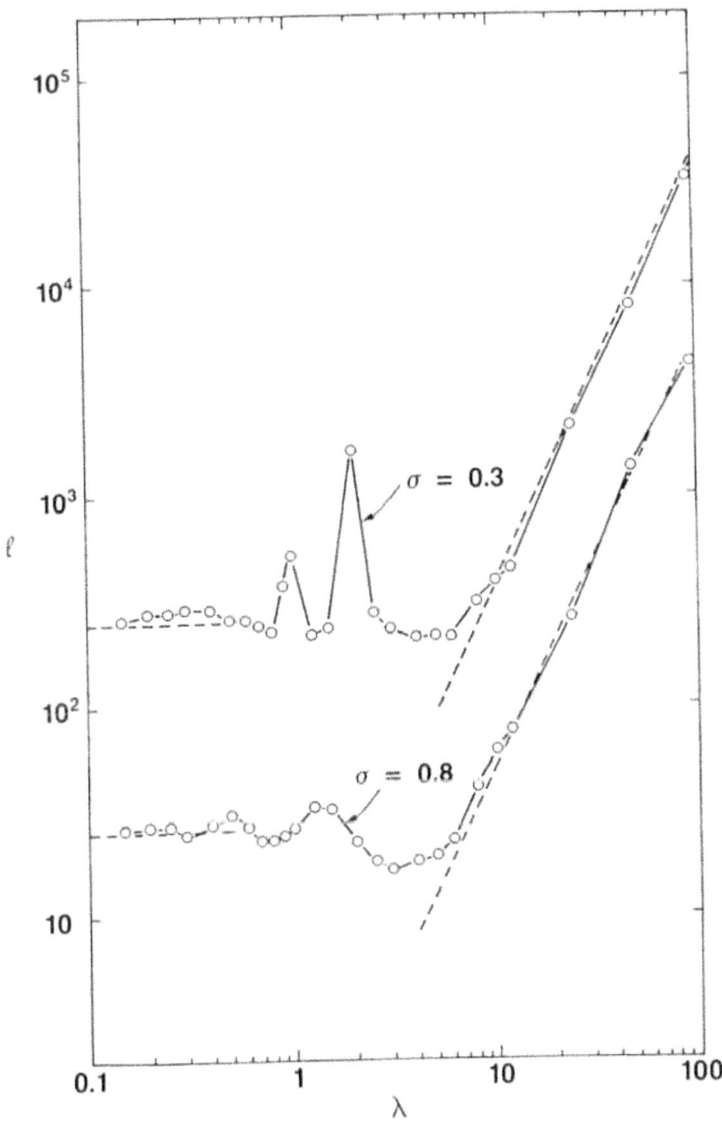



Fig. 4. Localization length plotted as a function of wavelength for model I, calculated at two values of the fluctuation parameter $\sigma = \sigma_\rho = \sigma_K = 0.3$, 0.8 and $\rho_0 = K_0 = 1$. Both $l$ and $\lambda$ are in units of the layer thickness $\overline{a}$. Dashed lines denote the predictions of the analytic theory in the high- and low-frequency limits. Each data point is an average of 40 configurations.

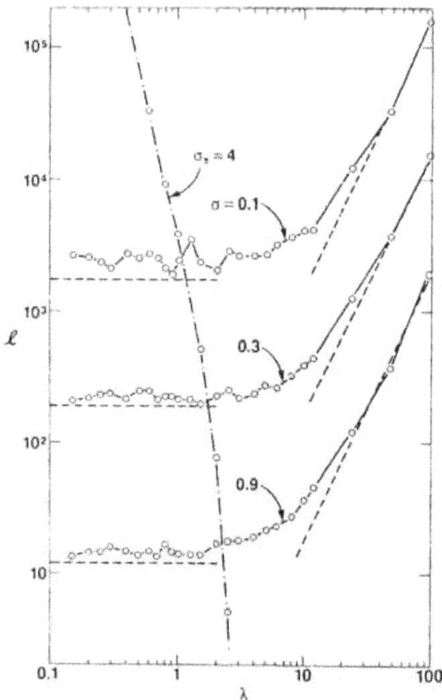

Fig. 5. Localization length plotted as a function of wavelength for model II, calculated at three different values of the parameter $\sigma = \sigma_\rho = \sigma_K = 0.1$, 0.3, 0.9 and $\rho_0 = K_0 = 1$. Both $l$ and $\lambda$ are in units of the average layer thickness $\bar{a}$. Dashed lines denote the predictions of the analytic theory in the high- and low-frequency limits. Each data point represents an average of 40 configurations. Comparison with Fig. 4 for the case of $\sigma = 0.3$ clearly shows a factor of 2 in the low-frequency value of $l$. The dashed-dot line denote the localization length for a quantum particle described by the Schrödinger equation.

Also shown in Fig. 5 by a dot-dashed line is the localization length behavior for a quantum particle whose behavior is governed by the Schrödinger equation with a random potential $V$. In this case we define $\lambda = 2\pi\hbar/\sqrt{2mE}$, where $\hbar$ is Planck's constant, $m$ the mass of the particle and $E$ its energy defined relative to the lower bound of $V$. The localization length, again calculated by the transfer-matrix method, clearly shows a frequency dependence that is opposite to the classical wave behavior. This is due to the fact that whereas the reflection coefficient at a sharp interface is independent of the frequency for the classical waves, the quantum particle always becomes more transmitting as its energy increases. Also, the classical wave scattering tends to become weaker as energy decreases whereas the quantum particle scattering tends to become stronger because the particle no longer has enough energy to overcome the potential barrier. As a result, there is no minimum value for the localization length of a quantum particle in random potentials.

For model III, where the material parameters vary continuously, the frequency dependence of the localization length is plotted in Fig. 6 for three cases where the material parameters are obtained from those of Fig. 5 through a convolution with an exponential filter. It is seen that $l(\omega)$ has a well-defined minimum value at $\lambda \sim 10\bar{a}$. The fact that $l(\omega)$ diverges at high frequencies is indeed verified, and the low-frequency behavior agrees with that of model II as expected. From our data the divergence at high frequencies seems to follow the form $l \sim \omega^2$.

Common to the localization length of all the three models is the fact that there is a well-defined minimum value. Furthermore, this minimum localization length is generally orders of magnitude larger than $\bar{a}$. Comparisons with the Schrödinger wave localization behavior indicates that there is no counterpart to the minimum localization length in the quantum case.

For oblique incidence of electromagnetic waves, we have evaluated the $E$- and $H$-polarized localization lengths (along the $z$ direction) in a model II-type medium. In Fig. 7, values of $l$ for the two polarizations are plotted as a function of angle of incidence at $\lambda/\bar{a} = 48$. In the present case we have assumed that $\sigma_\epsilon = 0.3$ and $\sigma_\mu = 0$, and the transfer matrix for the $H$-polarization has exactly the same form as Eq. (58b) except now $I_n = \varepsilon_n/\sqrt{(\varepsilon_n/\varepsilon_0) - \sin^2\theta}$ and $\nu_n = \omega a_n\sqrt{\varepsilon_n - \varepsilon_0 \sin^2\theta}/2c$, where $\theta$ is the angle of incidence at the interface between the homogeneous medium (dielectric constant $\varepsilon_0$) and the random medium (dielectric constant $\varepsilon_0[1 + 2\sigma_\epsilon\xi(z)]$). For the $E$-polarization the only difference in the transfer matrix is that $I_n = 1/\sqrt{(\varepsilon_n/\varepsilon_0) - \sin^2\theta}$. The agreement between theory and data is seen to be excellent for all angles up to the critical angle of total internal reflection (calculated in the effective-medium sense). The analytical formulas are then expected to

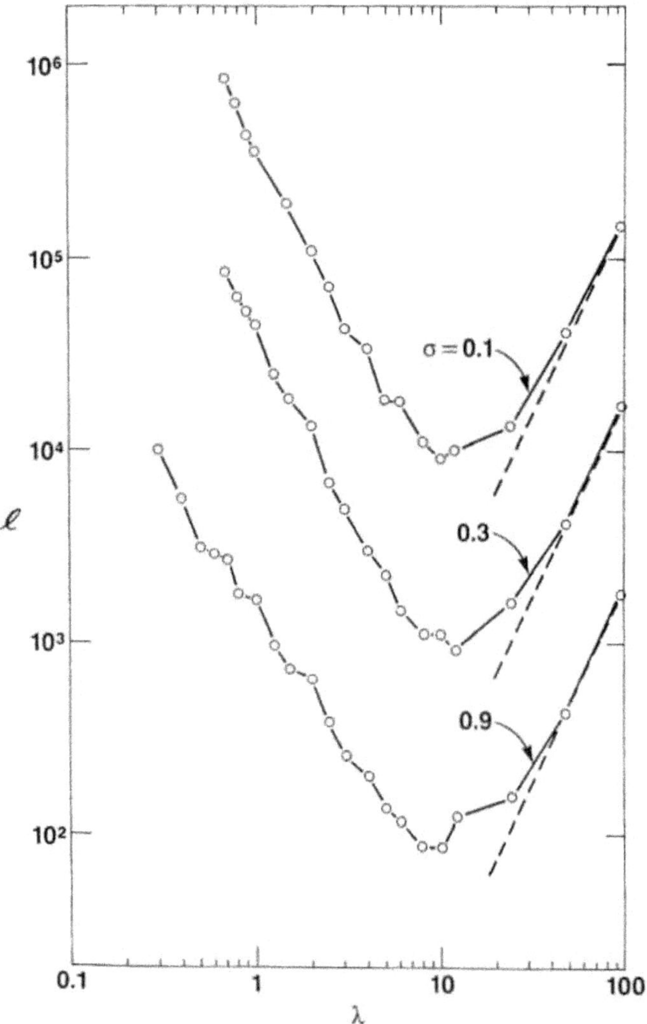



Fig. 6. Localization length plotted as a function of wavelength for model III, calculated for the materials obtained from Fig. 5 through the convolution with an exponential filter. Both *l* and $\lambda$ are in units of $\overline{a}$ defined by model II. Dashed lines denotes the prediction of the low frequency theory. Each data point is an average of 40 configurations.

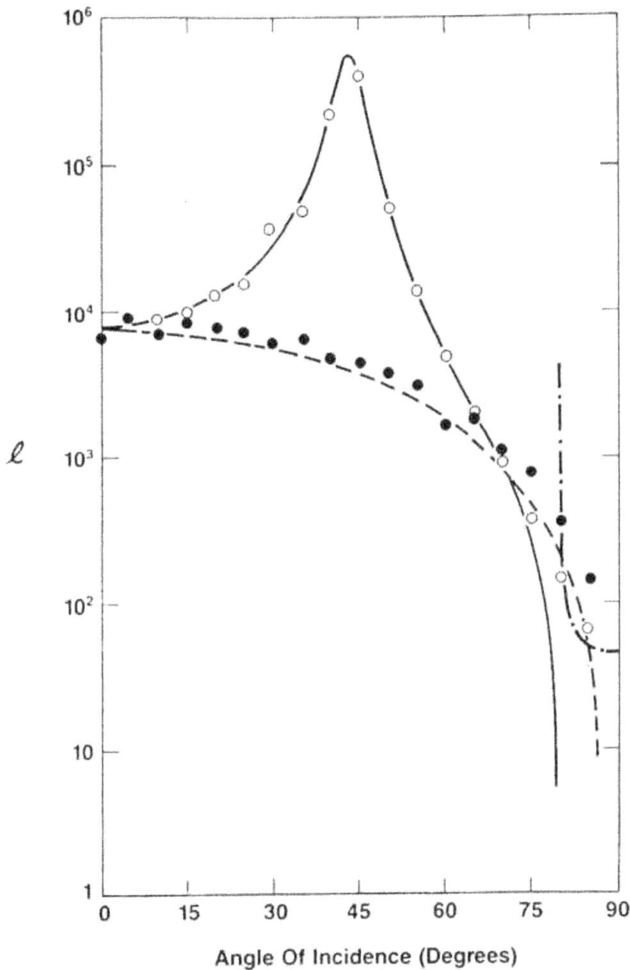



Fig. 7. Localization length in units of mean layer thickness $\overline{a}$ plotted as a function of incidence angle for $\lambda = 48\overline{a}$. The dielectric constant $\varepsilon(z)$ of the random medium is defined in the text. Open circles are results of Monte Carlo simulation for $H$-polarized light, and solid circles are for $E$-polarized light. Solid and dashed lines are calculated from the long-wavelength theory for the $H$- and the $E$-polarized waves, respectively. The dash-dot curve indicates the evanescent decay length beyond the critical angle of the $H$-polarized light calculated in the effective-medium theory.

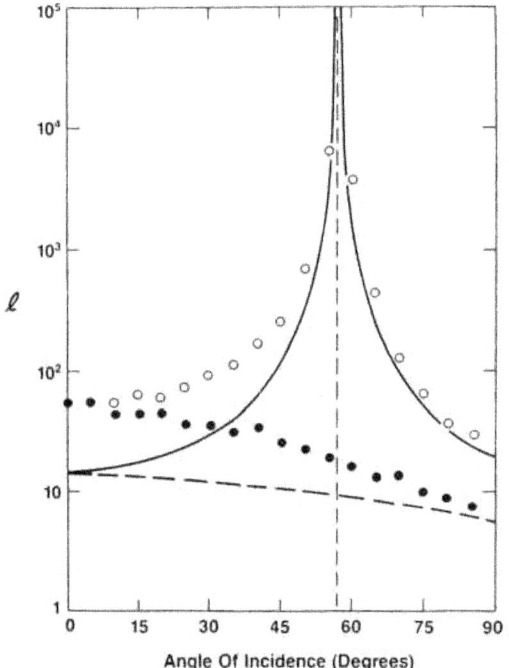

Fig. 8. Localization length in units of mean layer thickness $\bar{a}$ plotted as a function of incidence angle for $\lambda = 5\bar{a}$. The random medium is a binary layered system with $\varepsilon_a = \varepsilon_0$ and $\varepsilon_b = 2.42\varepsilon_0$. Open circles are results of simulation for $H$-polarized light, and solid circles are for $E$-polarized light. Solid and dashed curves are calculated from the long-wavelength theory for the $H$- and $E$-polarized waves, respectively.

fail due to the divergence of the fluctuations in the material parameters (e.g. see Eq. (52b)). However, an effective medium estimate of the evanescent decay length beyond the critical angle for the $H$-polarization yields reasonable agreement with the data. From the figure it is clear that the Brewster anomaly enhances the localization length of the $H$-polarized wave by two orders of magnitude with a fairly wide angular spread. The maximum occurs at $\theta \simeq 45°$ as predicted.

In Fig. 8 we show a case for binary layered system with $\varepsilon_a = \varepsilon_0$, $\varepsilon_b = 2.42\varepsilon_0$ and $\lambda = 5\bar{a}$. Since the wavelength is now comparable to $\bar{a}$, the low frequency theory is no longer expected to be quantitatively correct. Nevertheless, the divergence of the $H$-polarization localization length and the position of where it occurs for the binary system,

$$\theta_B = \sin^{-1} \sqrt{\frac{\varepsilon_a \varepsilon_b}{\varepsilon_0 (\varepsilon_a + \varepsilon_b)}} \, , \tag{61}$$

are still predicted accurately.

## 5. LOCALIZATION CHARACTERISTICS IN THE TIME DOMAIN

### 5.1. Formulation

We would now like to consider the response of a randomly-layered halfspace to a transient pressure pulse.[3-5] Since the problem is linear, for the case of the matched-medium boundary condition the interface pressure variation as a function of time $\tau$ can be written formally as

$$p(\tau) = (2\pi)^{-1} \int_{-\infty}^{\infty} d\omega \exp\left(i\omega\tau\right) f(\omega) R(\omega) \, , \tag{62}$$

where $f(\omega)$ is the pulse frequency spectrum and $R(\omega)$ the reflection coefficient. Based on the consideration of pulse multiple scattering, the function $p(\tau)$ is expected to have a fluctuating character. Therefore what we would like to study is the power spectrum of the response, $S(\tau, \omega)$. From the definition of the power spectrum we have

$$\langle p(\tau)p(\tau + t)\rangle = (2\pi)^{-2} \int_{-\infty}^{\infty} d\omega_1 \int_{-\infty}^{\infty} d\omega_2 f(\omega_1) f^*(\omega_2)$$

$$\times \exp\left(-i\omega_2 t\right) \langle R(\omega_1) R^*(\omega_2)\rangle \exp\left[i(\omega_1 - \omega_2)\tau\right]$$

$$= (2\pi)^{-1} \int_{-\infty}^{\infty} d\omega \exp\left(-i\omega t\right) S\left(\tau + \frac{t}{2}, \omega\right) \, , \tag{63a}$$

where

$$S(\tau, \omega) = (2\pi)^{-1} \int_{-\infty}^{\infty} d\Delta_\omega f\left(\omega - \frac{\Delta_\omega}{2}\right) f^*\left(\omega + \frac{\Delta_\omega}{2}\right)$$

$$\times \exp\left(-i\Delta_\omega \tau\right) U_1(\Delta_\omega, \omega) \, , \tag{63b}$$

$$U_1(\Delta_\omega,\omega) = \left\langle R\left(\omega - \frac{\Delta_\omega}{2}\right) R^*\left(\omega + \frac{\Delta_\omega}{2}\right)\right\rangle . \qquad (63c)$$

Here $\omega = (\omega_1 + \omega_2)/2$, $\Delta_\omega = \omega_2 - \omega_1$ denotes the frequency difference, and the subscript 1 denotes the case of the matched-medium interface boundary condition. From Eq. (63c) it is clear that $U_1(0,\omega) = 1$. One also intuitively expects $U_1(\infty,\omega) = 0$. If one assumes that $U_1(\Delta_\omega,\omega) \sim \delta(\Delta_\omega)$, i.e. $R(\omega_1)$ and $R(\omega_2)$ are not correlated unless $\omega_1 = \omega_2$, then

$$S(\tau,\omega) \propto |f(\omega)|^2$$

becomes independent of $\tau$ as one would expect for stationary noise. However, in reality we expect $R(\omega_1)$ and $R(\omega_2)$ to be correlated over a frequency range $\Delta\omega$ that is small compared to $\omega$. In that case $S$ may be written as

$$S(\tau,\omega) \cong |f(\omega)|^2 \cdot N(\tau,\omega) , \qquad (64a)$$

$$N(\tau,\omega) = (2\pi)^{-1} \int_{-\infty}^{\infty} d\Delta_\omega \, \exp\left(-i\Delta_\omega \tau\right)U_1(\Delta_\omega,\omega) , \qquad (64b)$$

where $N(\tau,\omega)$ denotes the noise spectrum and is seen to be just the Fourier transform of the correlation function for the reflection coefficient at two different frequencies. Since $\Delta_\omega$ and $\tau$ are conjugate variables as evidenced in Eq. (64b), a finite correlation range $\Delta\omega$ for the function $U_1(\Delta_\omega,\omega)$ would imply a definite time scale $(\Delta\omega)^{-1}$ for the non-stationarity of the noise spectrum $N(\tau,\omega)$. What we are going to show, both analytically and numerically, is that $U_1$ is a smoothly decaying function governed by only one width parameter $\Delta\omega$. Moreover, $\Delta\omega \sim l^{-1}(\omega)$ so that the time scale of non-stationarity in the multiple scattering noise is determined by none other than the localization length.[4]

From Eq. (64) it is clear that in order to make mathematical progress we must evaluate the function $U_1(\Delta_\omega,\omega)$ as defined by Eq. (63c). To this end let us first non-dimensionalize and transform the wave equation for $p$ and $w$, Eq. (2), into a form for the forward-

and backward-traveling wave amplitudes $A$ and $B$:

$$\frac{d}{dz}\begin{bmatrix} A \\ B \end{bmatrix} = i\omega \begin{bmatrix} g & -h\exp\left(2i\omega z\right) \\ h\exp\left(2i\omega z\right) & -g \end{bmatrix}\begin{bmatrix} A \\ B \end{bmatrix}, \tag{65a}$$

where we have assumed harmonic time dependence for $p$ and $w, g = (\widehat{\rho} + \widehat{K}^{-1})/2$, $h = (\widehat{\rho} - \widehat{K}^{-1})/2$ ($\widehat{\rho} = 2\sigma_\rho\xi_1$, $\widehat{K}^{-1} = 2\sigma_K\xi_2$ for our models I and II), and

$$p = \frac{1}{2}\left[A\exp\left(i\omega z\right) + B\exp(-i\omega z)\right], \tag{65b}$$

$$w = \frac{1}{2}\left[A\exp\left(i\omega z\right) - B\exp\left(-i\omega z\right)\right]. \tag{65c}$$

The variables in Eq. (65) are all expressed in terms of the natural units of the problem: $p \rightarrow p/K_0$, $w \rightarrow w/v_0$, $z \rightarrow z/\bar{a}$, $\omega \rightarrow \omega\bar{a}/v_0$. From Eq. (65a) it is easy to show that if one defines $R = B/A$, then $R$ satisfies a Riccati equation[5]

$$\frac{dR}{dz} = i\omega[h\exp\left(2i\omega z\right) - 2gR + hR^2\exp\left(-2i\omega z\right)]. \tag{66}$$

For a finite slab of length $L$, we have $R(L) = 0$ because of the radiation boundary condition at the end of the slab, and Eq. (66) is to be integrated backward to the front interface $z = 0$, where $R$ may be identified as the reflection coefficient. As $L \rightarrow \infty$, the boundary condition at $z = L$ may be determined by noting first that $|R| = 1$ because of the localizing nature of the random medium, then writing $R = \exp\left(i\psi\right)$ everywhere to obtain a corresponding equation[5] for $\psi$:

$$\frac{d\psi}{dz} = 2\omega[h\cos(\psi - 2\omega z) - g]. \tag{67}$$

Instead of a boundary condition at $z = L$ we now take the statistically stationary solution to Eq. (67). In terms of $\psi$ the function $U_1(\Delta_\omega, \omega)$ can be written as

$$U_1(\Delta_\omega, \omega) = \langle\exp\left[i(\psi_1 - \psi_2)\right]\rangle = \int_0^{2\pi} d\overline{\psi}\, P(\overline{\psi})\exp\left(i\overline{\psi}\right), \tag{68}$$

where $\psi_1$ corresponds to the phase angle at frequency $\omega - (\Delta_\omega/2)$ and $\psi_2$ corresponds to the phase angle at frequency $\omega + (\Delta_\omega/2)$, $\overline{\psi} = \psi_1 - \psi_2$, and $P(\overline{\psi})$ denotes the stationary distribution of $\overline{\psi}$.

Up to now we have assumed that the random medium is bordered by a homogeneous medium with $\rho = \rho_0$, $K = K_0$. It is equally possible to have the other case where the random medium interfaces with vacuum so that backscattered waves are totally returned into the random medium. Since $p = 0$ is now the boundary condition at the interface for all times except during the application of the pressure pulse, one has to monitor the medium response in terms of the displacement velocity $w$ at $z = 0$. In this case the expression corresponding to $U_1(\Delta_\omega, \omega)$ would be

$$U_2(\Delta_\omega, \omega) = \left\langle w\left(\omega - \frac{\Delta_\omega}{2}\right) w^*\left(\omega + \frac{\Delta_\omega}{2}\right)\right\rangle, \tag{69a}$$

where

$$w(\omega) = \frac{1 - R(\omega)}{1 + R(\omega)}. \tag{69b}$$

The particular form of $w$, Eq. (69b), may be deduced as follows. We write the pressure at $z = 0$ as a superposition of forward- and backward-going wave amplitudes in accordance with Eq. (65b):

$$p(\omega) = \frac{1}{2}(A + B), \quad z = 0.$$

If a $\delta(\tau)$ pulse is applied at $z = 0$, then $f(\omega) = 1$ for the pressure pulse and we have the condition of

$$\frac{A + B}{2} = 1. \tag{70a}$$

The displacement velocity, in accordance with Eq. (65c), is given by

$$w = \frac{1}{2}(A - B). \tag{70b}$$

By defining $R = B/A$ as a reflection coefficient and using Eq. (70a), we have $A = 2/(1 + R)$, $B = 2R/(1 + R)$, and

$$w = \frac{1 - R}{1 + R}. \tag{70c}$$

Since $R = \exp(i\psi)$, the right-hand side of Eq. (70c) can be singular. This ambiguity may be resolved by writing $R = r \exp(i\psi)$ in the definition of $U_2(\Delta_\omega, \omega)$ and let $r$ approach unity from below. That is,

$$
\begin{aligned}
U_2(\Delta_\omega, \omega) \\
= \lim_{r_1, r_2 \to 1^-} & \left\langle \left[\frac{1 - r_1 \exp(i\psi_1)}{1 + r_1 \exp(i\psi_1)}\right]\left[\frac{1 - r_2 \exp(-i\psi_2)}{1 + r_2 \exp(-i\psi_2)}\right]\right\rangle \\
= \lim_{r_1, r_2 \to 1^-} & \left\langle \frac{1 + r_1 r_2 \exp(i\overline{\psi}) - \exp(i\overline{\psi}/2)[r_1 \exp(i\psi_0) + r_2 \exp(-i\psi_0)]}{1 + r_1 r_2 \exp(i\overline{\psi}) + \exp(i\overline{\psi}/2)[r_1 \exp(i\psi_0) + r_2 \exp(-i\psi_0)]}\right\rangle,
\end{aligned}
$$

$$(71\text{a})$$

where $\psi_0 = (\psi_1 + \psi_2)/2$. The distribution of $\psi_0$ is expected to be uniform (this will be justified later) and therefore in the evaluation of $U_2$ we can first integrate $\psi_0$ from 0 to $2\pi$ with $P(\psi_0) = (2\pi)^{-1}$. A contour integration (with $\exp(i\psi_0)$ being the complex variable) yields

$$
U_2(\Delta_\omega, \omega) = \lim_{r \to 1^-} \int_0^{2\pi} \frac{1 + 3r \exp(i\overline{\psi})}{1 - r \exp(i\overline{\psi})} P(\overline{\psi})\, d\overline{\psi}, \qquad (71\text{b})
$$

where $r = r_1 r_2$. From $U_2(\Delta_\omega, \omega)$ the noise spectrum $N(\tau, \omega)$ may be obtained from Eq. (64b) with $U_2$ replacing $U_1$.

Equations (68) and (71b) tell us that the knowledge of $P(\overline{\psi})$ is necessary and sufficient for the evaluation of $U_{1,2}(\Delta_\omega, \omega)$ and $N(\tau, \omega)$ under both types of interface boundary conditions. To make further progress, we must derive the relevant stochastic differential equation for $P(\overline{\psi})$ and obtain its solution. The description of these tasks is presented in the next section.

## 5.2. Solution in the White Noise Limit

Up to now no approximation has been made in the derivation of the equations for $R$ (Eq. (66)) and $\psi$ (Eq. (67)). While the general solution of $P(\overline{\psi})$ based on these equations is not known, in this section we wish to show that the problem is analytically soluble in the white noise limit[5] of $\overline{a} \to 0$. More precisely, the limit involves the introduction of a dimensionless parameter $\overline{a}\omega/v_0 = \varepsilon \ll 1$ (not to be confused with the dielectric constant symbol $\varepsilon$ used in other sections) and the specification that $\bar{a}$ is on the order of $\varepsilon^2$ when measured in terms of a macroscale, taken to be 1. Physically, the macroscale may be thought of as the distance traveled by the pulse front during the time scale of non-stationarity $(\Delta\omega)^{-1}$ that we discussed before. The intermediate scale, $v_0/\omega$, may be identified as the pulse width. This can be seen from the fact that since $\bar{a}/1 \sim \varepsilon^2$ and $\bar{a}\omega/v_0 = \varepsilon$, we must have $\omega \to \omega/\varepsilon$ so that the wavelength is on the order of $v_0/(\omega/\varepsilon) \sim \varepsilon$. On the other hand, $\Delta_\omega$ is the difference frequency which determines the time scale for the non-stationarity and therefore $\Delta_\omega \to \Delta_\omega$ under the scale transformation. From Eq. (67), we can write a set of equations for $\psi_1$ and $\psi_2$ under the new scales:

$$\frac{d\psi_1^\varepsilon}{dz} = \frac{2\omega}{\varepsilon} \left[ h^\varepsilon \cos\left( \psi_1^\varepsilon - 2\omega\frac{z}{\varepsilon} + \Delta_\omega z \right) - g^\varepsilon \right]$$
$$- \Delta_\omega \left[ h^\varepsilon \cos\left( \psi_1^\varepsilon - 2\omega\frac{z}{\varepsilon} + \Delta_\omega z \right) - g^\varepsilon \right] , \qquad (72a)$$

$$\frac{d\psi_2^\omega}{dz} = \frac{2\omega}{\varepsilon} \left[ h^\varepsilon \cos\left( \psi_2^\varepsilon - 2\omega\frac{z}{\varepsilon} - \Delta_\omega z \right) - g^\varepsilon \right]$$
$$+ \Delta_\omega \left[ h^\varepsilon \cos\left( \psi_2^\varepsilon - 2\omega\frac{z}{\varepsilon} - \Delta_\omega z \right) - g^\varepsilon \right] . \qquad (72b)$$

Here we have assumed that random material fluctuations are generated by the random process $\boldsymbol{\xi}$ so that, $h^\varepsilon, g^\varepsilon$ can be written as $h^\varepsilon = h[\boldsymbol{\xi}(z/\varepsilon^2)]$, $g^\varepsilon = g[\boldsymbol{\xi}(z/\varepsilon^2)]$, i.e., material fluctuations occur on an extremely fast scale. What we want to examine is the effect of these fast material fluctuations integrated over the macroscale. Equation (72) can be written succinctly as

$$\frac{d\boldsymbol{\psi}^\varepsilon}{dz} = \frac{1}{\varepsilon}\mathbf{H}[z,\varsigma,\boldsymbol{\xi},\Delta_\omega] + \mathbf{G}[z,\varsigma,\boldsymbol{\xi},\Delta_\omega] , \qquad (72c)$$

where $\boldsymbol{\psi}^\varepsilon = (\psi_1^\varepsilon, \psi_2^\varepsilon)$, $\varsigma = z/\varepsilon$, and

$$\mathbf{H} = 2\omega \begin{bmatrix} h[\boldsymbol{\xi}]\cos(\psi_1 - 2\omega\varsigma + \Delta_\omega z) - g[\boldsymbol{\xi}] \\ h[\boldsymbol{\xi}]\cos(\psi_2 - 2\omega\varsigma - \Delta_\omega z) - g[\boldsymbol{\xi}] \end{bmatrix} , \qquad (72d)$$

$$\mathbf{G} = \Delta_\omega \begin{bmatrix} -h[\boldsymbol{\xi}]\cos(\psi_1 - 2\omega\varsigma + \Delta_\omega z) + g[\boldsymbol{\xi}] \\ h[\boldsymbol{\xi}]\cos(\psi_2 - 2\omega\varsigma - \Delta_\omega z) - g[\boldsymbol{\xi}] \end{bmatrix} . \qquad (72e)$$

The process $(\boldsymbol{\xi}, \boldsymbol{\psi}^\epsilon)$ is jointly Markovian with the infinitesimal generator

$$\begin{aligned}
L^\epsilon &= \frac{1}{\epsilon^2}Q + \frac{\partial \boldsymbol{\psi}^\epsilon}{\partial z} \cdot \boldsymbol{\nabla}_\psi \\
&= \frac{1}{\epsilon^2}Q + \frac{1}{\epsilon}\mathbf{H} \cdot \boldsymbol{\nabla}_\psi + \mathbf{G} \cdot \boldsymbol{\nabla}_\psi \ .
\end{aligned} \tag{73}$$

From $L^\epsilon$ one can write the backward Kolmogorov equation

$$-\frac{\partial V^\epsilon}{\partial z} + \frac{1}{\epsilon^2}QV^\epsilon + \frac{1}{\epsilon}\mathbf{H} \cdot \boldsymbol{\nabla}_\psi V^\epsilon + \mathbf{G} \cdot \boldsymbol{\nabla}_\psi V^\epsilon = 0 \tag{74a}$$

for any quantity $V^\epsilon(z, \boldsymbol{\xi}, \boldsymbol{\psi})$. We will consider the case where the conditions at $z = 0$ do not depend on $\boldsymbol{\xi}$ i.e.,

$$V^\epsilon(z, \boldsymbol{\xi}, \boldsymbol{\psi})|_{z=0} = V_0(\boldsymbol{\psi}) \ . \tag{74b}$$

To obtain the infinitesimal generator in the limit of $\epsilon \to 0$, let us expand Eq. (74a) in multiple scales of $z$ and $\varsigma = z/\epsilon$ so that $\partial/\partial z \to \partial/\partial z + \epsilon^{-1}\partial/\partial\varsigma$. Then Eq. (74a) becomes

$$\frac{1}{\epsilon^2}QV^\epsilon + \frac{1}{\epsilon}\left[\mathbf{H} \cdot \boldsymbol{\nabla}_\psi V^\epsilon - \frac{\partial V^\epsilon}{\partial\varsigma}\right] + \left[\mathbf{G} \cdot \boldsymbol{\nabla}_\psi V^\epsilon - \frac{\partial V^\epsilon}{\partial z}\right] = 0 \ . \tag{74c}$$

If $z$ is measured on the macroscale, then $\varsigma$ is measured on the pulse-width scale. By expressing $V^\epsilon$ as

$$V^\epsilon = V^{(0)} + \epsilon V^{(1)} + \epsilon^2 V^{(2)} + \cdots \ , \tag{75}$$

in Eq. (74c) and equating terms with the same powers of $\epsilon$ that results, we get

$$QV^{(0)} = 0 \ , \tag{76a}$$

$$QV^{(1)} + \mathbf{H} \cdot \boldsymbol{\nabla}_\psi V^{(0)} - \frac{\partial V^{(0)}}{\partial\varsigma} = 0 \ , \tag{76b}$$

$$QV^{(n)} + \mathbf{H} \cdot \boldsymbol{\nabla}_\psi V^{(n-1)} - \frac{\partial V^{(n-1)}}{\partial\varsigma}$$
$$+ \mathbf{G} \cdot \boldsymbol{\nabla}_\psi V^{(n-2)} - \frac{\partial V^{(n-2)}}{\partial z} = 0, n \geq 2 \ . \tag{76c}$$

The first equation tells us that $V^{(0)}$ is independent of $\xi$, i.e. $V^{(0)} = V^{(0)}(z, \psi)$. Now multiplying Eq. (76b) by $\overline{P}(\xi)$, the stationary distribution of $\xi$, and integrating with respect to $\xi$ gives

$$0 = \int d\xi\, \overline{P}(\xi) Q V^{(1)} + \int d\xi\, \overline{P}(\xi) \mathbf{H} \cdot \boldsymbol{\nabla}_\psi V^{(0)} + \frac{\partial V^{(0)}}{\partial \varsigma}$$

$$= \int d\xi\, V^{(1)} Q^* \overline{P}(\xi) + \frac{\partial V^{(0)}}{\partial \varsigma} = \frac{\partial V^{(0)}}{\partial \varsigma} \,, \tag{77a}$$

where we recognize that since $\mathbf{H}$ (and $\mathbf{G}$ as well) is linearly proportional to $\xi_1$ and $\xi_2$ ($h$, $g$ are functions of the fluctuating parts of $\rho$ and $K^{-1}$, see Eq. (65a)), its average must be 0. Also $Q^* \overline{P}(\xi) = 0$ (see Secs. 4.1 and 4.2). From Eq. (77a) and Eq. (76b) we get

$$V^{(1)} = -Q^{-1} \mathbf{H} \cdot \boldsymbol{\nabla} V_\psi^{(0)} + V_0^{(1)} \,. \tag{77b}$$

Since $Q$ has a nontrivial null space as demonstrated by Eq. (76a), its inverse $Q^{-1}$ must be defined in the sense of Eq. (29) and acts only on states that are perpendicular to the null space. The term $V_0^{(1)}$, on the other hand, is independent of $\xi$ and lies in the null space of $Q$. Now the substitution of Eq. (77b) into Eq. (76c) with $n = 2$ yields

$$QV^{(2)} - \mathbf{H} \cdot \boldsymbol{\nabla}_\psi Q^{-1} \mathbf{H} \cdot \boldsymbol{\nabla}_\psi V^{(0)} + \mathbf{H} \cdot \boldsymbol{\nabla}_\psi V_0^{(1)} - \frac{\partial V_0^{(1)}}{\partial \varsigma}$$

$$+ \mathbf{G} \cdot \boldsymbol{\nabla}_\psi V^{(0)} - \frac{\partial V^{(0)}}{\partial z} = 0 \,. \tag{78a}$$

By multiplying by $\overline{P}(\xi)$ and integrating with respect to $\xi$, the terms $QV^{(2)}$, $\mathbf{H} \cdot \boldsymbol{\nabla}_\psi V_0^{(1)}$, and $\mathbf{G} \cdot \boldsymbol{\nabla}_\psi V^{(0)}$ in Eq. (78a) all vanish, giving

$$\frac{\partial V^{(0)}}{\partial z} + \frac{\partial V_0^{(1)}}{\partial \varsigma} + E[\mathbf{H} \cdot \boldsymbol{\nabla}_\psi Q^{-1} \mathbf{H} \cdot \boldsymbol{\nabla}_\psi]_\xi V^{(0)} = 0 \,, \tag{78b}$$

where the notation $E[\ ]_\xi$ is used to denote taking the expectation value with respect to $\xi$, which is essentially averages over the material fluctuations on the $\varepsilon^2$ scale. We want to further average over the $\varepsilon$ scale by applying the operation

$$E[\ ]_\varsigma = \lim_{T \to \infty} \frac{1}{T} \int_0^T d\varsigma \,.$$

The term $\partial V_0^{(1)}/d\varsigma$ in Eq. (78b) would become $\lim_{T\to\infty}[V_0^{(1)}(\varsigma = T) - V_0^{(1)}(\varsigma = 0)]/T$ under the $\varsigma$ averaging. If $V_0^{(1)}$ is bounded as $T \to \infty$, then the $V_0^{(1)}$ term vanishes, giving us the final result

$$
\begin{aligned}
\frac{\partial V^{(0)}}{\partial z} &+ E[\mathbf{H}\cdot\boldsymbol{\nabla}_\psi Q^{-1}\mathbf{H}\cdot\boldsymbol{\nabla}_\psi]_{\boldsymbol{\epsilon},\varsigma}V^{(0)} \\
&= \frac{\partial V^{(0)}}{\partial z} - \int_0^\infty ds\, E[\mathbf{H}[z,\boldsymbol{\xi}(0)]\cdot\boldsymbol{\nabla}_\psi\mathbf{H}[z,\boldsymbol{\xi}(s)]\cdot\boldsymbol{\nabla}_\psi]_{\boldsymbol{\epsilon},\varsigma}V^{(0)} \\
&= 0 \;,
\end{aligned}
\tag{78c}
$$

where we have used Eq. (29) as the definition of $Q^{-1}$. It is immediately recognized that Eq. (78c) is exactly the backward Kolmogorov equation in the limit of $\varepsilon \to 0$ since now $\varepsilon$ has disappeared during the averaging process. The limiting infinitesimal generator is seen to be

$$
L = \int_0^\infty ds\, E[\mathbf{H}[z,\boldsymbol{\xi}(0)]\cdot\boldsymbol{\nabla}_\psi\mathbf{H}[z,\boldsymbol{\xi}(s)]\cdot\boldsymbol{\nabla}_\psi]_{\boldsymbol{\epsilon},\varsigma} \;.
\tag{79}
$$

By using the fact that

$$
\boldsymbol{\nabla}_\psi\mathbf{H}\cdot\boldsymbol{\nabla}_\psi = (\mathbf{H}\cdot\boldsymbol{\nabla}_\psi)\boldsymbol{\nabla}_\psi + (\boldsymbol{\nabla}_\psi\mathbf{H})\cdot\boldsymbol{\nabla}_\psi \;,
\tag{80}
$$

and that the $\varsigma$ averaging yields a factor of $1/2$ for the square of the cosine terms, $\cos(\psi_1 - \psi_2 + 2\Delta_\omega z)/2$ for the cross-cosine terms, and zero for everything else, the operator $L$ may be evaluated explicitly as

$$
\begin{aligned}
L = 4\omega^2 \Bigg\{ &\alpha_{gg}\left[\frac{\partial}{\partial\psi_1} + \frac{\partial}{\partial\psi_2}\right]^2 + \frac{1}{2}\alpha_{hh}\left[\frac{\partial^2}{\partial\psi_1^2} + \frac{\partial^2}{\partial\psi_2^2}\right. \\
&\left. + 2\cos(\psi_1 - \psi_2 + 2\Delta_\omega z)\frac{\partial^2}{\partial\psi_1\partial\psi_2}\right]\Bigg\} \;,
\end{aligned}
\tag{81a}
$$

where

$$
\alpha_{gg} = \int_0^\infty ds\langle g(0)g(s)\rangle \;,
\tag{81b}
$$

and

$$
\alpha_{hh} = \int_0^\infty ds\langle h(0)h(s)\rangle
\tag{81c}
$$

are the integrals of the correlation functions for $g$ and $h$. Here we have used the notations $g(s) = g[\xi(s)]$, $h(s) = h[\xi(s)]$, and $\langle\ \rangle$ for $E[\ ]_\xi$ (same as configurational averaging because of ergodicity). From the definition of $g$ and $h$ as $(\widehat{\rho} \pm \widehat{K}^{-1})/2$ and comparison with Eq. (32), we immediately obtain the equivalence of $\alpha_{gg} = (\alpha_{\rho\rho} + \alpha_{KK} + 2\alpha_{\rho K})/4$ and $\alpha_{hh} = (\alpha_{\rho\rho} + \alpha_{KK} - 2\alpha_{\rho K})/4$. Furthermore, the low-frequency localization length is simply given by

$$l(\omega) = \frac{v_0^2}{\alpha_{hh}\omega^2}, \quad \omega \to 0 \tag{82}$$

as indicated by Eq. (34b).

From Eq. (81) we can now get $L$ in terms of $\overline{\psi} = \psi_1 - \psi_2$ and $\psi_0 = (\psi_1 + \psi_2)/2$:

$$L = 4\omega^2 \left\{ \left( \alpha_{gg} + \frac{1}{4}\alpha_{hh}[1 + \cos(\overline{\psi} + 2\Delta_\omega z)] \right) \frac{\partial^2}{\partial\psi_0^2} \right.$$
$$\left. + \alpha_{hh}[1 - \cos(\overline{\psi} + 2\Delta_\omega z)]\frac{\partial^2}{\partial\overline{\psi}^2} \right\} . \tag{83}$$

Note that the coefficients in Eq. (83) do not depend on $\psi_0$. That means the knowledge of $\overline{\psi}$ at any position $z$ is sufficient to predict the state at $z + dz$, i.e. $\overline{\psi}$ is Markovian by itself. For $\psi_0$, on the other hand, the only stationary distribution form compatible with Eq. (83) would be $a\psi_0 + b$. Since on physical basis there is no reason to expect $-\psi$ to be any different from $+\psi$, we conclude that $P(\psi_0)$ must be uniform as asserted earlier. By taking into account these considerations, the Fokker-Planck equation $L^*P = \partial P/\partial z$ for the distribution $P(\overline{\psi})$ may be written in the following form:

$$4\omega^2\alpha_{hh}\frac{\partial^2}{\partial\overline{\psi}^2}[1 - \cos(\overline{\psi} + 2\Delta_\omega z)]P = \frac{\partial P}{\partial z} . \tag{84a}$$

By making a variable transformation $\overline{\psi} + 2\Delta_\omega z \to \overline{\psi}'$, $z \to z'$, we get $\partial/\partial z' \to \partial/\partial z' + (\partial\overline{\psi}'/\partial z')(\partial/\partial\overline{\psi}') = \partial/\partial z' + 2\Delta_\omega \partial/\partial\overline{\psi}'$ so that

$$\frac{\partial P}{\partial z} + 2\Delta_\omega \frac{\partial P}{\partial\overline{\psi}} = 4\omega^2\alpha_{hh}\frac{\partial^2}{\partial\overline{\psi}^2}[1 - \cos\overline{\psi}]P , \tag{84b}$$

where we have dropped the primes. In the new equation the argument of the cosine is no longer dependent on $z$. For stationary distribution, we may set $\partial P/\partial z = 0$ to get

$$\Delta_\omega \frac{\partial}{\partial s}(1 + s^2)P(s) + 2\omega^2 \alpha_{hh} \frac{\partial}{\partial s}(1 + s^2)\frac{\partial P(s)}{\partial s} = 0 \ , \qquad (85)$$

where $s = \cot(\overline{\psi}/2)$ so that $\partial/\partial\overline{\psi} = -[(1 + s)^2/2]\partial/\partial s$, $1 - \cos\overline{\psi} = 2/(1+s^2)$, and $P(\overline{\psi}) = P(s)(\partial s/\partial\overline{\psi}) = -(1+s)^2 P(s)/2$. Equation (85) may be solved by quadrature[5]:

$$P_{\Delta_\omega}(s) = \frac{\Delta_\omega}{2\pi\omega^2\alpha_{hh}} \int_0^\infty \frac{\exp\left(-\Delta_\omega q/2\omega^2\alpha_{hh}\right)}{1 + (q - s)^2} \, dq, \quad \Delta_\omega \geq 0 \ . \quad (86a)$$

If $\Delta_\omega < 0$, the change of $s \to -s$ in Eq. (85) is seen to take the equation back to its original state. Therefore,

$$P_{-\Delta_\omega}(s) = P_{\Delta_\omega}(-s) \ . \qquad (86b)$$

To evaluate $U_1(\Delta_\omega, \omega)$, we simply recognize that $\exp(i\overline{\psi}) = (s+i)/(s-i)$. For the matched-medium boundary condition that means

$$U_1(\Delta_\omega, \omega) = \int_{-\infty}^\infty P_{\Delta_\omega}(s) \ \frac{s + i}{s - i} ds = U_1(y)$$

$$= y \int_0^\infty \exp\left(-qy\right)\frac{q}{q + i} dq, \quad y > 0 \qquad (87)$$

and $U_1^*(-y) = U_1(y)$. Here $y = \Delta_\omega l(\omega)/v_0$, with $l(\omega)$ expressing the low-frequency localization length as given by Eq. (82).

The fact that $U_1$ is a function of only one variable $y$ has the following physical implications. In the last section we have predicted that the function $U_1(\Delta_\omega, \omega)$ must decay to zero over a finite width $|\Delta_\omega| < \Delta\omega$, where $(\Delta\omega)^{-1}$ gives the time scale of non-stationary for the multiple scattering noise. Numerical evaluation of Eq. (87) (see Fig. 9 in next section) now indeed tells us that $U_1(y)$ has a width of $\Delta y \simeq 1$. From the definition of $y = \Delta_\omega l(\omega)/v_0$, we get

$$(\Delta\omega)^{-1} \simeq \frac{l(\omega)}{v_0} \ . \qquad (88)$$

That is, the width of the function $U_1$ scales as $l^{-1}(\omega)$, and therefore the time scale of non-stationarity is directly proportional to $l(\omega)$. Since $l(\omega)$ increases as $\omega$ decreases, it follows that the low-frequency components of the multiple scattering noise should persist longer than the high-frequency components. This is intuitively reasonable because the high-frequency components of the pulse should be reflected into the homogeneous medium much earlier due to the shorter localization length. This phenomenon may indeed be detected by looking at numerically simulated coda as shown in Fig. 1.

For the case of totally-reflecting boundary condition at the interface, the injected pulse energy would be trapped by a reflecting boundary on the one end and by a nonpenetrable random medium on the other. The noise statistics is therefore expected to be stationary. The evaluation of $U_2(\Delta_\omega, \omega)$ means the calculation of the following integral:

$$
\begin{aligned}
U_2(\Delta_\omega, \omega) &= \lim_{r \to 1^-} \int_{-\infty}^{\infty} P_{\Delta_\omega}(s) \left[ \frac{(3r+1)s + i(3r-1)}{(1-r)s - i(1+r)} \right] ds \\
&= \lim_{r \to 1^-} \int_{-\infty}^{\infty} dq \, \frac{y \exp(-yq/2)}{2\pi} \\
&\quad \times \int_{-\infty}^{\infty} ds \, \frac{[(3r+1)s + i(3r-1)]}{[1 + (q-s)^2][(1-r)s - i(1+r)]} \\
&= 1 - i\frac{4}{y} \\
&= U_2(y) \; .
\end{aligned}
\tag{89}
$$

The noise spectrum $N(\tau, \omega)$ can now be obtained directly from $U_1$ and $U_2$ by a Fourier transform:

$$
N(\tau, \omega) = \frac{\chi}{2\pi\tau} \int_{-\infty}^{\infty} dy \exp(-iy\chi) U_{1,2}(y)
$$

$$
= \begin{cases} \dfrac{1}{\tau}\mu_1(\chi); & \text{matched-medium B.C.} \\[2em] \dfrac{v_0}{l(\omega)}\delta(\chi) + \dfrac{1}{\tau}\mu_2(\chi); & \text{reflecting B.C.} , \end{cases}
\tag{90a}
$$

where

$$\mu_1(\chi) = \frac{\chi}{(1+\chi)^2} H(\chi) , \tag{90b}$$

$$\mu_2(\chi) = 4\chi H(\chi) , \tag{90c}$$

$$\chi = \frac{\tau v_0}{l(\omega)} . \tag{90d}$$

Here $H(\chi)$ is the unit step function defined as $H(\chi) = 1$ for $\chi \geq 0$ and $H(\chi) = 0$ for $\chi < 0$. It is clear that the scaling in the $\Delta_\omega$ domain naturally implies scaling in the $\tau$ domain through the definition of the variable $\chi$. While the forms and the physical implications of $\mu_1(\chi)$ and $\mu_2(\chi)$ will be discussed in Sec. 5.4., we just wish to note here that for $\mu_1$, the condition of $U_1(y = 0) = 1$ implies a sum rule for $\mu_1$ in the form of

$$U_1(0) = \int_{-\infty}^{\infty} \chi^{-1}\mu_1(\chi)d\chi = 1 . \tag{91}$$

From the form of $\mu_1(\chi)$, Eq. (90b), this is indeed true. For $\mu_2(\chi)$, no such sum rule exists, but at $\tau > 0$, $N(\tau,\omega) = 4\chi/\tau = 4v_0/l(\omega)$ is independent of $\tau$ as expected.

## 5.3. Numerical Simulations

Since the analytical results obtained in the last section are based on the assumption of $\bar{a}\omega/v_0 = \varepsilon \to 0$, a natural use of numerical simulations would be to examine the prediction of the theory when the ratio $\bar{a}\omega/v_0$ is no longer small. What we find is that the theory is surprisingly accurate for all ranges of the ratio $\bar{a}\omega/v_0$ provided one uses not just the low frequency expression for $l(\omega)$ but the true value of $l(\omega)$ when $\omega$ is beyond the low frequency regime. This indicates to us that the analytical results have general validity.

To compare our analytical results with numerical simulation data which can extend beyond the white-noise limit, we choose to focus only on the set $U_1(y)$ and $\mu_1(\chi)$ since the same basic quantity $P(\bar{\psi})$ is used to evaluate both $U_1$ and $U_2$. To obtain $U_1(y)$, we use the transfer-matrix algorithm as described in Sec. 4.5 to calculate $R_{1,2} = \exp(i\psi_{1,2})$. $U_1 = \langle \exp(i\overline{\psi}) \rangle$ is then obtained by configurational averaging. Care is taken in using sufficient large sample length $L$ so that $|R| = 1$ and the value of $\overline{\psi}$ stabilizes. All three material models were simulated. For model I, we used $\overline{a} = 1$, $\rho_0 = 8$, $K_0 = 2$, $\sigma_\rho = 0.05$, and $\sigma_K = 0.4$. In model II the parameters are $\overline{a} = 1$, $\rho_0 = K_0 = 1$, $\sigma_\rho = 0.9$, and $\sigma_K = 0.9$. For model III we just get the material parameters from model II by convolving with an exponential filter. For a particular value of $\omega$ in a given model, $U_1$ is calculated as a function of $\Delta_\omega$. A total of 32 curves is obtained with frequencies ranging from $\overline{a}\omega/v_0 \sim 0.05$ to $\overline{a}\omega/v_0 \sim 10$. When $\mathrm{Re}(U_1)$ and $\mathrm{Im}(U_1)$ are plotted as a function of $\ln \Delta_\omega$, all the curves appear to be similar except that they are horizontally displaced from one another. This suggests the introduction of a scaling frequency $\Delta_s = v_0/l_s$ for each curve whose role is to scale $\Delta_\omega$ such that when the curves are plotted as $\ln(\Delta_\omega/\Delta_s) = \ln(\Delta_\omega l_s/v_0)$ all the data would collapse into a single functional form. The value of $l_s(\omega)$ obtained in this way can then be compared with independently calculated localization length $l(\omega)$ to see if $y = \Delta_\omega l(\omega)/v_0$ is indeed the scaling variable. In Fig. 9 we compare the collapsed simulation data with the prediction of the theory as evaluated from Eq. (87). The good agreement of the different curves with each other and with the theory indicates that $U_1$ is truly a function of only one variable, and in terms of this variable the functional form $U_1(y)$ is universal in the sense of being independent from model parameters and statistics. The function $U_1(y)$ is also noted to have a width of about $\Delta y \simeq 1$. In Fig. 10 we compare the values of $l_s(\omega)$ with that of numerically evaluated $l(\omega)$. For clarity, only those data for models I and II are shown. The remarkable tracking of $l_s(\omega)$ with $l(\omega)$ leaves no doubt that $l_s(\omega)$ is in fact the localization length (data for model III show equally good tracking). Also, the frequency range over which the scaling holds is seen to be very large. We therefore conclude that the $\omega$-dependence of the multiple scattering noise spectrum is completely determined by $l(\omega)$.

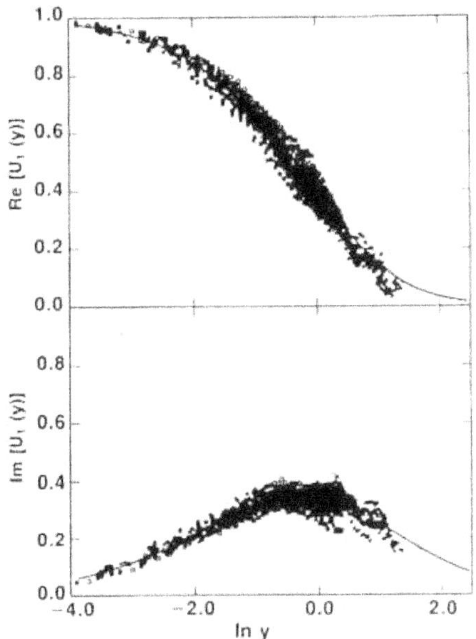

Fig. 9. The real and imaginary parts of $U_1(y)$ for the matched-medium case plotted as functions of ln $y$, where $y = \Delta_\omega l_s(\omega)/v_0$. Solid line denotes the prediction of the analytical theory as evaluated from Eq. (87). The symbols denote the simulation results for a total of 32 cases involving the three models and different values of the frequency $\omega$. Each symbol represents the average of over 40 configurations.

In the time domain, we simulate $\mu_1(\chi)$ directly by solving the wave equation

$$K^{-1}(z)\frac{\partial^2 p}{\partial t^2} = \frac{\partial}{\partial z}\rho^{-1}(z)\frac{\partial p}{\partial z}$$

numerically for 900 configurations of model I with a Gaussian incident pulse. By Fourier transformation of $\langle p(\tau)p(\tau+t)\rangle$ with respect to $t$, we get $S(\tau,\omega)$ for five different observation window times $\tau$. The values of $\mu_1(\chi)$ shown in Fig. 11 are evaluated as $S(\tau,\omega)\tau/|f(\omega)|^2$ and plotted as a function of $\sqrt{\chi}$ by using the numerically calculated values of the localization length. We use $\sqrt{\chi}$ as the plotting variable because at low frequencies $\sqrt{\chi}\,\alpha\,\omega$, which makes the plotted curve proportional to the noise spectrum in that limit. It is seen that the scaling property of $\mu_1(\chi)$ is indeed verified in that the data for different windows collapse

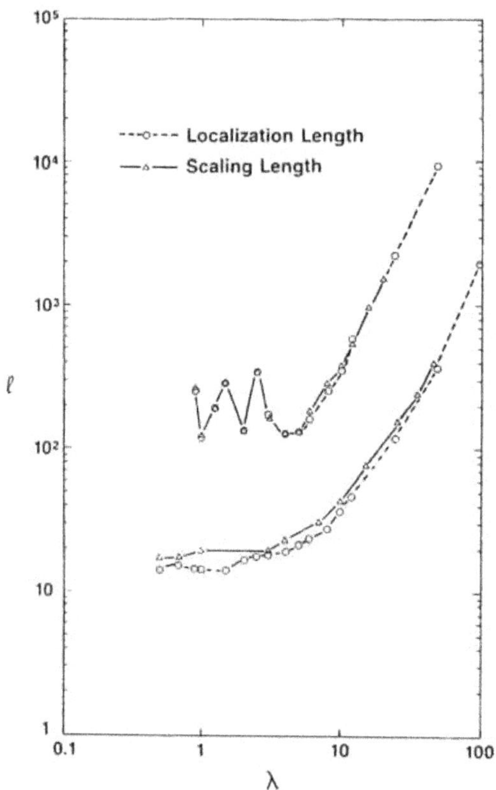



Fig. 10. The comparison between the scaling length obtained by collapsing the data in Fig. 9, and the independently calculated localization length. Both are in units of mean layer thickness $\overline{a}$. Results for model I are shown in the upper pair of curves, and those for model II are shown in the lower pair.

into a single curve when plotted in terms of the variable $\chi$. Comparison with the theoretical prediction also shows obvious agreement, although the numerical peak value at $\chi = 1$ is higher than predicted. This could be due to numerical inaccuracies arising from fluctuations in the data, which are largest near $\chi = 1$.

The excellent agreement between the simulation data and the analytical results does raise an interesting question. While the independence of $U_{1,2}(y)$ from the different types of model and model parameters is implicit in our derivation and therefore understandable, yet why

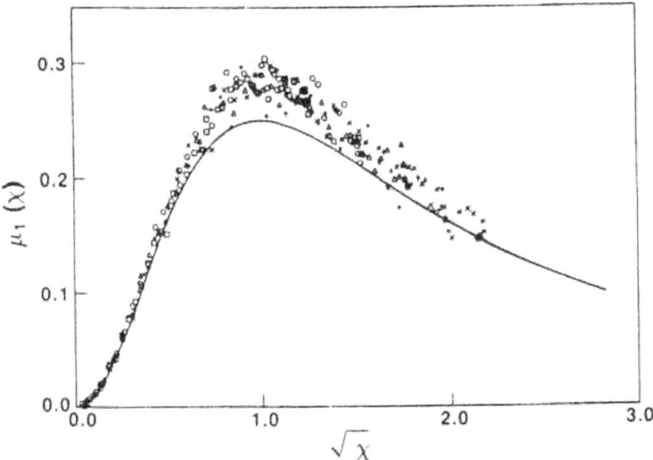

Fig. 11. The noise spectrum $\mu_1(\chi)$ plotted as a function of $\sqrt{\chi}$ where $\chi = \tau v_0/l(\omega)$. The simulation data, denoted by symbols, are obtained by solving the wave equation in the time domain for model I. In time units of $\bar{a}/v_0$, scaled noise data for $\tau = 200, 300, 400, 500$ and $600$ are shown. Solid line is the curve $\chi/(1+\chi)^2$.

should the results of the theory be valid at all frequency ranges when their derivation is based on the low-frequency assumption? We speculate that perhaps the present results are derivable from a much more general set of assumptions. One plausible general criterion for the validity of the theory is that the dimensionless parameter $Q^{-1} = 2v_0/\omega l(\omega)$ be small, where $Q$ is usually denoted as the "quality factor". Since we have seen that $l(\omega)$ is always bounded below by a fairly large minimum value (see Sec. 4.5) even for a medium with very large material parameter fluctuations, $Q^{-1}$ would therefore have a maximum value that is $\ll 1$ for all frequencies. Physically, that means the multiple-scattering attenuation is always a weak effect over a pulse-width. If that is the case, then $\omega\bar{a}/v_0 = \varepsilon \ll 1$ is just one realization of the general criterion and therefore would be a sufficient, but not necessary, condition for the derivation of the analytical results. In any case, our simulation data clearly demonstrate the general applicability of the analytical results.

## 5.4. Physical Implications

Knowledge of the form for $\mu_{1,2}(\chi)$ has direct implications for the measurement of the localization length in the time domain.[5] In the case of matched-medium boundary condition, it indicates that on measuring $\tau N(\tau,\omega) = \mu_1(\chi)$ at a certain frequency window, one should observe a peak at some time $\tau_0$ (with the time of pulse injection taken to be $\tau = 0$) corresponding to $\chi = \tau_0 \cdot v_0 / l(\omega) = 1$. Thus, knowing the mean speed of the random medium would yield $l(\omega) = \tau_0 v_0$. If $v_0$ is not known, one can still get a relative ratio of the localization lengths for different random media by comparing the peaking times. For the pressure-release boundary conditions, on the other hand, $N(\tau,\omega) = 4v_0/l(\omega)$ so that the localization length is proportional to the inverse of the noise spectrum and therefore can again be determined directly in the time domain.

A curious feature about $\mu_1(\chi)$ is that it has a universal maximum value of $1/4$. That means the maximum $\tau N(\tau,\omega)$ value for all randomly stratified media should be the same, regardless of the magnitude of the material-parameter fluctuations. This is somewhat anti-intuitive at first sight because one would normally expect the medium with large-magnitude fluctuations in $\rho$ or $K^{-1}$ to be strongly scattering and therefore to yield a higher magnitude for the maximum of $\mu_1$. But in reality what happens, as evidenced by our analytical and numerical results, is that for different media only the peaking times of $\tau N(\tau,\omega)$ are different. For a medium with small randomness one has to wait longer before the same peak magnitude of $\tau N(\tau,\omega)$ is achieved.

The validity of our analytical results for the pulse reflection problem for all values of $\omega$ means that if $l(\omega)$ is known, then $N(\tau,\omega)$ may be explicitly predicted. In the case of a model-II type medium, for example, we have $l(\omega) \cong c_1^{-1} + 6v_0^2/(\sigma^2\omega^2 a)$ (where $\sigma = \sigma_\rho = \sigma_K$) as an accurate representation for $l(\omega)$ over the entire frequency range. That means

$$
N(\tau,\omega) = \begin{cases} \dfrac{c_1^{-1} + \frac{6}{a}\left(\frac{v_0}{\sigma\omega}\right)^2}{\left[c_1^{-1} + \frac{6}{a}\left(\frac{v_0}{\sigma\omega}\right)^2 + \tau v_0\right]^2}\, v_0\,, & \text{matched-medium} \\[2em] \left[c_1^{-1} + \dfrac{6}{a}\left(\dfrac{v_0}{\sigma\omega}\right)^2\right]^{-1} v_0\,, & \text{reflecting}\,, \end{cases}
\tag{92}
$$

where $c_1$ can be calculated from Eq. (48). The knowledge of $N(\tau,\omega)$ may enable one to optimally detect reflections from target objects buried in a random medium. Of course, $N(\tau,\omega)$ represents only an expected value. Results of measurement from a single configuration are expected to show large fluctuations.

In this work we have not considered the parallel effect of dissipative attenuation. Of course, if the dissipative attenuation length $l_d(\omega)$ is significantly smaller than the localization length, then the attenuation effects of multiple scattering would obviously be secondary and we would not expect to observe true localization effects. On the hand, if the $l_d(\omega) > l(\omega)$, then we might view the dissipation and multiple scattering as two independent sources of attenuation, and in that case it can be shown that the power spectrum $\mu_{1,2}$ is multiplied by the factor $\exp(-2\gamma_0\tau/\rho_0)$, where $\gamma_0 = \langle\gamma_i\rangle$ and $\gamma_i$ (displacement velocity) represents the local dissipation in $i$th layer.

## 6. CONCLUDING REMARKS

In this paper we have delineated in some detail the recent theoretical developments on the problem of wave propagation in randomly stratified media. While a fair amount is already known, yet much remains to be done. For the pulse scattering problem, for example, there is obviously a general theory waiting to be uncovered that would relax the low-frequency assumption. Challenges also exist in understanding the statistics of pulse transmission through a finite slab of random medium, as well as in the application of the present approach to the problem of statistical data inversion. By the latter we mean a generalization of the model so that the mean of the material parameters, $\rho_0$ and $K_0$, are no longer constant but can actually vary over distances that are large as compared with $\bar{a}$. The aim of an inversion scheme would then be the optimal recovery of $\rho_0(z)$ and $K_0(z)$ from noisy data, where the noise arises from multiple scattering by material fluctuations on the scale of $\bar{a}$.

On a more general level, one may consider a model where each layer may contain lateral inhomogeneities. By adjusting the amount of lateral fluctuations, it is possible to cross over from one-dimensional randomness to fully three-dimensional randomness. The study of this more general model, however, would require mathematical methods very different from those considered here, and we leave it as a subject of future inquiry.

APPENDIX

## THE TRACE THEOREM

For differential equations of the type

$$\frac{d}{dz}Y = tY \; ,  \tag{A.1}$$

where $Y$ and $t$ are $n \times n$ matrices, it is simple to show that there is a relationship between the determinant of $Y$ and the trace of $t$. Equation (A.1) can be rewritten as

$$Y(z + \Delta z) = (1 + t\Delta z)Y(z) \; .  \tag{A.2}$$

By taking the determinant of both sides, we get

$$\begin{aligned}
\det(Y)_{z+\Delta z} &= \det(1 + t\Delta z)\det(Y)_z \\
&\cong \left[1 + \mathrm{Tr}\,(t)\Delta z\right]\det(Y)_z
\end{aligned}  \tag{A.3}$$

to the first order in $\Delta z$. That means

$$\frac{d}{dz}\det(Y) = \mathrm{Tr}\,(t)\det(Y) \; .  \tag{A.4}$$

If $\mathrm{Tr}(t) = 0$ as in the case of Eq. (5), then $\det(Y)$ is a constant independent of $z$. That means if one set of solution, say $(p_1, u_1)$, decays like $\exp(-\gamma z)$, then the second set of solution $(p_2, u_2)$ must grow exponentially as $\exp(\gamma z)$ with exactly the same $\gamma$ so that the determinant of the solution matrix would maintain its constancy as a function of $z$.

# REFERENCES

1. See for example, K. Aki and P. G. Richards, *Quantitative Seismology* (W. H. Freeman, San Francisco, California, 1980).

2. See, for example, B. Souillard, "Waves and Electrons in Inhomogeneous Media" in *Chance and Matter*, J. Souletie, J. Vannimenus and R. Stora, eds. (North-Holland, Amsterdam, 1987), and references therein.

3. R. Burridge, G. Papanicolaou and B. White, *SIAM J. Appl. Math.* **47** (1987) 146.

4. P. Sheng, Z. Q. Zhang, B. White and G. Papanicolaou, *Phys. Rev. Lett.* **57** (1986) 1000.

5. B. White, P. Sheng, Z. Q. Zhang and G. Papanicolaou, *Phys. Rev. Lett.* **59** (1987) 1918.

6. P. Sheng, B. White, Z. Q. Zhang and G. Papanicolaou, *Phys. Rev.* **B34** (1986) 4757.

7. P. G. Richards and W. Menke, *Bull. of Seismological Soc. of Am.* **73** (1983) 1005.

8. W. Menke, *Geophys. J. Res. Astr. Soc.* **75** (1983) 541.

9. J. E. Sipe, P. Sheng, B. W. White and M. H. Cohen, *Phys. Rev. Lett.* **60** (1988) 108.

10. S. John, H. Sompolinsky and M. J. Stephen, *Phys. Rev.* **B27** (1983) 5592.

11. M. Y. Azbel, *Phys. Rev.* **B28** (1983) 4106.

12. W. Kohler and G. Papanicoloau, *J. Math. Phys.* **14** (1973) 1733.

13. M. Schoenberger and F. K. Levin, *Geophysics* **39** (1974) 278.

14. M. Schoenberger and F. K. Levin, *Geophysics* **43** (1978) 730.

15. R. F. O'Doherty and N. A. Anstey, *Geophysical Prospecting* **19** (1971) 430.

16. C. W. Gardiner, *Handbook of Stochastic Methods for Physics, Chemistry and the Natural Sciences* (Springer-Verlag, New York, 1983).

17. M. Born and E. Wolf, *Principle of Optics*, 2nd edn. (Pergamon Press, New York, 1964) p. 43.

SUBJECT INDEX

1-photon Green's function    378
2-photon Green's function    379
    average    388
a.c. conductivity    76
absorption    70, 74, 132, 248, 347, 358
    coefficient    249
    effect on cone    104, 117
    length    251, 259
    length coefficient    241
    rate    214
    time    214, 301
acoustic waves    427
adjoint operator    573
Aharanov-Bohm effect    215
albedo    112
amplitude    409
    fluctuations    409
Anderson
    localization    375, 406, 427, 428
    model    217, 218
    transition    212, 235
angular dependence    590
anomalous
    absorption    73
    diffusion    167

approximation
  average-t-matrix   130
  coherent potential   130
  diffusion   102
  independent scattering   129
autocorrelation function   319, 580
  measured   339


backscattering   101
  coherent   102, 103
  enhanced (see also cone)   101, 102
  polarization   103
  time-solved   118, 124
backward Kolmogorov equation   573, 579, 605, 607
band edge trajectories   414
band gaps   414
band structure techniques   414
Bethe-Salpeter equation   150, 388
binary system   592
bistatic coefficient   113, 119
Boltzmann
  diffusion constant   155, 160
  equation   444
Born regime   184
boundary conditions   435
  radiation   223
  reflecting   226
  waveguide   231
Bragg reflection   545
Brewster
  anomaly   598
  effect   567, 589
Brownian motion   100
  analogy   149
bubbly fluid   490

built-in correlation    588

classical wave    406, 423
   localization    415
close packing concentration    414
coda    566
coherence
   area    215, 228, 242
   function    60
   length    213, 477
   volume    214, 229
Coherent Potential Approximation (CPA)    38, 227, 411, 416
coherent
   backscattering    25, 156, 210, 216, 226, 283, 285, 300, 346, 406
   scatterers    15
conductance    230
   distribution    220
   fluctuations    212, 214, 288, 292
cone (of enhanced backscattering)    101, 102
   difference technique    104, 107, 122
   diffusion theory    115
   polarization    104, 106, 107
   recording    104, 105
   rigorous theory    112
   shape    103, 121
   spatial anisotropy    104, 107, 110
   time-resolved    104, 106, 118
   width    101, 102, 103, 106, 121
configuration average    245, 262, 267
conjugate variables    600
continuous random models    586
continuum    411, 413
correlated disorder    43
correlation    293
   effects    190

frequency    211, 223, 264

   fluctuating parts    581

   length    166, 227, 229, 300

   reflection coefficients    600

Coulombic repulsion    368

critical

   angle    598

   exponents    235, 429

   point    409

   volume fraction    409

cross section    410

crossover exponent    531

cumulant    334

cylindrical scatterer    459, 463, 466


degree of correlation    228, 232

delocalization    471

density correlation function    433

density of states (DOS)    19, 211, 219, 226, 248

density propagator    433

dephasing time    235

diagonal disorder    30

diagrams

   ladder    24, 112, 125, 152, 376

   maximally crossed    112, 125, 157, 159, 377, 393
     447, 448

diagrammatic method    388

dielectric

   constant    242

   function    236, 406, 407

diffusing wave spectroscopy    199, 316

diffusion    133

   approximation    102, 115

   boundary condition    328

coefficient   21, 23, 45, 101, 133, 268, 409, 418, 419, 420, 434, 468, 469, 475, 482
   classical   223, 249
   collective   366
   photon   258
   renormalized   217, 225, 300, 302
   scale dependent   304
   scale dependent, renormalized   224
equation   214, 393
ladder diagram   191
diffusive regime   406
dimensionality   99
dimensionless
  conductance   214, 219, 231, 294
  level width   211, 221, 224, 228, 298
disorder   409
  average   437
  parameter   131
disordered medium   409
dispersion relation   140
dissipation   484, 485, 516, 556
dissipative attenuation   617
distribution of path lengths   324
dynamical correlation function   199

E-polarization   587, 595
effective medium   411
  speed   145, 460
  propagation constant   410
eigenmodes   211, 223, 544, 556
Einstein relation   214, 219
elastic mean free path   41
elastic wave equation   569
electrical polarizability   76
electromagnetic   406

localization    430, 489
  wave    416
electron diffusion    214
electron-electron interaction    427, 428, 532
electron-hole propagator    78
electronic localization    406
electronic tight-binding model    171
em energy density    391
energy density propagator    436, 504, 526
energy diffusion coefficient    35, 38
energy shift    218
enhancement factor    103, 106, 110, 124
ensemble average    379
  intensity    187
extended state    408

Faraday effect    27
fast material fluctuations (effect of)    604
Feynman
  diagrams    385
  paths    213, 231
field correlation function    223, 261, 263
field factorization approximation    281, 286
field theory    36, 526, 530
fixed point    530
Fokker-Planck equation    573, 575, 582, 608
forbidden band    545, 550
form factor    193, 195
forward scattering amplitude    411, 412
full potential theory (for surface gravity wave)    551

Gaussian
  distribution    384
  integrals    31
  noise    486
  wave packet    190

Goldstone mode    40, 53, 55, 62, 529
Green's
  function    433, 504
  operator    439

H-polarization    587, 595
hard (Neumann) scatterers    454
Hikami box    194
Hubbard-Stratonovitch transformation    37
Huygens' wavelets    186
hydrodynamic cutoff    449

independent scatterers    12
index of refraction, random variation of    430
inelastic
  scattering    428
  scattering length    210, 214, 235
infinitesimal generator    573, 575, 605, 607
infrared singularity    523
intensity
  correlation    187, 286
    function    191, 224, 262, 263, 267, 271
    sample configuration with time    277
    sample rotation    281
  cross correlation    298
  fluctuations    196, 215, 267, 494
interference    215
intermediate scale    604
inversion scheme    617
Ioffe-Regel    300
  criterion    13, 42, 99, 132, 211, 227, 247, 444, 452, 464, 469
irreducible
  diagrams    441, 442
  four-point vertex function    442

kinetic energy    435
Kronecker products    528


Lagrangian    33
lateral inhomogeneities    617
lattice
  model    411, 413, 414
  spacing    413
level
  spacing    211, 218, 225, 227
  width    211, 223, 225, 227, 294, 298
light path    100, 103, 107
    loop type    127
    self-avoiding    125
    time-reversed pair    100
local fluctuations    210
localization    408, 419, 541, 544, 548, 551, 555
  criterion    162, 394
  effect in the time domain    569
  electrons    218
  length    42, 168, 219, 409, 438, 450, 468, 474, 475, 482, 483, 487, 489, 509
    angular dependence of    567
    divergence    567, 580
    high frequency behavior    582
    low frequency behavior    575
    minimum value    595
    non-decreasing trend    583
    numerical simulation    590
    quantum particle    594
    time-domain measurement    616
  of light, strong    131
  of light waves    406
  of light, weak    100, 131
  parameter    420

   phase diagrams    168
   phenomenon    561
   photon    217
   scaling theory of    219, 224
   threshold    211, 212, 218, 220, 226, 277, 300
long range
  correlation    196, 210, 263, 285
  intensity    292
    correlation    294
lower critical dimension    25, 534
Lyapunov exponents    553

macroscale    604
magnetoresistance    215
Markov process    573, 574
matched-medium boundary condition    571, 599, 609
material fluctuations    606
maximally crossed diagrams    112, 125, 157, 159, 377, 393, 447, 448
maximum statistical information (from inversion)    569
mean free path    156, 210, 254, 409, 410, 420, 444, 464
  transport    106
medium contrast    141
memory effect    188, 193, 200, 281
MET    414, 415
micro spheres    103
Mie
  resonance    236, 302, 377, 415, 417
  scattering    236, 238, 259
Milne equation    113
minimum impedance ratio (for localization)    169, 170, 171
minimum metallic conductance    163, 164
mobility edge    10, 70, 163, 166, 212, 220, 225, 229, 235, 300, 376, 408
  trajectory    408, 416, 417

momentum
  conservation   59
  shell integration   84
multiple scattering   410, 411, 438
  formalism   439
  regime   184

near field transmission   204
negative exponential statistics   273, 274
noise spectrum   600, 610
nonlinear
  $\sigma$-model   61, 68
  wave   542, 545
null space   576, 578, 606

oblique incidence   586, 595
oceanography   545
off-diagonal disorder   30
optical
  band gap   17
  localization   420
  theorem   153, 160, 445, 455
orthogonal symmetries   527

pair correlation   383
parity breaking   28
particle diffusion   210
partition function   31, 36
path length distribution   210, 249
percolation   408, 409
periodic substrate   518
permeable scatterers   454
perturbation theory   410
phase   409
  boundary   468, 470
  coherence length   409

difference 283

velocity 409, 411

phonon localization 12, 234, 427

photomultiplier tube 243

photon

absorption time 288

diffusion 214, 248, 249

Green's function 386

photonic band gaps 7, 236, 247, 248

photonic band tail 19

polarization effects 107

polyballs 91

polydispersity 351

potential

theory 542, 545, 546, 551

well analogy (PWA) 410

power spectrum 568, 599

universal maximum value 616

probability

distribution intensity 272, 273, 291

stretched exponential 291

distribution of path lengths 222

distribution, time of flight 210, 222, 249, 257, 301

distribution transmission 221

of return 212, 282, 294

propagation constant 409, 411, 412

pseudosphere approximation 486, 487

pulse

localization 566, 568

propagation 173

quality factor 615

quanternion 528

quantum

particle 140

theory of solids  545

quasielastic light scattering  315


random

  layered media  566

  models  570

  phase  261, 284

  walks  186,

Rayleigh

  distribution  273

  scattering  257, 580

reducible diagrams  442

reflection  182, 238, 270, 289

  boundary condition  571

  coefficient  591, 599

  fluctuations  233

  speckle pattern  204

renormalization  76, 409

  diffusion constant  162

  resistance  89

  speed of sound  455, 504, 505

replica  46

  field representation  46

  method  32

resistance  221

  fluctuations  215

resonance  418, 419, 509, 510

  behavior  458

  modes  558, 561

  scattering  169, 171

retroreflection  102

Riccati equation  601


scalar wave  185

  equation  142

scale dependence
  diffusion coefficient    214, 300
scale of non-stationarity    600
scale-dependent
  conductivity    165
  diffusion coefficient    70, 240
scaling
  behavior    163, 429
  frequency    612
  function    165, 219
  property    613
  relation    163
  theory of localization    300
  variable    534
scatterer correlation    533
scattering    106
  amplitude    143, 144, 455
  correlated    129
  cross section    411
  dependent    130
    correction    130
  independent    129
  isotropic point    112
  length    387
  mean free path    465
  Mie    106
  potential    382
  Rayleigh    106
  resonances    490
  single    104, 120, 127
    contribution    104, 120
  t-matrix    398
Schottky diode    246, 267
Schrödinger
  equation    552

operator   56

self focussing nonlinearities   7

self-consistent

approach   533

diagrammatic approach   423, 429, 432

theory of localization   444, 446

self-energy   381, 440, 455

shallow water   546, 550

theory   550, 551, 553

Siegert relation   329

sigma model method   429

soft (Dirichlet) scatterers   454

solvability condition   576, 578

speckle pattern   228, 231, 233, 242, 263, 264, 289, 299, 315

transmission   182

sphere   455, 461, 465

spherical tensors   447

static structure factor   350

stationary distribution   574, 602, 609

function   573

stochastic differential equation   572

stochastic resonances   545, 556

modes   542

strip method   413

surface

enhanced Raman scattering   376

gravity waves   541

plasmon polariton   375

tension   554

T matrix   377

operator   142

third sound   500, 502, 518, 532

Thouless

criterion   211

number   194

tight-binding
  Hamiltonian    413
  model    197
time domain localization    566, 599
  analytical solution    603
  numerical simulations    611
time reversal
  invariance    522
  symmetry    376
time scale of non-stationarity    610
titania/polystyrene composites    243
total
  cross section    412
    scattering    455
  scattering matrix    381
  transmission    298
trace theorem    618
transfer-matrix    591, 612
transmission    230, 238, 250, 254, 299
  coefficient    591
  fluctuations    230, 233
  resonances    220, 512
transport
  mean-free path    106, 322
  theory    249

ultraviolet cutoff    449, 469
unitary symmetries    527
universal
  conductance fluctuations    197, 300
  function    568
  maximum    616
  scaling    613
universality    219, 221, 305
upper critical dimension    534

Urbach edge    10

vector
   spherical harmonic function    385
   waves    99, 107, 111, 127
velocity    210, 268, 301
vertex function    150, 442
viscosity    555
viscous dissipation    555

Ward identity    153, 446
wave
   diffusion    149
   guide    189
   length    413
   scalar    111
   time-reversed pair    110
   vector    99, 111, 127
weak localization    157, 185, 210, 215, 217, 281
white noise limit    603
wire method    413, 414, 415

www.ingramcontent.com/pod-product-compliance
Lightning Source LLC
Chambersburg PA
CBHW060748240726
48664CB00009BA/1660